Graduate Texts in Mathematics 53

For other titles in this series, go to
http://www.springer.com/series/136

Yu. I. Manin

A Course in Mathematical Logic for Mathematicians

Second Edition

Chapters I-VIII translated from the Russian
by Neal Koblitz

With new chapters by Boris Zilber and Yuri I. Manin

Author:
Yu. I. Manin
Max-Planck Institut für Mathematik
53111 Bonn
Germany
manin@mpim-bonn.mpg.de

Contributor:
B. Zilber
Mathematical Institute
University of Oxford
Oxford OX1 3LB
United Kingdom
zilber@maths.ox.ac.uk

First Edition Translated by:
Neal Koblitz
Department of Mathematics
University of Washington
Seattle, WA 98195
USA
koblitz@math.washington.edu

ISSN 0072-5285
ISBN 978-1-4614-2479-6 e-ISBN 978-1-4419-0615-1
DOI 10.1007/978-1-4419-0615-1
Springer New York Dordrecht Heidelberg London

Mathematics Subject Classification (2000): 03-XX, 03-01

Printed on acid-free paper

Springer is part of Springer Science+Business Media (www.springer.com)

To Nikita, Fedor and Mitya, with love

Preface to the Second Edition

1. The first edition of this book was published in 1977. The text has been well received and is still used, although it has been out of print for some time.

In the intervening three decades, a lot of interesting things have happened to mathematical logic:

(i) Model theory has shown that insights acquired in the study of formal languages could be used fruitfully in solving old problems of conventional mathematics.

(ii) Mathematics has been and is moving with growing acceleration from the set-theoretic language of structures to the language and intuition of (higher) categories, leaving behind old concerns about infinities: a new view of foundations is now emerging.

(iii) Computer science, a no-nonsense child of the abstract computability theory, has been creatively dealing with old challenges and providing new ones, such as the P/NP problem.

Planning additional chapters for this second edition, I have decided to focus on model theory, the conspicuous absence of which in the first edition was noted in several reviews, and the theory of computation, including its categorical and quantum aspects.

The whole Part IV: Model Theory, is new. I am very grateful to Boris I. Zilber, who kindly agreed to write it. It may be read directly after Chapter II.

The contents of the first edition are basically reproduced here as Chapters I–VIII. Section IV.7, on the cardinality of the continuum, is completed by Section IV.7.3, discussing H. Woodin's discovery.

The new Chapter IX: Constructive Universe and Computation, was written especially for this edition, and I tried to demonstrate in it some basics of categorical thinking in the context of mathematical logic. More detailed comments follow.

I am grateful to Ronald Brown and Noson Yanofsky, who read preliminary versions of new material and contributed much appreciated criticism and suggestions.

2. Model theory grew from the same roots as other branches of logic: proof theory, set theory, and recursion theory. From the start, it focused on language and formalism. But the attention to the foundations of mathematics in model

theory crystallized in an attempt to understand, classify, and study models of theories of real-life mathematics.

One of the first achievements of model theory was a sequence of *local theorems* of algebra proved by A. Maltsev in the late 1930s. They were based on the compactness theorem established by him for this purpose. The compactness theorem in many of its disguises remained a key model-theoretic instrument until the end of the 1950s. We follow these developments in the first two sections of Chapter X, which culminate with a general discussion of nonstandard analysis discovered by A. Robinson. The third section introduces basic tools and concepts of the model theory of the 1960s: types, saturated models, and modern techniques based on these.

We try to illustrate every new model-theoretic result with an application in "real" mathematics. In Section 4 we discuss an algebro-geometric theorem first proved by J. Ax model-theoretically and re-proved by G. Shimura and A. Borel. Moreover, we explain an application of the Tarski–Seidenberg quantifier elimination for $\mathbf{R}$ due to L. Hörmander. A real gem of model-theoretic techniques of the 1980s is the calculation by J. Denef of the Poincaré series counting p-adic points on a variety based on A. Macintyre's quantifier elimination theorem for $\mathbf{Q}_p$.

In the last two sections we present a survey of classification theory, which started with M. Morley's analysis of theories categorical in uncountable powers in 1964, and was later expanded by S. Shelah and others to a scale that no one could have envisaged.

The striking feature of these developments is the depth of the very abstract "pure" model theory underlying the classification, in combination with the diversity of mathematical theories affected by it, from algebraic and Diophantine geometry to real analysis and transcendental number theory.

3. The formal languages with which we work in the first, and in most of the second, edition of this book are exclusively *linear* in the following sense. Having chosen an alphabet consisting of letters, we proceed to define classes of well-formed *expressions* in this alphabet that are some *finite sequences of letters*. At the next level, there appear well-formed *sequences of words*, such as deductions and descriptions. Church's λ-calculus furnishes a good example of strictures imposed by linearity.

Nonlinear languages have existed for centuries. Geometers and composers could not perform without using the languages of drawings, resp. musical scores; when alchemy became chemistry, it also evolved its own two-dimensional language. For a logician, the basic problem about nonlinear languages is the difficulty of their formalization.

This problem is addressed nowadays by relegating nonlinear languages of contemporary mathematics to the realm of more conventional mathematical objects, and then formally describing such languages as one would describe any other structure, that is, linearly.

Such a strategy probably cannot be avoided. But one must be keenly aware that some basic mathematical structures *are* "linguistic" at their core. Recognition or otherwise of this fact influences the problems that are chosen, the questions that are asked, and the answers that are appreciated.

It would be difficult to dispute nowadays that category theory as a language is replacing set theory in its traditional role as *the* language of mathematics. Basic expressions of this language, *commutative diagrams*, are one-dimensional, but *nonlinear:* they are certain decorated graphs, whose topology is that of 1-dimensional triangulated spaces.

When one iterates the philosophy of category theory, replacing sets of morphisms by objects of a category of the next level, commutative diagrams become two-dimensional simplicial sets (or cell complexes), and so on. Arguably, in this way the whole of homotopy topology now develops into the language of contemporary mathematics, transcending its former role as an important and active, but reasonably narrow research domain. Much remains to be recognized and said about this emerging trend in foundations of mathematics.

The first part of Chapter IX in this edition is a very brief and tentative introduction to this way of thinking, oriented primarily to some reshuffling of classical computability theory, as was explained in the Part II of the first edition.

4. The second part of the new Chapter IX is dedicated to some theoretical problems of classical and quantum computing. It introduces the P/NP problem, classical and quantum Boolean circuits, and presents several celebrated results of this early stage of theoretical quantum computing, such as Shor's factoring and Grover's search algorithms.

The main reason to include these topics is my conviction that at least some theoretical achievements of modern computer science must constitute an organic part of contemporary mathematical logic.

Already in the first edition, the manuscript for which was completed in September 1974, "quantum logic" was discussed at some length; cf. Section II.12.

A Russian version of the Part II of first edition was published as a separate book, *Computable and Uncomputable*, by "Soviet Radio" in 1980. For this Russian publication, I had written a new introduction, in which, in particular, I suggested that quantum computers could be potentially much more powerful than classical ones, if one could use the exponential growth of a quantum phase space as a function of the number of degrees of freedom of the classical system.

When a mathematical implementation of this idea, massive quantum parallelism, made possible by quantum entanglement, gradually matured, I gave a talk at a Bourbaki seminar in June 1999, explaining the basic ideas and results.

Chapter IX is a revised and expanded version of this talk.

5. Finally, a few words about the last digression in Chapter II, "Truth as Value and Duty: Lessons of Mathematics."

“Mathematical truth” was the central concept of the first part of the book, “Provability.” Writing this part, I felt that if I did not compensate somehow the aridity and sheer technicality of the analysis of formal languages, I would not be able to convince people–the readers that I imagined, working mathematicians like me—that it is worth studying at all. The literary device I used to struggle with this feeling of helplessness was this: from time to time I allowed myself free associations, and wrote the outcome in a series of six digressions, with which the first two Chapters were interspersed.

By the end of the second chapter, I realized that I was finally on the fertile soil of “real mathematics,” and the need for digressions faded away.

Nevertheless, the whole of Part I was left without proper summary.

Its role is now played by the “Last Digression,” published here for the first time. It is a slightly revised text of the talk prepared for a Balzan Foundation International Symposium on “Truth in the Humanities, Science and Religion” (Lugano, 2008), where I was the only mathematician speaker among philosophers, historians, lawyers, theologians, and physicists. I was confronted with the task to explain to a distinguished “general audience” what is so different about mathematical truth, and what light the usage of this word in mathematics can throw on its meaning in totally foreign environments.

The main challenge was this: avoid sounding ponderous.

Yu. Manin, Bonn December 31, 2008

Preface to the First Edition

1. This book is above all addressed to mathematicians. It is intended to be a textbook of mathematical logic on a sophisticated level, presenting the reader with several of the most significant discoveries of the last ten or fifteen years. These include the independence of the continuum hypothesis, the Diophantine nature of enumerable sets, and the impossibility of finding an algorithmic solution for one or two old problems.

All the necessary preliminary material, including predicate logic and the fundamentals of recursive function theory, is presented systematically and with complete proofs. We assume only that the reader is familiar with "naive" set-theoretic arguments.

In this book mathematical logic is presented both as a part of mathematics and as the result of its self-perception. Thus, the substance of the book consists of difficult proofs of subtle theorems, and the spirit of the book consists of attempts to explain what these theorems say about the mathematical way of thought.

Foundational problems are for the most part passed over in silence. Most likely, logic is capable of justifying mathematics to no greater extent than biology is capable of justifying life.

2. The first two chapters are devoted to predicate logic. The presentation here is fairly standard, except that semantics occupies a very dominant position, truth is introduced before deducibility, and models of speech in formal languages precede the systematic study of syntax.

The material in the last four sections of Chapter II is not completely traditional. In the first place, we use Smullyan's method to prove Tarski's theorem on the undefinability of truth in arithmetic, long before the introduction of recursive functions. Later, in the seventh chapter, one of the proofs of the incompleteness theorem is based on Tarski's theorem. In the second place, a large section is devoted to the logic of quantum mechanics and to a proof of von Neumann's theorem on the absence of "hidden variables" in the quantum-mechanical picture of the world.

The first two chapters together may be considered as a short course in logic apart from the rest of the book. Since the predicate logic has received the widest dissemination outside the realm of professional mathematics, the author has not resisted the temptation to pursue certain aspects of its relation to linguistics, psychology, and common sense. This is all discussed in a series of digressions, which, unfortunately, too often end up trying to explain "the exact meaning of a proverb" (E. Baratynsky).[1] This series of digressions ends with the second chapter.

The third and fourth chapters are optional. They are devoted to complete proofs of the theorems of Gödel and Cohen on the independence of the continuum hypothesis. Cohen forcing is presented in terms of Boolean-valued models; Gödel's constructible sets are introduced as a subclass of von Neumann's universe. The number of omitted formal deductions does not exceed the accepted norm; due respects are paid to syntactic difficulties. This ends the first part of the book: "Provability."

The reader may skip the third and fourth chapters, and proceed immediately to the fifth. Here we present elements of the theory of recursive functions and enumerable sets, formulate Church's thesis, and discuss the notion of algorithmic undecidability.

The basic content of the sixth chapter is a recent result on the Diophantine nature of enumerable sets. We then use this result to prove the existence of versal families, the existence of undecidable enumerable sets, and, in the seventh chapter, Gödel's incompleteness theorem (as based on the definability of provability via an arithmetic formula). Although it is possible to disagree with this method of development, it has several advantages over earlier treatments. In this version the main technical effort is concentrated on proving the basic fact that all enumerable sets are Diophantine, and not on the more specialized and weaker results concerning the set of recursive descriptions or the Gödel numbers of proofs.

The last section of the sixth chapter stands somewhat apart from the rest. It contains an introduction to the Kolmogorov theory of complexity, which is of considerable general mathematical interest.

The fifth and sixth chapters are independent of the earlier chapters, and together make up a short course in recursive function theory. They form the second part of the book: "Computability."

The third part of the book, "Provability and Computability," relies heavily on the first and second parts. It also consists of two chapters. All of the seventh chapter is devoted to Gödel's incompleteness theorem. The theorem appears later in the text than is customary because of the belief that this central result can only be understood in its true light after a solid grounding both in formal mathematics and in the theory of computability. Hurried expositions, where

[1] Nineteenth century Russian poet (translator's note). The full poem is:

We diligently observe the world,
We diligently observe people,
And we hope to understand their deepest meaning.
But what is the fruit of long years of study?
What do the sharp eyes finally detect?
What does the haughty mind finally learn
At the height of all experience and thought,
What?—the exact meaning of an old proverb.

the proof that provability is definable is entirely omitted and the mathematical content of the theorem is reduced to some version of the "liar paradox," can only create a distorted impression of this remarkable discovery. The proof is considered from several points of view. We pay special attention to properties which do not depend on the choice of Gödel numbering. Separate sections are devoted to Feferman's recent theorem on Gödel formulas as axioms, and to the old but very beautiful result of Gödel on the length of proofs.

The eighth and final chapter is, in a way, removed from the theme of the book. In it we prove Higman's theorem on groups defined by enumerable sets of generators and relations. The study of recursive structures, especially in group theory, has attracted continual attention in recent years, and it seems worthwhile to give an example of a result which is remarkable for its beauty and completeness.

3. This book was written for very personal reasons. After several years or decades of working in mathematics, there almost inevitably arises the need to stand back and look at this research from the side. The study of logic is, to a certain extent, capable of fulfilling this need.

Formal mathematics has more than a slight touch of self-caricature. Its structure parodies the most characteristic, if not the most important, features of our science. The professional topologist or analyst experiences a strange feeling when he recognizes the familiar pattern glaring out at him in stark relief.

This book uses material arrived at through the efforts of many mathematicians. Several of the results and methods have not appeared in monograph form; their sources are given in the text. The author's point of view has formed under the influence the ideas of Hilbert, Gödel, Cohen, and especially John von Neumann, with his deep interest in the external world, his open-mindedness and spontaneity of thought.

Various parts of the manuscript have been discussed with Yu. V. Matiyasevič, G. V. Čudnovskiĭ, and S. G. Gindikin. I am deeply grateful to all of these colleagues for their criticism.

W. D. Goldfarb of Harvard University very kindly agreed to proofread the entire manuscript. For his detailed corrections and laborious rewriting of part of Chapter IV, I owe a special debt of gratitude.

I wish to thank Neal Koblitz for his meticulous translation.

Yu. I. Manin Moscow, September 1974

Interdependence of Chapters

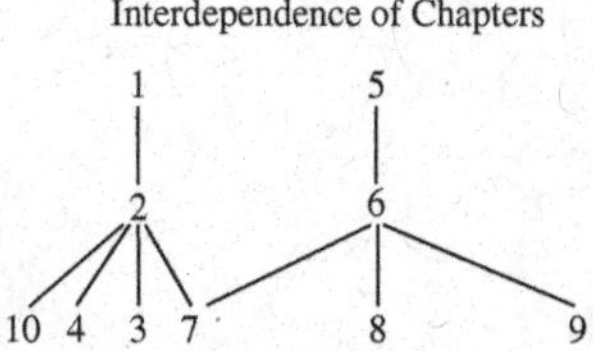

Contents

Part I
PROVABILITY

I

Introduction to Formal Languages

Gelegentlich ergreifen wir die Feder
Und schreiben Zeichen auf ein weisses Blatt,
Die sagen dies und das, es kennt sie jeder,
Es ist ein Spiel, das seine Regeln hat.
        H. Hesse, "Buchstaben"
We now and then take pen in hand
And make some marks on empty paper.
Just what they say, all understand.
It is a game with rules that matter.
        H. Hesse, "Alphabet"
        (translated by Prof. Richard S. Ellis)

1 General Information

1.1. Let A be any abstract set. We call A an *alphabet.* Finite sequences of elements of A are called *expressions* in A. Finite sequences of expressions are called *texts.*

We shall speak of a *language with alphabet A* if certain expressions and texts are distinguished (as being "correctly composed," "meaningful," etc.). Thus, in the Latin alphabet A we may distinguish English word forms and grammatically correct English sentences. The resulting set of expressions and texts is a working approximation to the intuitive notion of the "English language."

The language Algol 60 consists of distinguished expressions and texts in the alphabet {Latin letters} ∪ {digits} ∪ {logical signs} ∪ {separators}. *Programs* are among the most important distinguished texts.

In natural languages the set of distinguished expressions and texts usually has unsteady boundaries. The more formal the language, the more rigid these boundaries are.

The rules for forming distinguished expressions and texts make up the *syntax* of the language. The rules that tell how they correspond with reality make

Yu. I. Manin, *A Course in Mathematical Logic for Mathematicians, Second Edition*, Graduate Texts in Mathematics 53, DOI 10.1007/978-1-4419-0615-1_1,

up the *semantics* of the language. Syntax and semantics are described in a *metalanguage.*

1.2. "Reality" for the languages of mathematics consists of certain classes of (mathematical) arguments or certain computational processes using (abstract) automata. Corresponding to these designations, the languages are divided into formal and algorithmic languages. (Compare: in natural languages, the declarative versus imperative moods, or—on the level of texts—statement versus command.)

Different formal languages differ from one another, in the first place, by the scope of the formalizable types of arguments—their expressiveness; in the second place, by their orientation toward concrete mathematical theories; and in the third place, by their choice of elementary modes of expression (from which all others are then synthesized) and written forms for them.

In the first part of this book a certain class of formal languages is examined systematically. Algorithmic languages are brought in episodically.

The "language–parole" dichotomy, which goes back to Humboldt and Saussure, is as relevant to formal languages as to natural languages. In §3 of this chapter we give models of "speech" in two concrete languages, based on set theory and arithmetic, respectively, because, as many believe, habits of speech must precede the study of grammar.

The language of set theory is among the richest in expressive means, despite its extreme economy. In principle, a formal text can be written in this language corresponding to almost any segment of modern mathematics—topology, functional analysis, algebra, or logic.

The language of arithmetic is one of the poorest, but its expressive possibilities are sufficient for describing all of elementary arithmetic, and also for demonstrating the effects of self-reference à la Gödel and Tarski.

1.3. As a means of communication, discovery, and codification, no formal language can compete with the mixture of mathematical argot and formulas that is common to every working mathematician.

However, because they are so rigidly normalized, formal texts can themselves serve as an object for mathematical investigation. The results of this investigation are themselves *theorems of mathematics.* They arouse great interest (and strong emotions) because they can be interpreted as *theorems about mathematics.* But it is precisely the possibility of these and still broader interpretations that determines the general philosophical and human value of mathematical logic.

1.4. We have agreed that the expressions and texts of a language are elements of certain abstract sets. In order to work with these elements, we must somehow fix them materially. In the modern European tradition (as opposed to the ancient Babylonian tradition, or the latest American tradition, using computer memory), the following notation is customary. The elements of the alphabet are indicated by certain symbols on paper (letters of different kinds of type, digits, additional signs, and also combinations of these). An expression in an alphabet A is written in the form of a sequence of symbols, read from left to right, with hyphens when necessary. A text is written as a sequence of written expressions, with spaces or punctuation marks between them.

1.5. If written down, most of the interesting expressions and texts in a formal language either would be physically extremely long, or else would be psychologically difficult to decipher and learn in an acceptable amount of time, or both.

They are therefore replaced by "abbreviated notation" (which can sometimes turn out to be physically longer). The expression "$xxxxxx$" can be briefly written "$x \dots x$ (six times)" or "x^6." The expression "$\forall z(z \in x \Leftrightarrow z \in y)$" can be briefly written "$x = y$." Abbreviated notation can also be a way of denoting any expression of a definite type, not only a single such expression (any expression $101010\dots10$ can be briefly written "the sequence of length $2n$ with ones in odd places and zeros in even places" or "the binary expansion of $\frac{2}{3}(4^n - 1)$").

Ever since our tradition started, with Viète, Descartes, and Leibniz, abbreviated notation has served as an inexhaustible source of inspiration and errors. There is no sense in, or possibility of, trying to systematize its devices; they bear the indelible imprint of the fashion and spirit of the times, the artistry and pedantry of the authors. The symbols Σ, $\int$, $\in$ are classical models worthy of imitation. Frege's notation, now forgotten, for "P and Q" (actually "not [if P, then not Q]" whence the asymmetry):

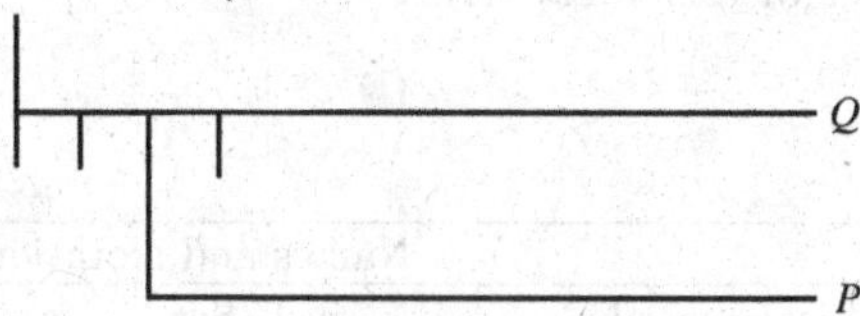

shows what should be avoided. In any case, abbreviated notation permeates mathematics.

The reader should become used to the trinity

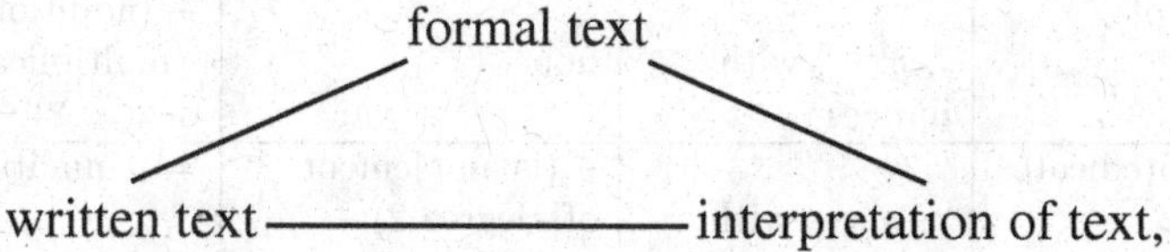

which replaces the unconscious identification of a statement with its form and its sense, as one of the first priorities in his study of logic.

2 First-Order Languages

In this section we describe the most important class of formal languages $\mathfrak{L}_1$—the first-order languages—and give two concrete representatives of this class: the Zermelo–Fraenkel language of set theory $\mathrm{L}_1\mathrm{Set}$, and the Peano language of arithmetic $\mathrm{L}_1\mathrm{Ar}$. Another name for $\mathfrak{L}_1$ is *predicate languages.*

2.1. The alphabet of any language in the class $\mathfrak{L}_1$ is divided into six disjoint subsets. The following table lists the generic name for the elements in each subset, the standard notation for these elements in the general case, the special notation used in this book for the languages $\mathrm{L}_1\mathrm{Set}$ and $\mathrm{L}_1\mathrm{Ar}$. We then describe the rules for forming distinguished expressions and briefly discuss semantics.

The distinguished expressions of any language L in the class $\mathfrak{L}_1$ are divided into two types: *terms* and *formulas.* Both types are defined recursively.

2.2. **Definition.** *Terms* are the elements of the least subset of the expressions of the language that satisfies the following two conditions:

(a) Variables and constants are (atomic) terms.
(b) If f is an operation of degree r and $t_1, \dots, t_r$ are terms, then $f(t_1, \dots, t_r)$ is a term.

In (a) we identify an element with a sequence of length one. The alphabet does not include commas, which are part of our abbreviated notation: $f(t_1,\ t_2,\ t_3)$ means the same as $f(t_1t_2t_3)$. In §1 of Chapter II we explain how a sequence of terms can be uniquely deciphered despite the absence of commas.

If two sets of expressions in the language satisfy conditions (a) and (b), then the intersection of the two sets also satisfies these conditions. Therefore the definition of the set of terms is correct.

Language Alphabets

Subsets of the Alphabet	Names and Notation		
	General	in $\mathrm{L}_1\mathrm{Set}$	in $\mathrm{L}_1\mathrm{Ar}$
connectives and quantifiers	$\Leftrightarrow$(equivalent); $\Rightarrow$(implies); $\vee$(inclusive or); $\wedge$ (and); $\neg$(not); $\forall$ (universal quantifier); $\exists$ (existential quantifier)		
variables	$x,\ y,\ z,\ u,\ v, \dots$ with indices		
constants	c $\cdots$ with indices	$\emptyset$ (empty set)	$\bar{0}$ (zero); $\bar{1}$ (one)
operations of degree 1, 2, 3, ...	$f,\ g, \dots$ with indices	none	$+$ (addition, degree 2); $\cdot$(multiplication, degree 2)
relations (predicates) of degree 1, 2, 3, ...	$p,\ q, \dots$ with indices	$\in$ (is an element of, degree 2); $=$ (equals, degree 2)	$=$ (equality, degree 2)
parentheses	((left parenthesis);)(right parenthesis)		

2.3. **Definition.** *Formulas* are the elements of the least subset of the expressions of the language that satisfies the following two conditions:

(a) If p is a relation of degree r and $t_1, \dots, t_r$ are terms, then $p(t_1, \dots, t_r)$ is an (atomic) formula.

(b) If P and Q are formulas (abbreviated notation!), and x is a variable, then the expressions

$$(P) \Leftrightarrow (Q),\ (P) \Rightarrow (Q),\ (P) \vee (Q),\ (P) \wedge (Q),$$
$$\neg(P),\ \forall x(P),\ \exists x(P)$$

are formulas.

It is clear from the definitions that any term is obtained from atomic terms in a finite number of steps, each of which consists in "applying an operation symbol" to the earlier terms. The same is true for formulas. In Chapter II, §1 we make this remark more precise.

The following initial interpretations of terms and formulas are given for the purpose of orientation and belong to the so-called "standard models" (see Chapter II, §2 for the precise definitions).

2.4. Examples and interpretations

(a) The terms stand for (are notation for) the objects of the theory. Atomic terms stand for indeterminate objects (variables) or concrete objects (constants). The term $f(t_1, \ldots, t_r)$ is the notation for the object obtained by applying the operation denoted by f to the objects denoted by $t_1, \ldots, t_r$. Here are some examples from $\mathrm{L}_1\mathrm{Ar}$:

$$\begin{aligned} \bar{0}\ & \text{ denotes zero;} \\ \bar{1}\ & \text{ denotes one;} \\ +(\bar{1}, \bar{1})\ & \text{ denotes two } (1 + 1 = 2 \text{ in the usual notation});\\ +\Big(\bar{1} + (\bar{1}, \bar{1})\Big)\ & \text{ denotes three;} \\ \cdot\Big(+ (\bar{1}, \bar{1}) + (\bar{1}, \bar{1})\Big)\ & \text{ denotes four } (2 \times 2 = 4). \end{aligned}$$

Since this normalized notation is different from what we are used to in arithmetic, in $\mathrm{L}_1\mathrm{Ar}$ we shall usually write simply $t_1 + t_2$ instead of $+(t_1, t_2)$ and $t_1 \cdot t_2$ instead of $\cdot(t_1, t_2)$. This convention may be considered as another use of abbreviated notation:

$$\begin{aligned} x &\text{ stands for an indeterminate integer;} \\ x + \bar{1} \text{ (or } + (x, \bar{1})) &\text{ stands for the next integer.} \end{aligned}$$

In the language $\mathrm{L}_1\mathrm{Set}$ all terms are atomic:

$$\begin{aligned} x &\text{ stands for an indeterminate set;} \\ \varnothing &\text{ stands for the empty set.} \end{aligned}$$

(b) The formulas stand for statements (arguments, propositions, ...) of the theory. When translated into formal language, a statement may be either true, false, or indeterminate (if it concerns indeterminate objects); see Chapter II for the precise definitions. In the general case the atomic formula $p\,(t_1, \ldots, t_r)$ has roughly the following meaning: "The ordered r-tuple of objects denoted by $t_1, \ldots, t_r$ has the property denoted by p." Here are some examples of atomic formulas in $\mathrm{L}_1\mathrm{Ar}$. Their general structure is $= (t_1, t_2)$, or, in nonnormalized notation, $t_1 = t_2$:

$$\bar{0} = \bar{1}, \qquad x + \bar{1} = y.$$

Here are some examples of formulas which are not atomic:

$$\begin{aligned}&\neg(\bar{0} = \bar{1}),\\ &(x = \bar{0}) \Leftrightarrow (x + \bar{1} = \bar{1}),\\ &\forall\, x\Big((x = \bar{0}) \vee (\neg(x \cdot x = \bar{0}))\Big).\end{aligned}$$

Some atomic formulas in L_1 Set

$$y \in x \qquad (y \text{ is an element of } x),$$

and also $\varnothing \in y,\ x \in \varnothing$, etc. Of course, normalized notation must have the form $\in (xy)$, and so on.

Some nonatomic formulas:

$$\exists\, x\big(\forall y(\neg(y \in x))\big): \qquad \text{there exists an } x \text{ of which no } y \text{ is an element.}$$

Informally this means: "The empty set exists." We once again recall that an informal interpretation presupposes some standard interpretive system, which will be introduced explicitly in Chapter II.

$$\forall\, y(y \in z \Rightarrow y \in x): \qquad z \text{ is a subset of } x.$$

This is an example of a very useful type of abbreviated notation: four parentheses are omitted in the formula on the left. We shall not specify precisely when parentheses may be omitted; in any case, it must be possible to reinsert them in a way that is unique or is clear from the context without any special effort.

We again emphasize: the abbreviated notation for formulas are only material designations. Abbreviated notation is chosen for the most part with psychological goals in mind: speed of reading (possibly with a loss in formal uniqueness), tendency to encourage useful associations and discourage harmful ones, suitability to the habits of the author and reader, and so on. The mathematical objects in the theory of formal languages are the formulas themselves, and not any particular designations.

Digression: Names

On several occasions we have said that a certain object (a sign on paper, an element of an alphabet as an abstract set, etc.) is a notation for, or denotes, another element. A convenient general term for this relationship is naming.

The letter x is the name of an element of the alphabet; when it appears in a formula, it becomes the name of a set or a number; the notation $x \in y$ is the name of an expression in the alphabet A, and this expression, in turn, is the name of an assertion about indeterminate sets; and so on.

When we form words, we often identify the names of objects with the objects themselves: we say "the variable x," "the formula P," "the set z." This can sometimes be dangerous. The following passage from Rosser's book *Logic for Mathematicians* points up certain hidden pitfalls:

> The gist of the matter is that, if we have a statement such as "3 is greater than $\frac{9}{12}$" about the rational number $\frac{9}{12}$ and containing a name "$\frac{9}{12}$" of this rational number, one can replace this name by any other name of the same rational number, for instance, "$\frac{3}{4}$." If we have a statement such as "3 divides the denominator of '$\frac{9}{12}$' " about a name of a rational number and containing a name of this name, one can replace this name of the name by some other name of the same name, but not in general by the name of some other name, if it is a name of some other name of the same rational number.

Rosser adds that "failure to observe such distinctions carefully can seldom lead to confusion in logic and still more seldom in mathematics." However, these distinctions play a significant role in philosophy and in mathematical practice.

"A rose by any other name would smell as sweet"—this is true because roses exist outside of us and smell in and of themselves. But, for example, it seems that Hilbert spaces "exist" only insofar as we talk about them, and the choice of terminology here makes a difference. The word "space" for the set of equivalence classes of square integrable functions was at the same time a codeword for an entire circle of intuitive ideas concerning "real" spaces. This word helped organize the concept and led it in the right direction.

A successfully chosen name is a bridge between scientific knowledge and common sense, between new experience and old habits. The conceptual foundation of any science consists of a complicated network of names of things, names of ideas, and names of names. It evolves itself, and its projection on reality changes.

3 Beginners' Course in Translation

3.1. We recall that the formulas in $\mathrm{L_1Set}$ stand for statements about sets; the formulas in $\mathrm{L_1Ar}$ stand for statements about natural numbers; these formulas contain names of sets and numbers, which may be indeterminate.

In this section we give the first basic examples of two-way translation "argot $\Leftrightarrow$ formal language." One of our purposes will be to indicate the great expressive possibilities in $\mathrm{L}_1\mathrm{Set}$ and $\mathrm{L}_1\mathrm{Ar}$, despite the extremely limited modes of expression.

As in the case of natural languages, this translation cannot be given by rigid rules, is not uniquely determined, and is a creative process. Compare Hesse's quatrain with its translation in the epigraph to this book: the most important aim of translation is to "understand ... just what they say."

Before reading further, the reader should look through the appendix to Chapter II: "The von Neumann Universe." The semantics implicit in $\mathrm{L}_1\mathrm{Set}$ relates to this universe, and not to arbitrary "Cantor" sets.

A more complete picture of the meaning of the formulas can be obtained from §2 of Chapter II.

Translation from $\mathrm{L}_1\mathrm{Set}$ *to argot.*

3.2. $\forall\, x(\neg(x \in \varnothing))$: "for all (sets) x it is false that x is an element of (the set) $\varnothing$" (or "$\varnothing$ is the empty set").

The second assertion is equivalent to the first only in the von Neumann universe, where the elements of sets can only be sets, and not real numbers, chairs, or atoms.

3.3. $\forall\, z(z \in x \Leftrightarrow z \in y) \Leftrightarrow x = y$: "if for all z it is true that z is an element of x if and only if z is an element of y, then it is true that x coincides with y; and conversely," or "a set is uniquely determined by its elements."

In the expression 3.3 at least six parentheses have been omitted; and the subformulas $z \in x$, $z \in y$, $x = y$ have not been normalized according to the rules of $\mathfrak{L}_1$.

3.4. $\forall u\ \forall v\ \exists x\ \forall z(z \in x \Leftrightarrow (z = u \vee z = v))$: "for any two sets u, v there exists a third set x such that u and v are its only elements."

This is one of the axioms of Zermelo–Fraenkel. The set x is called the "unordered pair of sets u, v" and is denoted $\{u,\ v\}$ in the appendix.

3.5. $\forall y\ \forall z(((z \in y \wedge y \in x) \Rightarrow z \in x) \wedge (y \in x \Rightarrow \neg(y \in y)))$: "the set x is partially ordered by the relation $\in$ between its elements."

We mechanically copied the condition $y \in x \Rightarrow \neg(y \in y)$ from the definition of partial ordering. This condition is automatically fulfilled in the von Neumann universe, where no set is an element of itself.

A useful exercise would be to write the following formulas:

"x is totally ordered by the relation $\in$";

"x is linearly ordered by the relation $\in$";

"x is an ordinal."

3.6. $\forall x(y \in z)$: The literal translation "for all x it is true that y is an element of z" sounds a little strange. The formula $\forall x\, \exists x(y \in z)$, which agrees with the rules for constructing formulas, looks even worse. It would be possible to make the rules somewhat more complicated, in order to rule out such formulas, but in general they cause no harm. In Chapter II we shall see that from the point of view of "truth" or "deducibility," such a formula is equivalent to the formula $y \in z$. It is in this way that they must be understood.

Translation from argot to $\mathrm{L_1Set}$.

We choose several basic constructions having general mathematical significance and show how they are realized in the von Neumann universe, which contains only sets obtained from $\varnothing$ by the process of "collecting into a set," and in which all relations must be constructed from $\in$.

3.7. "x is the direct product $y \times z$."

This means that the elements of x are the ordered pairs of elements of y and z, respectively. The definition of an unordered pair is obvious: the formula

$$\forall u\ (u \in x \Leftrightarrow (u = y_1 \vee u = z_1))$$

"means," or may be briefly written in the form, $x = \{y_1, z_1\}$ (compare 3.4). The ordered pair y_1 and z_1 is introduced using a device of Kuratowski and Wiener: this is the set x_1 whose elements are the unordered pairs $\{y_1, y_1\}$ and $\{y_1, z_1\}$.

We thus arrive at the formula

$$\exists y_2\ \exists z_2(\text{“}x_1 = \{y_2, z_2\}\text{”} \wedge \text{“}y_2 = \{y_1, y_1\}\text{”} \wedge \text{“}z_2 = \{y_1, z_1\}\text{”}),$$

which will be abbreviated

$$x_1 = \langle y_1, z_1 \rangle$$

and will be read "x_1 is the ordered pair with first element y_1 and second element z_1." The abbreviated notation for the subformulas is in quotes; we shall later omit the quotation marks.

Finally, the statement "$x = y \times z$" may be written in the form

$$\forall x_1(x_1 \in x \Leftrightarrow \exists y_1\ \exists z_1(y_1 \in y \wedge z_1 \in z \wedge \text{“}x_1 = \langle y_1, z_1 \rangle\text{”})).$$

In order to remind the reader for the last time of the liberties taken in abbreviated notation, we write this same formula adhering to all the canons of $\mathfrak{L}_1$:

$$\forall x_1 \Bigg[(\in (x_1 x))$$
$$\Leftrightarrow \Bigg[\exists y_1 \Bigg(\exists z_1 \Bigg(\Big(((\in (y_1 y)) \wedge (\in (z_1 z)) \Big) \wedge \Bigg(\exists y_2 \Bigg(\exists z_2 \Bigg(\Bigg(\Big(\forall u ((\in (u x_1))$$
$$\Leftrightarrow ((= (u y_2) \vee (= (u z_2)) \Big) \Big) \Big) \wedge (\forall u ((\in (u y_2))$$
$$\Leftrightarrow (= (u y_1))))) \wedge (\forall u ((\in (u z_2) \Leftrightarrow ((= (u y_1)) \vee (= (u z_1)))))) \Bigg) \Bigg) \Bigg) \Bigg) \Bigg] \Bigg]$$

EXERCISE: Find the open parenthesis corresponding to the fifth closed parenthesis from the end. In §1 of Chapter II we give an algorithm for solving such problems.

3.8. "*f is a mapping from the set u to the set v.*"

First of all, mappings, or functions, are identified with their graphs; otherwise, we would not be able to consider them as elements of the universe. The following formula successively imposes three conditions on f: f is a subset of $u \times v$; the projection of f onto u coincides with all of u; and each element of u corresponds to exactly one element of v:

$$\begin{aligned}&\forall z\big(z \in f \Rightarrow (\exists u_1\, \exists v_1 (u_1 \in u \wedge v_1 \in v \wedge \text{“}z = \langle u_1, v_1\rangle\text{”}))\big)\\ &\wedge \forall u_1 (u_1 \in u \Rightarrow \exists v_1\, \exists z (v_1 \in v \wedge \text{“}z = \langle u_1, v_1\rangle\text{”} \wedge z \in f))\\ &\wedge \forall u_1\, \forall v_1\, \forall v_2 (\exists z_1\, \exists z_1 (z_1 \in f \wedge z_2 \in f \wedge \text{“}z_1 = \langle u_1, v_1\rangle\text{”} \wedge \text{“}z_2 = \langle u_1, v_2\rangle\text{”}))\\ &\qquad\qquad\qquad\qquad\qquad\qquad\qquad\qquad\qquad\qquad\qquad\qquad\qquad \Rightarrow v_1 = v_2\big).\end{aligned}$$

EXERCISE: Write the formula "f is the projection of $y \times z$ onto z."

3.9. "*x is a finite set.*"

Finiteness is far from being a primitive concept. Here is Dedekind's definition: "there does not exist a one-to-one mapping f of the set x onto a proper subset." The formula:

$$\begin{aligned}&\neg\exists f\big(\text{“}f \text{ is a mapping from } x \text{ to } x\text{”} \wedge \forall u_1\, \forall u_2\, \forall v_1\, \forall v_2 ((\text{“}\langle u_1, v_1\rangle \in f\text{”}\\ &\wedge \text{“}\langle u_2, v_2\rangle \in f\text{”} \wedge \neg(u_1 = u_2)) \Rightarrow \neg(v_1 = v_2) \wedge \exists v_1 (v_1 \in x \wedge \neg\exists u_1\\ &\qquad\qquad\qquad\qquad\qquad\qquad\qquad\qquad\qquad\qquad (\text{“}\langle u_1, v_1\rangle \in f\text{”}))\big).\end{aligned}$$

The abbreviation "$\langle u_1, v_1\rangle \in f$" means, of course, $\exists y(\text{“}y = \langle u_1, v_1\rangle)\text{”} \wedge y \in f)$.

3.10. "*x is a nonnegative integer.*"

The natural numbers are represented in the von Neumann universe by the finite ordinals, so that the required formula has the form

$$\text{“}x \text{ is totally ordered by the relation } \in\text{”} \wedge \text{“}x \text{ is finite.”}$$

EXERCISE: Figure out how to write the formulas "$x + y = z$" and "$x \cdot y = z$" where x, y, z are integers $\geqslant 0$.

After this it is possible in the usual way to write the formulas "x is an integer," "x is a rational number," "x is a real number" (following Cantor or Dedekind), etc., and then construct a formal version of analysis. The written statements will have acceptable length only if we periodically extend the language $\mathrm{L_1Set}$ (see §8 of Chapter II). For example, in $\mathrm{L_1Set}$ we are not allowed to write term-names for the numbers 1, 2, 3, ... ($\varnothing$ is the name for 0), although we may construct the formulas "x is the finite ordinal containing 1 element," "x is the finite ordinal containing 2 elements," etc. If we use such roundabout methods of expression, the simplest numerical identities become incredibly long; but of course, in logic we are mainly concerned with the theoretical possibility of writing them.

3.11. "*x is a topological space.*"

In the formula we must give the topology of x explicitly. We define the topology, for example, in terms of the set y of all open subsets of x. We first write that y consists of subsets of x and contains x and the empty set:

$$P_1 : \quad \forall z(z \in y \Rightarrow \forall u(u \in z \Rightarrow u \in x)) \wedge x \in y \wedge \varnothing \in y.$$

The intersection w of any two elements u, v in y is open, i.e., belongs to y:

$$P_2 : \quad \forall u\, \forall v\, \forall w((u \in y \wedge v \in y \wedge \forall z((z \in u \wedge z \in v) \Leftrightarrow z \in w)) \Rightarrow w \in y).$$

It is harder to write "the union of any set of open subsets is open." We first write

$$P_3 : \quad \forall u(u \in z \Leftrightarrow \forall v(v \in u \Rightarrow v \in y)),$$

that is, "z is the set of all subsets of y." Then

$$P_4 : \quad \forall u\, \forall w((u \in z \wedge \forall v_1(v_1 \in w \Leftrightarrow \exists v(v \in u \wedge v_1 \in v))) \Rightarrow w \in y).$$

This means (taking into account P_3, which defines z); "If u is any subset of y, i.e., a set of open subsets of x, then the union w of all these subsets belongs to y, i.e., is open." Now the final formula may be written as follows:

$$P_1 \wedge P_2 \wedge \forall z(P_3 \Rightarrow P_4).$$

The following comments on this formula will be reflected in precise definitions in Chapter II, §§1 and 2. The letters x, y have the same meaning in all the P_i, while z plays different roles: in P_1 it is a subset of x, and in P_3 and P_4 it is the set of subsets of x. We are allowed to do this because as soon as we "bind" z by the quantifier $\forall$, say in P_1, z no longer stands for an (indeterminate) individual set, and becomes a temporary designation for "any set." Where the "scope of action" of $\forall$ ended, z can be given a new meaning. In order to "free" z for later use, $\forall z$ was also put before $P_3 \Rightarrow P_4$.

Translation from argot to L₁Ar.

3.12. "$x < y$": $\exists z(y = (x + z) + \bar{1})$. Recall that the variables are names for nonnegative integers.

3.13. "*x is a divisor of y*": $\exists z(y = x \cdot z)$.

3.14. "*x is a prime number*": "$\bar{1} < x$" $\wedge$ ("*y is a divisor of x*" $\Rightarrow (y = \bar{1} \vee y = x)$).

3.15. "*Fermat's last theorem*": $\forall x_1\, \forall x_2\, \forall x_3\, \forall u($"$\bar{2} < u$" $\wedge$ "$x_1^u + x_2^u = x_3^u$" $\Rightarrow$ "$x_1 x_2 x_3 = \bar{0}$"$)$. It is not clear how to write the formula $x_1^u + x_2^u = x_3^u$

in L_1Ar. Of course, for any *concrete* $u = 1, 2, 3$ there is a corresponding atomic formula in L_1Ar, but how do we make u into a variable? This is not a trivial problem. In the second part of the book we show how to find an atomic formula $p(x, u, y, z_1, \ldots, z_n)$ such that the assertion that $\exists z_1 \cdots \exists z_n p\,(x, u, y, z_1, \ldots, z_n)$ in the domain of natural numbers is equivalent $y = x^u$. Then $x_1^u + x_2^u = x_3^u$ can be translated as follows:

$$\exists y_1\, \exists y_2\, \exists y_3\, (\text{“}x_1^u = y_1\text{”} \wedge \text{“}x_2^u = y_2\text{”} \wedge \text{“}x_3^u = y_3\text{”} \wedge y_1 + y_2 = y_3).$$

The existence of such a p is a nontrivial number-theoretic fact, so that here the very possibility of performing a translation becomes a mathematical problem.

3.16. “*The Riemann hypothesis.*” The Riemann zeta function $\zeta\,(s)$ is defined by the series $\Sigma_{n=1}^{\infty}\, n^{-s}$ in the half-plane Re $s \geq 1$. It can be continued meromorphically onto the entire complex s-plane. The Riemann hypothesis is the assertion that the nontrivial zeros of $\zeta(s)$ lie on the line Re $s = \frac{1}{2}$. Of course, in this form the Riemann hypothesis cannot be translated into L_1Ar. However, there are several purely arithmetic assertions that are demonstrably equivalent to the Riemann hypothesis. Perhaps the simplest of them is the following.

Let $\mu(n)$ be the Möbius function on the set of integers $\geqslant 1$: it equals 0 if n is divisible by a square, and equals $(-1)^r$, where r is the number of prime divisors of n, if n is square-free. We then have

$$\text{Riemann hypothesis } \Leftrightarrow \forall\varepsilon > 0\ \exists x\ \forall y \left[y > x \Rightarrow \left[\left| \sum_{n=1}^{y} \mu(n) \right| < y^{1/2+\varepsilon} \right] \right].$$

Only the exponent is not an integer on the right; but ε need only run through numbers of the form $1/z$, z an integer $\geqslant 1$, and then we can raise the inequality to the $(2z)$th power. The formula

$$\left(\sum_{n=1}^{y} \mu(n) \right)^{2z} < y^{z+2}$$

can then be translated into L_1Ar, although not completely trivially. The necessary techniques will be developed in the second part of the book.

The last two examples were given in order to show the complexity that is possible in problems that can be stated in L_1Ar, despite the apparent simplicity of the modes of expression and the semantics of the language.

We conclude this section with some remarks concerning higher-order languages.

3.17. *Higher-order languages.* Let L be any first-order language. Its modes of expression are limited in principle by one important consideration: we are not allowed to speak of arbitrary properties of objects of the theory, that is, arbitrary subsets of the set of all objects. Syntactically, this is reflected in the prohibition against forming expressions such as $\forall p(p(x))$, where p is a relation of degree 1; relations must stand for fixed rather than variable properties.

Of course, certain properties can be defined using nonatomic formulas. For example, in $\mathrm{L_1Ar}$ instead of "x is even" we may write $\exists y(x = (\bar{1} + \bar{1}) \cdot y)$. However, there is a continuum of subsets of the integers but only a countable set of definable properties (see §2 of Chapter II), so there are automatically properties that cannot be defined by formulas. Thus, it is impossible to replace the forbidden expression $\forall p(p(x))$ by a sequence of expressions $P_1(x),\ P_2(x),\ P_3(x), \ldots$.

Languages in which quantifiers may be applied to properties and/or functions (and also, possibly, to properties of properties, and so on) are called higher-order languages. One such language—$\mathrm{L_2Real}$—will be considered in Chapter III for the purpose of illustrating a simplified version of Cohen forcing.

On the other hand, the same extension of expressive possibilities can be obtained without leaving $\mathfrak{L}_1$. In fact, in the first-order language $\mathrm{L_1Set}$ we may quantify over all subsets of any set, over all subsets of the set of subsets, and so on. Informally this means that we are speaking of all properties, all properties of properties, ... (with transfinite extension). In addition, any higher-order language with a "standard interpretation" in some type of structured sets can be translated into $\mathrm{L_1Set}$ so as to preserve the meanings and truth values in this standard interpretation. (An apparent exception is the languages for describing Gödel–Bernays classes and "large" categories; but it seems, based on our present understanding of paradoxes, that no higher-order languages can be constructed from such a language.)

The attentive reader will notice the contrast between the *possibility* of writing a formula in $\mathrm{L_1Set}$ in which $\forall$ is applied to all subsets (informally, to all properties) of finite ordinals (informally, of integers) and the *impossibility* of writing a formula in $\mathrm{L_1Set}$ that would define any concrete subset in the continuum of undefinable subsets. (There are fewer such subsets in $\mathrm{L_1Set}$ than in $\mathrm{L_1Ar}$, but still a continuum.) We shall examine these problems more closely in Chapter II when we discuss "Skolem's paradox."

Let us summarize. Almost all the basic logical and set-theoretic principles used in the day-to-day work of the mathematician are contained in the first-order languages and, in particular, in $\mathrm{L_1Set}$. Hence, those languages will be the subject of study in the first and third parts of the book. But concrete oriented languages can be formed in other ways, with various degrees of deviation from the rules of $\mathfrak{L}_1$. In addition to $\mathrm{L_2Real}$, examples of such languages examined in Chapter II include SELF (Smullyan's language for self-description) and SAr, which is a language of arithmetic convenient for proving Tarski's theorem on the undefinability of truth.

Digression: Syntax

1. The most important feature that most artificial languages have in common is the ability to encompass a rich spectrum of modes of expression starting with a small finite number of generating principles.

In each concrete case the choice of these principles (including the alphabet and syntax) is based on a compromise between two extremes. Economical use of modes of expression leads to unified notation and simplified mechanical analysis of the text. But then the texts become much longer and farther removed from natural language texts. Enriching the modes of expression brings the artificial texts closer to the natural language texts, but complicates the syntax and the formal analysis. (Compare machine languages with such programming languages as Algol, Fortran, Cobol, etc.)

We now give several examples based on our material.

2. *Dialects of* $\mathfrak{L}_1$

(a) Without changing the logic in $\mathfrak{L}_1$, it is possible to discard parentheses and either of the two quantifiers from the alphabet, and to replace all the connectives by one, namely $\downarrow$ (conjunction of negations). (In addition, constants could be declared to be functions of degree 0, and functions could be interpreted as relations.)

This is accomplished by the following change in the definitions. If $t_1, \dots, t_r$ are terms, f is an operation of degree r, and p is a relation of degree r, then $ft_1 \dots t_r$ is a term, and $pt_1 \dots t_r$ is an atomic formula. If P and Q are formulas, then $\downarrow PQ$ and $\forall x P$ are formulas. The content of $\downarrow PQ$ is "not P and not Q" so that we have the following expressions in this dialect:

$$\begin{aligned} \neg(P): &\quad \downarrow PP, \\ (P) \wedge (Q): &\quad \downarrow\downarrow PP \downarrow QQ, \\ (P) \vee (Q): &\quad \downarrow\downarrow PQ \downarrow PQ. \end{aligned}$$

Clearly, economizing on parentheses and connectives leads to much repetition of the same formula. Nevertheless, it may become simpler to prove theorems about such a language because of the shorter list of syntactic norms.

(b) Bourbaki's language of set theory has an alphabet consisting of the signs $\square$, τ, $\vee$, $\neg$, $=$, $\in$ and the letters. Expressions in this language are not simply sequences of signs in the alphabet, but sequences in which certain elements are paired together by superlinear connectives. For example:

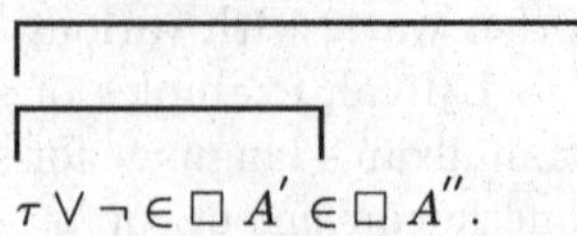

$$\tau \vee \neg \in \square A' \in \square A''.$$

The main difference between Bourbaki's language and $\mathrm{L}_1\mathrm{Set}$ is the use of the "Hilbert choice symbol." If, for example, $\in xy$ is the formula "x is an element of y," then

$$\tau \in \square\, y$$

is a term meaning "some element of the set y."

Bourbaki's language is not very convenient and is not widely used. It became known in the popular literature thanks to an example of a very long abbreviated notation for the term "one," which the authors imprudently introduced:

$$\tau_z\Big((\exists u)(\exists U)(u = (U,\ \{\varnothing\}, Z) \wedge U \subset \{\varnothing\} \times Z \wedge (\forall k)((x \in \{\varnothing\})$$
$$\Rightarrow (\exists y)((x,y) \in U)) \wedge (\forall x)(\forall y)(\forall y')(((x,y \in U \wedge (x,y') \in U)$$
$$\Rightarrow (y = y')) \wedge (\forall y)((y \in Z) \Rightarrow (\exists x)x((x,y) \in U))))\Big).$$

It would take several tens of thousands of symbols to write out this term completely; this seems a little too much for "one."

(c) A way to greatly extend the expressive possibilities of almost any language in $\mathfrak{L}_1$ is to allow "class terms" of the type $\{x|P(x)\}$, meaning "the class of all objects x having the property P." This idea was used by Morse in his language of set theory and by Smullyan in his language of arithmetic; see §10 of Chapter II.

3. *General remarks.* Most natural and artificial languages are characteristically discrete and linear (one-dimensional). On the one hand, our perception of the external world is not felt by us to be either discrete or linear, although these characteristics are observed on the level of physiological mechanisms (coding by impulses in the nervous system). On the other hand, the languages in which we communicate tend to transmit information in a sequence of distinguishable elementary signs. The main reason for this is probably the much greater (theoretically unlimited) uniqueness and reproducibility of information than is possible with other methods of conveyance. Compare with the well-known advantages of digital over analog computers.

The human brain clearly uses both principles. The perception of images as a whole, along with emotions, are more closely connected with nonlinear and nondiscrete processes—perhaps of a wave nature. It is interesting to examine from this point of view the nonlinear fragments in various languages.

In mathematics this includes, first of all, the use of drawings. But this use does not lend itself to formal description, with the exception of the separate and formalized theory of graphs. Graphs are especially popular objects, because they are as close as possible both to their visual image as a whole and to their description using all the rules of set theory. Every time we are able to connect a problem with a graph, it becomes much simpler to discuss it, and large sections of verbal description are replaced by manipulation with pictures.

A less well-known class of examples is the commutative diagrams and spectral sequences of homological algebra. A typical example is the "snake lemma." Here is its precise formulation.

Suppose we are given a commutative diagram of abelian groups and homomorphisms between them (in the box below), in which the rows are exact sequences:

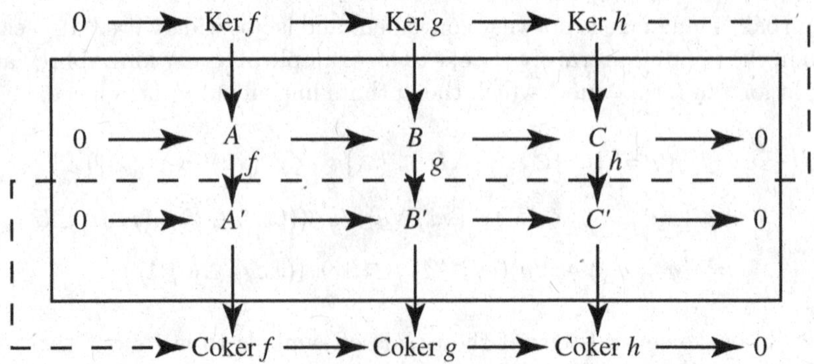

Then the kernels and cokernels of the "vertical" homomorphisms f, g, h form a six-term exact sequence, as shown in the drawing, and the entire diagram of solid arrows is commutative. The "snake" morphism Ker $h \to$ Coker f, which is denoted by the dotted arrow, is the basic object constructed in the lemma.

Of course, it is easy to describe the snake diagram sequentially in a suitable, more or less formal, linear language. However, such a procedure requires an artificial and not uniquely determined breaking up of a clearly two-dimensional picture (as in scanning a television image). Moreover, without having the overall image in mind, it becomes harder to recognize the analogous situation in other contexts and to bring the information together into a single block.

The beginnings of homological algebra saw the enthusiastic recognition of useful classes of diagrams. At first this interest was even exaggerated; see the editor's appendix to the Russian translation of *Homological Algebra* by Cartan and Eilenberg.

There is one striking example of an entire book with an intentional two-dimensional (block) structure: C. H. Lindsey and S. G. van der Meulen, *Informal Introduction to Algol 68* (North-Holland, Amsterdam, 1971). It consists of eight chapters, each of which is divided into seven sections (eight of the 56 sections are empty, to make the system work!). Let (i, j) be the name of the jth section of the ith chapter; then the book can be studied either "row by row" or "column by column" in the (i, j) matrix, depending on the reader's intentions.

As with all great undertakings, this is the fruit of an attempt to solve what is in all likelihood an insoluble problem, since, as the authors remark, Algol 68 "is quite impossible to describe ... until it has been described."

II

Truth and Deducibility

1 Unique Reading Lemma

The basic content of this section is Lemma 1.4 and Definitions 1.5 and 1.6. The lemma guarantees that the terms and formulas of any language in $\mathfrak{L}_1$ can be deciphered in a unique way, and it serves as a basis for most inductive arguments. (The reader may take the lemma on faith for the time being, provided that he was able independently to verify the last formula in 3.7 of Chapter I. However, the proof of the lemma will be needed in (§4 of Chapter VII.) It is important to remember that the theory of any formal language begins by checking that the syntactic rules are free of ambiguity.

We begin with the standard combinatoric definitions, in order to fix the terminology.

1.1. Let A be a set. By a sequence of length n of elements of A we mean a mapping from the set $\{1, \ldots, n\}$ to A. The image of i is called the ith term of the sequence. Corresponding to $n = 0$ we have the empty sequence. Sequences of length 1 will sometimes be identified with elements of A.

A sequence of length n can also be written in the form $a_1, \ldots, a_i, \ldots, a_n$, where a_i is its ith term. The number i is called the index of the term a_i. If $P = (a_1, \ldots, a_n)$ and $Q = (b_1, \ldots, b_m)$ are two sequences, their concatenation PQ is the sequence $(a_1, \ldots, a_n, b_1, \ldots, b_m)$ of length $m + n$ whose ith term is a_i for $i \leqslant n$ and b_{i-n} for $n+1 \leqslant i \leqslant n+m$. We similarly define the concatenation of a finite sequence of sequences.

An *occurrence* of the sequence Q in P is any representation of P as a concatenation P_1QP_2. Substituting a sequence R in place of a given occurrence of Q in P amounts to constructing the sequence P_1RP_2.

Let Π^+, Π^- be two disjoint subsets of $(1, \ldots, n)$. A map $c : \Pi^+ \to \Pi^-$ is called a *parentheses bijection* if it is bijective and satisfies the following conditions:

(a) $c(i) > i$ for all $i \in \Pi^+$;
(b) for every i and $j, j \in [i, c(i)]$ if and only if $c(j) \in [i, c(i)]$.

Yu. I. Manin, *A Course in Mathematical Logic for Mathematicians, Second Edition*, Graduate Texts in Mathematics 53, DOI 10.1007/978-1-4419-0615-1_2,

1.2. **Lemma**. *Given Π^+ and Π^-, if a parentheses bijection exists, then it is unique.*

This lemma will be applied to expressions in languages in $\mathfrak{L}_1$: Π^+ will consist of the indices of the places in the expression at which "(" occurs, Π^- will consist of the indices of the places at which ")" occurs, and the map c correlates to each left parenthesis the corresponding right parenthesis.

PROOF OF THE LEMMA. Let the function $\varepsilon : \{1, \dots, n\} \to \{0, \pm 1\}$ take the value 1 on Π^+ , -1 on Π^-, and 0 everywhere else. We claim that for every $i \in \Pi^+$, for any parentheses bijection $c : \Pi^+ \to \Pi^-$, and for any $k, 1 \leqslant k \leqslant c(i) - i$, we have the relations

$$\sum_{j=1}^{c(i)} \varepsilon(j) = 0, \qquad \sum_{j=1}^{c(i)-k} \varepsilon(j) > 0.$$

The lemma follows immediately from these relations, since we obtain the following recipe for determining c from Π^+ and Π^-; $c(i)$ is the least $l > i$ for which $\sum_{j=i}^{l} \varepsilon(j) = 0$.

The first relation holds because the elements of Π^+ and Π^- that appear in the interval $[i, c(i)]$ do so in pairs $(j, c(j))$, and $\varepsilon(j) + \varepsilon(c(j))= 0$.

To prove the second relation, suppose that for some i and k we have $\sum_{j=i}^{c(i)-k} \varepsilon(j) \leqslant 0$. Since $\varepsilon(i) = 1$, it follows that $\sum_{j=i+1}^{c(i)-k} \varepsilon(j) < 0$. Hence, the number of elements of Π^- in the interval $[i + 1, c(i) - k]$ is strictly greater than the number from Π^+. Let $c(j_0) \in \Pi^-$ be an element in the interval such that $j_0 \notin [i + 1, c(i) - k]$. Then $j_0 \leqslant i$, and in fact, $j_0 < i$, since $c(i)$ is outside the interval. But then only one element of the pair $j_0, c(j_0)$ lies in $[i, c(i)]$, which contradicts the definition of c. □

1.3. Now let A be the alphabet of a language L in $\mathfrak{L}_1$ (see §2 of Chapter I). Finite sequences of elements of A are the expressions in this language. Certain expressions have been distinguished as formulas or terms. We recall that the definitions in §2 of Chapter I imply that:

(a) Any term in L either is a constant, is a variable, or is represented in the form $f(t_1, \dots, t_r)$, where f is an operation of degree r, and $t_1, \dots, t_r$ are terms shorter in length.

(b) Any formula in L is represented either in the form $p(t_1, \dots, t_r)$, where p is a relation of degree r and $t_1, \dots, t_r$ are terms shorter in length, or in one of the seven forms

$$(P) \Leftrightarrow Q, \quad (P) \Rightarrow (Q), \quad (P) \vee (Q), \quad (P) \wedge (Q),$$
$$\neg(P), \quad \forall x(P), \quad \exists x(P),$$

where P and Q are formulas shorter in length, and x is a variable.

The following result is then obtained by induction on the length of the expression: *if E is a term or a formula, then there exists a parentheses bijection between the set Π^+ of indices of left parentheses in E and the set Π^- of indices of right parentheses.* In fact, the new parentheses in 1.3(a) and (b) have a natural bijection, while the old ones (which might be contained in the terms $t_1, \ldots, t_r$ or the formulas P, Q) have such a bijection by the induction assumption. In addition, the new parentheses never come between two paired old parentheses.

We can now state the basic result of this section:

1.4. Unique Reading Lemma. *Every expression in L is either a term, or a formula, or neither. These alternatives, as well as all of the alternatives listed in* 1.3(a) *and* (b), *are mutually exclusive. Every term (resp. formula) can be represented in exactly one of the forms in* 1.3(a) (*resp.*1.3(b)), *and in a unique way.*

In addition, in the course of the proof we show that *if an expression is the concatenation of a finite sequence of terms, then it is uniquely representable as such a concatenation.*

PROOF. Using induction on the length of the expression E, we describe an informal algorithm for syntactic analysis, which uniquely determines which alternative holds.

(a) If there are no parentheses in E, then E is either a constant term, a variable term, or neither a term nor a formula.
(b) If E contains parentheses, but there is no parentheses bijection between the left and right parentheses, then E is neither a term nor a formula.
(c) Suppose E contains parentheses with a parentheses bijection. Then either E is uniquely represented in one of the nine forms

$$f(E_0) \quad \text{(where } f \text{ is an operation)},$$

$$p(E_0) \quad \text{(where } p \text{ is a relation)},$$

$$(E_1) \Leftrightarrow (E_2), \quad (E_1) \Rightarrow (E_2), \quad (E_1) \vee (E_2), \quad (E_1) \wedge (E_2),$$

$$\neg(E_3), \quad \forall x(E_3), \quad \exists x(E_3),$$

or else E is neither a term nor a formula. Here the pairs of parentheses we have written out are connected by the unique parentheses bijection that is assumed to exist in E; this is what ensures uniqueness. In fact, we obtain the form $f(E_0)$ if and only if the first element of the expression is a function, the second element is "(", and the last element is the ")" that corresponds under the bijection: and similarly for the other forms.

We have thereby reduced the problem to the syntactic analysis of the expressions E_0, E_1, E_2, E_3, which are shorter in length. This almost completes our description of the algorithm, since what remains to be determined about E_1, E_2, E_3 is whether they are formulas. However, for E_0 we must determine whether this expression is a concatenation of the right number of terms, and we must ask whether such a representation must be unique.

The answer to the latter question is positive. We have the following recipe for breaking off terms from left to right in a union of terms.

(d) Let E_0 be an expression having a parentheses bijection between its left and right parentheses. If E_0 can be represented in the form tE_0', where t is a term, then this representation is unique. In fact, either E_0 can be uniquely represented in one of the forms

$$xE_0', \quad cE_0', \quad f(E_0'')E_0'$$

(where x is a variable, c is a constant, and f is an operation whose parentheses correspond under the unique parentheses bijection in E_0), or else E_0 cannot be represented in the form tE_0', where t is a term. In the cases $E_0 = xE_0'$ or $E_0 = cE_0'$, this is obviously the only way to break off a term from the left. In the case $E_0 = f(E_0'')E_0'$, the question reduces to whether E_0'' is a concatenation of degree-(f) terms. By induction on the length of E_0, we may assume that either E_0'' is not such a concatenation, or else it is uniquely representable as a concatenation of terms. The lemma is proved. □

EXERCISE: State and prove a unique reading lemma for the "parentheses-less" dialect of $\mathfrak{L}_1$ described in 2(a) of "Digression: Syntax" in Chapter I.

Here is the first inductive description of the difference between free and bound occurrences of a variable in terms and formulas. The correctness of the following definitions is ensured by Lemma 1.4.

1.5. **Definition.**

(a) Every occurrence of a variable in an atomic formula or term is free.
(b) Every occurrence of a variable in $\neg(P)$ or in $(P_1) * (P_2)$ (where $*$ is any of the connectives "$\vee$", "$\wedge$", "$\Rightarrow$", "$\Leftrightarrow$") is free (respectively bound) if and only if the corresponding occurrence in P, P_1, or P_2 is free (respectively bound).
(c) Every occurrence of the variable x in $\forall x(P)$ and $\exists x(P)$ is bound. The occurrences of other variables in $\forall x(P)$ and $\exists x(P)$ are the same as the corresponding occurrences in P.

Suppose the quantifier $\forall$ (or $\exists$) occurs in the formula P. It follows from the definitions that it must be followed in P by a variable and a left parenthesis. The expression that begins with this variable and ends with the corresponding right parenthesis is called the scope of the given (occurrence of the) quantifier.

1.6. **Definition.** Suppose we are given a formula P, a free occurrence of the variable x in P, and a term t. We say that t is free for the given occurrence of x in P if the occurrence does not lie in the scope of any quantifier of the form $\exists y$ or $\forall y$, where y is a variable occurring in t.

In other words, if t is substituted in place of the given occurrence of x, all free occurrences of variables in t remain free in P.

We usually have to substitute a term for each free occurrence of a given variable. It is important to note that this operation takes terms into terms and formulas into formulas (induction on the length). If t is free for each free occurrence of x in P we simply say that t is free for x in P.

1.7. We shall start working with Definitions 1.5 and 1.6 in the next section. Here we shall only give some intuitive explanations.

Definition 1.5 allows us to introduce the important class of *closed* formulas. By definition, this consists of formulas without free variables. (They are also called sentences.) The intuitive meaning of the concept of a closed formula is as follows. A closed formula corresponds to an assertion that is completely determined (in particular, regarding truth or falsity); indeterminate objects of the theory are mentioned only in the context "all objects x satisfy the condition ..." or "there exists an object y with the property ... ." Conversely, a formula that is not closed, such as $x \in y$ or $\exists x(x \in y)$, may be true or false depending on what sets are being designated by the names x and y (for the first) or by the name y (for the second). Here truth or falsity is understood to mean for a fixed interpretation of the language, as will be explained in §2.

In particular, Definition 1.6 gives the rules of hygiene for changing notation. If we want to call an indeterminate object x by another name y in a given formula, we must be sure that x does not appear in the parts of the formula where this name y is already being used to denote an *arbitrary* indeterminate object (after a quantifier). In other words, y must be free for x. Moreover, if we want to say that x is obtained from certain operations on other indeterminate objects (x = a term containing $y_1, \dots, y_n$), then the variables $y_1, \dots, y_n$ must not be bound.

There is a close parallel to these rules in the language of analysis: instead of $\int_1^x f(y)\,dy$ we may confidently write $\int_1^x f(z)\,dz$ but we must not write $\int_1^x f(x)\,dx$; the variable y is bound, in the scope of $\int f(y)\,dy$.

2 Interpretation: Truth, Definability

2.1. Suppose we are given a language L in $\mathfrak{L}_1$ and a set (or class) M. To give an *interpretation* of L in M means to tell how a formula in L can be given a meaning as a statement about the elements of M.

More precisely, an interpretation ϕ of the language L in M consists of a collection of mappings that correlate terms and formulas of the language to elements of M and structures over M (in the sense of Bourbaki). These mappings are divided into *primary* mappings, which actually determine the interpretation, and *secondary* mappings, which are constructed in a natural and unique way from the primary mappings. We shall use the term interpretation to refer to the mappings themselves, and sometimes also to the values they take.

Let us proceed to the systematic definitions. We shall sometimes call the elements of the alphabet of L *symbols.* The notation ϕ for the interpretation will either be included when the mappings are written or omitted, depending on the context.

2.2. *Primary mappings*

(a) An interpretation of the constants is a map from the set of symbols for constants (in the alphabet of L) to M that takes a symbol c to $\phi(c) \in M$.
(b) An interpretation of the operations is a map from the set of symbols for operations (in the alphabet of L) that takes a symbol f of degree r to a function $\phi(f)$ on $M \times \cdots \times M = M^r$ with values in M.
(c) An interpretation of the relations is a map from the set of symbols for relations (in the alphabet of L) that takes a symbol p of degree r to a subset $\phi(p) \subset M^r$.

Secondary mappings. Intuitively, we would like to interpret variables as names for the "generic element" of the set M, which can be given specific values in M. We would like to interpret the term $f(x_1, \ldots, x_r)$ as a function $\phi(f)$ of r arguments that run through values in M, and so on.

In order to give a precise definition, we introduce the *interpretation class* $\overline{M}$:

$\overline{M}$ = the set of all maps to M from the set of symbols for variables in the alphabet of L.

Thus, every point $\xi \in \overline{M}$ correlates to any variable x a value $\phi(x)(\xi) \in M$, which we shall usually denote simply by x^ξ. This allows us to consider variables as *functions on* $\overline{M}$ *with values in* M. More generally:

2.3. The interpretation of terms correlates to each term t a function $\phi(t)$ on $\overline{M}$ with values in M. This correspondence is defined inductively by the following compatibilities:

(a) If c is a constant, then $\phi(c)$ is the constant function whose value is defined by the primary mapping.
(b) If x is a variable, then $\phi(x)$ is $\phi(x)(\xi)$ as a function of ξ.
(c) If $t = f(t_1, \ldots, t_r)$, then for all $\xi \in \overline{M}$,

$$\phi(t)(\xi) = \phi(f)(\phi(t_1)(\xi), \ldots, \phi(t_r)(\xi)),$$

where the $\phi(t_i)(\xi)$ are defined by the induction assumption, and $\phi(f) : M^r \to M$ is given by the primary mapping. Instead of $\phi(t)(\xi)$ we shall sometimes write simply t^ξ.

2.4. *Interpretation of atomic formulas.* An interpretation ϕ assigns to every formula P in L a *truth* function $|P|_\phi$. This is a function on the interpretation class $\overline{M}$ that takes only the values 0 ("false") and 1 ("true"). It is defined for atomic formulas as follows:

$$|p(t_1, \ldots, t_r)|_\phi(\xi) = \begin{cases} 1, & \text{if } \langle t_1^\xi, \ldots, t_r^\xi \rangle \in \phi(p), \\ 0, & \text{otherwise.} \end{cases}$$

Intuitively, a statement p about the names $t_1, \ldots, t_r$ for objects in M becomes true if the objects named by $t_1, \ldots, t_r$ satisfy the relation named by p.

2.5. *Interpretation of formulas.* The truth function for nonatomic formulas is defined inductively by means of the following relations (for brevity, we have omitted parentheses and explicit mention of ϕ and ξ):

$$|P \Leftrightarrow Q| = |P||Q| + (1 - |P|)(1 - |Q|):$$

$P \Leftrightarrow Q$ is true when either P and Q are both true or P and Q are both false.

$$|P \Rightarrow Q| = 1 - |P| + |P||Q|:$$

$P \Leftrightarrow Q$ is false only when P is true and Q is false.

$$|P \vee Q| = \max(|P|, |Q|):$$

$P \vee Q$ is false only when P and Q are both false.

$$|P \wedge Q| = \min(|P|, |Q|):$$

$P \wedge Q$ is true only when P and Q are both true.

$$|\neg P| = 1 - |P|:$$

$\neg P$ is false only when P is true.

Finally, we must describe what happens when quantifiers are introduced. Suppose that $\xi \in \overline{M}$ and x is a variable. By a *variation of ξ along x* we mean any point $\xi' \in \overline{M}$ for which $y^{\xi} = y^{\xi'}$ whenever y is a variable different from x. Then

$$|\forall x P|(\xi) = \min_{\xi'} |P|(\xi'),$$

$$|\exists x P|(\xi) = \max_{\xi'} |P|(\xi'),$$

where ξ' runs through all variations of ξ along x.

A formula P is called ϕ-true if $|P|_{\phi}(\xi) = 1$ for all $\xi \in \overline{M}$. The interpretation ϕ (or M) is called a *model* for a set of formulas $\mathcal{E}$ if all the elements of $\mathcal{E}$ are ϕ-true.

2.6. Example: Standard Interpretation of L_1Ar. This is the interpretation in the set N of nonnegative integers, in which $\bar{0}, \bar{1}$ are interpreted as 0, 1, respectively, and $+, \cdot, =$ are interpreted as addition, multiplication, and equality, respectively.

2.7. Example: Standard Interpretation of L_1Set. This is the interpretation in the von Neumann universe V, in which $\varnothing$ is interpreted as the empty set, $\in$ is interpreted as the relation "is an element in," and $=$ is interpreted as equality.

All of the examples of translations in Chapter I were based on these standard interpretations. The relationship between those examples and the above definitions is as follows. Let $\Pi(x, y, z)$ be a statement in argot about the indeterminate sets x, y, z in V; and let $P(x, y, z)$ be a translation of Π into the language $\mathrm{L_1Set}$. Then for any point ξ interpreting x, y, z as the names of sets x^ξ, y^ξ, z^ξ in the von Neumann universe, we have:

$$\Pi(x^\xi, y^\xi, z^\xi) \text{ is true} \Leftrightarrow |P(x, y, z)|(\xi) = 1.$$

Thus, every formula expresses, or defines, a property of objects in the interpretation set:

2.8. **Definition**. A set $S \subset M^r, r \geqslant 1$, is called ϕ-definable (by the formula P in L with the interpretation ϕ) if there exist variables $x_1, \dots, x_r$ such that

$$|P|_\phi(\xi) = 1 \Leftrightarrow \langle x_1^\xi, \dots, x_r^\xi \rangle \in S$$

for all ξ in $\overline{M}$.

One of the most important problems concerning formal languages is to understand the structure of the sets of

$$\phi\text{-true formulas in } L;$$

$$\phi\text{-definable sets in } \bigcup_{r \geqslant 1} M^r.$$

2.9. Example. The sets definable by means of $\mathrm{L_1Ar}$ with the standard interpretation constitute the smallest class of sets in $\bigcup_{r \geqslant 1} N^r$ that

(a) contains all sets of the form

$$\{\langle k_1, \dots, k_r \rangle | F(k_1, \dots, k_r) = 0\} \subset N^r,$$

where F runs through all polynomials with integral coefficients;

(b) is closed relative to finite intersections, unions, and complements (in the appropriate N^r);

(c) is closed relative to the projections $\overline{\mathrm{pr}}_i : N^r \to N^{r-1}$:

$$\overline{\mathrm{pr}}_i \langle k_1, \dots, k_r \rangle = \langle k_1, \dots, k_{i-1}, k_{i+1}, \dots, k_r \rangle.$$

In fact, sets of type (a) are defined by atomic formulas of the form $t_1^F = t_2^F$, where t_1^F is a term corresponding to the sum of the monomials in F with positive coefficients, and t_2^F corresponds to the sum of the monomials with negative coefficients. Further, if $S_1, S_2 \subset N^r$ are definable by formulas P_1, P_2 (with the same variables), then $S_1 \cap S_2$ is definable by $P_1 \wedge P_2$, $S_1 \cup S_2$ is definable by $P_1 \vee P_2$, and $N^r \setminus S_1$ is definable by $\neg P_1$. Finally, the set $\mathrm{pr}_i(S_1)$ is definable by the formula $\exists x_i(P_1)$. The connectives $\Rightarrow$ and $\Leftrightarrow$ and the quantifier $\forall$ give nothing new, since without changing the set being defined, we may replace them by combinations of the logical operations already discussed: $\forall x$ may be replaced by $\neg \exists x \neg$, and so on.

This first description of *arithmetical* sets, i.e., $\mathrm{L_1Ar}$-definable sets, will be greatly amplified in the second and third parts of the book. At this point it is not immediately clear how to develop the subtler properties of definability, such as the definability of the set of prime numbers in N (see Example 3.14 in Chapter I), the definability of the set of partial fractions in the continued fraction expansion of $\sqrt[3]{2}$, or the definability of the set of pairs

$$\{\langle i, i\text{th digit in the decimal expansion of } \pi\rangle\} \subset N^2.$$

However, as we shall see in §11 and in Chapter VII, the "Gödel numbers of the true formulas of arithmetic" form a much more complicated set, and this set is not definable.

We now give several simple technical results.

2.10. **Proposition.** *Let P be a formula in L, ϕ an interpretation in M, and $\xi, \xi' \in \overline{M}$. Suppose that x^ξ coincides with $x^{\xi'}$ for all variables x occurring freely in P. Then $|P|_\phi(\xi) = |P|_\phi(\xi')$.*

2.11. **Corollary.** *In any interpretation the closed formulas P have well-defined truth values: $|P|_\phi(\xi')$ does not depend on (ξ).*

PROOF.
(a) Let t be a term, and suppose that for any variable x in t we have $x^\xi = x^{\xi'}$. Then Lemma 1.4 and induction on the length of t give $t^\xi = t^{\xi'}$.
(b) Assertion 2.10 holds for atomic formulas P of the form $p(t_1, \ldots, t_r)$. In fact,

$$|P|(\xi) = \begin{cases} 1, & \text{if} \langle t_1^\xi, \ldots, t_r^\xi \rangle \in \phi(P), \\ 0, & \text{otherwise,} \end{cases}$$

and similarly for $|P|(\xi')$. But if ξ and ξ' coincide on all the variables in P (all of which occur freely), then a fortiori they coincide on all the variables in t_i, and by part (a), we have $t_i^\xi = t_i^{\xi'}, i = 1, \ldots, r$. Therefore $|P|(\xi) = |P|(\xi')$.

(c) We now use induction on the total number of connectives and quantifiers in P. If P has the form $\neg Q$ or $Q_1 * Q_2$, then 2.10 for P follows trivially from 2.10 for Q, Q_1, Q_2. Now suppose that P has the form $\forall x(Q)$, and that 2.10 holds for Q. (The case $\exists x(Q)$ can be treated analogously or can be reduced to the case $\forall x$ by replacing $\exists x$ by $\neg \forall x \neg$.) By definition, we have

$$|\forall xQ|(\xi) = \begin{cases} 1, & \text{if} |Q|(\eta) = 1 \text{ for variations } \eta \text{ of } \xi \text{ along } x, \\ 0, & \text{otherwise;} \end{cases}$$

$$|\forall xQ|(\xi') = \begin{cases} 1, & \text{if} |Q|(\eta') = 1 \text{ for variations } \eta' \text{ of } \xi' \text{ along } x, \\ 0, & \text{otherwise.} \end{cases}$$

On the right we may let η and η' vary in addition on all variables that do not occur freely in Q. The assertions after the word "if" remain true or false in this wider range of values if they were true or false before, by the induction hypothesis on Q. But then η and η' run through the same values, because ξ

and ξ' differ only on variables that do not occur freely in Q, and on x. The proposition is proved. □

The following almost obvious fact is the basis for many phenomena that attest to the inadequacy of formal languages for completely describing intuitive concepts (see "Skolem's paradox" below):

2.12. **Proposition**. *The cardinality of the class of ϕ-definable sets does not exceed*

$$\mathrm{card}(\text{alphabet of}\, L) + \aleph_0.$$

Here and below, by "card(alphabet of L)" we mean the cardinality of the alphabet of L *without the set of variables.*

PROOF. If the language has $\leqslant \aleph_0$ variables, then there are at most

$$\mathrm{card}(\text{alphabet of}\, L) + \aleph_0 \text{ formulas.}$$

If, on the other hand, it has an uncountable set of variables, then we note that every definable set can be defined by a formula whose variables belong to a fixed countable subset of the variables that is chosen once and for all. □

2.13. **Corollary**. *If M is infinite and* card(alphabet of L) $< 2^{\mathrm{card}\, M}$, *then "almost all" sets are undefinable.*

Thus, the only way to define all subsets of M is to include a tremendous number of names in the language. For languages that are to describe actual mathematical reasoning this is an unrealistic program. Essentially, any finitely describable collection of modes of expression allows us to define only a countable number of sets. However, it is often technically useful to include in the alphabet, for example, names for all the elements of M.

In the following sections we proceed to study systematically sets of true formulas.

3 Syntactic Properties of Truth

Let L be a language in $\mathfrak{L}_1$, let ϕ be an interpretation of L, and let $T_\phi L$ be the set of ϕ-true formulas. In this section we list some properties of $T_\phi L$ that reflect the logic inherent in languages of $\mathfrak{L}_1$, regardless of the specific nature of the interpretation ϕ.

3.1. *The set $T_\phi L$ is complete.* By definition, this means that for any closed formula P, either P or $\neg P$ lies in $T_\phi L$. This property follows from Corollary 2.11 above.

3.2. *The set $T_\phi L$ does not contain a contradiction*, that is, there is no formula P for which P and $\neg P$ both lie in $T_\phi L$. In fact, $T_\phi L = \{P \mid |P|_\phi = 1\}$, while $|\neg P|_\phi = 1 - |P|_\phi$.

3.3 *The set $T_\phi L$ is closed under the rules of deduction* MP (modus ponens) *and* Gen (*generalization*). By definition, this means that if P and $P \Rightarrow Q$ lie in $T_\phi L$, then Q also lies in $T_\phi L$, and that if P lies in $T_\phi L$, then $\forall x P$ lies in $T_\phi L$ for any variable x. The verification is immediate: if $|P|_\phi = 1$ and $|P \Rightarrow Q|_\phi = 1$, then we must have $|Q|_\phi = 1$; if $|P|_\phi(\xi) = 1$ for all ξ, then also $|\forall x P|_\phi(\xi) = 1$. The formula Q is called a *direct consequence of the formulas P and $P \Rightarrow Q$ using the rule of deduction* MP. The formula $\forall x P$ is called a *direct consequence of the formula P using the rule of deduction* Gen.

The intuitive meaning of these rules of deduction is as follows. The rule MP corresponds to the following type of argument: "If P is true, and if the truth of P implies the truth of Q, then Q is true." Thus, one might say that the semantics of the expression "if ... then" in natural languages is divided between the semantics of the *connective* $\Rightarrow$ and the semantics of the *rule of deduction* MP in languages of $\mathcal{L}_1$. Neglecting this point of view often leads to confusion when one attempts to explain the rules for assigning truth values to the formula $P \Rightarrow Q$.

The rule Gen corresponds to the practice in mathematics of writing "identities" or universally true assertions. When we write $(a+b)^2 = a^2+2ab+b^2$ or "in a right triangle the square of the hypotenuse is equal to the sum of the squares of the other two sides," the quantifiers $\forall a$ $\forall b$ and $\forall$ triangles are omitted. Putting the quantifiers back in does not change the truth values, and has the advantage of freeing the notation for later use.

3.4. *The set $T_\phi L$ contains all tautologies.* To define what a tautology is, we first introduce the notion of a *logical polynomial* over a set of formulas $\mathcal{E}$. This is an element in the minimal set of formulas that contains $\mathcal{E}$ and is closed with respect to constructing formulas from shorter formulas using logical connectives.

A sequence of formulas $P_1, \ldots, P_n$ and representations of each P_i, either in the form Q, where $Q \in \mathcal{E}$, or in the form $\neg Q$ or $Q_1 * Q_2$, where Q, Q_1, Q_2 lie in $\{P_i, \ldots, P_{i-1}\}$, is called a *representation* of P_n as a logical polynomial over $\mathcal{E}$. The representation of P_n is not necessarily unique: for example, if $\mathcal{E} = \{P, Q, P \Rightarrow Q\}$, then $P \Rightarrow Q$ has two representations.

Let $|| : \mathcal{E} \to \{0, 1\}$ be any map. If we are given a representation r of the formula P_n as a logical polynomial over $\mathcal{E}$, then we can use the formulas in 2.5 to determine $|P_n|_r$ recursively.

A formula P is called a tautology if there exist a set of formulas $\mathcal{E}$ and a representation r of P as a logical polynomial over $\mathcal{E}$ such that $|P|_r = 1$ for all maps $|| : \mathcal{E} \to \{0, 1\}$. The property of being a tautology is effectively decidable, since; by syntactically analyzing P we can enumerate all representations of P as a logical polynomial. All tautologies obviously belong to $T_\phi L$.

Here are our first examples of tautologies:

A0. $P \Rightarrow P$;

A1. $P \Rightarrow (Q \Rightarrow P)$;

A2. $(P \Rightarrow (Q \Rightarrow R)) \Rightarrow ((P \Rightarrow Q) \Rightarrow (P \Rightarrow R))$;

A3. $(\neg Q \Rightarrow \neg P) \Rightarrow ((\neg Q \Rightarrow P) \Rightarrow Q)$;

B1. $\neg\neg P \Rightarrow P, P \Rightarrow \neg\neg P$;
B2. $\neg P \Rightarrow (P \Rightarrow Q)$.

Here P, Q, and R are arbitrary formulas in L; the form in which these tautologies are written makes it clear what representation as a logical polynomial over $\{P, Q, R\}$ is intended.

Thus, tautologies are formulas that are true regardless of the truth or falsity of the component parts (if the notion of component is suitably chosen). Bl is the law of the excluded middle: a double negation is equivalent to the original assertion. B2 is the mechanism by which a contradiction in a set of formulas $\mathcal{E}$ in L leads to the deducibility of any formula, and thereby destroys the entire system. (See Proposition 4.2 below.)

Example of how a tautology is verified. We give three versions of how to verify that the simple formula Al is a tautology.

Version (a). By the formulas in 2.5, we have

$$|P \Rightarrow (Q \Rightarrow P)| = 1 - |P| + |P|\,|Q \Rightarrow P|$$
$$= 1 - |P| + |P|(1 - |Q| + |P|\,|Q|) = 1,$$

since $|P|^2 = |P|$.

Version (b). We tabulate $|P \Rightarrow (Q \Rightarrow P)|$ as a function of $|P|$ and $|Q|$:

$\|P\|$	$\|Q\|$	$\|Q \Rightarrow P\|$	$\|P \Rightarrow (Q \Rightarrow P)\|$
0	0	1	1
0	1	0	1
1	0	1	1
1	1	1	1

This is an example of a "truth table."

Version (c). The basic property of the connective $\Rightarrow$ is that $P \Rightarrow Q$ is false only if P is true and Q is false. If $P \Rightarrow (Q \Rightarrow P)$ were false, then P would be true and $Q \Rightarrow P$ would be false; then, in turn, Q would be true and P would be false, a contradiction.

The reader would do well to verify that the more complicated axioms, for example A2, are tautologies, and to decide which of the three versions he prefers.

3.5. *The set $T_\phi L$ contains the "logical quantifier axioms," that is, the formulas*

(a) $\forall x(P \Rightarrow Q) \Rightarrow (P \Rightarrow \forall xQ)$, if all the occurrences of x in P are bound.
(b) $\forall x \neg P \Leftrightarrow \neg\exists xP$.
(c) $\forall xP(x) \Rightarrow P(t)$, if t is free for x in P (*axiom of specialization*). Here we use the notation $P(t)$ for the result of substituting t for each free occurrence of x in P. In all other respects P and Q are arbitrary formulas.

In 3.7 we verify that the formulas in 3.5 are ϕ-true. The intuitive meaning of these formulas is more or less clear. For example, the axiom of specialization means that if $P(x)$ is true for all x, then $P(t)$ is also true, where t is the name of any object. The condition that t must be free for x is the rule of hygiene for changing notation.

The set

$$\mathrm{Ax}\, L = \{\text{tautologies of}\, L\} \cup \{\text{quantifier axioms}\}$$

is called the *set of logical axioms in the language L.*

A set of formulas $\mathcal{E}$ in L will be called *Gödelian* if it is complete, does not contain a contradiction, is closed with respect to the rules of deduction MP and Gen, and contains all the logical axioms of L. The basic conclusion of our discussion is then the following:

3.6. Proposition. *The set of true formulas of L (in any interpretation) is Gödelian.*

In §6 we prove that conversely, any Gödelian set is a set of true formulas in a suitable interpretation. Thus, the concept of a Gödelian set is the closest approximation to the concept of truth that can be attained "without regard to meaning."

3.7. *Verification that axioms* 3.5 *are true.*

(a) Let R be the formula 3.5(a). We suppose that $|R|(\xi) = 0$ for some $\xi \in \overline{M}$ and show that this leads to a contradiction.

In fact, then $|\forall x(P \Rightarrow Q)|(\xi) = 1$ and $|P \Rightarrow \forall x\, Q|(\xi) = 0$. The second equation implies that $|P|(\xi) = 1$ and $|\forall x\, Q|(\xi) = 0$. Let ξ' be a variation of ξ along x for which $|Q|(\xi') = 0$. Then $|P|(\xi') = |P|(\xi) = 1$ by Proposition 2.10, since x does not occur freely in P. Hence, $|P \to Q|(\xi') = 0$, which contradicts the relation $|\forall x(P \Rightarrow Q|(\xi) = 1$.

(b) For all $\xi \in \overline{M}$ and for all variations ξ' of ξ along x, we have

$$|\forall x \neg P|(\xi) = \max_{\xi'} |\neg P|(\xi') = 1 - \min_{\xi'} |P|(\xi');$$

$$|\neg \exists x\, P|(\xi) = 1 - \min_{\xi'} |P|(\xi').$$

Hence, the truth values of $\forall x \neg P$ and $\neg \exists x P$ coincide, so that $\forall x \neg P \Leftrightarrow \neg \exists x\, P$ is identically true.

(c) Suppose that $|\forall x\, P(x) \Rightarrow P(t)|(\xi) = 0$ for some point $\xi \in \overline{M}$. We show that this leads to a contradiction. In fact, then

$$|\forall x\, P(x)|(\xi) = 1, \qquad |P(t)|(\xi) = 0.$$

The first equation implies that $|P(x)|(\xi') = 1$ for all variations ξ' or ξ along x. For ξ' we take the variation such that $x^{\xi'} = t^{\xi}$. If we prove that $|P(t)|(\xi) = |P(x)|(\xi')$, then we obtain the desired contradiction.

We prove this by induction on the total number of connectives and quantifiers in P.

(c$_1$) Let P be an atomic formula $p(t_1, \ldots, t_n)$. Letting $\bar{t}_i$ denote the result of substituting t for each occurrence of x in t_i, we successively obtain

$$t^{\xi} = x^{\xi'} \quad \text{(by the definition of } \xi'\text{)},$$

$$\bar{t}_i^{\xi} = t_i^{\xi'} \quad \text{(by induction on the length of } t_i\text{)},$$

$$|P(x)|(\xi') = |P(t_1, \ldots, t_n)|(\xi') = |P(\bar{t}_1, \ldots, \bar{t}_n)|(\xi) = |P(t)|(\xi).$$

(c$_2$) Let P have the form $\neg Q$ or $Q_1 \mapsto Q_2$, where $\mapsto$ is a connective. Since x does not bind t in P by assumption, the same is true for Q, Q_1, and Q_2, and the necessary induction step is automatic.

(c$_3$) Finally, let P have the form $\exists y\, Q$ or $\forall y\, Q$. We shall examine the first case; the proof for the second case is analogous.

Subcase 1. $y = x$. Then x is bound in P; therefore, $P(x) = P(t)$, and $|P|(\xi) = |P|(\xi')$ by Proposition 2.10.

Subcase 2. $y \neq x$. The induction assumption has the form $|Q(t)|(\eta) = |Q(x)|(\eta')$ if η is any point in $\overline{M}$ and η' is a variation of η along x for which $x^{\eta'} = t^{\eta}$. We must show that the following two truth values coincide (where ξ and ξ' are defined as above):

$$|\exists y\, Q(x)|(\xi') = \begin{cases} 1, & \text{if } |Q(x)|(\eta') = 1 \text{ for some variation } \eta' \text{ of } \xi' \text{ along } y, \\ 0, & \text{otherwise.} \end{cases}$$

$$|\exists y\, Q(t)|(\xi) = \begin{cases} 1, & \text{if } |Q(t)|(\eta) = 1 \text{ for some variation } \eta \text{ of } \xi \text{ along } y, \\ 0, & \text{otherwise.} \end{cases}$$

We recall that ξ' is the variation of ξ along x for which $x^{\xi'} = t^{\xi}$.

We first suppose that the second truth value is 1. We choose $\eta \in \overline{M}$ such that $|Q(t)|(\eta) = 1$, and then construct the variation η' of η along x for which $x^{\eta'} = t^{\eta}$. Then, by the induction assumption, $1 = |Q(t)|(\eta) = |Q(x)|(\eta')$. We show that η' is a variation of ξ' along y; this will imply that the first truth value is also 1. In fact, η' was obtained by varying η along x, η was obtained by varying ξ along y, and ξ was obtained by varying ξ' along x. Hence, η' is a variation of ξ' along x and y; we must show the variation along x did not actually take place:

$$x^{\eta'} = x^{\xi}.$$

But the left-hand side is t^{η} by the definition of η'; the right-hand side is t^{ξ} by the definition of ξ'; and η was obtained by varying ξ along y. Since t is free for x in $P = \exists y\, Q$, it follows that y does not occur in t.

It remains to verify that if the second truth value is 0, then the first is also 0. The argument is almost the same. If the second truth value is 0, then $|Q(t)|(\eta) = 0$ for all variations η of ξ along y. For each such η we construct η' as in the first part of the proof. As before, we verify that η' is a variation of ξ' along y and, moreover, η' runs through all such variations when η runs through all variations of ξ along y. Hence, the first truth value is also 0.

The proposition is proved. □

Digression: Natural Logic

1. Logic does not concern itself with the external world, but only with systems for trying to understand it. The logic of one such system—mathematics—is normalized to such an extent that it resembles a rigid stencil, which we can attempt to impose on any other system. But whether this stencil fits the system should not be seen as the criterion of suitability or the measure of worth of the system. The physicist's descriptions do not have to form a consistent or coherent whole; his job is to describe nature effectively on certain levels. Natural languages and the spontaneous workings of the mind are even less logical. In general, adherence to logical principles is only a condition for effectiveness in certain narrowly specialized spheres of human endeavor.

Although comparisons between the logic of predicates and the logic of natural languages or their subsystems have no normative force, such comparisons may be interesting and enlightening. Here we give some selected material from linguistics and psychology.

2. B. Russell, K. Döhmann, H. Reichenbach, U. Weinreich, and many others have studied the problem of finding parallels in natural languages for categories that can be formalized in languages of $\mathfrak{L}_1$ and of cataloguing the methods of transmitting these categories. This leads to the grouping of words into so-called *logico-semantic classes*, instead of the traditional division into verbs, nouns, articles, etc. (A. V. Gladkii and I. A. Mel'čuk, *Éléments de linguistique mathématique*, Paris, Dunod, 1972, §6).

For example, the words *sleeps, smart, crybaby* are parallel to relation symbols (predicates) of rank 1; the words *loves, friendly, sister* correspond to relations of rank 2. For each of them we have atomic formulas, such as "N sleeps," "X is friendly to Y," and so on.

"All, sometimes, something" are quantifier words; while "and, or, but, if ... then" are, of course, connectives. "The nose, le cadeau" are constants. Nouns are made into constants by using the definite article or its semantic equivalent. In Russian, which does not have definite articles, one must either use the demonstrative articles *etot* (this), *tot* (that), or make it clear from the context that the noun is meant as a constant. The words *nos* (nose), *podarok* (gift) are more like variables that stand for any object satisfying the simple predicate "is a nose," "is a gift." Incidentally, there are other possible interpretations.

The pronoun "he" is, without doubt, a variable. The pronouns "I" and "you" have much more complicated semantics, involving a correlation with who is speaking that does not exist in the speakerless languages of $\mathfrak{L}_1$. Certain aspects of the first person pronoun are included in the semantics of algorithmic languages. The right type of "memory key" in a program for the IBM 360 will allow the program to change what is contained in any byte in the basic memory region. The memory guard asks "Who is there?", and the program answers, "It is I." Finally, it is even possible in languages of $\mathfrak{L}_1$ to find models for certain types of self-description; see 9–11 and the digression on self-reference.

In Russian, "ili" (or) can be used not only to express the logical $\vee$, but also to express the exclusive "or" and even to express conjunction $\wedge$, as in the sentence "$x^2 > 0$ for $x > 0$ or for $x < 0$" (E. V. Padučeva). In Latin, the functions of exclusive and inclusive "or" are expressed by two different words, *aut* and *vel.* "And" can sometimes express a time sequence: compare the sentences "Jane got married and had a baby" with "Jane had a baby and got married" (S. Kleene). The conjunction $\wedge$ can be expressed in different languages by

juxtaposition: Chinese: ma mo—horse and donkey
Swahili: shika kitabu usome—take a book and read

a preposition: Russian: Petya s Mašeĭ—Peter and Masha

a conjunction: and, i, et

a postpositional particle: Latin: senatus populusque—the senate and the people

two conjunctions: Russian: kak ... tak.

Döhmann has catalogued the ways of expressing 16 logical polynomials in two variables in several languages of the world.

3. Curious as all this material may be, it should be regarded critically; in such comparisons with logic, the subtleties of usage often elude us. As an example, let us analyze the natural semantics of "if ... then." We have already mentioned that in languages of $\mathfrak{L}_1$ this connective corresponds not only to "$\Rightarrow$" but also to the rule of deduction modus ponens. Moreover, MP more adequately represents the meaning of "if ... then."

Actually, the rule that any conditional is true if its antecedent is known to be false has almost no parallel in natural logic. Examples of the type "if snow is black, then $2 \times 2 = 5$," which keep cropping up in textbooks, are capable only of confusing the student, since no natural subsystem in our language has expressions with this semantics. A possible exception is certain poetic and expressive formulas with extremely limited usage ("If she be false, O, then heaven mocks itself!"). Formal mathematics, in which a single contradiction destroys the entire system, clearly has the features of poetic hyperbole.

Finally, in the logic of predicates there is no place at all for the modal aspect of the use of "if... then" in instructions of the type "if this happens, do that." On the other hand, this aspect can easily be expressed by the semantics of the connective "if ... then ... else" in algorithmic languages such as Algol. Unless one uses techniques suggested by algorithmic languages, any attempt to find a model for modality in languages based on $\mathfrak{L}_1$ is doomed to failure (compare: A. A. Ivin, *The Logic of Norms*, MGU Press, 1973).

4. We have mentioned several times that the choice of the primitive modes of expression in the logic of predicates does not reflect psychological reality. Elementary logical operations, even one-step deductions, may require a highly trained intellect; yet, logically complicated operations can often be performed as a single elementary act of thought even by a damaged brain.

Sublieutenant Zasetsky, aged twenty-three, suffered a head injury 2 March 1943 that penetrated the left parieto-occipital area of the cranium. The

> injury... was further complicated, by inflammation that resulted in adhesions of the brain to the meninges and marked changes in the adjacent tissues.

Professor A. R. Luria met Zasetsky at the end of May 1943, and observed his condition for the next 26 years. In this time Zasetsky wrote nearly 3000 pages, describing with agonizing effort his life and illness as he struggled to regain his reason. His notebooks, which provided the material for Luria's book *The Man with a Shattered World* (Basic Books, Inc., New York, 1972, translated by L. Solotaroff), not only show his perseverance and determination, but are also revealing from a psychological point of view.

At first, the destruction of Zasetsky's psyche was overwhelming. The predominant disorder was asemia, the inability to connect symbols with their meaning. Luria describes his first meeting with Zasetsky:

> "'Try reading this page," I suggested to him.
>
> "What's this?... No, I don't know... don't understand... what *is* this?...."
>
> I suggested he try to do something simple with numbers, like add six and seven.
>
> "Seven ... six ... what's it? No, I can't ... just don't know."
>
> The ability to understand the simplest predicates was lost: "What season is there before winter? Before winter? After winter?... Summer?... Or *something!* No, I can't get it. Before spring? It's spring now ... and ... and before ... I've already forgotten, just can't remember."
>
> Zasetsky lost the ability to interpret the syntactic devices for organizing meaning: "In the school where Dunya studied a woman worker from the factory came to give a report." What did this mean to him? Who gave the report—Dunya or the factory worker? And where was Dunya studying? Who came from the factory? Where did she speak?

This is a fairly difficult example composed by Professor Luria, but here is what Zasetsky himself writes:

> I also had trouble with expressions like: "Is an elephant bigger than a fly?" and "Is a fly bigger than an elephant?" All I could figure out was that a fly is small and an elephant is big, but I didn't understand the words *bigger* and *smaller*. The main problem was I couldn't understand which word they referred to.

What attracts our attention is the complexity of Zasetsky's metalinguistic text describing his linguistic difficulties. The subtlety of the analysis seems incompatible with the crude errors being analyzed. This could be explained by the retrospective nature of the analysis, but the following even more complicated description was written concurrently with the experience of the mental defect being described:

> Sometimes I'll try to make sense out of those simple questions about the elephant and the fly, decide which is right or wrong. I know that when you rearrange the words, the meaning changes. At first I didn't think it did, it didn't seem to make any difference whether or not you rearranged the words. But after I thought about it a while I noticed that the sense of the four words (*elephant, fly, smaller, larger*) did change when the words were in a different order. But my brain, my memory, can't figure out right away what the word *smaller* (or *larger*) refers to. So I always have to think about them for a while ... So sometimes ridiculous expressions like "a fly is bigger than an elephant" seem right to me, and I have to think about it a while longer.

We can also see how complicated mental abilities were preserved while "simple" ones were lost from examples of Zasetsky's creative imagination, which resemble literary-psychological studies:

> Say I'm a doctor examining a patient who is seriously ill. I'm terribly worried about him, grieve for him with all my heart. (After all, he's human too, and helpless. I might become ill and also need help. But right now it's him I'm worried about—I'm the sort of person who can't help caring.) But say I'm another kind of doctor—someone who is bored to death with patients and their complaints. I don't know why I took up medicine in the first place, because I don't really want to work and help anyone. I'll do it if there's something in it for me, but what do I care if a patient dies? It's not the first time people have died, and it won't be the last.

All of this shows that there is no basis whatsoever for Rosser's opinion that "once the proof is discovered, and stated in symbolic logic, it can be checked by a moron." The human mind is not at all well suited for analyzing formal texts.

4 Deducibility

4.1. **Definition.** A deduction of a formula P from a set of formulas $\mathcal{E}$ (in a language L in $\mathfrak{L}_1$) is a finite sequence of formulas $P_1, \ldots, P_n = P$ with the property that for each $i = 1, \ldots, n$ at least one of the following alternatives holds:

(a) $P_i \in \mathcal{E}$;
(b) $\exists j < i$ such that P_i is a direct consequence of P_j using Gen;
(c) $\exists j, k < i$ such that P_i is a direct consequence of P_j and P_k using MP.

We shall write $\mathcal{E} \vdash P$ to abbreviate "there exists a deduction of P from $\mathcal{E}$." A deduction of P, together with a precise indication for each $i \leqslant n$ of which of the alternatives (a), (b), (c) and which indices j in case (b) or j, k in case (c) are used to obtain P_i, is called a description of a deduction. A single deduction may have several descriptions.

We usually consider deductions from sets $\mathcal{E}$ that contain Ax L, the logical axioms of L. The other elements of $\mathcal{E}$ may be formulas of L that are "guessed" to be true in the standard interpretation; these are called *special axioms* of L.

(Examples will be given later in 4.6–4.9.) Such deductions may be considered the formal equivalents of *mathematical proofs* (of a formula $P = P_n$ from the hypotheses $\mathcal{E}$). This identification is justified for the following reasons:

(a) As shown in 3.3, if $\mathcal{E} \subset T_\phi L$ for some interpretation ϕ, and if $\mathcal{E} \vdash P$, then $P \in T_\phi L$; only true formulas can be deduced from true formulas.
(b) A large amount of experimental work has been done on formalizing mathematical proofs, that is, replacing them by deductions in suitable languages of $\mathfrak{L}_1$, especially $\mathrm{L_1Set}$. This work has shown that for large segments of mathematics, including the foundations of the theory of integers and real numbers, set theory, and so on, proofs can successfully be formalized as deductions within the framework of $\mathfrak{L}_1$. There is much material on this theme in the literature on mathematical logic; see, in particular, Mendelson's book.
(c) Gödel's completeness theorem for the logical modes of expression in $\mathfrak{L}_1$ (see §6) shows that any formula that is not deducible from $\mathcal{E}$ must be false in some model (interpretation) of $\mathcal{E}$.

For further discussion, see "Digression: Proof."

We occasionally consider deductions from another type of sets $\mathcal{E}$. For example, we might remove from $\mathcal{E}$ certain logical axioms, such as the "law of the excluded middle" (B_1 in Section 3.4), in order to investigate formally intuitionistic principles. Or we might add to $\mathcal{E}$ a formula that we think is false in order to deduce a contradiction from $\mathcal{E}$; this is the so-called "proof by contradiction."

We now prove some formal aspects of contradiction.

4.2. **Proposition**. *Suppose that $\mathcal{E}$ contains all tautologies of type* B.2 *in Subsection* 3.4. *Then the following two properties of $\mathcal{E}$ are equivalent*:

(a) *There exists a formula P such that $\mathcal{E} \vdash P$ and $\mathcal{E} \vdash \neg P$.*
(b) *$\mathcal{E} \vdash Q$ for any formula Q.*

A set $\mathcal{E}$ with these properties is called inconsistent.

PROOF. (b) $\Rightarrow$(a) is obvious. Conversely, suppose $\mathcal{E} \vdash P$ and $\mathcal{E} \vdash \neg P$. We first add the formula $\neg P \to (P \to Q)$, which is assumed to lie in $\mathcal{E}$, to the descriptions of the two deductions. Then, applying MP twice (to this formula and $\neg P$; then to $P \Rightarrow Q$ and P), we obtain a description of a deduction $\mathcal{E} \vdash Q$.

4.3. A large part of the theorems of logic consists in proving assertions of the type "$\mathcal{E} \vdash P$" or "t is not true that $\mathcal{E} \vdash P$" for various languages L, sets $\mathcal{E}$, and (classes of) formulas P.

A result of the form $\mathcal{E} \vdash P$ may be proved by presenting a description of a deduction of P from $\mathcal{E}$. However, even in slightly complicated cases, this procedure becomes so long that it is replaced by more or less complete instructions on how to compose such a description. Finally, "$\mathcal{E} \vdash P$" may be proved without presenting even an incomplete description of a deduction of P from $\mathcal{E}$. In this case we "are not proving P, but are proving that a proof of P exists"; see the example in §8 concerning language extensions.

In rare cases a result of the form "it is not true that $\mathcal{E} \vdash P$" can be proved by a purely syntactic argument. But usually such a result is obtained by constructing a model, i.e., an interpretation, in which $\mathcal{E}$ is true and P is false; see the discussion of the continuum problem in Chapters III–IV. If it is true neither that $\mathcal{E} \vdash P$ nor that $\mathcal{E} \vdash \neg P$, we say that P is *independent* of $\mathcal{E}$.

We now give two useful elementary results concerning deductions. It is clear that compared with usual proofs, deductions are made up of very minor details. The mathematician, as if wearing seven-league boots, covers entire fields of formal deductions in one step.

4.4. **Lemma**. *Suppose that $\mathcal{E}$ contains all tautologies. If $\mathcal{E} \vdash P$ and $\mathcal{E} \vdash Q$, then $\mathcal{E} \vdash P \wedge Q$.*

PROOF. If $P_1, \dots, P_m$ and $Q_1, \dots, Q_n$ are deductions of P and Q, respectively, then

$$P_1, \dots, P_m, Q_1, \dots, Q_n, P \Rightarrow (Q \Rightarrow (P \wedge Q)), Q \Rightarrow (P \wedge Q), P \wedge Q$$

is a deduction of $P \wedge Q$. The third formula from the end is a tautology; the second formula from the end is a direct consequence of this tautology and $P_m = P$ using MP; and the last formula is a direct consequence of the second to last and $Q_n = Q$ using MP. □

4.5. **Deduction Lemma**. *Suppose that $\mathcal{E} \supset \mathrm{Ax}\, L$ and P is a closed formula. If $\mathcal{E} \cup \{P\} \vdash Q$, then $\mathcal{E} \vdash P \Rightarrow Q$.*

PROOF. Let $Q_1, \dots, Q_n = Q$ be a deduction of Q from $\mathcal{E} \cup \{P\}$. We show by induction on n that there exists a deduction of $P \Rightarrow Q$ from $\mathcal{E}$.

(a) $n = 1$. Then either $Q \in \mathcal{E}$, or else $Q = P$. In the first case $P \Rightarrow Q$ is deduced from Q and the tautology $Q \Rightarrow (P \Rightarrow Q)$ using MP. In the second case $P \Rightarrow P$ is a tautology.

(b) $n \geqslant 2$. We assume that the lemma holds for deductions of length $\leqslant n - 1$. Then $\mathcal{E} \vdash P \Rightarrow Q_i$ for all $i \leqslant n - 1$. Further, we have the following possibilities for $Q_n = Q$: $(\mathrm{b}_1)\, Q \in \mathcal{E}$; $(\mathrm{b}_2)\, Q = P$; $(\mathrm{b}_3)\, Q$ is deduced from Q_i and $Q_j = (Q_i \Rightarrow Q)$ using MP; and $(\mathrm{b}_4)\, Q$ has the form $\forall x\, Q$; for $j \leqslant n - 1$. The first two cases are handled in exactly the same way as for $n = 1$.

In case (b_3), $P \Rightarrow Q$ can be deduced from $\mathcal{E}$ in the following way:

(1) deduction of $P \Rightarrow Q$ (induction assumption);
(2) deduction of $P \Rightarrow (Q_i \Rightarrow Q)$ (induction assumption);
(3) $(P \Rightarrow (Q_i \Rightarrow Q)) \Rightarrow ((P \Rightarrow Q_i) \Rightarrow (P \Rightarrow Q))$ (tautology);
(4) $(P \Rightarrow Q_i) \Rightarrow (P \Rightarrow Q)$ (from (2) and (3) using MP);
(5) $P \Rightarrow Q$ (from (1) and (4) using MP).

From now on, arguments of this sort will be presented more briefly, with explicit mention of only the last steps of the induction (here (3), (4), and (5)).

Finally, in case (b_4), we obtain a deduction of $P \Rightarrow \forall x\, Q_j$ from $\mathcal{E}$ if we add the following formulas to the deduction of $P \Rightarrow Q_j$ from $\mathcal{E}$ (which exists by the induction assumption):

$$\forall x(P \Rightarrow Q_j) \qquad \text{(Gen)}$$

$$\forall x(P \Rightarrow Q_j) \Rightarrow (P \Rightarrow \forall x Q_j) \qquad \text{(logical quantifier axiom, since } P \text{ is closed)}$$

$$P \Rightarrow \forall x\, Q_j \qquad \text{(MP applied to the two preceding formulas).}$$

The lemma is proved. □

We record for future reference that in the parts of deductions constructed in Lemmas 4.4 and 4.5, only tautologies of the types A0, A1, and A2 in Section 3.4 were used.

We now give some basic examples of special axioms.

Axioms of equality

Let L be a language in $\mathfrak{L}_1$ whose alphabet includes a relation $=$ of rank two. We shall write t_1, t_2 instead of $= (t_1, t_2)$. If P is a formula, x is a variable, and t is a term, we let $P(x, t)$ denote the result of substituting t in P in place of *any or all* of the free occurrences of x in P for which t is free.

4.6. Proposition.

(a) *The formulas*

$$t = t; \qquad t_1 = t_2 \Rightarrow t_2 = t_1; \qquad t_1 = t_2 \wedge t_2 = t_3 \Rightarrow t_1 = t_3;$$
$$x = t \Rightarrow (P(x, x) \Rightarrow P(x, t))$$

are ϕ-true for any interpretation of L in which $\phi(=)$ is equality.

(b) *All the formulas in* (a) *are deducible from the set*

$$\text{Ax } L \cup \{x = x | x \text{ is a variable}\}$$
$$\cup \{x = y \Rightarrow (P(x, x) \Rightarrow P(x, y)) | P \text{ is an atomic formula}\}.$$

The formulas in this list, except for Ax L, *are called the axioms of equality.*

(c) *Let ϕ be any interpretation of L in a set M for which the axioms of equality are true. Then $\phi(=)$ is an equivalence relation in M that is compatible with the interpretations of all the relations and operations of L in M. If ϕ' denotes the obvious interpretation of L in the quotient set $M' = M/\phi(=)$, then $\phi'(=)$ is equality, and $T_\phi L = T_{\phi'} L$.*

PROOF (SKETCH)

(a) The ϕ-truth is easily established. We illustrate this by showing that the last formula is ϕ-true. Suppose it were false at a point $\xi \in \overline{M}$. Then

$|x = t|(\xi) = 1, |P|(\xi) = 1$ and $|P(x,t)|(\xi) = 0$. The first assertion means that $x^\xi = t^\xi$. But then $|P|(\xi) = |P(x,t)|(\xi)$ by Proposition 2.10, contradicting the second and third assertions.

(b) Deduction of $t = t$: $x = x$ (axiom of equality); $\forall x(x = x)$ (Gen); $\forall x(x = x) \Rightarrow t = t$ (logical axiom of specialization); $t = t$ (MP).
Deduction of $t_1 = t_2 \Leftrightarrow t_2 = t_1$:

(1) $x = y \Rightarrow (x = x \Rightarrow y = x)$ (axiom of equality with $=$ for P);
(2) $Q \Rightarrow ((P \Rightarrow (Q \Rightarrow R)) \Leftrightarrow (P \Rightarrow R))$, where P is $x = y, Q$ is $x = x, R$ is $y = x$ (tautology);
(3) $x = x$ (axiom of equality);
(4) $(P \Rightarrow (Q \Rightarrow R)) \Rightarrow (P \Rightarrow R)$ (MP is applied to (2) and (3));
(5) $x = y \Rightarrow y = x$ (MP applied to (1) and (4)).

We then twice apply Gen, the axiom of specialization, and MP, in order to deduce the formula $t_1 = t_2 \Rightarrow t_2 = t_1$ from (5); we replace t_1 by t_2 and t_2 by t_1 to deduce $t_2 = t_1 \Rightarrow t_1 = t_2$; we use Lemma 4.4 to deduce the conjunction of these two formulas; and, finally, the tautology $(t_1 = t_2 \Rightarrow t_2 = t_1) \wedge (t_2 = t_1 \Rightarrow t_1 = t_2) \Rightarrow (t_1 = t_2 \Leftrightarrow t_2 = t_1)$, together with MP, gives the required formula.

The deduction of the third and fourth formulas in (a) will be left to the reader. The existence of a deduction of the fourth formula can be proved by induction on the number of connectives and quantifiers in P. P is represented in the form $\neg Q, Q_1 * Q_2, \forall x\, Q$, or $\exists x\, Q$; we assume that the formula with Q, Q_1, and Q_2 in place of P has already been deduced, and we complete the deduction for P (see Mendelson, Chapter 2, Proposition 2.25).

(c) If the axioms of equality are ϕ-true, then so are the formulas in (a), since they are deducible. The first three formulas in (a), applied to three different variables x, y, and z, then show that the relation $\phi(=)$ on M is reflexive, symmetric, and transitive. In fact, let X, Y, and Z be any three elements of M, let $\xi \in \overline{M}$ be a point such that $x^\xi = X, y^\xi = Y$; and $z^\xi = Z$ and let $\sim$ be the relation $\phi(=)$ on M. The ϕ-truth of the formulas in (a) means that

$$X \sim X; \qquad X \sim Y \Leftrightarrow Y \sim X; \qquad X \sim Y; \text{ and } Y \sim Z \Rightarrow X \sim Z.$$

By definition, to say that $\sim$ is compatible with the ϕ-interpretation of all relations and operations on M means the following. Let p be a relation, and let $\phi(p) \subset M^r$ be its interpretation. If $\langle X_1, \dots, X_r\rangle \in \phi(p)$ and $X_i' \sim X_i$, then $\langle X_1, \dots, X_i', \dots, X_r\rangle \in \phi(p)$. Now let f be an operation, and let $\phi(f) : M^r \Rightarrow M$ be its interpretation. If $\phi(f)(X_1, \dots, X_r) = Y$ and $X_i' \sim X_i$, then $\phi(f)(X_1, \dots, X_i', \dots, X_r) = Y' \sim Y$.

We verify this compatibility by using the ϕ-truth of the last formula in 4.6(a) at a suitable point $\xi \in \overline{M}$. Here we take the formulas $p(x_1, \dots, x_r)$ and $f(x_1, \dots, x_r) = y$, respectively, for P; we take the variable x_i' for t and the variable x_i for x; and we set $x_i^\xi = X_i, x_i'^\xi = X_i'$, and $y^\xi = Y$.

It follows from the compatibility that we can construct an interpretation ϕ' of L in $M' = M/\sim$ such that $\phi'(p) = \phi(p) \bmod \sim, \phi'(f) = \phi(f) \bmod \sim$, and

$\phi'(=)$ is equality. The last formula in 4.6(a) will then imply that all the ϕ-true formulas remain ϕ'-true, and conversely. □

From now on, when we speak of the special axioms for any language in $\mathfrak{L}_1$ having the symbol $=$, we shall without explicit mention always include among them the axioms of equality for $=$. Models in which $=$ is interpreted as equality are called *normal* models.

Special axioms of arithmetic

4.7. **Proposition.** *The following formulas are true in the standard interpretation of* $\mathrm{L}_1\mathrm{Ar}$, *and are called the special axioms of* $\mathrm{L}_1\mathrm{Ar}$:

(a) *The axioms of equality.*
(b) *The axioms of addition:*

$$x + \bar{0} = x; \qquad x + y = y + x; (x + y) + z = x + (y + z);$$
$$x + z = y + z \Rightarrow x = y.$$

(c) *The axioms of multiplication:*

$$x \cdot \bar{0} = \bar{0}; \qquad x \cdot 1 = x; \qquad x \cdot y = y \cdot x; \qquad (x \cdot y) \cdot z = x \cdot (y \cdot z).$$

(d) *The distributive axiom:*

$$x \cdot (y + z) = x \cdot y + x \cdot z.$$

(e) *The axioms of induction:*

$$P(\bar{0}) \wedge \forall x (P(x) \Rightarrow P(x + \bar{1})) \Rightarrow \forall x\, P(x),$$

where P *is any formula in* $\mathrm{L}_1\mathrm{Ar}$ *having one free variable.*

The proof is trivial and will be left to the reader. We note only that the "proof" that the induction axioms are true itself uses induction.

Remarks

(a) In (b), (c), and (d) above, we have written the usual axioms for a commutative (semi) ring in order to shorten the formal deductions; any informal computation that uses only these axioms can easily be transformed into a formal deduction of the result of the computation in $\mathrm{L}_1\mathrm{Ar}$. In Chapter 3 of Mendelson's textbook, he gives an apparently weaker set of axioms, and then shows how to deduce our formulas from them. This takes up 5–6 pages of text, and is basically a tribute to a historical tradition going back to Peano.

(b) The induction axioms are a countable set of formulas in $\mathrm{L}_1\mathrm{Ar}$; it is customary to say that 4.7(e) is an *axiom schema*. The corresponding fact in intuitive mathematics is stated as follows; "For any property P of nonnegative integers, if 0 has the property P, and, whenever x has the property P, $x + 1$ also has the property P, then all nonnegative integers have the property P."

Here "property of nonnegative integers" means the same as "any subset of the nonnegative integers."

However, in the means of expression of $\mathrm{L_1Ar}$ there is no way to say "any subset." Neither is there any way to say "all properties"; we can only list one by one the properties that are definable by formulas in the language. We recall that there are only countably many such properties, while the intuitive interpretation refers to a continuum of properties. Thus, the formal axiom of induction is weaker than the informal one, and is also weaker than the version of this axiom that is obtained by embedding $\mathrm{L_1Ar}$ in $\mathrm{L_1Set}$.

Special axioms of Zermelo–Fraenkel set theory

(*see the description of* V *in the appendix to Chapter* II)

4.8. **Proposition.** *The following formulas are true in the standard interpretation of* $\mathrm{L_1Set}$ *in the von Neumann universe* V:

(a) *Axiom of the empty set*: $\forall x\, \neg(x \in \varnothing)$.
(b) *Axiom of extensionality*: $|\forall z(z \in x \Leftrightarrow z \in y) \Leftrightarrow x = y$.
(c) *Axiom of pairing*: $\forall u\, \forall w\, \exists x\, \forall z(z \in x \Leftrightarrow z = u \vee z = w)$.
(d) *Axiom of the union*: $\forall x\, \exists y\, \forall u(\exists z(u \in z \wedge z \in x) \Leftrightarrow u \in y)$.
(e) *Axiom of the power set*: $\forall x\, \exists y\, \forall z(z \subset x \Leftrightarrow x \in y)$, *where* $z \subset x$ *is abbreviated notation for the formula* $\forall u(u \in z \Rightarrow u \in x)$.
(f) *Axiom of regularity*: $\forall x(\neg x = \varnothing \Rightarrow \exists y(y \in x \wedge y \cap x = \varnothing))$, *where* $y \cap x = \varnothing$ *is abbreviated notation for* $\neg\exists z(z \in y \wedge z \in x)$.

Proof and explanations. This is not a complete list of the axioms of Zermelo–Fraenkel; the axiom of infinity, axiom of replacement, and also the axiom of choice, which are more subtle, will be discussed in the next subsection.

(a) The truth of these formulas must, of course, be proved by computing the function $|\ |$ using the rules in 2.4 and 2.5. We do this, for example, for the axiom of extensionality. Let ξ be any point in the interpretation class, and let $X = x^{\xi}, Y = y^{\xi}$. We must show that

$$|\forall z(z \in x \Leftrightarrow z \in y)|(\xi) = |x = y|(\xi),$$

i.e., that

$$\min_{Z \in V}(|Z \in X|\,|Z \in Y| + (1 - |Z \in X|)(1 - |Z \in Y|)) = |X = Y|,$$

where we have written $|Z \in X|$ instead of $|z \in x|(\xi')$ with $z^{\xi'} = Z, x^{\xi'} = X$, and so on. But the left-hand side equals 1 if and only if for every $Z \in V$ either both $Z \in X$ and $Z \in Y$, or else both $Z \notin Y$ and $Z \notin Y$, that is, if and only if $X = Y$.

More generally, if we replace V by any subclass $M \subset V$ and restrict the standard interpretation of $\mathrm{L_1Set}$ to M, then the same reasoning shows that *The axiom of extensionality is true in* M *if and only if for any elements* $X, Y \in M$ *we have*

$$X = Y \Leftrightarrow X \cap M = Y \cap M,$$

i.e., if and only if every element of M is uniquely determined by its elements which lie in M. This result will be used later.

The analogous computations for all the other axioms will be given systematically in a much more difficult context in Chapter III. Hence, at this point we shall only explain how to translate them into argot, as in Chapter I, and why they are fulfilled in V.

(b) The axiom of the empty set does not need special comment. We only remark that if we interpret L$_1$Set in a subclass $M \subset V$, then the constant $\varnothing$ may be interpreted as any element $X \in M$ with the property that $X \cap M = \varnothing$, and this axiom will still hold.

(c) The axiom of pairing is true, because if $U, W \in V_\alpha$, then $\{U, W\} \in \mathcal{P}(V_\alpha)$, so that all pairs lie in V.

(d) The axiom of the union is true, because if $X \in V$, then the set $Y = \bigcup_{Z \in X} Z$ also lies in V. In fact, if $X \in V_{\alpha+1} = \mathcal{P}(V_\alpha)$, then the elements of X are subsets of V_α, and their union therefore lies in $V_{\alpha+1}$.

(e) The axiom of the power set is true, because if $X \in V$, then $\mathcal{P}(X) \in V$. In fact, if $X \in V_\alpha$, then $X \subset V_\alpha$, and hence $\mathcal{P}(X) \subset \mathcal{P}(V_\alpha) = V_{\alpha+1}$, so that $\mathcal{P}(X) \in V_{\alpha+2}$.

(f) The axiom of regularity is true, because any nonempty set $X \in V$ has an empty intersection with at least one of its elements; in this form the axiom is proved in the appendix to this chapter.

4.9. The axioms of L$_1$Set in Section 4.8 have one property in common: their simplest model in the standard interpretation is precisely the union $V_{\omega_0} = \bigcup_{n=0}^{\infty} V_n$ of the first ω_0 levels of the von Neumann universe. In other words, this is the set of hereditarily finite sets $X \in V$, i.e., those such that if $X_n \in X_{n-1} \in \cdots \in X_0 = X$ then all the X_i are finite.

V_{ω_0} is the reliable, familiar world of combinatorics and number theory. Additional principles are needed to force us out of this world. There are two such principles: the axiom of infinity and the axiom schema of replacement.

(a) *Axiom of infinity:*

$$\exists x(\varnothing \in x \wedge \forall y(y \in x \Rightarrow \{y\} \in x)).$$

Here $\{y\} \in x$ is abbreviated notation for $\exists z(z = \{y, y\} \wedge z \in x)$, where the meaning of $z = \{y, y\}$ was explained in 3.7 of Chapter I. This axiom requires that we add to V_{ω_0} some set containing the elements $\varnothing, \{\varnothing\}, \{\{\varnothing\}\}, \ldots$ (a countable sequence). Then, in order to preserve the intuitive version of the axiom of the power set, we must add $\mathcal{P}(X), \mathcal{P}^2(X), \ldots$, thereby hopelessly leaving the realm of finite sets, countable sets, continua, and so on.

It is a striking fact that none of this is necessary in the formal, as opposed to intuitive, version of set theory, where we can always limit ourselves to hereditarily countable submodels of V. This important fact will be discussed in detail in §7.

(b) *Axiom schema of replacement.* We introduce the following convenient abbreviated notation (in any language of $\mathfrak{L}_1$ having the notion of equality): $\exists! y\, P(y)$ means $\exists y\, P(y) \wedge \forall x \forall y(P(x) \wedge P(y) \Rightarrow x = y)$. Thus, this formula is read; "There exists a unique object y with the property P," where we assume that $=$ is interpreted as equality. When other variables besides y occur freely in P, the formula $\exists! yP(y)$ is true precisely when P determines y as an "implicit function" of the other variables.

We can now write the replacement axioms. In the formula P below we list all the variables that occur freely in P:

$$\forall z_1 \cdots \forall z_n \forall u(\forall x(x \in u \Rightarrow \exists! y\, P(x, y, z_1, \ldots, z_n)) \\ \Rightarrow \exists w\, \forall y(y \in w \Leftrightarrow \exists x(x \in u \wedge P(x, y, z_1, \ldots, z_n)))).$$

The hypothesis says that "P gives y as a function of $x \in u$ (for given values of the parameters $z_1, \ldots, z_n$)"; the conclusion says that "the image of the set u under this function is some set w."

From the standpoint of the formal theory it is worthwhile to note that from this axiom and the axioms of equality are deducible the so-called separation axioms, namely

$$\forall z_1 \cdots \forall z_n\, \forall x\, \exists y\, \forall u(u \in y \Leftrightarrow u \in x \wedge P(u, z_1, \ldots, z_n)).$$

This says that if we take the class of sets having a property P and intersect it with a set x, we obtain a set.

The replacement axioms should be looked at very carefully. They go beyond the usual, "intuitively obvious" working tools of the topologist and analyst. The axioms assert that, for example, it is impossible to "stretch" an ordinal α too far by means of a function f; for any f we choose, there is always an ordinal β such that all the values $f(\gamma), \gamma \leqslant \alpha$, lie in V_β. In other words, the universe V is incomparably more infinite than any of its levels V_α.

Even if we adopt this axiom, questions remain that are very similar in style, that are beyond the reach of our intuition, and that are not solvable using this and the other axioms. For example, do there exist so-called *inaccessible cardinals* γ? One of the properties of an inaccessible cardinal γ is the following: if f is a function from V_α to V_γ (with $\alpha < \gamma$), then the set of values of f is an element of V_γ. In particular, there is an "upper bound" beyond which ordinals not exceeding γ cannot be "stretched." Do such infinities exist or not?

After thinking about this and related problems, many specialists on the foundations of mathematics have come to the conclusion that such languages of set theory as $\mathrm{L_1Set}$ with a suitable axiom system are the only reality one should work with, and any attempt to make intrinsic sense out of the universe V or similar models is in principle doomed to failure. In particular, the set of formulas in $\mathrm{L_1Set}$ that are true in the standard interpretation is not defined, and we can only talk about formulas that are deducible from the axioms.

But we shall not entirely adopt this point of view for several reasons. The simplest reason is the feeling that a language without an interpretation not only loses its intrinsic justification, but also cannot be used for anything. We cannot even play the "formal game" well unless we master the intuitive concepts that give meaning to the symbols. A language (along with the external world) helps bring order and precision to these intuitive concepts, which, in turn, make us change the language or at least revise our earlier linguistic constructions. But we can never assume that we have achieved complete clarity.

We should understand the need for certain types of self-restraint. However, intellectual asceticism (like all other forms of asceticism) cannot be the lot of many.

(a) *Axiom of choice*:

$$\forall x(\neg x = \varnothing \Rightarrow \exists y(\text{“}y \text{ is a function with domain of definition } x\text{”}$$
$$\wedge \forall u(u \in x \wedge \neg u = \varnothing \Rightarrow \exists w(w \in u \wedge \text{“}\langle u, w\rangle \in y\text{”})))).$$

That is, y chooses one element from each nonempty element $u \in x$.

The belief that this axiom is true in V is at least as justified as the belief in the existence of V itself. Over the past fifty years it has become customary for every working mathematician to accept this axiom, and the heated controversies about it at the beginning of the century are now all but forgotten. The interested reader is referred to Chapter II of *Foundations of Set Theory* by Fraenkel and Bar-Hillel (North-Holland, Amsterdam, 1958).

4.10. *General properties of axioms*. Despite the wide variety of concepts reflected in these axioms, each of our sets of axioms for languages in $\mathfrak{L}_1$ (tautologies; Ax L; special axioms of $\mathrm{L_1Ar}$ and $\mathrm{L_1Set}$) have the following informal syntactic characteristics:

(a) An algorithm can be given that tells whether any given expression is an axiom (compare the syntactic analysis in §1 and the verification of the tautologies in Section 3.4).
(b) A finite number of rules can be given for generating the axioms.

It is clear that a priori, property (b) is less restrictive than (a). In fact, an algorithm as in (a) can be transformed into a rule for generating the axioms: "Write out all possible expressions one by one in some order, and take those for which the algorithm gives a positive answer."

It is actually natural to suppose that property (a) should characterize axioms, and property (b) should characterize deducible formulas, no matter how we explicitly describe the axioms and the deducible formulas in a given language. In Part III we make these intuitive ideas into precise definitions and show that (b) is strictly weaker than (a). See also the discussion in Section 11.6(c) of this chapter.

Digression: Proof

1. A proof becomes a proof only after the social act of "accepting it as a proof." This is as true for mathematics as it is for physics, linguistics, or biology. The evolution of commonly accepted criteria for an argument's being a proof is an almost untouched theme in the history of science. In any case, the ideal for what constitutes a mathematical demonstration of a "nonobvious truth" has remained unchanged since the time of Euclid: we must arrive at such a truth from "obvious" hypotheses, or assertions that have already been proved, by means of a series of explicitly described, "obviously valid" elementary deductions.

Thus, the method of deduction is a method of mathematics *par excellence.* ("Mathematical induction" clearly comes out of the same tradition. Peano's induction principle allows us to write only the first step and the general step of a proof, and is thereby in some sense the first metamathematical principle. This point is observed by the tradition of listing Peano's axiom among the special axioms (see 4.7(e)), but one way or another, it is one of the archetypes of mathematical thought.)

The longer the deductive argument, the more important it is for all its elementary components to be written in an explicit and normalized fashion. In the last analysis, the amount of initial data in formal mathematics is so small that failure to observe the rules of hygiene in long deductions would lead to the collapse of the system if we did not have external checks on the system. In induction, on the other hand, relatively short deductions are based on a vast amount of initial information. Darwin's theory of evolution is explained to school children, but life is not long enough to judge how persuasive the proofs are. We see a similar situation in comparative linguistics when the features of the so-called protolanguages are reconstructed. In such uses of induction, the "rules of deduction" cannot be so very rigid, despite the critical viewpoint of the neo-grammarians.

2. The above observations concerning the method of deduction are supported by the fact that the notion of a formal deduction in languages of $\mathfrak{L}_1$ is a close approximation to the concept of an ideal mathematical proof. It is therefore enlightening to examine the differences between deductions and the arguments we use in day-to-day practice.

(a) *Reliability of the principles.* Not only the mathematics implicit in the special axioms of $\mathrm{L}_1\mathrm{Set}$ and $\mathrm{L}_1\mathrm{Ar}$, but even the logic of the languages of $\mathfrak{L}_1$ is not accepted by everyone. In particular, Brouwer and others have called into question the law of the excluded middle. From their extremely critical perspective, our "proofs" are at best harmless deductions of nonsense out of falsehood.

The mathematician cannot permit himself to be completely deaf to these criticisms. After thinking about them for a while, he should at least be willing to admit that proofs can have objectively different "degrees of proofness."

(b) *Levels of "proofness."* Every proof that is written must be approved and accepted by other mathematicians, sometimes by several generations of mathematicians. In the meantime, both the result and the proof itself are liable to be refined and improved. Usually the proof is more or less an outline of a formal deduction in a suitable language. But, as mentioned before, an assertion P is sometimes established by proving that a proof of P exists. This hierarchy of proofs of the existence of proofs can, in principle, be continued indefinitely. We can take down the hierarchy using sophisticated logical and set-theoretic principles; however, not everyone might agree with these principles. Papers on constructive mathematics abound with assertions of the type, "there cannot not exist an algorithm that computes x," whereas a classical mathematician would simply say "x exists," or even "x exists and is effectively computable."

(c) *Errors.* The peculiarities of the human mind make it impossible in practice to verify formal deductions, even if we agree that in principle, such a verification is the ideal form for a proof. Two circumstances act together with perilous effect: formal deductions are much longer than texts in argot, and humans are much slower at reading and comprehending such formal arguments than texts in natural languages.

A proof of a single theorem may take up five, fifteen, or even fifty pages. In the theory of finite groups, the proofs of the two Burnside conjectures occupy nearly five hundred pages apiece. Deligne has estimated that a complete proof of Ramanujan's conjecture assuming only set theory and elementary analysis would take about two thousand pages. The length of the corresponding formal deductions staggers the imagination.

Hence, the absence of errors in a mathematical paper (assuming that none are discovered), as in other natural sciences, is often established indirectly: how well the results correspond to what was generally expected, the use of similar arguments in other papers, examination of small sections of the proof "under the microscope," even the reputation of the author—in short, its reproducibility in the broadest sense of the word. "Incomprehensible" proofs can play a very useful role, since they stimulate the search for more accessible arguments.

The last two decades have seen the appearance of a very powerful method for performing long formal deductions, namely the use of computers. At first glance, it would seem that the status of formal deductions might greatly improve, so that the Leibnizian ideal of being able to verify truth mechanically would become attainable. But the state of affairs is actually much less trivial.

We first give two authoritative opinions on this question by C. L. Siegel and H. P. F. Swinnerton-Dyer. Both opinions relate to the solution by computer of concrete number-theoretic problems.

3. The present level of knowledge concerning Fermat's last theorem is as follows. Let p be a prime. It is called regular if it does not divide the numerator of any of the Bernoulli numbers $B_2 = \frac{1}{6}, B_4 = \frac{1}{30}, \ldots, B_{p-3}$. Fermat's theorem was proved for regular prime exponents by Kummer. For irregular p there is a series of criteria for Fermat's theorem to hold. These criteria reduce to checking that certain divisibility properties do not hold; if they hold, we must try certain other divisibility properties, and so on. The verification for each p requires extensive computer computations. As of 1955, this was successfully done for all $p < 4002$ (J. L. Selfridge, C. A. Nicol, H. S. Vandiver, *Proc. Nat. Acad. Sci. USA*, 41, 970-973 (1955)).

Let $v(x)$ denote the ratio of the number of irregular primes $\leqslant x$ to the number of regular primes $\leqslant x$. Kummer conjectured that $v(x) \to \frac{1}{2}$ as $x \to \infty$. Siegel *(Nachrichten Ak. Wiss. Göttingen, Math. Phys. Klasse*, 1964, No. 6, 51–57) suggests that $\sqrt{e} - 1$ is a more likely value for the limit, supports this opinion with probabilistic arguments, compares with the data of Selfridge–Nicol–Vandiver, and concludes this discussion with the following unexpected sentence: "In addition, it must be taken into account that the above numerical values for $v(x)$ were obtained using computers, and therefore, strictly speaking, cannot be considered proved"!

4. Siegel's point of view can be explained as a natural reaction to information received at second hand. But the excerpts below are from an article by a professional mathematician and experienced computer programmer (Acta Arithmetica, XVIII, 1971, 371–385). The article is devoted to the following problem:

Let L_1, L_2, L_3 be three homogeneous linear forms in u, v, w with real coefficients and determinant Δ; and suppose that the lower bound of $|L_1 L_2 L_3|$ for integer values of u, v, w not all zero is 1. What can be said about the possible value for Δ?

The corresponding problem for the product of two linear forms is much easier, and was essentially completely solved by Markov. There are countably many possible values of Δ less than 3, each of which has the form

$$\Delta = (9 - 4n^{-2})^{1/2}$$

for some integer n; the first few values of n are 1, 2, 5, 13, 29, and there is an algorithm for constructing all the permissible values of n.

For three forms Davenport (1943) proved that $\Delta = 7$ or $\Delta = 9$ or $\Delta > 9.1$. In Swinnerton–Dyer's paper, all values of $\Delta \leqslant 17$ are computed *under the assumption* that there are only finitely many such values and he gives a list of them: the third value is 148, and the last (the eighteenth) is $\sqrt{2597/9}$. Discussing this result, he makes a very interesting comment:

> When a theorem has been proved with the help of a computer, it is impossible to give an exposition of the proof which meets the traditional test—that a sufficiently patient reader should be able to work through the proof and verify that it is correct. Even if one were to print all the programs and all the sets of data used (which in this case would occupy some forty very dull pages) there can be no assurance that a data tape has not been mispunched or misread. Moreover, every modern computer has obscure faults in its software and hardware—which so seldom cause errors that they go undetected for years—and every computer is liable to transient faults. Such errors are rare, but a few of them have probably occurred in the course of the calculations reported here.

The arguments on the positive side are also very curious:

> However, the calculation consists in effect of looking for a rather small number of needles in a six-dimensional haystack; almost all the calculation is concerned with parts of the haystack which in fact contain no needles, and an error in those parts of the calculation will have no effect on the final results. Despite the possibilities of error, I therefore think it almost certain that the list of permissible $\Delta \leqslant 17$ is complete; and it is inconceivable that an infinity of permissible $\Delta \leqslant 17$ have been overlooked.

His conclusion:

> Nevertheless, the only way to verify these results (if this were thought worth while) is for the problem to be attacked quite independently, by a different machine. This corresponds exactly to the situation in most experimental sciences.

We note that it is becoming more and more apparent that the processing, and also the storage, of large quantities of information outside the human brain leads to social problems that go far beyond questions of the reliability of mathematical deductions.

5. In conclusion, we quote an impression concerning mechanical proofs, even ones done by hand, which is experienced by many.

After stating a proposition to the effect that "the function $T_{W,\eta_0}\tilde{\theta}$ is correctly defined," a gifted and active young mathematician writes (*Inventiones Math.*, vol. 3, f.3 (1967), 230):

> The proof of this Proposition is a ghastly but wholly straightforward set of computations. It took me several hours to do every bit and as I was no wiser at the end—except that I knew the definition was correct—I shall omit details here.

The moral: a good proof is one that makes us wiser.

5 Tautologies and Boolean Algebras

5.1 **Proposition.** *A finite list, or "basis," of tautologies—logical polynomials in three variables P, Q, R—can be given with the following property.*

Let L be any language in $\mathfrak{L}_1$, and let $\mathcal{F}$ be the set of all formulas in L that can be obtained from the basis tautologies by substituting all possible formulas in place of P, Q, R. Then any tautology in L is deducible from $\mathcal{F}$ using only the rule of deduction MP.

The choice of the basis tautologies is by no means unique. Our list will consist of the tautologies A0, Al, A2, A3, Bl, B2 in Section 3.4 and the following tautologies:

C1 $\neg(P \Rightarrow \neg Q) \Rightarrow (P \wedge Q), (P \wedge Q) \Rightarrow \neg(P \Rightarrow \neg Q)$.
C2 $(\neg P \Rightarrow Q) \Rightarrow (P \vee Q), (P \vee Q) \Rightarrow (\neg P \Rightarrow Q)$.
C3 $P \Rightarrow (\neg Q \Rightarrow \neg(P \Rightarrow Q))$.
C4 $(P \Rightarrow Q) \Rightarrow ((\neg P \Rightarrow Q) \Rightarrow Q)$.
C5 $(P \Rightarrow Q) \Rightarrow (\neg Q \Rightarrow \neg P)$.
C6 $(P \Rightarrow Q) \Rightarrow ((Q \Rightarrow P) \Rightarrow (P \Leftrightarrow Q))$.
C7 $(P \Leftrightarrow Q) \Rightarrow (P \Rightarrow Q), (P \Leftrightarrow Q) \Rightarrow (Q \Rightarrow P)$.

We are not trying to economize on the size of the basis, but rather on the length of the proof of Proposition 5.1; hence, A0–C7 is not the shortest possible list. This does not make any difference for studying the logic of $\mathfrak{L}_1$; but the study of modified logical systems, for example those of the intuitionist type, requires more careful analysis of this list.

PROOF OF PROPOSITION 5.1. Let $\mathcal{E}$ be a finite set of formulas in L, and let P be a logical polynomial (with a fixed representation) over $\mathcal{E}$. For any map $v : \mathcal{E} \to \{0, 1\}$, we extend v to P using the same rules that defined the truth function $|\ |$ in Section 2.5. We set

$$P^v = \begin{cases} P, & \text{if } v(P) = 1, \\ \neg P, & \text{if } v(P) = 0. \end{cases}$$

5.2. **Fundamental Lemma**. *Let* $\mathcal{E}^v = \{Q^v | Q \in \mathcal{E}\}$. *Then for any* v *we have* $\mathcal{F} \cup \mathcal{E} \vdash P^v$ (*using* MP).

This lemma expresses the following idea. It is natural to prove Proposition 5.1 by induction on the length of the tautology. However, the component parts of a tautology themselves might not be tautologies. The operation of taking P to P^v forces any formula to be "v-true" and makes it possible for us to use induction.

5.3. PROOF OF 5.1 ASSUMING THE FUNDAMENTAL LEMMA. Let P be a tautology, so that $P^v = P$ for all v, Set $\mathcal{E} = \{P_1, \dots, P_r\}$. By the fundamental lemma, $\mathcal{F} \cup \{P_1^v, \dots, P_r^v\} \vdash P$ using MP for any v: We show that then $\mathcal{F} \cup \{P_1^v, \dots, P_{r-1}^v\} \vdash P$ using MP. Descending induction on r then gives the required assertion (the assumption that P is a logical polynomial in $P_1, \dots, P_r$ is not used in the induction step).

The Deduction Lemma 4.5 shows that $\mathcal{F} \cup \{P_1^v, \dots, P_{r-1}^v\} \vdash (P_r^v \Rightarrow P)$ using MP; to see this we only need examine the proof and notice that the deduction used only MP and the tautologies in $\mathcal{F}$, since the rule of deduction Gen was not needed.

Since for any v there exists a v' that coincides with v on $P_1, \dots, P_{r-1}$ but takes a different value on P_r, it follows that $P_r \Rightarrow P$ and $\neg P_r \Rightarrow P$ are deducible from $\mathcal{F} \cup \{P_1^v, \dots, P_{r-1}^v\}$ using MP. On the other hand, the tautology C4: $(P_r \Rightarrow P) \Rightarrow ((\neg P_r \Rightarrow P) \Rightarrow P)$ lies in $\mathcal{F}$. Applying MP twice, we deduce P. □

5.4. PROOF OF THE FUNDAMENTAL LEMMA. We use induction on the number of connectives in the representation of P as a logical polynomial over $\mathcal{E}$. If there are no connectives, that is, $P \in \mathcal{E}$, then the assertion is obvious. Otherwise, P has the form $\neg Q$ or $Q_1 * Q_2$, where $*$ is one of the binary connectives.

(a) *The case* $P = \neg Q$. If $v(Q) = 0$, then $Q^v = \neg Q = P = P^v$. That $Q^v = P^v$ is deducible from $\mathcal{F} \cup \mathcal{E}^v$ is precisely the induction assumption.

On the other hand, if $v(Q) = 1$, then $Q^v = Q, P^v = \neg\neg Q$. Here Q is deducible from $\mathcal{F} \cup \mathcal{E}^v$ by the induction assumption, and then the tautology $Q \Rightarrow \neg\neg Q$ in $\mathcal{F}$ along with MP gives a deduction of P^v.

(b) *The case* $P = Q_1 * Q_2$. For the different connectives and possible values of $v(Q_1)$ and $v(Q_2)$ we first tabulate the formulas for which deductions exist by the induction assumption and the formulas for which we must find deductions. In the columns under $\wedge$ and $\vee$ we give formulas from which $(Q_1 \wedge Q_2)^v$ and $(Q_1 \vee Q_2)^v$, respectively, are deducible using MP and the tautologies in $\mathcal{F}$ (tautologies Cl, C2, and C5). Hence it suffices to find deductions of each of formulas 1–16 from $\mathcal{F}$ and the pair of formulas in the appropriate row in the second column using MP.

Deduction of formulas 1–16.

	Given: deductions of	Must Find: Deduction of $(Q_1 * Q_2)^v$	
$v(Q_1) v(Q_2)$	Q_1^v and Q_2^v	$\Rightarrow$	$\wedge$
0 0	$\neg Q_1, \neg Q_2$	1. $Q_1 \Rightarrow Q_2$	5. $\neg\neg(Q_1 \Rightarrow \neg Q_2)$
0 1	$\neg Q_1, Q_2$	2. $Q_1 \Rightarrow Q_2$	6. $\neg\neg(Q_1 \Rightarrow \neg Q_2)$
1 0	$Q_1, \neg Q_2$	3. $\neg(Q_1 \Rightarrow Q_2)$	7. $\neg\neg(Q_1 \Rightarrow \neg Q_2)$
1 1	Q_1, Q_2	4. $Q_1 \Rightarrow Q_2$	8. $\neg(Q_1 \Rightarrow \neg Q_2)$

$v(Q_1)\ v(Q_2)$	Q_1^v and Q_2^v	$\vee$	$\Leftrightarrow$
0 0	$\neg Q_1, \neg Q_2$	9. $\neg(\neg Q_1 \Rightarrow Q_2)$	13. $Q_1 \Leftrightarrow Q_2$
0 1	$\neg Q_1, Q_2$	10. $\neg Q_1 \Rightarrow Q_2$	14. $\neg(Q_1 \Leftrightarrow Q_2)$
1 0	$Q_1, \neg Q_2$	11. $\neg Q_1 \Rightarrow Q_2$	15. $\neg(Q_1 \Leftrightarrow Q_2)$
1 1	Q_1, Q_2	12. $\neg Q_1 \Rightarrow Q_2$	16. $Q_1 \Leftrightarrow Q_2$

Note that if P is deducible then for any Q the formula $Q \Rightarrow P$ is also deducible (tautology A1 and MP) and if $\neg P$ is deducible then for any Q the formula $P \Rightarrow Q$ is deducible (tautology B2 and MP). This immediately yields deductions of 1, 2, 4, 10, and 12. If we remove the double negations in the $\wedge$ column using tautology B1 and MP, we obtain deductions of 5, 6, and 7. And 11 is deducible since by B1 the second column yields a deduction of $\neg\neg Q_1$. In the first and last rows the deductions of 1 and 4 yield deductions of $Q_2 \Rightarrow Q_1$ by symmetry; tautology C6 and MP twice give a deduction of 13 and 16 from $Q_1 \Rightarrow Q_2$ and $Q_2 \Rightarrow Q_1$.

3 is deduced from C3: $Q_1 \Rightarrow (\neg Q_2 \Rightarrow \neg(Q_1 \Rightarrow Q_2))$ and the second column using MP twice.

8 is deduced from C3: $Q_1 \Rightarrow (\neg\neg Q_2 \Rightarrow \neg(Q_1 \Rightarrow \neg Q_2))$ and the second column using MP, applying B1 to Q_2, and again using MP.

9 is deduced from C3: $\neg Q_1 \Rightarrow (\neg Q_2 \Rightarrow \neg(\neg Q_1 \Rightarrow Q_2))$ using MP twice.

15 is deduced from 3 by C7 and C5 and MP twice.

Finally, the deduction of 3 from Q_1 and $\neg Q_2$ yields by symmetry a deduction of $\neg(Q_2 \Rightarrow Q_1)$from $\neg Q_2$ and Q_2. Hence on the second row the deduction of 14 is analogous to that of 15.

Proposition 5.1 is proved. □

5.5. *Tautologies and probability.* Tautologies are statements that are true independently of the truth or falsity of their "component parts." This assertion still holds even if the components of a tautology are assigned probabilistic truth values $\|P\|$ in the algebra of measurable sets in some probability space.

An example: the tautology $R \vee S \vee \neg R \vee \neg S$—"either it will rain, or it will snow, or it won't rain, or it won't snow"[1]—is a reliable weather forecast despite the great complexity of the meteorological probability space.

For a precise result, it is convenient to use the terminology of Boolean algebras.

5.6. *Boolean algebras.* A Boolean algebra B is a set with an operation of rank one, with two operations $\vee$ and $\wedge$ of rank two, and with two distinguished elements 0 and 1, such that the following axioms hold:

(a) $(A')' = A$ for all $A \in B$;
(b) $\wedge$ and $\vee$ are each associative and commutative;
(c) $\wedge$ and $\vee$ are distributive with respect to one another;
(d) $(a \vee b)' = a' \wedge b', (a \wedge b)' = a' \vee b'$;
(e) $a \vee a = a \wedge a = a$;
(f) $1 \wedge a = a; 0 \vee a = a$.

EXAMPLES.

(a) B is the set of all subsets of a set M, $'$ is complement, $\wedge$ is intersection, $\vee$ is union, 0 is the empty subset, and 1 is all of M.
(b) B is the set of open-and-closed subsets of a topological space M with the same operations.
(c) B is the algebra of measurable subsets (modulo measure-zero subsets) of a probability space M with the same operations.

In all of these cases B can be identified with the space of characteristic functions of the corresponding subsets of M (taking the value 1 on the subset and 0 on the complement).

5.7. *Boolean truth functions.* Let B be a Boolean algebra, and let $\mathcal{E}$ be a set of formulas in a language L. Let $\|\ \ \| : \mathcal{E} \to B$ be any map. We extend this map to the logical polynomials over $\mathcal{E}$ (more precisely, to their representations) by means of the recursive formulas

$$\begin{aligned} \|P \Leftrightarrow Q\| &= (\|P\| \wedge \|Q\|) \vee (\|P\|' \wedge \|Q\|'), \\ \|P \Rightarrow Q\| &= \|P\|' \vee \|Q\|, \\ \|P \vee Q\| &= \|P\| \vee \|Q\|, \\ \|P \wedge Q\| &= \|P\| \wedge \|Q\|, \\ \|\neg P\| &= \|P\|'. \end{aligned}$$

[1] A Russian proverb (translator's note).

In the case $B = \{0, 1\}$, these formulas coincide with the definitions in 2.5. We note that $\vee$ and $\wedge$ have different meanings in the left- and right-hand sides.

5.8. **Proposition.** *Let the logical polynomial P be a tautology over $\mathcal{E}$. Then for any map $\| \quad \| : \mathcal{E} \to B$ to any Boolean algebra B we have $\|P\| = 1$.*

PROOF. An example of a natural map $\| \quad \|$ can be obtained as follows: if we are given an interpretation of L in a set M, then the truth functions $|P|(\xi)$ can be considered as the characteristic functions of the definable subsets of the interpretation class $\overline{M}$ (compare §2). Hence, our usual truth functions are essentially Boolean-valued. They are embedded in the Boolean algebra of all subsets of $\overline{M}$, which decomposes as a direct product of two-point Boolean algebras $\{0, 1\}$. Hence the proposition follows trivially in this case.

In the general case one could use Stone's structure theorem for Boolean algebras. However, instead of this we shall indicate how to reduce the problem to some simple computations using Proposition 5.1. Because of Proposition 5.1, it suffices to verify that the basis tautologies are $\| \quad \|$-true and that $\| \quad \|$-truth is preserved when we use MP. For example, if $\|P\| = 1$ and $\|P \Rightarrow Q\| = 1$, then $\|P\|' = 0$ while $\|P\|' \vee \|Q\| = 1$, so that $\|Q\| = 1$ by 5.6(f); this answers the question about MP. The truth values of the basis tautologies are computed in a similar manner using the axioms in 5.6. □

Boolean truth functions will be the basic tool in the presentation of Cohen forcing in Chapter III.

Digression: Kennings

1. The process in §5 generates all possible tautologies starting with a finite number of tautologies and using a finite number of rules. It has become very popular in modern linguistics to attempt to find a suitable description of natural languages by means of such generating rules (N. Chomsky and others; see, for example, the book *Éléments de linguistique mathématique* by A. V. Gladkiĭ and I. A. Mel'čuk, Paris, Dunod, 1972).

However, many psychologists consider that this conception has little to do with the actual process of speech. According to one such opinion, real speech has more in common with a game of chance, chasing a fugitive, or a river current near a jagged shoreline. The choice of the next word in a sentence is determined statistically both by a formulating principle (an idea, situation, or psychological state) and by the peculiarities of semantics, grammar, phonetics, and the associative cloud formed by the earlier words.

There is reason to hope that formal grammars are more closely suited to describing special fragments of natural languages that are in some sense more rigidly defined, such as certain language fragments in poetry or law. In these

[1] A metaphorical compound word or phase used specially in Old English and Old Norse poetry, e.g. 'swan-road' for 'ocean'—Webstar's New Collegiate Dictionary (translator's note).

fragments an essential role is played by "prohibitions," which weed out, say, all texts not having a certain rhythmic pattern. Even the most casual attempt at writing poetry reveals the psychological reality of prohibitions in versification. But it is much less obvious that there is a set of generating rules that also has a psychological reality.

2. Yet there has been at least one poetic system in which generating rules occupied an important place. One of the basic elements of skaldic (ancient Icelandic) poetry consisted of special formulas called *kennings.* A kenning is an expression that can replace a single word. For example,

"storm of spears" is a kenning for "battle"

"tree of battle"
"bush of the helmet"
"thrower of swords"
"giver of gold"
} are kennings for warrior or man

"sea of the wagon" is a kenning for "earth"

"fire of war" is a "kenning for "gold"

"sky of sand"
"field of seals"
} are kennings for "sea," and so on.

A *simple kenning* is a kenning no part of which is a kenning. The examples above are all simple kennings. They play the role of axioms; obviously, only very great poets have the right to create new simple kennings. It falls to the lot of the lesser poets to create new kennings using the rules of deduction. The *rule of deduction* of a new kenning from earlier kennings is as follows: any word in a kenning may be replaced by a (not necessarily simple) kenning for that word. Here is a complicated example of a kenning together with its decomposition into simple kennings (an actual example):

"thrower of the fire of the storm of the witch of the moon of the steed of the ship stables"

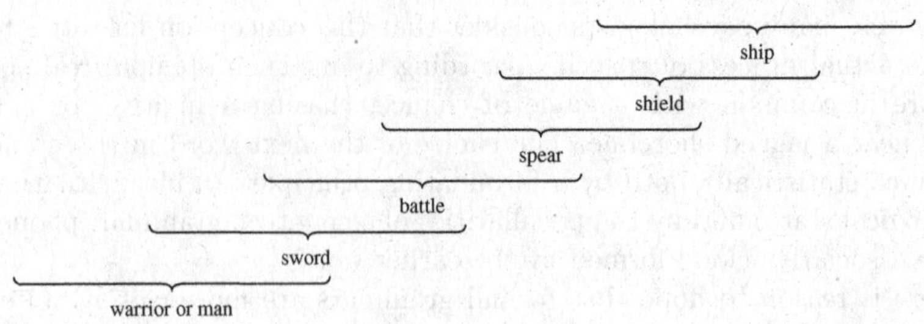

The Soviet poet Leonid Martynov thought of kennings as metaphors (a fundamental error, although an understandable one—kennings and metaphors play completely different structural roles in different poetic systems), and he wrote a poem "Songs of the Skalds" which ends as follows:

...But perhaps the translators have gotten a bit carried away?

No!
In our times, too,
might there not live
some *throwers*
of the fire
of the storm
of the witch
of the moon
of the steed
of the ship stables,
squanderers
of the amber
of the cold earth
of the great boar?

or

Anything is possible !!
And who can be so very sure
That there are no longer songs
which could be called

Surf
of yeast
of the people
of the bones
of the fjord?

Perhaps there really are such songs now,
Who can tell??

After all this, the professional opinion of M. I. Steblin-Kamenskii, whose book *Icelandic Culture* (Leningrad, Nauka, 1967) provided us with the above examples, sounds a little anticlimactic: "As a rule, any kenning for a man or warrior was no richer in content than the pronoun 'he.'"

EXERCISES:

(a) Find the simple kennings from which the last two kennings in Martynov's poem are deduced.
(b) Construct the kennings of maximum length that are deducible from all the simple kennings in the above text. Prove that it is impossible to deduce longer kennings.

6 Godel's Completeness Theorem

6.1. Let L be a language in $\mathfrak{L}_1$, let ϕ be an interpretation of L, and let $T_\phi L$ be the set of ϕ-true formulas. In §3 it was shown that the set $T_\phi L$ is *Gödelian*: it is complete, does not contain a contradiction, is closed with respect to deduction, and contains all the logical axioms Ax L. We say that a set of formulas $\mathcal{E}$

in L is *consistent* if the set of formulas deducible from $\mathcal{E}$ does not contain a contradiction, i.e., if there is no P such that $\mathcal{E} \vdash P$ and $\mathcal{E} \vdash \neg P$; otherwise, we say that $\mathcal{E}$ is *inconsistent.* The basic purpose of this section is to prove the following converse of the result in §3:

6.2. **Theorem** (Gödel)

(a) *Any Gödelian set T is the set of ϕ-true formulas $T_\phi L$ for a suitable interpretation of L in some set M having cardinality* $\leqslant$ card *(alphabet of L)* + $\aleph_0$. *(Here and below we always mean the cardinality of the alphabet without the variables.)*
(b) *Any set of formulas $\mathcal{E}$ which contains* Ax L *and is consistent can be imbedded in a Godelian set.*

The model M which is constructed in the proof consists of expressions in some extension of the alphabet of L, and thus has a somewhat artificial character. In the next section we show that, if we are given some natural interpretation (M, ϕ) of L, then we can find a submodel having cardinality $\leqslant$ card (alphabet of L) + $\aleph_0$.

6.3 **Corollary**. (Deducibility criterion). *Let* $\mathcal{E} \supset$ Ax L.

(a) *A formula P is deducible from $\mathcal{E}$ if and only if either $\mathcal{E}$ is inconsistent, or P is ϕ-true for all models ϕ of the set $\mathcal{E}$ having cardinality* $\leqslant$ card *(alphabet of L)* + $\aleph_0$.
(b) *A formula P is independent of $\mathcal{E}$ if and only if both $\mathcal{E} \cup \{P\}$ and $\mathcal{E} \cup \{\neg P\}$ are consistent; by Theorem* 6.2, *this is true if and only if $\mathcal{E} \cup \{P\}$ and $\mathcal{E} \cup \{\neg P\}$ have models.*

In what follows we shall often omit the verification that various formal deductions exist. If the reader wants to fill in such a verification, this can almost always be done more easily using deducibility criterion 6.3 than directly.

PROOF OF THE COROLLARY

(a) If $\mathcal{E}$ is inconsistent, then any formula can be deduced from $\mathcal{E}$ (Proposition 4.2). Suppose $\mathcal{E}$ is consistent and P is ϕ-true for all models of $\mathcal{E}$. Let $\bar{P} = \forall x_1 \cdots \forall x_n P$ be the "closure" of P. To prove that $\mathcal{E} \vdash P$. we consider two cases.
(a_1) $\mathcal{E} \cup \{\neg \bar{P}\}$ is *inconsistent.* Then $\mathcal{E} \cup \{\neg \bar{P}\} \vdash \bar{P}$, so that, by the Deduction lemma, $\mathcal{E} \vdash \neg \bar{P} \Rightarrow \bar{P}$. The tautology $(\neg \bar{P} \Rightarrow \bar{P}) \Rightarrow \bar{P}$ and MP give $\mathcal{E} \vdash \bar{P}$, and then the axiom of specialization and MP give $\mathcal{E} \vdash P$.
(a_2) $\mathcal{E} \cup \{\neg \bar{P}\}$ is *consistent.* Then, by Theorem 6.2, the set $\mathcal{E} \cup \{\neg \bar{P}\}$ has a model. In this model $\mathcal{E}$ is true and P is false, so that this case is impossible.

(b) Suppose that P is independent of $\mathcal{E}$, i.e., neither P nor $\neg P$ is deducible. Then, by part (a), there exists a model of $\mathcal{E}$ in which P is true and a model of $\mathcal{E}$ in which P is false. The converse is obvious. □

We now proceed to the proof of Gödel's completeness theorem.

6.4. **Definition.** Let $\mathcal{E}$ be a set of formulas in a language L. The alphabet of L is said to be *sufficient* for $\mathcal{E}$ if, for each closed formula $\neg\forall x P(x)$ in $\mathcal{E}$ there exists a constant c_P (depending on P) such that the formula

$$R_P\colon \neg\forall x\, P(x) \Rightarrow \neg P(c_P)$$

belongs to $\mathcal{E}$.

The intuitive meaning of R_P is; "If not all x have the property P, then some concrete object c_P can be found that does not have this property." We say that the *alphabet* (rather than $\mathcal{E}$) is "sufficient" or "insufficient" because if $\mathcal{E}$ does not contain enough formulas of the type R_P, we can simply add all the R_p to $\mathcal{E}$, while if there are not enough constants c_P, we then have to add them to the alphabet of the language.

The plan for proving Theorem 6.2 is as follows. We first prove the fundamental lemma:

6.5. **Fundamental Lemma.** *If a set of formulas $\mathcal{E}$ in a language L is consistent and complete and contains* Ax L, *and if the alphabet of L is sufficient for $\mathcal{E}$, then $\mathcal{E}$ has a model with cardinality $\leqslant$* card(*alphabet of L*) $+ \aleph_0$.

The next two lemmas allow us to embed any consistent $\mathcal{E}$ in a complete set, or in one for which the alphabet is sufficient.

6.6. **Lemma.** *If $\mathcal{E}$ is consistent and contains* Ax L, *then there exists a consistent and complete set of formulas $\mathcal{E}' \supset \mathcal{E}$.*

6.7. **Lemma.** *If $\mathcal{E}$ is consistent and contains* Ax L, *then there exist:*

(a) *a language L' whose alphabet is obtained from the alphabet of L by adding a set of new constants having cardinality $\leqslant$* card(alphabet of L) $+ \aleph_0$.
(b) *a set of formulas $\mathcal{E}'$ in L' that is consistent, contains $\mathcal{E}$ and* Ax L', *and has the property that the alphabet of L' is sufficient for $\mathcal{E}'$.*

However, these constructions get in each other's way. If we complete a set $\mathcal{E}$ for which the alphabet is sufficient, we might obtain a set with an insufficient alphabet; if we add new constants, we increase the overall supply of formulas in the language, and thereby lose the completeness of $\mathcal{E}$. Hence, we have to alternate the constructions in 6.6 and 6.7 a countable number of times in order to prove our last lemma:

6.8. **Lemma.** *If $\mathcal{E} \supset$* Ax L *is consistent, then there exist:*

(a) *a language $L^{(\infty)}$ whose alphabet is obtained from the alphabet of L by adding a set of new constants having cardinality $\leqslant$* card(alphabet of L) $+ \aleph_0$.
(b) *a set of formulas $\mathcal{E}^{(\infty)}$ in $L^{(\infty)}$ that is complete and consistent, contains $\mathcal{E}$ and* Ax $L^{(\infty)}$, *and has the property that the alphabet of $L^{(\infty)}$ is sufficient for $\mathcal{E}^{(\infty)}$.*

After Lemma 6.8 is proved, Theorem 6.2 is obtained from the fundamental lemma applied to $\mathcal{E}^{(\infty)}$ if we restrict the resulting model to L and $\mathcal{E}$.

We now prove the lemmas. The fundamental lemma is proved in 6.9, and Lemmas 6.5, 6.6, and 6.7 are proved in Sections 6.10, 6.11, and 6.12, respectively.

6.9. PROOF OF THE FUNDAMENTAL LEMMA. We begin by explicitly constructing the interpretation ϕ of L that will be our model for $\mathcal{E}$.
(a) By a *constant term* we mean a term in L that does not contain any symbols for variables. We let $M = \{\bar{t} \mid t \text{ is a constant term}\}$ be a "second copy" of the set of constant terms, and we define the *primary mappings* of the interpretation ϕ of L in M as follows:

$\phi(c) = \bar{c}$ (for any constant c);

$\phi(f)(\bar{t}_1, \ldots, \bar{t}_r) = \overline{f(t_1, \ldots, t_r)}$ (for each operation symbol f of degree r and all constant terms $t_1, \ldots, t_r$);

$\langle \bar{t}_1, \ldots \bar{t}_r \rangle \in \phi(p)$ if and only if $p(t_1, \ldots, t_r) \in \mathcal{E}$ (for each relation p of degree r and all constant terms $t_1, \ldots, t_r$).

We now prove the following claim:

(b) **Claim**. Let P be a closed formula. Then $|P|_\phi = 1$ if and only if $P \in \mathcal{E}$. (This claim implies that ϕ is a model for $\mathcal{E}$. In fact, if $P \in \mathcal{E}$ is not closed, then its closure $\forall x_1 \cdots \forall x_n P$ is deducible from $\mathcal{E}$ using Gen, and hence, since $\mathcal{E}$ is complete and consistent, $\forall x_1 \cdots \forall x_n P \in \mathcal{E}$. By the claim, $|\forall x_1 \cdots \forall x_n P|_\phi = 1$, so that $|P|_\phi = 1$.)

PROOF OF THE CLAIM. We use induction on the total number of quantifiers and connectives in P. We shall write $|P|$ instead of $|P|_\phi$.

(b_1) P is an atomic formula $p(t_1, \ldots, t_n)$. The claim follows from the definition of $|P|$ and the list of primary mappings, since the t_i are constant terms (or else P would not be closed).

(b_2) $P = \neg Q$. If $|P| = 1$, then $|Q| = 0$ and $Q \notin \mathcal{E}$ by the induction assumption applied to Q; since $\mathcal{E}$ is complete, we have $\neg Q \in \mathcal{E}$, i.e., $P \in \mathcal{E}$. On the other hand, if $|P| = 0$, then $|Q| = 1$ and $Q \in \mathcal{E}$, so that $\neg Q \notin \mathcal{E}$ since $\mathcal{E}$ is consistent.

(b_3) $P = (Q_1 \Rightarrow Q_2)$. We first show that if $|P| = 0$ then $P \notin \mathcal{E}$. In fact, in this case $|Q_1| = 1$ and $|Q_2| = 0$; by the induction assumption, $Q_1 \in \mathcal{E}, Q_2 \notin \mathcal{E}$; since $\mathcal{E}$ is complete, $\neg Q_2 \in \mathcal{E}$; using the tautology $Q_1 \Rightarrow (\neg Q_2 \Rightarrow \neg(Q_1 \Rightarrow Q_2))$ and using MP twice yields $\mathcal{E} \vdash (Q_1 \Rightarrow Q_2)$. Since $\mathcal{E}$ is complete and consistent, all closed formulas that are deducible from $\mathcal{E}$ belong to $\mathcal{E}$; hence, $\neg(Q_1 \Rightarrow Q_2) = \neg P \in \mathcal{E}$, so that $P \notin \mathcal{E}$.

We now show that if $P \notin \mathcal{E}$, then $|P| = 0$. In fact, since $\mathcal{E}$ is complete, we then have $\neg P = \neg(Q_1 \Rightarrow Q_2) \in \mathcal{E}$. The tautologies $\neg(Q_1 \Rightarrow Q_2) \Rightarrow Q_1$ and $\neg(Q_1 \Rightarrow Q_2) \Rightarrow \neg Q_2$ and MP give $\mathcal{E} \vdash Q_1$ and $\mathcal{E} \neg Q_2$, so that since $\mathcal{E}$ is complete and consistent, $Q_1 \in \mathcal{E}$ and $\neg Q_2 \in \mathcal{E}$. By the induction assumption, $|Q_1| = 1$ and $|Q_2| = 0$, so that $|P| = |Q_1 \Rightarrow Q_2| = 0$.

(b_4) $P = Q_1 \vee Q_2$ or $Q_1 \wedge Q_2$. Using the tautologies that express $\wedge$ and $\vee$ in terms of $\Rightarrow$ and $\neg$, we can reduce to the previous cases; we omit the details.

(b_5) $P = \forall x Q$. If x does not occur freely in Q, then $|P| = 1$ is equivalent to $|Q| = 1$, i.e., by the induction assumption, to $Q \in \mathcal{E}$. But $Q \in \mathcal{E}$ is equivalent to $\forall x\, Q \in \mathcal{E}$, in one direction using Gen and in the other direction using the axiom of specialization with $t = x$ and then MP.

We now assume that x occurs freely in Q. We first suppose that $|P| = 1$ but $P \notin \mathcal{E}$, and obtain a contradiction. If $P \notin \mathcal{E}$, then $\neg P \in \mathcal{E}$, i.e., $\neg\forall x\, Q(x) \in \mathcal{E}$. Since the alphabet of L is sufficient for $\mathcal{E}$, it follows that $\mathcal{E}$ contains the formula $\neg\forall x\, Q(x) \Rightarrow \neg Q(c_Q)$. Applying MP, we obtain $\mathcal{E} \vdash \neg Q(c_Q)$; since $\mathcal{E}$ is consistent, we have $Q(c_Q) \notin \mathcal{E}$. By the induction assumption, $|Q(c_Q)| = 0$ ($Q(c_Q)$ is closed!). This means that $|Q(x)|(\xi) = 0$ for $\xi \in \overline{M}$ if $x^\xi = c_Q$, contradicting the assumption that $|P| = 1$.

We now suppose that $|P| = 0$ but $P \in \mathcal{E}$, and obtain a contradiction. Since $|P| = 0$, for some $\xi \in \overline{M}$ we have $|Q(x)|(\xi) = 0$. Let t be the constant term for which $x^\xi = t$. Clearly t is free for x in Q, so that $0 = |Q(x)|(\xi) = |Q(t)|$. Hence $Q(t) \notin \mathcal{E}$ by the induction assumption, and $\neg Q(t) \in \mathcal{E}$ since $\mathcal{E}$ is complete. On the other hand, if $P \in \mathcal{E}$, i.e., $\forall x\, Q_x \in \mathcal{E}$, then the axiom of specialization $\forall x\; Q(x) \Rightarrow Q(t)$ gives us $\mathcal{E} \vdash Q(t)$. But since $\neg Q(t) \in \mathcal{E}$, this contradicts the consistency of $\mathcal{E}$.

(b_6) $P = \exists x\, Q$. This reduces to the previous case using the axiom that expresses $\exists$ in terms of $\forall$ and negation; we omit the details. □

6.10. Proof of Lemma 6.6. In order to embed $\mathcal{E}$ in a complete and consistent set $\mathcal{E}'$, we shall have to use Zorn's lemma and the deduction lemma for L (see Section 4.5 of Chapter II). Zorn's lemma will be applied to the set $C\mathcal{E}$ = the set of sets of formulas $\mathcal{E}'$ in L that contain $\mathcal{E}$ and are consistent. The set $C\mathcal{E}$ is ordered by inclusion.

Verification of the hypothesis of Zorn's lemma. Let$\{\mathcal{E}'_\alpha\}_{\alpha\in I}$ be a linearly ordered subset of $C\mathcal{E}$, i.e., for any α and β we have either $\mathcal{E}'_\alpha \leqslant \mathcal{E}'_\beta$ or $\mathcal{E}'_\beta \leqslant \mathcal{E}'_\alpha$. Then the union $\cup\mathcal{E}'_\alpha$ a belongs to $C\mathcal{E}$. In fact, otherwise $\cup\mathcal{E}'_\alpha$ would be inconsistent, and there would exist a deduction of a contradiction from a finite number of formulas. Suppose these formulas are contained in $\mathcal{E}'_{\alpha_1}, \ldots, \mathcal{E}'_{\alpha_n}$. But one of these sets contains the remaining $n-1$; this set would be inconsistent, contrary to the definition of $C\mathcal{E}$.

Proof of lemma 6.6 from Zorn's lemma. The set $C\mathcal{E}$ has a maximal element, i.e., a consistent set $\mathcal{E}' \supset \mathcal{E}$ such that if $Q \notin \mathcal{E}'$ then $\mathcal{E}' \cup \{Q\}$ is inconsistent. We claim that $\mathcal{E}'$ is complete. In fact, suppose that there were a closed formula P such that $P \notin \mathcal{E}'$ and $\neg P \notin \mathcal{E}'$. Since $\mathcal{E}'$ is maximal, it follows that $\mathcal{E}' \cup \{P\} \vdash R$ and $\mathcal{E}' \cup \{\neg P\} \vdash R$ for any formula R. By the deduction lemma, $\mathcal{E}' \vdash P \Rightarrow R$ and $\mathcal{E}' \vdash \neg P \Rightarrow R$. Using the tautology $(P \Rightarrow R) \Rightarrow ((\neg P \Rightarrow R) \Rightarrow R))$ and MP, we have $\mathcal{E}' \vdash R$, contradicting the consistency of $\mathcal{E}'$. □

6.11. Proof of Lemma 6.7. In constructing a language with a sufficient alphabet for a consistent set of formulas $\mathcal{E}'$ that contains $\mathcal{E}$ and Ax L', we proceed in the most natural way.

(a) We add to the alphabet of L a set of new constants whose cardinality is that of the alphabet of $L + \aleph_0$. We obtain a language L'.

(b) We consider the set of formulas $\mathcal{E} \cup \text{Ax}\, L'$ in the language L', where $Ax\, L'$ consists of all the logical axioms of L'. We claim that this set of formulas is consistent. In fact, if there were a deduction of a contradiction from $\mathcal{E} \cup \text{Ax}\, L'$ in L', then the following procedure would transform it into a deduction of a contradiction from $\mathcal{E}$ in L: take the finite set consisting of all the new constants that occur in the formulas in the deduction and replace these constants by old variables (in L) that do not occur in the formulas in the deduction. It is easily verified that the deduction of a contradiction remains a deduction of a contradiction, and now lies entirely in L.

(c) We consider the set S of formulas $P(x)$ containing one free variable x and such that $\neg\forall_x P(x) \in \mathcal{E} \cup \text{Ax}\, L'$. For each $P(x)$ in S we choose a new constant c_P subject to the following restriction: each c_P can be assigned a natural number, its *rank*, in such a way that if a constant of rank n occurs in $P(x)$ then c_P has rank $> n$. This can be done since card$(S) \leqslant$ card(alphabet of L') = card(alphabet of L) + $\aleph_0$. For each $P(x)$ in S define the formula

$$R_p : \neg\forall x\, P(x) \Rightarrow \neg P(c_P)$$

and finally let

$$\mathcal{E}' = \mathcal{E} \cup \text{Ax}\, L' \cup \{R_P | P(x) \in S\}.$$

Call any R_P an R-formula. Note that no R-formula has the form $\neg\forall x\ P(x)$, so that L' is sufficient for $\mathcal{E}'$. It remains only to verify that $\mathcal{E}'$ is consistent. If a contradiction were deducible from $\mathcal{E}'$ then it would be deducible using finitely many R-formulas. At least one R_P among these must be such that c_P does not occur in any of the others: namely, pick c_P of maximal rank. Hence it suffices to verify that if $\mathcal{E} \cup \text{Ax}\, L' \cup \mathcal{R}$ is consistent, where $\mathcal{R}$ is a set of formulas not containing c_P, then the addition of R_P does not lead to a contradiction.

Suppose $\mathcal{E} \cup \text{Ax}\, L' \cup \mathcal{R} \cup \{R_p\}$ were inconsistent. Then, in particular, we would have a deduction of $\neg R_P$ and, by the deduction lemma, $\mathcal{E} \cup \text{Ax}\, L' \cup \mathcal{R} \vdash R_P \Rightarrow \neg R_P$. The tautology $(R_P \Rightarrow \neg R_P) \Rightarrow \neg R_P$ and MP would yield a deduction of $\neg R_P$; that is,

$$\mathcal{E} \cup Ax\, L' \cup \mathcal{R} \vdash (\neg\forall x\, P(x) \Rightarrow \neg P(c_P)).$$

Then the tautology $\neg(P \Rightarrow \neg Q) \Rightarrow Q$ and MP would yield a deduction of $P(c_p)$. Transform this deduction by replacing the constant c_P with a variable y that does not occur in the formulas in the deduction. Since c_P does not occur in $\mathcal{R}$ it is easily verified that the transformation yields a deduction of $P(y)$ from $\mathcal{E} \cup \text{Ax}\, L' \cup \mathcal{R}$. Using Gen, $\mathcal{E} \cup \text{Ax}\, L' \cup \mathcal{R} \vdash \forall y\, P(y)$. But since $\neg\forall x\, P(x) \in \mathcal{E} \cup \text{Ax}\, L'$, we have $\mathcal{E} \cup \text{Ax}\, L' \vdash \neg\forall y\, P(y)$. Hence $\mathcal{E} \cup \text{Ax}\, L' \cup \mathcal{R}$ is inconsistent, contrary to hypothesis. □

6.12. PROOF OF LEMMA 6.8. Let L be a language in the class $\mathcal{L}_1$, and let $\mathcal{E}$ be a set of formulas in L. We embed $\mathcal{E}$ in a complete and consistent set $\mathcal{E}'$, and then apply Lemma 6.7 to $(L, \mathcal{E}')$. We let L^* and $\mathcal{E}^*$ denote the resulting language and set of formulas. We further define inductively

$$(L^{(0)}, \mathcal{E}^{(0)}) = (L, \mathcal{E}), \qquad (L^{(i+1)}, \mathcal{E}^{(i+1)}) = (L^{(i)^*}, \mathcal{E}^{(i)^*}),$$

and finally

$$L^{(\infty)} = \bigcup_{i=0}^{\infty} L^{(i)}, \qquad \mathcal{E}^{(\infty)} = \bigcup_{i=0}^{\infty} \mathcal{E}^{(i)}.$$

The set $\mathcal{E}^{(\infty)}$ is consistent, since any deduction of a contradiction would be obtained "at some finite level," and all the $\mathcal{E}^{(i)}$ are consistent. It is complete, since every closed formula in $L^{(\infty)}$ is written in the alphabet of $L^{(i)}$ for some i, and $\mathcal{E}^{(i+1)}$ contains the completion of $\mathcal{E}^{(i)}$ in $L^{(i)}$. Finally, the alphabet of $L^{(\infty)}$ is sufficient for $\mathcal{E}^{(\infty)}$ by the same argument.

This completes the proof of the lemmas. □

6.13. Deduction of Theorem 6.2 from the lemmas. Let T be a Gödelian set of formulas in L. Applying Lemma 6.8 to T, we embed (L, T) in $(L^{(\infty)}, T^{(\infty)})$, where the pair $(L^{\infty)}, T^{(\infty)})$ satisfies Lemma 6.5. Let $\phi^{(\infty)}$ be an interpretation of $L^{(\infty)}$ such as must exist by Lemma 6.5. The cardinality of $M^{(\infty)}$ does not exceed card(alphabet of L)$+\aleph_0$. The restriction ϕ of $\phi^{(\infty)}$ to L satisfies the condition $T \subset T_\phi L$. We prove that $T = T_\phi L$. In fact, let $P \in T_\phi L$. If P is closed, then $P \in T$, since either P or $\neg P$ lies in T by completeness, and $\neg P \notin T$ because P is ϕ-true. If P is not closed, and $x_1, \ldots, x_n$ are the variables that occur freely in P, then $\forall x_n\, P$ is closed and belongs to T. By the axiom of specialization, P is deducible from $T \cup \{\forall x_1 \cdots \forall x_n P\}$, so that $P \in T$, since T is closed under deduction. This proves the first assertion of the theorem.

The second assertion follows from the analogous argument applied to $\mathcal{E}$ instead of T. We find a model ϕ for $\mathcal{E}$; then $\mathcal{E} \subset T_\phi L$ and $T_\phi L$ is Gödelian. □

6.14. In conclusion, we note that if the alphabet of L contains a symbol $=$ for which the axioms of equality are included in $\mathcal{E}$ (or T), then there exists a normal interpretation that satisfies Theorem 6.2 and takes $=$ into equality. To prove this, we take the above model M and divide out by the equivalence relation $\phi(=)$, as in Section 4.6.

7 Countable Models and Skolem's Paradox

> "I know what you're thinking about," said Tweedledum: "but it isn't so, nohow."
> "Contrariwise," continued Tweedledee, "if it was so, it might be; and if it were so, it would be: but as it isn't, it ain't. That's logic."
>
> Lewis Carroll, *Through the Looking Glass*

7.1. In this section we discuss the technique of "cutting down" models, in particular, models for $\mathrm{L_1Set}$. Let L be a language in $\mathfrak{L}_1$, let $M \subset N$ be two sets (or classes in V), and let ϕ and ψ be interpretations of L in M and N, respectively, that are compatible in the obvious sense, so that ψ is an extension of ϕ. We have a natural embedding of interpretation classes $\overline{M} \subset \overline{N}$.

7.2. **Definition**. A formula P in L is called (M,N)-absolute if for all $\xi \in \overline{M}$ we have

$$|P|_M(\xi) = |P|_N(\xi).$$

(We write $|\ |_M$ instead of $|\ |_\phi$, and so on.)

The property of being absolute is usually used as follows: if P is absolute, and is also N-true, then it is automatically M-true. A formula P often fails to be absolute for the following reason: a formula $P = \exists x\, Q(x)$ can be N-true, so that N has an object with the property Q, but not M-true, because no such object lies in M. The proof of the following assertion shows how to handle this situation.

7.3. **Proposition**. *Let $\mathcal{E}$ be a set of formulas in L, let ψ be an interpretation of L in N, and let $M_0 \subset N$ be a subset. Then there exists a set $M, M_0 \subset M \subset N$, having cardinality $\leqslant$ card M_0 + card $\mathcal{E} + \aleph_0$, such that all the formulas in $\mathcal{E}$ are (M,N)-absolute.*

7.4. **Corollary** (Löwenheim–Skolem). *If the alphabet of L is countable and N is a model for $\mathcal{E}$, then N has a countable submodel for $\mathcal{E}$.*

The corollary follows from Proposition 7.3 if we construct a countable submodel with respect to which *all* the formulas of L are absolute, and in particular, in which all formulas that were true before remain true.

Proof of 7.3. Suppose the set $M_i \subset N, i \geqslant 0$, has already been defined. Set

$$M_{i+1} = M_i \cup \{x^{\xi'} \,|\, \xi' = \xi'(x, P, \xi)\},$$

where x runs through the variables in L, P runs through the subformulas of the formulas in $\mathcal{E}$, and ξ runs through the points of $\overline{M}_i$, and where for each fixed triple $(x, P, \xi), \xi'(x, P, \xi)$ is any one variation of ξ along x for which $|P|_N(\xi') = 1$ if such a variation exists; otherwise, the triple does not make any contribution to M_{i+1}.

Further, set $M = \cup_{i=0}^{\infty} M_i$. M clearly has the desired cardinality. We now show that all subformulas of the formulas in $\mathcal{E}$ are (M,N)-absolute. We use induction on the number of quantifiers and connectives in the formula. The result is obvious for atomic formulas; the inductive step when a new formula is constructed using a connective is also clear. The quantifier $\forall$ reduces to $\exists$ in the usual way.

Thus, suppose P is absolute. We show that $\exists x\, P$ is also absolute. It suffices to consider the case that x occurs freely in P. For $\xi \in \overline{M}$ we have

$$|\exists x\ P|_N(\xi) = \begin{cases} 1, & \text{if there exists a variation } \xi' \in \bar{N} \text{ of } \xi \text{ along } x \\ & \text{with } |P|_N(\xi') = 1, \\ 0, & \text{otherwise;} \end{cases}$$

$$|\exists x\ P|_M(\xi) = \begin{cases} 1, & \text{if if there exists a variation } \xi'' \in \overline{M} \text{ of } \xi \text{ along } x \\ & \text{with } |P|_M(\xi'') = 1, \\ 0, & \text{otherwise.} \end{cases}$$

But the conditions on the right are equivalent. In fact, there exists a variation η of the point ξ along variables that do not occur freely in P, such that $\eta \in \overline{M}_i$ for some i. Then in the case $|\exists x\ P|_N(\xi) = |\exists x\ P|_N(\eta) = 1$ there is a $\xi' \in \overline{N}$ with $|P|_N(\xi') = 1 \Rightarrow$ there is an $\eta' \in \overline{M}_{i+1}$ with $|P|_N(\eta') = 1$, where η' is a variation of η along x, by the construction of M_{i+1}. This completes the proof. □

7.5. We now apply Corollary 7.4 to the standard interpretation of $\mathrm{L_1Set}$ in the von Neumann universe V and the set $\mathcal{E}$ of Zermelo–Fraenkel axioms. We obtain a countable model N for this axiom system, but this model has one defect: if $X \in N$, some elements in X might not themselves belong to N, i.e., $\in$ is not necessarily transitive. The following result of Mostowski shows how to replace N by a transitive countable model.

Let $N \subset V$ be a subclass, and let $\varepsilon \subset N \times N$ be a binary relation. We shall write $X \varepsilon Y$ instead of $\langle X, Y\rangle \in \varepsilon$. For any $X \in N$ we set

$$[X] = \{Y | Y \varepsilon X\}.$$

Suppose that $[X]\varepsilon V$ for all $X \in N$, i.e., each $[X]$ is a set rather than a class. We consider the interpretation ϕ of $\mathrm{L_1Set}$ in the class N for which $\phi(\in)$ is ε and $\phi(=)$ is equality.

7.6. **Proposition** (Mostowski). *Suppose that the axiom of extensionality and the axiom of the empty set are ϕ-true, and that N does not contain any infinite chain $\cdots X_n \varepsilon X_{n-1} \varepsilon \cdots \varepsilon X_1 \varepsilon X_0$. Then there exist a unique transitive class $M \subset V$ and a unique isomorphism $f : (N, \varepsilon) \xrightarrow{\sim} (M, \in)$.*

If we apply this proposition to the countable model $(N, \in)$ for the Zermelo–Fraenkel axioms in Section 7.5, we obtain a transitive countable model $(M, \in)$, that is, a "small-universe." (The condition that all ε-chains are finite holds even in V, as well as in N; $[X]$ is the subset $X \cap N \subset X$, and hence is an element of V.)

7.7. Proof of Proposition 7.6. Using transfinite induction, for every ordinal α we construct sets $N_\alpha \subset N, M_\alpha \subset V$ and compatible isomorphisms $f_\alpha : (N_\alpha, \varepsilon|_{N_\alpha}) \xrightarrow{\sim} (M_\alpha, \in|_{M_\alpha})$, and we show that $\cup N_\alpha = N$.

(a) Since the axiom of extensionality is ϕ-true and $\phi(=)$ is equality, we easily obtain $X_1 = X_2 \Leftrightarrow [X_1] = [X_2]$ for all $X_1, X_2 \in N$. Let $\varnothing_N \in N$ be the interpretation of the constant $\varnothing$ of the language $\mathrm{L_1Set}$. Since the axiom of the empty set is ϕ-true, we may conclude that $\varnothing_N$ is the unique element of N for which $[\varnothing_N] = \varnothing \in V$. We set

$$N_0 = \{\varnothing_N\}, \qquad M_0 = \{\varnothing\}, \qquad f_0(\varnothing_N) = \varnothing.$$

(b) *Recursive construction.* Let α be an ordinal. Suppose that N_α, M_α, and f_α have already been constructed. We set

$$N_{\alpha+1} = \{X \in N | [X] \subset N_\alpha \wedge X \notin N_\alpha\} \cup N_\alpha;$$

$$f_{\alpha+1}(X) = \{f_\alpha(Y) | Y \in [X]\}, \quad \text{for } X \in N_{\alpha+1} \backslash N_\alpha; \quad f_{\alpha+1}|N_\alpha = f_\alpha;$$

$$M_{\alpha+1} = \text{image of } f_{\alpha+1} = \text{range of } f_{\alpha+1}.$$

If β is a limiting ordinal, we set $N_\beta = \cup_{\alpha<\beta} N_\alpha, M_\beta = \cup_{\alpha<\beta} M_\alpha$, and $f_\beta = \cup_{\alpha<\beta} f_\alpha$. Finally, we set $M = \cup M_\alpha$ and $f = \cup f_\alpha$, where the union is taken over all the ordinals.

(c) *Inductive proof.* We verify that for each α,

(c_1) *N_α is a set, i.e., $N_\alpha \in V$.*
(c_2) *M_α is a transitive subset of V.*
(c_3) *f_α is an isomorphism of N_α with M_α taking ε to $\in$.*
(c_4) $N = \cup_\alpha N_\alpha$.

Assertions (c_1)–(c_3) are obvious for $\alpha = 0$. If they hold for all $\alpha < \beta$ and if β is a limiting ordinal, then they also hold for β. It remains to check the step from α to $\alpha + 1$.

(c_1) $[\ \ ]$ is obviously a function from $N_{\alpha+1} \setminus N_\alpha$ to $\mathcal{P}(N_\alpha)$; since the axiom of extensionality is true, there exists an inverse function. Its image $N_{\alpha+1} \setminus N_\alpha$ is a set, since N_α, and therefore $\mathcal{P}(N_\alpha)$, are sets by the induction assumption.

(c_2) Any element in $M_{\alpha+l} \setminus M_\alpha$ has the form $\{f_\alpha(Y)|Y \in [X]\}$, where $X \in N_{\alpha+1} \setminus N_\alpha$. But then $[X] \subset N_\alpha$. Hence, an element $f_\alpha(Y)$ of this element of $M_{\alpha+1} \setminus M_\alpha$ belongs to the image of f_α, i.e., to the set $M_\alpha \subset M_{\alpha+1}$. This proves the transitivity of $M_{\alpha+1}$.

(c_3) We first verify that $f_{\alpha+1}$ is a bijection. The surjectivity is obvious; using the induction assumption, we see that it suffices to verify injectivity on $N_{\alpha+1} \setminus N_\alpha$. But if $X_1, X_2 \in N_{\alpha+1} \setminus N_\alpha$ and $f_{\alpha+1}(X_1) = f_{\alpha+1}(X_2)$, then

$$\{f_\alpha(Y)|Y \in [X_1]\} = \{f_\alpha(Y)|Y \in [X_2]\}.$$

Since f_α is injective, we obtain $[X_1] = [X_2]$, so that $X_1 = X_2$.

We then obtain

$$Y \varepsilon X \Leftrightarrow Y \in [X] \Leftrightarrow f_\alpha(Y) \in f_{\alpha+1}(X),$$

so that for $X \in N_{\alpha+1} \setminus N_\alpha$ the relation $Y \varepsilon X$ goes to $f_{\alpha+1}(Y) \in f_{\alpha+1}(X)$. This is clearly sufficient to complete the induction.

(c_4) Finally, we verify that $N = \cup N_\alpha$. Let $N' = N \setminus \cup N_\alpha$; we suppose that N' is nonempty and show that this leads to a contradiction. If there existed an $X \in N'$ such that $[X] \cap N' = \varnothing$, then we would have $[X] \cap N \subset \cup N_\alpha$; then $[X] \subset N_{\alpha_0}$ for some α_0, so that $X \in N_{\alpha_0+1}$, contradicting the assumption that $X \in N \setminus \cup N_\alpha$. On the other hand, if we had $[X_0] \cap N' \neq \varnothing$ for all $X_0 \in N'$, then, successively choosing $X_{n+1} \in [X_n] \cap N'$, we would obtain an infinite chain $X_{n+1} \varepsilon X_n \varepsilon X_{n-1} \varepsilon \cdots \varepsilon X_0$, contradicting the hypothesis of the theorem.

(d) Suppose we have two transitive subclasses M and M', and an isomorphism $g : (M, \in) \xrightarrow{\sim} (M', \in)$. We set $M_\alpha = V_\alpha \cap M$ and $M'_\alpha = V_\alpha \cap M'$. An obvious induction on α then shows that g is the identity map. The proposition is proved. □

7.8. *Skolem's paradox.* Let M be a transitive countable model for the Zermelo–Fraenkel axioms. Then the following formulas are M-true:

the axiom of infinity;
the power set axiom;

Cantor's theorem that there is no mapping of x onto $\mathcal{P}(x)$ for any set x (this theorem is deducible from the Zermelo–Fraenkel axioms).

Since $\mathcal{P}(X)$ is uncountable when X is countably infinite, the content of the assertion that the power set axiom is true in the countable model M must be very different from the content of the assertion that this axiom is V-true. In fact, in L$_1$Set let "$y = \mathcal{P}(x)$" be abbreviated notation for the formula $\forall z($"$z \subset x$" $\Leftrightarrow z \in y)$. Let $\xi \in \overline{M}, x^\xi = X \in M$, and $y^\xi = Y \in M$. Then we easily see that

$$|``y = \mathcal{P}(x)"|_M(\xi) = 1 \Leftrightarrow Y = \{Z | Z \subset X \wedge Z \in M\},$$

i.e., $\mathcal{P}(X)_M = \mathcal{P}(X) \cap M$ plays the role of $\mathcal{P}(X)$ in M. Here $\mathcal{P}(X)_M$ is at most countably infinite, since M is countable; so, from the usual point of view, there exists a mapping of a countably infinite set X onto $\mathcal{P}(X)_M$. This does not contradict Cantor's theorem, because the M-truth of Cantor's theorem merely means that there are no (graphs of) such mappings in the model M. Such graphs may exist outside of M, but if we add such a graph to M (along with everything that must be added for the axioms to remain true), we thereby increase M, and at the same time $\mathcal{P}(X)_M$, and the mapping stops being onto.

All such ways in which statements of set theory change their meaning in countable models are customarily referred to as Skolem's paradox.

Cohen was the first who was able to use the properties of countable models to prove the nondeducibility of the continuum hypothesis. In his models sets of "M-intermediate" cardinality lie between ω_0 and $\mathcal{P}(\omega_0)_M$, although from an external point of view both ω_0 and $\mathcal{P}(\omega_0)_M$, along with all the other sets, are simply countable. Cohen introduced fundamentally new ideas of relativizing the very notion of truth, and it is only with the benefit of hindsight that we can so easily understand the situation in his models. For details, see Chapter III.

Skolem himself, and other specialists on the foundations of mathematics, were willing to work with countably infinite sets, but not with larger infinities. They considered Skolem's paradox to be a manifestation of the relative character of set-theoretic concepts. In particular, they considered that there exist "different continua" $\mathcal{P}(\omega_0)_M$, none of which coincides with the "real" $\mathcal{P}(\omega_0)$.

From the point of view of the topologist or analyst, for whom the continuum is a working reality, the existence of countable models means that formal language has limitations as a means of imitating intuitive reasoning. We encountered similar limitations when discussing the formal axioms of induction in §4.

For the psychologist or philosopher, perhaps the most interesting aspect of the situation is that any mathematician can understand the viewpoint of another mathematician (without having to agree with it). This means that what mathematician A says, though demonstrably incapable of conveying unambiguous information about the continuum, nevertheless is capable of bringing the brain of mathematician B to the point where it forms an idea of the continuum that adequately represents the idea in A's brain. Then B is still free to reject this idea.

"I know what you're thinking about," said Tweedledum: "but it isn't so, nohow."

8 Language Extensions

8.1. In this section we study the formal version of "introducing new notation." Here we consider only names of new functions and constants that are "demonstrably definable" in the language. Adding such names to the alphabet shortens formulas and formal deductions, but does not increase the set of deducible formulas—this will be the fundamental theorem of this section.

Of course, in practice, abbreviated notation and well-chosen new names can immediately make accessible to our intuition entire areas of mathematical facts that were previously inaccessible. One of the best-known examples is the groups introduced by Galois to study equations. In 1924, commenting on the attempt to curb the inflation in Germany by introducing a new unit of currency, the Rentenmark, Hilbert remarked skeptically, "A problem cannot be solved by renaming the independent variable." But as his biographer Constance Reid noted, Hilbert was wrong: the economic situation gradually stabilized.

We start with the following data.

8.2. Let L' be a language in $\mathfrak{L}_1$ with equality and with an infinite set of variables, and let $P'(x)$ be a formula in L' in which x occurs freely. We recall that the abbreviated notation $\exists!x\, P'(x)$ (read: "there exists a unique x with the property P'") stands for the formula

$$\exists x\, P'(x) \wedge \forall x \forall y (P'(x) \wedge P'(y) \Rightarrow x = y).$$

Let $\mathcal{E}'$ be a set of formulas in L' that contains Ax L', the axioms of equality, and perhaps some special axioms. Suppose that the formula $\exists!x\, P'(x, y_1, \dots, y_n)$ is deducible from $\mathcal{E}'$, where P' has no free variables other than $x, y_1, \dots, y_n$. Intuitively, this means that P' defines x as an implicit function of $y_1, \dots, y_n$, and in the informal text we can introduce a new notation for this notation for this function, say, $x = f(y_1, \dots, y_n)$, and then always use that notation. Now we give the formal version of this procedure.

8.3. **Proposition.** *Under the conditions in 8.2, let L denote the language in $\mathfrak{L}_1$ whose alphabet is obtained from the alphabet of L' by adding a new operation symbol f of degree n if $n \geqslant 1$, or a constant f if $n = 0$. Let $\mathcal{E}$ be the smallest set of formulas in L containing* Ax L, *the axioms of equality, $\mathcal{E}'$, and the formula $P'(f(y_1, \dots, y_n), y_1, \dots, y_n)$.*

Then there exists an explicitly describable map from the set of formulas of the (richer) language L to the set of formulas of the (poorer) language L' that correlates with each Q a translation Q' and that has the following properties:

(a) *If f does not occur in Q, then the translation of Q coincides with Q.*
(b) *If Q is deducible from $\mathcal{E}$ in L, then Q' is deducible from $\mathcal{E}'$ in L'. In particular, the set of formulas in L' that are deducible from $\mathcal{E}'$ in L' coincides*

with the set of formulas in L that do not contain f and are deducible from $\mathcal{E}$ in L.

PROOF.

Translation of formulas. Suppose $n \geqslant 1$. (The case $n = 0$ is analogous, and is simpler, so we shall omit it.) The first effect of adding f is to increase the set of terms: L includes terms of the form $f(t_1, \ldots, t_n)$, where f can occur in $t_1, \ldots, t_n$, and so on. In order to decrease the number of references to f, we must say "$f(t_1, \ldots, t_n)$" in a roundabout way: "that x for which $P(x, t_1, \ldots, t_n)$." This is the basic idea behind the translation of formulas. We now give a precise inductive definition.

(a) A term $f(t_1, \ldots, t_n)$ is called a simple f-term if f does not occur in $t_1, \ldots, t_n$.
(b) Let Q be an atomic formula in L. If f does not occur in Q, we let Q be its own translation. If f occurs in Q, then there exists a simple f-term $f(t_1, \ldots, t_n)$ that occurs in Q. We take the very first occurrence of a simple f-term in Q, then take a variable symbol x that does not occur in Q, substitute it in place of this occurrence, thereby obtaining a formula Q^*, and finally construct the formula

$$Q'_{(1)}: \exists x(P(x, t_1, \ldots, t_n) \wedge Q^*(x)).$$

We apply this procedure to $Q'_{(1)}$ to obtain $Q'_{(2)}$, and so on. After a finite number of steps we obtain a formula $Q'_{(i)} = Q'$ in which f does not occur. This Q' is the translation of Q.
(c) If Q is not an atomic formula, it has the form $\neg Q_1$ or $Q_1 * Q_2$ (where $*$ is a connective), or else $\forall y\, Q_1$ or $\exists y\, Q_1$. In all cases Q is translated automatically using the translations of Q, Q_1, Q_2, i.e., by "from Q produce Q'" to the component parts.

Translation of deductions. The problem is the following: Let $Q_1, \ldots, Q_n = Q$ be a deduction of Q from $\mathcal{E}$, and let Q' be the translation of Q. We must construct a deduction of Q' from $\mathcal{E}'$. The most obvious idea is to write the sequence of translations $Q'_1, \ldots, Q'_n$. Why isn't this a deduction of Q' from $\mathcal{E}'$, since MP and Gen are translated in a trivial way, and tautologies are translated as tautologies? Because, for example, the logical axiom $\forall x\, R(x) \Rightarrow R(f)$ might appear in this sequence, and this formula stops being an axiom after it is translated, if f occurs in R. Hence, we must fill in the sequence $Q'_1, \ldots, Q'_n$ by adding deductions from $\mathcal{E}'$ of certain of its terms. This is a rather cumbersome combinatoric procedure, which one can read in §74 of Kleene's book *Introduction to Metamathematics* (Van Nostrand, New York–Toronto, 1952). (The moral of the story is that new notation really does economize on time and space.)

Instead of using this procedure, we shall give an ineffective proof that $\mathcal{E}' \vdash Q'$ using the deducibility criterion in 6.3. We state this criterion once more:

(a) *If Q' is true in any model of $\mathcal{E}'$, then $\mathcal{E}' \vdash Q'$.* Since $\mathcal{E}'$ contains the axioms of equality, we can slightly strengthen this as follows:

(b) *If Q' is true in any normal model of $\mathcal{E}'$ then Q' is true in any model of $\mathcal{E}'$.*

Recall that $=$ is interpreted as equality in a normal model. On the other hand, in §4 we showed that in any model $=$ is interpreted as an equivalence relation that is compatible with the interpretation of all the constants, functions, and relations. Factoring out by this equivalence relation leads to a normal model, in which the truth values of all the formulas remain as before.

(c) *The normal models of $\mathcal{E}'$ (in the language L') coincide with the normal models of $\mathcal{E}$ (in the language L).*

More precisely, we can give the following natural one-to-one correspondence between them that preserves the truth function. We shall limit ourselves to the case $n \geqslant 1$. Let ϕ be a normal interpretation of L' in M for which $|Q'|_\phi = 1$ for all $Q' \in \mathcal{E}'$. In particular, since $\mathcal{E}' \vdash \exists!\, x\, P'$, we have

$$|\exists! x\, P'(x, y_1, \ldots, y_n)|_\phi = 1.$$

Computing the truth value on the left at a point $\xi \in \overline{M}$ and using the normality of the model, we then find that to every n-tuple $\langle y_1^\xi, \ldots, y_n^\xi \rangle \in M^n$ there corresponds a unique $x^{\xi'} \in M$ such that $|P'(x^{\xi'}, y_1^\xi, \ldots, y_n^\xi)|_\phi = 1$ (this is not the standard notation, but the meaning is clear). We now interpret the symbol f (which is the new symbol in the language L) as the function $M^n \to M$ that takes $\langle y_1^\xi, \ldots, y_n^\xi \rangle$ to $x^{\xi'}$. We obviously obtain a normal model for $\mathcal{E}$ in L.

Conversely, any normal model for $\mathcal{E}$ can be restricted to L' to obtain a normal model for $\mathcal{E}'$.

(d) *If Q is deducible from $\mathcal{E}$ in L, then Q' is true in any normal model for $\mathcal{E}'$.*

Proof. Q is true in any model ϕ for $\mathcal{E}$. To prove that Q' is true, we begin with atomic formulas Q that contain f. In the notation in the first part of the proof (translation of formulas), we construct Q^* and then $Q'_{(1)} = \exists x(P(x, t_1, \ldots, t_n) \wedge Q^*(x))$. To verify that $|Q'_{(1)}|_\phi = 1$, for each point $\xi \in \overline{M}$ we must find a variation ξ' of ξ along x for which

$$|P|_\phi(\xi') = 1 \quad \text{and} \quad |Q^*(x)|_\phi(\xi') = 1.$$

We determine $x^{\xi'}$ from the condition $|P(x^{\xi'}, t_1^\xi, \ldots, t_n^\xi)|_\phi = 1$. The description in (c) of the interpretation of f shows that we now have $|Q^*|_\phi(\xi') = |Q|_\phi(\xi) = 1$.

Thus, truth is preserved in going from Q to $Q'_{(1)}$. Repeating this procedure, we find that Q' is true for atomic formulas Q. Finally, the truth of Q' in the general case is proved by induction on the number of connectives and quantifiers. Combining the results (a)–(d), we then obtain $\mathcal{E}' \mapsto Q'$, which which completes the proof of Proposition 8.3. □

8.4. Examples

(a) In $\mathrm{L_1Set}$ the following formula is deducible from the axioms of extensionality and pairing (and also the axioms of equality and the logical axioms):

$$\exists! x \forall z (z \in x \Leftrightarrow z = u \vee z = v).$$

Using Proposition 8.3, we see that we may add to L_1Set a new degree 2 function symbol $\{\}$, "unordered pair," without changing the set of formulas in L_1Set that are deducible from the Zermelo–Fraenkel axioms. Therefore, without hesitation we may use not only the abbreviated notation "$x = \{u, w\}$" as before, but also terms that are put together using the symbol $\{\}$. In particular (here the use of $\{\}$ is not normalized, but is in agreement with tradition): (b) We can introduce notation for the finite ordinals

$$\varnothing, \{\varnothing\}, \{\varnothing, \{\varnothing\}\}, \ldots$$

as terms in their own right in our language extension, and then embed formal arithmetic in formal set theory.

(c) After deducing the formula

$$\exists! x(\text{"}x \text{ is an ordinal"} \wedge \text{"}x \text{ is not finite"} \wedge \text{"}\forall \text{ ordinal} y < x, y \text{ is finite"})$$

from the Zermelo–Fraenkel axioms, we can introduce a new constant ω_0, and then continue to introduce names of more and more ordinals that are demonstrably uniquely characterized by formulas in L_1Set (or in language extensions that are formed in the same way).

We shall make use of this new freedom of action in Chapter III.

9 Undefinability of Truth: The Language *SELF*

9.1. When modeled in formal languages, arguments of the "liar paradox" type lead to important theorems on the limitations of the modes of expression and proof in these languages. The best known of these theorems are Tarski's theorem on the undefinability of the set of true formulas and Gödel's theorem on the impossibility of effectively axiomatizing arithmetic.

The next three sections are devoted to Tarski's theorem. Our presentation is based on an excellent article by Smullyan (Languages in which self-reference is possible, *J. Symb. Logic.* vol. 22, no. 1 (1957), 55–67).

In this section we describe the extremely elementary language SELF (which does not belong to $\mathfrak{L}_1$), which was designed to illustrate self-reference and which graphically demonstrates the idea of such a construction. In §10 we introduce the language SAr, which is just as expressive as L_1Ar, but does not belong to $\mathfrak{L}_1$. Its syntax is close to that of SELF, which greatly simplifies proofs. Finally, in §11 we use a method of Smullyan to prove Tarski's theorem for SAr.

9.2. *The language SELF* (*Smullyan's Easy Language For self-reference*)

The alphabet of SELF E, $*$ (symmetric quotes), r (relation of degree 1), $\neg$ (negation).

The *syntax* of SELF. The distinguished expressions are labels, displays, formulas, and names. The *label* of any expression P is $*P*$ ("P in quotes"). The *display* of any expression P is $P * P*$ ("something with a label"). *Formulas* are expressions of the form $rE \ldots E * P*$ or $\neg rE \ldots E * P*$, where E appears $k \geqslant 0$

times after r. We use the abbreviated notation $rE^k * P*$ and $\neg rE^k * P*$ for formulas. Finally, we introduce the binary relation "is the name of" on the set of all distinguished expressions. This relation is defined recursively:

(a) The label of P is a name of P.
(b) If P is a name of Q, then EP is a name of the display of Q, i.e., a name of the expression $Q * Q*$.

9.3. *Remarks*

(a) If P is a name of Q, then the display of Q has at least two different names: EP and $*Q*Q**$. Thus, an expression can have several names. But conversely, an expression is uniquely determined if we know its name; names all have the form $E^k * P*, k \geqslant 0$. We shall write $N(Q)$ in place of "one of the names of Q."
(b) Every formula has the form $rN(Q)$ or $\neg rNQ$. In 9.4 we interpret such a formula as the statement, "The expression Q has (or does not have) the property R," and it is natural that the formula, in saying something about Q, "calls Q by name."
(c) The expression $E * E*$ is one of two possible names for itself. In exactly the same way, the formula $rE * rE*$ "says something about itself" (see 9.5). The language SELF was constructed precisely in order to produce these effects of self-reference with the fewest possible modes of expression.

9.4. *The standard interpretations.* In order to give one of the standard interpretations of the language SELF, we choose any set (property) R of expressions of the language and introduce the truth function $|\ |_R$ on the formulas by stipulating

$$1 - |\neg rN(Q)|_R = |rN(Q)|_R = \begin{cases} 1, & \text{if } Q \in R, \\ 0, & \text{otherwise.} \end{cases}$$

We say that a formula is R-true (R-false) if the value of $|\ |_R$ in the formula equals 1 (resp. 0).

9.5. **Undefinability Theorem**. *For any property R,*

$$R \cap \{formulas\} \neq \begin{cases} R\text{-}true formulas, \\ R\text{-}false formulas. \end{cases}$$

PROOF.
(a) The formula $Q = \neg rE * \neg rE*$ is R-true $\Leftrightarrow rE * \neg rE*$ is R-false $\Leftrightarrow Q \notin R$, since $E * \neg rE*$ is a name of the display of $\neg rE$, i.e., a name of Q. Thus, Q cannot both lie in R and be true, which proves the first part of the theorem. The connection with the liar paradox becomes clear if we note that Q says about itself, "I do not have the property R."
(b) Analogously, the formula $rE * rE*$ says about itself, "I have the property R," and so cannot both lie in R and be R-false. □

10 Smullyan's Language of Arithmetic

10.1. In this section we describe the language of arithmetic SAr and its standard interpretation. The main difference between SAr and $\mathrm{L_1Ar}$ is that in SAr we are allowed to form "class terms"—names of certain sets of natural numbers. More precisely, if $P(x)$ is a formula in SAr with one free variable x, then the expression $x(P(x))$ in SAr names the set $\{x \in N | P(x) \text{ is true}\}$, and the expression $x(P(x))\bar{k}$, where the term $\bar{k}$ is a name for an integer $k \geqslant 1$, is a name for the statement "k satisfies P." The greater richness of the modes of expression in SAr, as opposed to $\mathrm{L_1Ar}$, does not increase the class of subsets in $\cup_{r\geqslant 1} N^r$ that are definable by formulas. But it brings the syntax of SAr so close to that of SELF that we can imitate the proof of Theorem 9.5.

In addition, the alphabet of SAr is somewhat altered and shortened in comparison with the alphabet of $\mathrm{L_1Ar}$, but this is done only in order to simplify the description of the syntax. These changes do not make the logic of SAr any poorer.

10.2. *The alphabet of* SAr: x (a variable); $'$ (used to form a countable set of variables $x, x', x'', \ldots$); $\cdot$ (multiplication, a degree-2 operation); $\uparrow$ (raising to a power, a degree-2 operation, as in Algol); $=$ (equality); $\downarrow$ (a connective, the conjunction of negations); $(,)$ (parentheses); and $\bar{1}$ (the constant one).

10.3. *The syntax and interpretation of* SAr. Because we are allowed to form the class terms $x(P(x))$ and the formulas $x(P(x))\bar{k}$, the syntax is more complicated than in languages of $\mathfrak{L}_1$. We use induction on the integer $i \geqslant 0$ to define two sequences of sets of expressions: Tm_{2i} (terms of rank $\leqslant 2i$) and Fl_{2i+1} (formulas of rank $\leqslant 2i+1$). (Using double induction—on the rank of the term or formula, and, within the set Tm_{2i} or Fl_{2i+1}, on the length of the term or formula—one can prove a unique reading lemma; this lemma is the basis for defining free and bound occurrences of variables and truth functions. However, since there is nothing new here beyond what was done in §1, we leave the details to the reader.)

Along with our description of the syntax, we give a parallel description of the standard interpretation of SAr in N. In order to interpret expressions with free variables, we must fix a point $\xi \in N^N = N \times N \times N \times \cdots$, which we shall identify with the corresponding infinite vector with natural number coordinates. Here the value of the kth variable $(x'^{\cdots}{}')^\xi$ ($k-1$ primes) is in the kth place in the vector.

(a_0) Tm_0 *is the set of numerical terms i.e., the least set of expressions that contains the variables* $x, x', x'', \ldots$ *and the names of the natural numbers* $\bar{1}, \bar{1}\bar{1}, \bar{1}\bar{1}\bar{1}, \ldots$ *and is closed with respect to forming the expressions* $(t_1) \cdot (t_2)$ *and* $(t_1) \uparrow (t_2)$, *where* $t_i \in Tm_0$.

Instead of $x'^{\cdots}{}'$($k-1$ primes) we shall write x_k, and instead of $\bar{1}\cdots\bar{1}$ ($k \geqslant 1$ ones) we shall write $\bar{k}$. The term $\bar{k}$ is interpreted as k (not depending on ξ); x_k^ξ is interpreted as the kth coordinate of ξ; and if $t_1^\xi, t_2^\xi \in N$ have already been determined, then $[(t_1) \cdot (t_2)]^\xi = t_1^\xi t_2^\xi$ and $[(t_1) \uparrow (t_2)]^\xi = (t_1^\xi)^{t_2^\xi}$. The occurrences of the expressions $x_k = x'^{\cdots\prime}$ in any term in Tm_0 are obviously independent of one another. All such occurrences are considered free.

(b$_0$) *Fl_1 is the least set of expressions that contains all expressions of the form $t_1 = t_2$ (where $t_i \in Tm_0$) and is closed with respect to forming the expressions $(P_1) \downarrow (P_2)$, where $P_i \in Fl_1$.* In other words, Fl_1 is the logical closure of the set of atomic formulas $\{t_1 = t_2 | t_i \in Tm_0\}$.

Choosing a point ξ determines a truth value for any formula $P \in Fl_1$ by induction on the number of times $\downarrow$ occurs:

$$|t_1 = t_2|(\xi) = \begin{cases} 1, & \text{if } t_1^\xi = t_2^\xi; \\ 0, & \text{otherwise;} \end{cases}$$

$$|(P_1) \downarrow (P_2)|(\xi) = \begin{cases} 1, & \text{if } |P|_1(\xi) = |P_2|(\xi) = 0, \\ 0, & \text{otherwise.} \end{cases}$$

All occurrences of variables in elements of Fl_1 are independent of one another, and are considered free.

Now let $i \geqslant 1$, and suppose that the sets Tm_{2k-2}, Fl_{2k-1} are already defined for $k \leqslant i$ along with the interpretations and the division into free and bound occurrences of variables. We define the next sets Tm_{2i} and Fl_{2i+1} as follows.

(a$_i$) Tm_{2i} consists of the class terms of rank $\leqslant 2i$:

$$Tm_{2i-2} \cup \{x_k(P) | k \geqslant 1, P \in Fl_{2i-1}\}$$

(Tm_0 need not be included when $i = 1$). These elements have the following interpretation:

$$(x_k(P))^\xi = \left\{ \begin{matrix} x_k^{\xi'} | \xi' \text{ runs through the variations of } \xi \text{ along } x_k \\ \text{for which } |P|(\xi') = 1 \end{matrix} \right\}.$$

All occurrences of the variable x_k in $x_k(P)$ are considered bound, and the occurrences of other variables remain the same (free or bound) as in P.

(b$_i$) *Fl_{2i+1} is the logical closure of the set of expressions*

$$Fl_{2i-1} \cup \{x_k(P) = x_k(Q) | k \geqslant 1; P, Q \in Fl_{2i-1}\} \cup \{T\bar{k} | k \geqslant 1, T \in Tm_{2i}\}$$

The *truth function* is defined as follows: if we set $x_k(P) = T_1$ and $x_k(Q) = T_2$, then

$$|x_k(P) = x_k(Q)|(\xi) = \begin{cases} 1, & \text{if } T_1^\xi = T_2^\xi \text{ as subsets of } N, \\ 0, & \text{otherwise;} \end{cases}$$

$$|T|\bar{k}(\xi) = \begin{cases} 1, & \text{if } k \in T^\xi, \\ 0, & \text{otherwise.} \end{cases}$$

The function $||$ is extended to the logical closure in the same way as in (b$_0$). All occurrences of variables in $x_k(P) = x_k Q$ and in $T\bar{k}$ are the same (free or bound) as in the corresponding class term. Composition using the connective $\downarrow$ does not change the nature of the occurrence. As in Section 2.10, one can prove that $|P|(\xi)$ depends only on the ξ-values of the variables that have free occurrences in the formula $P \in \cup_{i=0}^{\infty} Fl_{2i+1}$.

This finishes the description of the syntax and semantics of SAr.

In conclusion, we show that the classes of sets in $\cup_{r \geqslant 1} N^r$ that are definable by formulas in $\mathrm{L_1Ar}$ and in SAr coincide. This result is not used in the proof of Tarski's theorem in the next section. However, the result itself and the method of proof are instructive, and we shall return to these ideas in Part III of the book.

Let $\mathrm{L_1Ar}$ have a countable set of variables. If we denote them by $x_1, x_2, \dots, x_n, \dots$ and identify x_i with $x'^{\cdots}{}'$($i-1$ primes), we can also identify the interpretation classes for $\mathrm{L_1Ar}$ and SAr in the obvious way. Our claim that the classes of definable sets coincide is then an immediate consequence of the following stronger fact:

10.4. **Proposition.** *Two translation mappings*

$$\{\text{formulas of } \mathrm{L_1Ar}\} \rightleftarrows \{\text{formulas of SAr}\}$$

can be explicitly defined with the following properties:

(a) *At every point ξ the truth values of any formula and its translation coincide.*
(b) *The sets of free variables of any formula and its translation coincide.*

We note that the mappings we define will not be inverse to each other!

PROOF.

(a) *The translation from* $\mathrm{L_1Ar}$ to SAr. The translation of a formula P will be denoted by "P". We first translate atomic formulas, and then use induction on the length. The alphabet of SAr does not have addition, but it has both multiplication and raising to a power, so that in place of $z = x+y$ we can write $2^z = 2^x \cdot 2^y$.

(a_1) *Atomic formulas.* They have the form $t_1 = t_2$. By "carrying out the operations," we replace every nonzero term in $\mathrm{L_1Ar}$ by a "normalized term," i.e., a polynomial of the form $\Sigma x_i^{i_1} \cdots x_n^{i_n}$, where the monomials are written in the form $(\cdots(x_1 \cdot x_1)\cdots x_1)\cdot x_2)\dots)$, then arranged in lexicographic order, and finally separated by parentheses: $(\cdots((m_1+m_2)+m_3)+\cdots)$. It is clear how to correlate such a term t to the term "$\bar{2} \uparrow t$" in SAr. For example, "$\bar{2} \uparrow ((x_1)\cdot(x_1)+x_2)$" is $(\bar{2}\uparrow(x_1)\cdot(x_1))\cdot(\bar{2}\uparrow(x_2))$. By definition, the translation "$\bar{2} \uparrow \bar{0}$" is $\bar{1}$. Then we define the translation of the formula $t_1 = t_2$ to be "$2 \uparrow t_1$"= "$2 \uparrow t_2$". It is clear that such a formula and its translation have the same variables and are true at the same points ξ.

(a_2) If "Q", "Q_1", and "Q_2", have already been defined, then "$\neg Q$" is defined as "Q"$\downarrow$ "Q". We similarly construct "$Q_1 * Q_2$" for the other connectives (see "Digression: Syntax" in Chapter I).

(a_3) If "Q" has already been defined, then "$\forall x_k\, Q$" is defined as

$$x_k(\text{“}Q\text{”}) = x_k(x_k = x_k).$$

Both the formula and its translation are true at a point ξ if and only if Q (and "Q") are true at all variations ξ' of ξ along x_k. They also have the same free variables, since by induction, we may assume that this is the case for Q and "Q".

(a$_4$) By definition, "$\exists x_k Q$" coincides with "$\neg\forall x_k \neg Q$".

(b) *The translation from* SAr *to* L$_1$Ar. As before, we let "P" denote the translation of a formula P, although this time P will be a formula in SAr and "P" will be a formula in L$_1$Ar.

There is a subtle point here, namely, how to translate $x_1 = x_2 \uparrow x_3$. It will be shown in Part II of the book that such a translation exists, and can even be taken in the form $\exists x_4 \cdots \exists x_n p(x_1, x_2, x_3, x_4, \ldots, x_n)$, where p is an atomic formula in L$_1$Ar. Here we shall take this fact on faith, and choose a translation "$x_1 = x_2 \uparrow x_3$" once and for all.

(b$_1$) *Translation of formulas* in Fl_0. The following rules give an inductive definition:

"$t_1 = t_2$" has exactly the same form if $t_1, t_2 \in \{\text{variables}\} \cup \{\bar{1}, \bar{1}\bar{1}, \ldots\}$ (of course, in the sense that $x'^{\cdots\prime}$ is replaced by x_k and $\bar{1}\cdots\bar{1}$ is replaced by $(\cdots(\bar{1}+\bar{1})+\bar{1})+\cdots)$). "$x_k = t_1 \cdot t_2$" has the form $\exists x_i \exists x_j$("$x_i = t_1$" $\wedge$ "$x_j = t_2$" $\wedge\, x_k = x_i \cdot x_j$) and "$x_k = t_1 \uparrow t_2$" has the form $\exists x_i \exists x_j$("$x_i = t_1$"$\wedge$"$x_j = t_2$" $\wedge x_k = x_i \uparrow x_j$), where x_i and x_j are the first two variables not occurring in t_1 or t_2. We similarly translate formulas with the left- and right-hand sides permuted, and also with $\bar{1}\cdots\bar{1}$ instead of x_k. We further stipulate that "$t_1 = t_2$" has the form $\exists x_i$("$x_i = t_1$" $\wedge$ "$x_i = t_2$"), where x_i is the first variable not occurring in t_1 or t_2, and where we assume only that neither t_1 nor t_2 is a variable or $\bar{1}\cdots\bar{1}$. It is clear that the truth function and the set of free variables are preserved under these translations.

(b$_2$) Suppose that the formulas in Fl_{2i-1} have already been translated. Let

$$\text{"}x_k(P_1) = x_k(P_2)\text{" be } \forall x_k(\text{"}P_1\text{"} \Leftrightarrow \text{"}P_2\text{"}), \quad \text{and}$$
$$\text{"}x_k(P)\bar{n}\text{" be "}P\text{"}(\bar{n}),$$

where on the right $\bar{n} = (\cdots(\bar{1}+\bar{1})+\bar{1})+\cdots)$ is substituted in place of all free occurrences of x_k in "P". This completes the proof. □

11 Undefinability of Truth: Tarski's Theorem

11.1. The language SAr is interpreted in N, and not in the set of its own formulas the way SELF is. In order to be able to determine the set of definable formulas, we number formulas by (certain) integers as follows.

We number the symbols of the alphabet (of which there are nine) from 1 to 9 in any order, *as long as* $\bar{1}$ *corresponds to* 9. We then set (here $a_i \in$ {alphabet of SAr} and $v(a_i)$ is the number of a_i)

$$\text{number } (a_1 \cdots a_k) = n(a_1 \cdots a_k) = \sum_{i=1}^{k} v(a_i) 10^{k-i} + 1.$$

In other words, we obtain the number of an expression by replacing all of its symbols by the corresponding decimal digits ($\bar{1}$ is replaced by 9), then reading the resulting number in the decimal system and adding 1. It is clear that an expression can be reconstructed in a unique way if we know its number.

The name in SAr of the number of an expression P, i.e., $\bar{1}\cdots\bar{1}$ ($n(P)$ times), is called the *label* of P. As in SELF, we shall denote the label of P by $*P*$ (but now this is abbreviated notation). We call the expression $P*P*$ the *display* of P.

11.2. **Definition.** Let $P(x)$ be a formula in SAr with one free variable x.

(a) An expression Q satisfies P if the number of Q lies in the set $\{k|P(\bar{k})$ is true$\}$.

(b) An expression Q is displayed in P if the display of Q satisfies P.

11.3. **Lemma.** *Let* $P(x)$ *be as in* 11.2. *Let* $P_E(x)$ *denote the formula* $P((x)\cdot((\bar{1}\bar{0})\uparrow(x)))$ (*i.e., the term "*$x10^x$*" is substituted in place of all free occurrences of* x). *Then the set of expressions satisfying* P_E *coincides with the set of expressions displayed in* P.

PROOF. If Q has number k, then the display of Q has number $k\cdot 10^k$ (which is why $\bar{1}$ has number nine!):

$$n(Q*Q*) = n(Q\ \underbrace{1\cdots 1}_{n(Q)\text{times}})$$

$$= (n(Q)-1)10^{n(Q)} + \underbrace{9\cdots 9}_{n(Q)\text{times}} + n(Q)10^{n(Q)}.$$

Hence, $n(Q)$ satisfies P_E if and only if $n(Q*Q*)$ satisfies P. □

11.4. **Theorem.** *For any formula* $P(x)$ *as in* 11.2, *we have*

$$\textit{the set of formulas satisfying } P \neq \begin{cases}\textit{the set of true formulas}\\ \textit{the set of false formulas.}\end{cases}$$

PROOF. We consider the Tarski–Smullyan formula $S : xP_E * xP_E*$. According to the definitions, we have (recall that xP_E is a class term and $*xP_E*$ is the name of a number) S is true $\Leftrightarrow$ xP_E satisfies P_E $\Leftrightarrow$ xP_E is displayed in P (by Lemma 11.3)$\Leftrightarrow$ the display of xP_E satisfies $P \Leftrightarrow S$ satisfies P. Hence, S is either not false and satisfies P, or else is false and does not satisfy P. Therefore, the set of formulas satisfying P cannot coincide with the set of false formulas. As in §9, the formula S says, "I satisfy P."

Similarly, the formula

$$x((P)\downarrow(P))_E{}^*x((P)\downarrow(P))_E{}^*$$

says, "I do not satisfy P," and thus either satisfies P or is true, but not both. The theorem is proved. □

11.5. Of course, Lemma 11.3 is pure magic. The decimal system really has nothing to do with all this, and $\bar{1}$ did not really have to be number nine, but this way everything is much prettier.

More generally, let ?Ar be any language of arithmetic with a finite alphabet containing the alphabet of SAr. Let the rules for forming distinguished expressions and the standard interpretation of formulas in ?Ar be an arbitrary extension of the rules in SAr. We require only that the terms and formulas in SAr keep their earlier meaning, and that for any formula $P(x)$ in ?Ar with a free variable x, the expression $x(P(x))\bar{k}$ must be a formula in ?Ar and be interpreted by the same recipe as in SAr. (For example, we might add to SAr the $+$ sign, the connectives, and the quantifiers, and then allow formulas to be constructed by the rules of $\mathfrak{L}_1$ as well, thereby embedding $\mathrm{L}_1\mathrm{Ar}$ in ?Ar.)

Then the undefinability theorem 11.4 *holds for* ?Ar.

We must choose the numbering as follows: if m is the number of elements in the alphabet of ?Ar and v is a numbering of the symbols for which $v(\bar{1}) = m$, then

$$n(a_1 \cdots a_k) = \sum_{i=1}^{k} v(a_i)(m+1)^{k-i} + 1.$$

Then, using the same conventions as before, we have

$$\begin{aligned} n(Q * Q*) &= n(Q \underbrace{\bar{1}\cdots\bar{1}}_{n(Q)\,\text{times}}) \\ &= (n(Q)-1)(m+1)^{n(Q)} + m \sum_{j=0}^{n(Q)-1} (m+1)^j + 1 \\ &= n(Q)(m+1)^{n(Q)}. \end{aligned}$$

Defining $P_E(x)$ as $P((x)\cdot((\overline{m+1}) \uparrow (x)))$, without any further alterations we obtain Lemma 11.3 and Tarski's theorem for ?Ar.

11.6. *Remarks*

(a) If Tarski's theorem were not true, and there were a formula $P(x)$ such that $\{Q|Q$ is a formula and $P(\overline{n(Q)})$ is true$\}$ coincided with the set of all true formulas of arithmetic, then this would mean that all number-theoretic questions would reduce to a series of problems all of the same type. Instead of asking, "Is assertion number n true?" we could ask, "Is $P(\bar{n})$ true?" Although such an all-encompassing problem could still be rather complicated (in a certain sense even "infinitely complicated," see Part III), Tarski's theorem says that arithmetic has much more diversity than could be contained in any such single problem.

(b) We still have reason to suspect that perhaps everything worked out this way because we could "cleverly" number the formulas. This is not the case; the results in Part III will imply that Tarski's theorem remains true for any numbering in which a formula and its number can be effectively reconstructed from one another.

(c) It is natural to ask whether the set of numbers of *provable*, or *deducible*, formulas is definable (for some set of axioms and rules of deduction, for example in SAr). The answer is *yes*, this set is *definable.* We shall give some intuitive considerations in this direction, which anticipate the systematic theory in Part III.

However we define the notion of provability, it is natural to expect it to have the following property: *there exists an algorithm (for example, a computer program) that for any text of the given language determines whether this text is a proof and, if so, of what formula.*

We now write a program that constructs the texts in the language in lexicographic order, verifies whether each one is a proof, and, when it is, computes the number of the formula it proves. Roughly speaking, the graph of the function (number of a proof) $\mapsto$ (number of the formula proved) is definable in $\mathrm{L_1Ar}$ because machine logic and arithmetic are embedded in $\mathrm{L_1Ar}$. Hence, the set of numbers of provable formulas is definable in $\mathrm{L_1Ar}$, in SAr, or in any language ?Ar as in 11.5.

Combining this discussion with Tarski's theorem, we obtain the following form of Gödel's theorem:

11.7. **Gödel's Incompleteness Theorem for Arithmetic**. *In any language of arithmetic of type* ?Ar, *and for any definition of deducibility in which the set of (numbers of) deducible formulas is definable,*

$$\{\textit{true formulas}\} \neq \{\textit{deducible formulas}\}.$$

In Part III we discuss more general formulations of this theorem and other versions of the proof, and we give a detailed verification of the principle in 11.6(c) for deductions in $\mathrm{L_1Ar}$.

Digression: Self-Reference

In natural languages it is only recently that linguists have taken note of the so-called "performative" statements. The characteristic feature of such a statement is *self-reference*, which can be defined as the ability to "refer to a reality that it creates itself, because it is stated under circumstances which make it into an act" (E. Benveniste, La Philosophi analytique et le langage, *Les Et. Philos.*, No. 1 (1963) 9). Examples of performative statements include, "I solemnly swear," the saying of which constitutes the act of swearing; "I proclaim a general mobilization," and "I appoint you director," when these two statements come from an authority that has the power to carry out the respective acts. If we look carefully at the semantics of performative statements, we find an imperative nuance, even though it is expressed by the declarative mood of the verb.

In this connection, it is interesting to compare the role of self-reference in formal and algorithmic languages (see also Section 1.2 of Chapter I). In formal languages (and, in general, in descriptive languages), self-reference leads to logical circles, to paradoxes, or, if we try to avoid logical circles, to demonstrations of certain inadequacies of the language. On the other hand, in algorithmic languages (and in general, in control languages and systems), self-reference is the most important device for turning a finite program into a process that is potentially arbitrarily long ("loops"); it takes part in the control instructions (feedback), and is among the fundamental possibilities of the system.

A similar dichotomy can also be found in psychological behavior—compare with the distinction between introspection and self-improvement.

Finally, self-reference can play a role in the genetic causality of aging processes (of biological and social systems). A self-regenerating cycle, when repeated many times, leads to erosion at the place of generation.

12 Quantum Logic

12.1. The last section of this chapter is devoted to certain physical facts and to the mathematical constructions that have been developed to describe them. In particular, we discuss von Neumann's theorem that it is impossible to introduce hidden variables into the quantum-mechanical picture of the world. This material, while not completely traditional for a course in logic, is relevant here for two reasons.

In the first place, von Neumann's theorem is a vivid example of a metaphysical assertion. It is concerned with properties of the language, rather than with the subatomic world described by the language, and thus is analogous to, for example, Tarski's theorem in metamathematics. This is why it occupies an isolated position in physics, and why we are interested in it here.

In the second place, analyzing quantum-mechanical phenomena reveals a profound divergence between the internal logical structures of the macroworld and the microworld. Although explanations of these differences by means of natural language and natural logic are agonizingly difficult and, in the last analysis, always leave one feeling unsatisfied, these attempts to explain continue. The development of the foundations of physics in the twentieth century has taught us a serious lesson. Creating and understanding these foundations turned out to have very little to do with the epistemological abstractions that were of such importance to the twentieth-century critics of the foundations of mathematics: finiteness, consistency, constructibility, and in general, the Cartesian notion of intuitive clarity. Instead, completely unforeseen principles moved into the spotlight: complementarity, and a nonclassical, probabilistic truth function. The electron is infinite, capricious, and free, and does not at all share our love for algorithms.

The following exposition is based on the article by S. Kochen and E. P. Specker in *J. Math. Mech.*, vol. 17, no. 1 (1967), 59–87. Sections 12.9–12.16 contain pure algebra and formally do not depend on the preceding semiphysical considerations.

12.2. *The atom of orthohelium.* We now describe certain characteristics of the behavior of the physical system "an atom of orthohelium in the state $n = 2$, $l = 0, s = 1$." Such a helium atom is in an excited state: its two electrons are on the second energy level, and their spin is pointed in the same direction. Nevertheless, the state is metastable, because in order to fall to the first energy level, the electrons must turn their spins in opposite directions (parahelium); this creates a certain stability.

Spin is a physical quantity that is expressed in the same units as the "angular momentum." The total spin of our system (in atomic units: $h = 2\pi$) is represented by a unit vector in physical three-dimensional space. As a first approximation we may think of it as changing with time but having instantaneous values that can be measured. (The inadequacy of this picture will soon be demonstrated.)

An experiment for the purpose of measuring the instantaneous value of the spin of our system could consist in turning on a magnetic field having a specified geometry and registering the shift in energy levels (spectral lines) of the atom. Each outcome of such an experiment can be precisely interpreted as a measurement of the projection of the spin on some axis, which is uniquely determined by the geometry of the field. We shall identify these directions with points of the unit sphere S^2.

Quantum mechanics makes the following positive assertions concerning measurements of the spin of orthohelium. The following quantities are measurable:

(a) the projection $s(\alpha, t)$ of the spin in the direction $\alpha \in S^2$ at the moment of time t;

(b) the lengths $|s|(\alpha_i, t), i = 1, 2, 3$, of three projections of the spin in three pairwise orthogonal directions $\{\alpha_1, \alpha_1, \alpha_3\} \subset S^2$ (a "frame") at the time t.

The predictions concerning the results of these measurements are as follows:

(c) $s(\alpha, t)$ is a random variable that can take only the values -1, 0, 1. (The probabilities of these values can be predicted from the results of the previous measurements, but this is not essential for us here.)

(d) $\sum_{i=1}^{3} |s|(\alpha_i, t) = 2$ for any frame $\{\alpha_1, \alpha_2, \alpha_3\}$ and any t.

12.3. *Attempt at a classical interpretation.* This could consist in adopting the following hypotheses A and B:

A. There is a certain space Ω of "hidden variables" or "internal states" of the system and a function $s(\alpha, t; \omega)$, $\omega \in \Omega$, such that if the system is in the state ω at time t, then $s(\alpha, t; \omega)$ is the "true value of the projection of the spin on the α-axis" at this moment.

B. The probabilistic aspect of the predictions in 12.2(c) results from our not knowing the exact values of $\omega = \omega(t)$, so that for some measure $d\mu(\omega)$ we have

$$\text{mathematical expectation of } s(\alpha, t) = \int_{\Omega} s(\alpha, t; \omega) d\mu(\omega),$$

and similarly for $|s|$.

Generalizing, we might suppose that Ω does not depend only on the system itself but also on the arrangement for measuring the spin; μ may depend on the time, and so on. However, all of these possibilities actually contradict the predictions in 12.2(c) for the following startling reason.

12.4. **Proposition** (Kochen, Specker). *There does not exist a mapping $S^2 \to \{0, 1\}$ such that for every frame $\{\alpha_1, \alpha_2, \alpha_3\}$ this mapping takes the value zero* *on precisely one of the directions* α_i. *Moreover, it is possible to construct a finite system* $\Gamma \subset S^2$ *of* 117 *points with the following property. For any mapping* $k : \Gamma \to \{0,1\}$ *either there is a frame* $\{\alpha_1, \alpha_2, \alpha_3\} \subset \Gamma$ *in which* k *does take the value* 0 *exactly once, or else there is a pair of perpendicular directions* $\{\alpha_1, \alpha_2\} \subset \Gamma$ *on which* k *equals* 0.

Here we note that adopting both the assertions in 12.2 and the hypotheses in 12.3 would allow us to construct such a mapping of the sphere. In fact, it would be sufficient to consider

$$S^2 \to \{0,1\} : \alpha \mapsto |s|(\alpha, t; \omega)$$

for fixed t and ω. By 12(c), $|s|$ takes only the values 0 and 1, and by 12(d), it takes the value 1 twice and 0 once on any frame $\{\alpha_1, \alpha_2, \alpha_3\}$.

We prove Proposition 12.4 in Sections 12.12–12.15, and now proceed to a more systematic study of "quantum logic." We shall adhere to our customary and useful dualism between "language and interpretation," although these categories are much less formalized and are harder to distinguish from each other in physics.

12.5. *The language of nonrelativistic quantum mechanics.* We have a somewhat unusual situation in that quantum mechanics does not really have its own language. More precisely, to describe a physical system S such as a "free electron" or "atom of helium in a magnetic field," quantum mechanics uses a certain fragment of the language of functional analysis, "oriented on describing S." Assuming that the reader is familiar with functional analysis, we shall limit ourselves to a glossary of the most frequently used terms. We also give some synonyms used by physicists to indicate the "physical sense," i.e., the interpretation, which will be considered separately in our text.

(a) *A separable complex Hilbert space* $\mathcal{H}_S$. Here we are also interested in its one-dimensional subspaces and its vectors of length one. A synonym for the former is the (pure) states, and for the latter is the (normalized) ψ-functions, or, more precisely, the instantaneous values of the ψ-functions.
(b) *Unitary representations of* $\mathbf{R}$ *in* $\mathcal{H}_S$: $t \mapsto U_t = e^{-iH_st}$. For synonyms we have $t \mapsto U_t$ is the dynamic group; t is the time; and the infinitesimal generator H_S (which is a self-adjoint operator) is the dynamic operator, or Hamiltonian, of S.
(c) *Schrödinger equation*: $\partial\psi_t/\partial t = -iH_S\psi_t$. It is satisfied by the ψ-functions $\psi_t = e^{-H_St}$, which evolve with time.
(d) *Self-adjoint operators in* $\mathcal{H}_S$. Synonym: the observables of the system. The operator H_S is an energy observable. The discrete spectrum of H_S gives us the energy levels of S. We shall be especially interested in the orthogonal projection observables. Here the pure states $\mathbf{C}_\psi \subset \mathcal{H}_S$ are in one-to-one correspondence with the projections P_ψ onto the corresponding subspace.

    Another important class of projections is constructed using the spectral decomposition theorem. Let $A = \int_{-\infty}^{\infty} \lambda dP_A(\lambda)$. Then the projection $P_A(U)$ is defined for any Borel subset $U \subset \mathbf{R}$. In the simplest cases its image is spanned by the vectors in $\mathcal{H}_S$ that are eigenvectors for A with eigenvalues in U.

Projection observables are also called “questions” (Mackey) or “Eigenschaften” (von Neumann).

(e) *Commuting operators.* Synonym: compatible (or simultaneously measurable) observables. For unbounded operators A and B, whose formal commutator may have an empty domain of definition, we define commutativity to mean that $P_A(U_1)$ and $P_B(U_2)$ commute for all Borel sets $U_1, U_2 \subset \mathbf{R}$.

(f) *Unitary representations in $\mathcal{H}_S$ of various groups, such as $SO(3)$, $SU(2)$, S_n,* Synonym: symmetries of the system S (if the representations commute with the Hamiltonian H_S), or approximate symmetries (if $H_S = H_0 + H_i$, where the representations commute with H_0 and H_1, is a “small perturbation”).

12.7. EXAMPLE. Let S be “an electron in the electric field of a proton” (where we disregard the motion of the proton, the spin, and the relativistic effects). Here $\mathcal{H}_S = L^2(E^3)$ consists of the square integrable complex functions in the Euclidean “physical coordinate space of the electron.”

H_S is the self-adjoint extension of the operator

$$-\frac{h}{4\pi m}\Delta - \frac{1}{h}\frac{e^2}{r},$$

where h is Planck's constant, m is the mass of the electron, e is its charge, and r is its distance from the origin (where the proton is).

The energy levels (the discrete spectrum of H_S) are $E_n = -(2\pi^2 me^4/h^2)/(1/n^2), n = 1, 2, 3, \ldots$. The eigenfunctions ψ corresponding to the points of this spectrum are the states of an electron in a hydrogen atom. The energy level $n = 1$ corresponds to the unexcited state, and the other values of n correspond to excited states. The positive semiaxis is the continuous spectrum of H_S; in states with positive electron energy, “the hydrogen atom is ionized.”

The most important observables of the electron are the operators of multiplication by the three coordinate functions x_j (the coordinate observables), and the self-adjoint extension of the operators $p_j = (h/2\pi i)(\partial/\partial x_j)$ (the momentum projection observables). The operators x_j and p_j do not commute, so that the x_j-coordinate and the projection of the momentum on the x_j-axis are not simultaneously measurable.

The system S is spherically symmetric. The natural representation of SO(3) in $L^2(E^3)$ commutes with H_S. The restriction of this representation to the subspace of $\mathcal{H}_S$ corresponding to the discrete spectrum of H_S in a natural way splits into a direct sum of representations corresponding to a given energy level E_n. This E_n-subspace, in turn, splits into a direct sum of representations of SO(3) on spherical polynomials of degree $j = 0, 1, 2, \ldots, n-1$ with multiplicity one. If the ψ-function of the electron belongs to the level E_n and the subspace corresponding to the representation of SO(3) on spherical polynomials of degree j, we say that n and j are the principal and orbital quantum numbers, respectively, of the electron's state in the hydrogen atom.

The above text is typical of what might be found in a physics textbook. The “language” is mixed with the “metalanguage” that gives the standard interpretation of the language. We now describe them separately and more systematically.

12.8. *The interpretation.* A very important aspect of the interpretation that we shall not discuss here is the list of informal recipes for choosing $\mathcal{H}_S$, H_S, and the observables corresponding to a given system S. These "units of expression" are often chosen in two stages: a classical description is chosen, and then the "rules of quantization" are applied to it. This procedure might be "approximate" in the sense that certain circumstances are not taken into account (such as the spin in 12.7).

Suppose that $\mathcal{H}_S$ and H_S have already been chosen. The most characteristic peculiarity of the interpretation of quantum language is that it is "two-layered." Part of the mathematical statements are interpreted as assertions about a "freely evolving system," and part are interpreted as assertions about the results of observations on this system.

(a) *Freely evolving system.* It is generally believed that the system's ψ-function $\psi_t \in \mathcal{H}_S$ gives (within the framework of a given approximation) maximally complete information about the state of the system at time t. As long as no one looks in on the system, ψ_t evolves as $e^{-iH_St}\psi_0$, starting from the initial state ψ_0. (How do we know ψ_0? See Section 12.8(c) below.)

(b) *Observation.* Suppose we want to measure the instantaneous value of some physical quantity for our system S at the moment t. This quantity corresponds to an observable A. (How do we know the form of A? See the beginning of 12.8.) For simplicity we suppose that A has a discrete spectrum with all multiplicities one. The predictions of what will be observed are as follows.

If $A\psi_t = a\psi_t$, then a will be the value of the observable A at the time t for the system S in the state with ψ-function ψ_t.

In the general case, let $\psi_A^{(i)}, i = 1, 2, \ldots$, be an orthonormal basis for $\mathcal{H}_S$ consisting of eigenvectors for A. We expand ψ_t with respect to this basis: $\psi_t = \sum_{i=1}^{\infty} \alpha^{(i)}(t)\psi_A^{(i)}$. Let $A\psi_A^{(i)} = a_i\psi_A^{(i)}$. Then the result of measuring A will be a random variable taking the value a_i with probability $|\alpha^{(i)}(t)|^2$. (It is easy to see that the mathematical expectation of this random variable is $(A\psi_t, \psi_t)$. This formula holds for all A. More generally, the probability of A falling in a Borel subset $U \subset \mathbf{R}$ is equal to $(P_A(U)\psi_t, \psi_t)$, where $P_A(U)$ was defined in 12.5(d).)

(c) *System evolving after observation.* With the same assumptions as before, the ψ-function of the system after the observation is determined by the result of the observation. If we registered the value a_i for A at the time t_0, then, starting from $\psi_A^{(i)}$ at t_0, S evolves until the next observation completely independently of how it evolved before.

Thus, the result of the observation lets us know the form of the ψ-function *after* the observation, but it tells us nothing about the ψ-function *before* the observation. Hence, physicists often say that registering the value $\psi_A^{(i)}$ *prepares* the system in the state $\psi_A^{(i)}$ at the time t_0. Another synonym: at the moment of observation the ψ-function of the system *reduces* to $\psi_A^{(i)}$.

If we were able simultaneously to register the values of two observables, then we would prepare the system with a ψ-function that is an eigenfunction for both observables. Since noncommuting observables always have different eigenvectors, in general the values of such variables are not simultaneously measurable.

12.9 *Quantum logic.* We now investigate the algebraic framework of quantum logic. We start with the following analogous situation.

Suppose we are given a formal language in $\mathfrak{L}_1$ having one variable and an interpretation of this language in a set M where this variable takes values. Then we can distinguish the Boolean algebra B of definable sets in M (see §3). The conjunction of formulas corresponds to the Boolean intersection of the sets that define them, and so on. By definition, $N \in B$ if we can ask in the language, "Does the value of the variable belong to N?" The algebra B is the most important invariant of the pair {language, interpretation}.

We now consider the language of quantum mechanics, oriented on describing a system S. We shall exclude the time aspect by *fixing* a moment of time to which all statements about the state of the system refer. Then the "state of the system" will be the only variable in the language. It takes values in the set of lines in the Hilbert space $\mathcal{H}_S$. The only questions to which we can give a yes or no answer are those of the form; "Does the state of the system belong to a given closed subspace of $\mathcal{H}_S$?" It is the closed subspaces of $\mathcal{H}_S$ that form the analogy of the Boolean algebra B. The conjunction of questions corresponds to the intersection of subspaces, and the disjunction corresponds to their sum, but both operations can be performed only when the corresponding projection observables commute. Only in this case are the Boolean identities fulfilled.

We axiomatize the situation as follows:

12.10. **Definition**. A partial Boolean algebra is a set B together with the following structures on B:

(a) A reflexive and symmetric binary relation $*$ called "compatible measurability." Instead of $(a, b) \in *$ we write $a * b$.
(b) Partial binary operations $\vee$ and $\wedge$ and a unary operation $'$.
(c) Two elements 0 and $1 \in B$.

These structures must satisfy the following axioms:

(d) The relation $*$ is closed with respect $\wedge, \vee$, and $'$: if a_1, a_2, and a_3 are pairwise compatibly measurable, then $(a_1 \wedge a_2) * a_3, (a_1 \vee a_2) * a_3$, and $a_1' * a_3$; in addition, $a * 0$ and $a * 1$ for all $a \in B$.
(e) If a_1, a_2, and a_3 are pairwise compatibly measurable, then together with 0 and 1 they generate a Boolean algebra relative to the operations $\vee, \wedge$, and $'$.

12.11. Example. Let $\mathcal{H}$ be a Hilbert space (possibly real and finite-dimensional). The partial Boolean algebra $B(\mathcal{H})$ is defined as the set of closed subspaces of $\mathcal{H}$ with the following structures:

(a) $a * b$ if and only if there exist three pairwise orthogonal closed subspaces $c, d, e \in \mathcal{H}$ such that $a = c \oplus d$ and $b = e \oplus d$. The motivation for this definition is that this condition is equivalent to commutativity of the projections onto a and b.
(b) $a \wedge b =$ the intersection of a and b.
(c) $a \vee b =$ the sum of a and b.
(d) $a' =$ the orthogonal complement of a.
(e) $0 = \{0\}$ and $1 = \mathcal{H}$.

One form for the theorem that there are no hidden variables is as follows.

12.12. **Theorem:** *If* $\dim \mathcal{H} \geqslant 3$, *then* $B(\mathcal{H})$ *cannot be embedded in a Boolean algebra in such a way that the operations are preserved.*

This result can be strengthened formally in various ways: see §5 of Kochen and Specker, and also N. Fierier, M. Schlessinger, *Duke Math. J.*, vol. 32, no. 2 (1965), 251–262. We shall not dwell on this here.

PROOF. We choose a real Euclidean space $E^3 \subset \mathcal{H}$ and show that even $B(E^3)$ cannot be embedded in a Boolean algebra. Otherwise there would exist a homomorphism of the partial Boolean algebra $B(E^3)$ onto the two-element Boolean algebra $\{0, 1\}$, since for any pair of elements in any Boolean algebra, there exists a homomorphism onto $\{0, 1\}$ that separates them.

Let h be such a homomorphism. If $a_1, a_2, a_3 \in E^3$ are pairwise orthogonal lines, then $h(a_i \wedge a_j) = h(a_i) \wedge (a_j) = 0$ for $i \neq j$. Hence, in any pair of orthogonal lines, at least one of the pair must go to 0 under h. Furthermore, $h(a_1 \vee a_2 \vee a_3) = h(a_1) \vee h(a_2) \vee h(a_3) = h(E^3) = 1$. Hence, in any frame exactly one of the lines goes to 1.

If we map the points of the unit sphere S^2 onto the lines joining them to the origin and then apply h, we obtain a mapping of S^2 with the property in Proposition 12.4 (where we have only to switch the roles of 0 and 1). We prove that no such map exists even on a certain subset consisting of 117 points on S^2. The latter stronger result is combinatorially elegant and physically meaningful: a physicist might raise objections to asking to be able to measure the projection of the spin of orthohelium simultaneously in *all* directions, independently of the question whether hidden variables are possible. In fact, we need only finitely many directions to show the futility of such an attempted measurement.

Consider a finite graph. By a *realization* of the graph on S^2 we mean any embedding of the set of its vertices in S^2 for which the distance between the endpoints of any edge equals 90°.

12.13. **Lemma.** *Let* α *and* β *be points on* S^2 *such that the sine of the angle between them* $\in [0, \frac{1}{3}]$. *Then there exists a realization of the following graph* Γ_1 *in which* a_0 *goes to* α *and* a_9 *goes to* β.

PROOF. Let $\bar{x}, \bar{y}, \bar{z}$ be a triple of pairwise orthogonal vectors on S^2. We take a_5 to $\bar{x}$ and a_6 to $\bar{z}$. For certain $\xi, \eta \in \mathbf{R}$ (to be chosen later), we set

$$a_1 \mapsto \frac{\bar{y} + \xi \bar{z}}{\sqrt{1+\xi^2}}, \qquad a_2 \mapsto \frac{\bar{x} + \eta \bar{y}}{\sqrt{1+\eta^2}}.$$

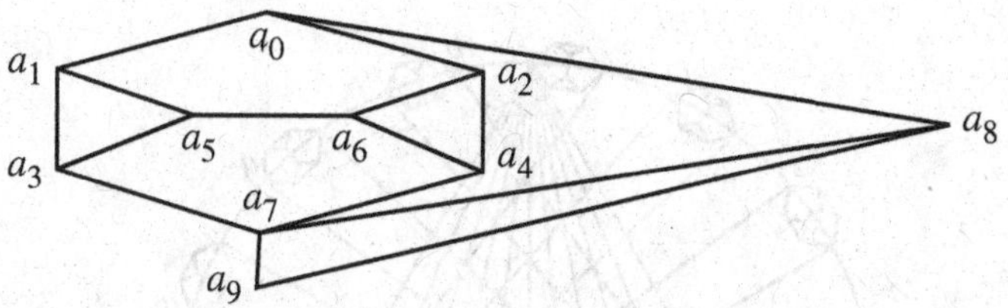

Then the images of a_3 and a_4 are determined up to a sign by the property of being orthogonal to (a_1, a_5) and (a_2, a_6), and we choose

$$a_3 \mapsto \frac{\xi\bar{y} - \bar{z}}{\sqrt{1+\xi^2}}, \qquad a_4 \mapsto \frac{\eta\bar{x} - \bar{y}}{\sqrt{1+\eta^2}}.$$

We similarly set

$$a_0 \mapsto \frac{\xi\eta\bar{x} - \xi\bar{y} + \bar{z}}{\sqrt{1+\xi^2+\xi^2\eta^2}}, \qquad a_7 \mapsto \frac{\bar{x} + \eta\bar{y} + \xi\eta\bar{z}}{\sqrt{1+\eta^2+\xi^2\eta^2}},$$

and finally, a_8 and a_9 are determined up to sign. The sine of the angle between a_0 and a_9 is easy to compute: it equals

$$\xi\eta/\sqrt{(1+\xi^2+\xi^2\eta^2)(1+\eta^2+\xi^2\eta^2)}.$$

As ξ and η vary, this expression takes on all values in $[0, \frac{1}{3}]$. □

12.14. **Lemma**. *Consider the graph Γ_2 that is obtained from Figure 1 by identifying the vertices $a = p_0, b = q_0$, and $c = r_0$ (the apparent intersections of the edges inside the circle are not vertices). This graph is realized on S^2.*

PROOF. For $0 \leqslant k \leqslant 4$ set

$$\begin{aligned} p_k &\mapsto \cos\frac{\pi k}{10}\cdot\bar{x} + \sin\frac{\pi k}{10}\cdot\bar{y}, \\ q_k &\mapsto \cos\frac{\pi k}{10}\cdot\bar{y} + \sin\frac{\pi k}{10}\cdot\bar{z}, \\ r_k &\mapsto \sin\frac{\pi k}{10}\cdot\bar{x} + \cos\frac{\pi k}{10}\cdot\bar{z}. \end{aligned}$$

Since $\sin(\pi/10) < \frac{1}{3}$, we can first extend this map to a realization of the subgraph between the points p_0, p_1, and r_0 using the preceding lemma. Rotating the resulting realization around r_0 so as to take (p_0, p_1) to $(p_1, p_2), (p_2, p_3), \ldots$, we obtain a realization of the "lower arc" and r_0. By similarly rotating around the images of p_0 and q_0, we obtain a realization of the other two arcs as well. □

12.15. END OF THE PROOF OF PROPOSITION 12.4 AND THEOREM 12.12. Consider an arbitrary map k of the vertices of the graph Γ_2 to $\{0, 1\}$. Suppose that exactly one vertex in each triangle goes to 1 and at least one of the two vertices on each edge goes to 0. In the triangle $\{p_0, r_0, q_0\}$ suppose that p_0 goes to 1. We consider the copy of the graph Γ_1 between the vertices p_0, r_0, and p_1, which we identify with a_0, a_8, and a_9, respectively.

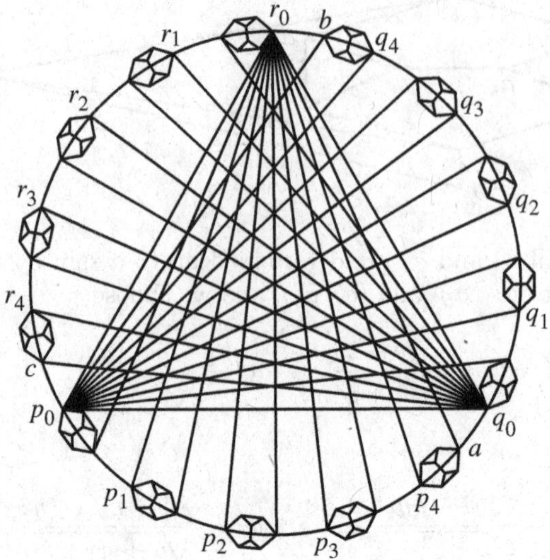

Figure 1.

We must have $k(p_1) = k(a_9) = 1$. In fact, if we had $k(a_9) = 0$, then we would also have $k(a_7) = 1$, and then $k(a_1) = k(a_2) = k(a_3) = k(a_4) = 0$, and $k(a_5) = k(a_6) = 1$, which is a contradiction.

We now return to Γ_2. Since $k(p_0) = k(p_1) = 1$, we similarly find that $k(p_2) = 1$, and then $k(p_3) = k(p_4) = k(q_0) = 1$. But $k(q_0) = 1$ contradicts the fact that $k(p_0) = 1$. This completes the proof. □

12.6. *Quantum tautologies.* This theme has been largely neglected. We give a counterexample due to Kochen and Specker and formulate some recent results of Gelfand and Ponomarev.

(a) *Counterexample.* This consists of the following: it is possible to give a logical polynomial in 117 variables that represents a classical tautology but that is defined and takes the value 0 in the partial Boolean algebra $B(E^3)$ for some values of the variables. This is simply another aspect of the impossibility of embedding $B(E^3)$ in a Boolean algebra.

In fact, let $P(p,q,r)$ be a logical polynomial in three variables that takes the truth value 1 when exactly one of $|p|$, $|q|$, and $|r|$ is 1. We may assume that only the connectives $\vee, \wedge$ and $\neg$ occur in P. Similarly, let $Q(p,q) = \neg p \vee \neg q$. Then Q takes the value 1 when at least one of $|p|, |q|$ is 0. We index the vertices of Γ_2 from 1 to 117 and set

$$R(p_1, \ldots, p_{117}) = \neg\left(\bigwedge_{\{i,j,k\}} P(p_i, p_j, p_k) \bigwedge_{\{r,s\}} Q(p_r, p_s)\right).$$

The first $\bigwedge$ is taken over all triples $\{i,j,k\}$ corresponding to triangles in Γ_2, and the second $\bigwedge$ is taken over all pairs $\{r,s\}$ corresponding to edges. The argument in 12.15 shows that for any mapping $\{p_1, \ldots, p_{117}\} \to \{0,1\}$ at least one of the Boolean factors takes the value 0. Hence R is a classical tautology.

But if we substitute for p_i the line from the origin to the image of the ith vertex in a fixed realization of Γ_2, then we obtain for the value of R the element $0 \in B(E^3)$. In fact, if p_r and p_s are orthogonal, then $p_r^{'} \vee p_s^{'} = E^3$. Similarly, if p_i, p_j, and p_k are orthogonal, then $P(p_i, p_j, p_k) = 1 \in B(E^3)$. The latter assertion is verified as follows: if we set

$$a + b = (a \wedge b^{'}) \vee (a^{'} \wedge b),$$

then we may take

$$P(p, q, r) = p + q + r + p \wedge q \wedge r$$

(for any arrangement of parentheses on the right), so that

$$P(p_i, p_j, p_k) = p_i \oplus p_j \oplus p_k = E^3.$$

(b) *Results of Gelfand and Ponomarev.* We start with the following observation. The operations $\wedge, \vee$, and $^{'}$ are actually defined everywhere on the set $B(\mathcal{H})$ of closed subspaces of the Hilbert space $\mathcal{H}$, although they do not satisfy the Boolean axioms, and if we ignore the compatible measurability relation $*$, it seems as if they no longer have physical meaning.

Nevertheless, it is also natural to investigate these structures, which were first introduced into the logic of quantum mechanics by G. Birkhoff and J. von Neumann (*Annals of Math*. vol. 37 (1936), 823–843). Here is how these structures are axiomatized:

Definition. A *modular structure* L is a set with binary operations $\wedge$ and $\vee$ that satisfy the following conditions:
(a) $\wedge$ and $\vee$ are associative and commutative;
(b) $a \wedge a = a \vee a = a$ for all $a \in L$;
(c) If $a \wedge b = b$, then $(a \vee c) \wedge b = b \vee (c \wedge b)$ (the "modular identity").

Birkhoff and von Neumann also require an "orthogonal complement" operation to exist with the usual axioms, but we shall omit this here.

We note that the modular identity is fulfilled universally in $B(\mathcal{H})$ only if $\mathcal{H}$ is finite-dimensional. It is also fulfilled for triples a, b, c whose elements have finite-dimension or codimension in $\mathcal{H}$.

I. M. Gelfand and V. A. Ponomarev (*Uspehi mat. nauk*, vol. XXIX (1974), No. 6 (180), 3–58) have studied the linear representations of free modular structures with r generators in $B(\mathcal{H})$ for finite-dimensional spaces over arbitrary fields. Such a representation is called indecomposable if it does not split into a direct sum of representations in $B(\mathcal{H}_1) \oplus B(\mathcal{H}_2)$.

Definition. A modular question is an element of a free modular structure that takes the value 0 or 1 for any indecomposable finite-dimensional representation.

One of the main results of Gelfand and Ponomarev is the construction of a very nontrivial countable series of modular questions. We shall only formulate these results here.

Let L^n be a free modular structure with n generators $\{a_1, \ldots, a_n\}$. We set $I = \{1, \ldots, n\}$. A sequence $\alpha = (i_1, \ldots, i_l)$ of length $l \geqslant 1$ of elements of I is called *admissible* if it does not have any identical neighboring entries.

A sequence $\beta = (k_1, \dots, k_{l-1})$ of length $l-1$ of elements of I is called subordinate to α if it is admissible and if $\forall_j \leqslant l-1, k_j \notin \{i_j, i_{j+1}\}$. For admissible α we inductively define

$$a_\alpha = a_{i_1} \dots \alpha_{i_l} = a_{i_1} \wedge (\vee_\beta \alpha_\beta),$$

where β runs through all sequences subordinate to α. Further, for $t \in \{1, \dots, n\}$ we define

$$A_t(l) = \bigvee{}_\alpha a_\alpha,$$

where α runs through all admissible sequences of length l with last entry t. Finally, we set

$$H_t(l) = \bigvee{}_{j \neq t} A_j(l).$$

The substructure in L^n generated by the elements $H_1(l), \dots, H_n(l)$ consists entirely of modular questions for all $l \geqslant 1$.

This is a difficult result. It is relatively easy to prove that this substructure is a Boolean algebra consisting of 2^n elements. If we substitute the elements in this Boolean algebra for the variables in the usual Boolean tautologies, we obtain "quantum tautologies," but to see this we must consider structures with complements.

It is not yet clear whether this algebra leads to nontrivial physics. Perhaps one should combine it with the techniques in the representation theory of symmetry groups.

12.17. *The orthohelium atom revisited.* In conclusion, we return to the orthohelium atom S and show how the material in 12.2 looks from a more general vantage point.

(a) *Choice of* $\mathcal{H}_S$. As explained in 12.7, an electron without spin corresponds to the space $L^2(E^3)$. If we want to take the spin into account, we must introduce a "two-component" ψ-function, i.e., use the space $L^2(E^3) \otimes \mathbf{C}^2$. The system of two electrons in helium is described by ψ-functions in the tensor square of this space. However, by Pauli's principle, the ψ-function of this system must behave antisymmetrically when the electrons corresponding to the two parts of the tensor square are permuted. Hence, we finally obtain $\mathcal{H}_S = \Lambda^2(L^2(E^3) \otimes \mathbf{C}^2)$.

(b) *Choice of* H_S. This is a difficult problem, because each electron moves in the variable electromagnetic field created by the nucleus and the other electron. The principal term in the Hamiltonian corresponds to the spherically symmetric constant potential obtained by averaging over time. The remainder is treated as a small perturbation. We give the approximate form of the ψ-function of orthohelium, more precisely, of the element in $\Lambda^2(L^2(E^3))$ corresponding to the projection of $\mathcal{H}_S$ onto the subspace of the unit projection of the spin:

$$\begin{aligned}\psi \approx e^{-k(r_1+r_2)} &+ [(C_1 + C_2(r_1 + r_2) + C_4 r_{12} + C_5 r_{12}(r_1 + r_2) \sinh C_0(r_1 - r_2) \\ &+ (r_1 - r_2)(C_3 + C_0 r_{12}) \cosh C_0(r_1 - r_2))],\end{aligned}$$

where $r_i = (\sum_{j=1}^3 x_{ij}^2)^{1/2}, i = 1, 2$; $r_{12} = (\sum_{j=1}^3 (x_{1j} - x_{2j})^2)^{1/2}$, and the constants $k, C_1, \dots, C_6$ are found experimentally. (E. U. Condon and G. H. Shortley, *The Theory of Atomic Spectra*, Cambridge University Press, London, 1935.)

(c) *Approximate symmetries.* The group SU(2) acts on the space $\mathcal{H}_S$: on $L^2(E^3)$ through the quotient group SO(3), and on $\mathbf{C}^2$ by the standard representation. This is the group of approximate symmetries of the system. The ψ-function of orthohelium is "not too far" from the subspace corresponding to a suitable representation of SU(2), so we may speak of the principal (n), orbital (j), and other quantum numbers of the state, as in the case of a hydrogen atom.

(d) *Spin.* The total angular momentum operator $\mathcal{J}$ commutes with the Hamiltonian H_S. In the state $n = 2$ and $j = 1$, its eigenvalue is 2 (in atomic units). The eigensubspace $N \subset \mathcal{H}_S$ corresponding to this eigenvalue is three-dimensional. Further, the squared spin projection operators $\mathcal{J}_x^2, \mathcal{J}_y^2, \mathcal{J}_z^2$ commute in pairs (this is a peculiarity of spin 1). Letting P denote the projection of $\mathcal{H}_S$ onto N, we are then able to embed the partial Boolean algebra $B(E^3)$ in $B(\mathcal{H}_S)$ by letting a line $\alpha \subset E^3$ correspond to the image in $\mathcal{H}_S$ of the operator $P\mathcal{J}_\alpha^2$. This takes the place of the somewhat naive picture in 12.2.

Appendix: The Von Neumann Universe

1. The premises of "naive" Cantorian set theory reduce to the following: a set may consist of any distinguishable elements (of the physical or intellectual world); a set is uniquely determined by its elements, and any property determines a set, namely, the set of objects that have this property.

However, the formal language of set theory $\mathrm{L_1Set}$ was introduced in order to describe a more restricted class of sets (a *universe*). Part of these restrictions come from considerations of convenience, and part come from the desire to avoid the so-called paradoxes. This gives an "upper bound" for our classes. We give a "lower bound" by asking that the class of sets be closed with respect to all mathematical constructions needed for certain (ideally, "all") parts of intuitive mathematics.

2. Following Zermelo, von Neumann, and others, we consider two basic restrictions on sets.

(a) All elements of sets must themselves be sets. In particular, since any chain $X_0 \in X_1 \in X_2 \in \cdots$ in the von Neumann universe V must terminate (see below), it follows that the last element in such a chain must be the empty set. Thus, all the sets in V are constructed "from nothing."

(b) The assumption that every collection of sets, even sets as in (a), is again a set in V, immediately leads to contradictions (Burali–Forti, Russell, and others). In particular, the collection of all sets in the universe is not itself an element of V. Hence, we must give a sufficiently complete description of which operations do not take us outside of V. The two basic formal languages of set theory—that of Gödel–Bernays and that of Zermelo–Fraenkel—differ in the choice of objects over which the variable symbols are to range under the standard interpretation of the language in V. In the Zermelo–Fraenkel language (our $\mathrm{L_1Set}$), they range over the sets in V. In the Gödel–Bernays language, they name classes (collections of sets in V) that "are not necessarily sets," and the property of "being a set" is specifically defined as the property of "being an element of another class." The Gödel–Bernays language is studied in Chapter 4 of Mendelson's book.

In this section we describe the von Neumann universe using the customary terminology of intuitive mathematics. The relationship of this construction to formalism will be discussed in Section 18.

3. *The first levels.* The von Neumann universe is constructed inductively, starting from the empty set, by successively applying the "set of all subsets" or "power set" operation $\mathcal{P}$. In this way,

$$\begin{aligned} V_0 &= \varnothing, \\ V_1 &= \mathcal{P}\{\varnothing\} = \{\varnothing\}, \\ V_2 &= \mathcal{P}(V_1) = \{\varnothing, \{\varnothing\}\}, \\ &\vdots \\ V_{n+1} &= \mathcal{P}(V_n), \\ &\vdots \end{aligned}$$

It is easy to see that $V_n \subset V_{n+1}$ (later this will be proved in complete generality). The level V_n consists of

$$2^{2^{\cdot^{\cdot^{2}}}} \quad (n-1 \text{ twos})$$

finite sets, whose elements are also finite sets, and so on.

We cannot go beyond finite sets unless we regard all the V_n as "already constructed" and apply $\mathcal{P}$ to the union of the V_n. We set

$$\begin{aligned} V_{\omega_0} &= \bigcup_{n=0}^{\infty} V_n, \\ V_{\omega_0} + 1 &= \mathcal{P}(V_{\omega_0}) \\ &\vdots \end{aligned}$$

The indices that we now use for the levels are the names of the first infinite ordinals. This remarkable idea of transfinite iteration of such constructions is due to Cantor, who first applied it to study trigonometric series, and then investigated it systematically, finding in it the key to the infinite.

In the next two subsections our sets will temporarily be Cantorian sets. We shall return to V after developing some properties of ordinals.

4. *Ordinals.* Let X be any set on which we are given a binary relation $<$. We consider the following properties of this relation:

(a) *$Y \not< Y$ for all $Y \in X$; if $Y_1 < Y_2$ and $Y_2 < Y_3$, then $Y_1 < Y_3$.*
(b) *For any $Y, Z \in X$, either $Y < Z$* or *$Z < Y$, or else $Y = Z$.*
(c) *Every nonempty subset of X has a least element (in the sense of $<$).*

The relation $<$ is a *partial ordering* of X if it satisfies (a), a *linear ordering* of X if it satisfies (a) and (b), and a *well-ordering* of X if it satisfies all three conditions (a), (b), and (c).

Let $(X, <)$ be a well-ordering. The *initial segment* $\hat{Y}$ determined by an element $Y \in X$ is the well-ordered set $(Z, <)$, where $Z = \{Y' | Y' < Y\}$. As is customary when speaking about a well-ordered set, we shall omit the explicit indication of the ordering if it is clear from the context.

5. **Lemma.** *Let X and Y be two well-ordered sets. Then exactly one of the following alternatives holds:*

(a) *X and Y are isomorphic.*
(b) *X is isomorphic to an initial segment in Y.*
(c) *Y is isomorphic to an initial segment in X.*

In each case the isomorphism is uniquely determined.

PROOF. We divide the argument into several steps.

(a) Let X be well-ordered, and let $f\colon X \to X$ be a monotonic map, i.e., $Z_1 < Z_2 \Rightarrow f(Z_1) < f(Z_2)$. Then for all $Z \in X$ we have $f(Z) \geqslant Z$. In fact, among the elements not having this property there would have to be a least element Z_0. But $f(Z_0) < Z_0$ and the monotonicity of f imply that $f(f(Z_0)) < f(Z_0)$, so that we would have an even smaller element in the set of elements not having the desired property.

(b) Therefore X is not isomorphic to any of its initial segments $\hat{X}_1$: if $f : X \xrightarrow{\sim} \hat{X}_1$, then $f(X_1) < X_1$.

(c) Now let X and Y be well-ordered. We set $f = \{\langle X_1, Y_1\rangle | X_1 \in X, Y_1 \in Y$, and there exists an isomorphism of $\hat{X}_1$ with $\hat{Y}_1\}$. First of all, f is the graph of a one-to-one mapping of $\mathrm{pr}_1 f$ onto $\mathrm{pr}_2 f$. In fact, if $X_1 \neq X_2$, say $X_1 < X_2$, then by (b), $\hat{X}_1$ is not isomorphic to $\hat{X}_2$; by symmetry, the same holds for f^{-1}. It is also clear from this that f and f^{-1} are monotonic. Further, if $X_1 \in \mathrm{pr}_1 f$ and $X_2 < X_1$, then $X_2 \in \mathrm{pr}_1 f$ and similarly for $\mathrm{pr}_2 f$. Finally, we show that either $\mathrm{pr}_1 f = X$, or else $\mathrm{pr}_2 f = Y$. Otherwise, there would exist a minimal element X_1 in $X \setminus \mathrm{pr}_1 f$ and a minimal element Y_1 in $Y \setminus \mathrm{pr}_2 f$. But by the preceding paragraph, f induces an isomorphism of $\hat{X}_1$ with $\hat{Y}_1$. By the definition of f, we then have $\langle X_1, Y_1\rangle \in f$, a contradiction.

(d) All of this means that either f is an isomorphism (more precisely, the graph of an isomorphism) of the set X onto Y or an initial segment in Y, or else f^{-1} is an isomorphism of Y onto X or an initial segment of X. It is clear from the definition of f that the graph of any other isomorphism must be contained in the graph of f, so we have uniqueness. The lemma is proved. □

As a preliminary definition, we can now consider the class of all well-ordered sets isomorphic to some fixed totally ordered set X, and call that class an ordinal. Two ordinals α and β satisfy the relation $\alpha = \beta, \alpha < \beta$, or $\alpha > \beta$ depending on which of the alternatives in Lemma 5 holds for representatives $X \in \alpha$ and $Y \in \beta$ (this obviously does not depend on the choice of representatives).

The next step is, naturally, to consider "all" ordinals as a class and show that $<$ induces a well-ordering on this class, thereby giving a universal well-ordering. However, an unnecessary difficulty arises here: the class of well-ordered sets isomorphic to a fixed X is extremely large, and so the class of ordinals must be a "class of classes," which needlessly complicates matters. An elegant technical discovery, due to von Neumann, removes this difficulty: instead of a vast number of possible orderings imposed on X from outside, we consider a single relation given by internal properties. Recall that a set X is transitive if $Z \in X$ whenever $Z \in Y \in X$ for some Y.

6. **Definition**. An ordinal is a transitive set X of sets that is well-ordered by the relation $\in$ between its elements.

7. **Theorem.**

(a) *The class of ordinals* On *is well-ordered by the relation* $\alpha \in \beta$ (*which we shall also write* $\alpha < \beta$).

(b) *Any well-ordered set is isomorphic to a unique ordinal* α, *and also to a unique initial segment of ordinals* (*those less than* $\alpha \cup \{\alpha\}$).

PROOF.

(a) We must verify conditions (a), (b), and (c) of Section 4. The first of them follows immediately from the definition.

To prove the second condition, we consider two ordinals α and β. By Lemma 5, there exists an isomorphism f of one of them, say α, onto either β or an initial segment of β. We show that then $\alpha = \beta$ or $\alpha \in \beta$. To do this, we prove that $f(\gamma) = \gamma$ for all $\gamma \in \alpha$. In fact, if γ_1 is the minimal element with $f(\gamma_1) \neq \gamma_1$, then $f(\gamma_2) = (\gamma_2)$ for all $\gamma_2 \in \gamma_1$. Since f is an isomorphic embedding of α with respect to the ordering $\in$, and since γ_1 and $f(\gamma_1)$ are sets, we have $f(\gamma_1) = \{f(\gamma_2)|\gamma_2 \in \gamma_1\} = \{\gamma_2|\gamma_2 \in \gamma_1\} = \gamma_1$, which contradicts the choice of γ_1. The same argument shows that $f(\alpha) = \alpha$, from which the condition follows.

Finally, let C be a nonempty class of ordinals, and let $\alpha \in C$. If α is not the least element in C, then the least element in the intersection $\alpha \cap C$ will be the least element in C.

(b) Let X be a well-ordered set. Let S denote the set of ordinals that are isomorphic to some initial segment in X. S is nonempty, since, for example, the ordinal $\{\varnothing\}$ is isomorphic to the segment consisting of the least element of X. It is easy to see that the set $\beta = \cup_{\alpha \in S}\alpha$ is an ordinal. We claim that β is isomorphic to X. In fact, if this were not the case, then β would be isomorphic to an initial segment in X, say $\hat{X}_1$. But then the ordinals $\beta \cup \{\beta\}$, which is larger than β, would be isomorphic to the initial segment $\hat{X}_1 \cup \{X_1\}$, contradicting the definition of β. □

We now give the elementary properties of ordinals.

8. (a) The finite ordinals are the "natural numbers" (and zero) in the first levels of the universe V. Thus, we shall write

$$0 = \varnothing, \quad 1 = \{\varnothing\}, \quad 2 = \{\varnothing, \{\varnothing\}\}, \quad 3 = \{\varnothing, \{\varnothing\}, \{\varnothing, \{\varnothing\}\}\}, \dots .$$

(b) The ordinal that immediately follows a given α is $\alpha \cup \{\alpha\}$. It is also denoted by $\alpha+1$, which agrees with the notation in (a) in the case of finite α.

(c) An ordinal β is called a *limit* ordinal if $\beta \neq \emptyset$ and $\beta \neq \alpha+1$ for any α. The first limit ordinal ω_0 is isomorphic as a totally ordered set to $\{0, 1, 2, 3, \ldots, n, \ldots\}$. If α is a limit ordinal, then $\alpha = \bigcup_{\beta<\alpha} \beta$. The converse is also true.

Ordinals are mainly used for three purposes: proofs using (transfinite) induction, constructions using (transfinite) recursion, and measuring cardinalities. Here are the basic principles.

9. *Transfinite induction.* Let C be a class of ordinals for which

(a) $\varnothing \in C$.
(b) If $\alpha \in C$, then $\alpha+1 \in C$.
(c) If a set of ordinals $\{\alpha_i\}$ is contained as a subset in C, then $\cup\alpha_i \in C$.

Then C contains all ordinals.

In fact, otherwise there would exist a least ordinal not in C, but this could not be the empty set by (a), a limit ordinal by (c), or any other ordinal by (b). In concrete applications, the verifications of (a) and (c) are often trivial and are omitted.

10. *Transfinite recursion.* Let G be a function of sets (it will actually be sufficient to assume that G is defined on all sets in the universe) whose values are sets. Then there exists a unique function F on the ordinals such that

$$F(a) = G(\text{the set of values of } F \text{ on the elements of } \alpha).$$

In fact, this equality uniquely determines $F(0) = G(\varnothing)$, and then $F(1) = G(\{F(0)\}), F(2) = G(\{F(0), F(1)\})$, and so on. Thus, if we consider the class C of ordinals α for which we can define F with the required property on the initial segment of ordinals $< \alpha$, then C satisfies the conditions 9(a)–(c), and therefore contains all the ordinals. Uniqueness follows similarly (if $F \neq F'$, consider the least α with $F(\alpha) \neq F'(\alpha)$).

11. *Measuring cardinalities.* Different ordinals can have the same cardinality. For example, all the ordinals $\omega_0, \omega_0+1, \omega_0+2, \ldots$ (and many more after them!) are countable. However, jumps in cardinality occur arbitrarily far out.

An ordinal that does not have the same cardinality as any lower ordinal is called a cardinal. All finite ordinals and ω_0 are cardinals. Clearly, any infinite cardinal is a limit ordinal. Further, any set has the same cardinality as some cardinal, and in fact, a unique one (see §1 of Chapter III). The infinite cardinals form a totally ordered class, which is naturally indexed by ordinals. Thus

$$\begin{aligned} \omega_0 &= \text{the first countable ordinal,} \\ \omega_1 &= \text{the first ordinal of cardinality} > \omega_0 \\ &= \text{the set of all finite and countable ordinals,} \\ \omega_2 &= \text{the first ordinal of cardinality} > \omega_1 \\ &= \text{the set of all ordinals of cardinality} \leqslant \omega_1, \end{aligned}$$

and so on.

We can now give our fundamental definition.

12. **Definition.** The (von Neumann) universe V is the class of sets $\bigcup_{\alpha\in\mathrm{On}} V_\alpha$, where the set V_a is defined by the following transfinite recursion:

$$V_0 = \varnothing,$$
$$V_{\alpha+1} = \mathcal{P}(V_\alpha),$$
$$V_\alpha = \bigcup\nolimits_{\beta<\alpha} V_\beta, \text{ if } \alpha \text{ is a limit ordinal.}$$

We give some elementary properties of the universe V.

13. *Each of the sets V_α is transitive: if $Y \in X \in V_\alpha$ then $Y \in V_\alpha$. (In other words, $V_\alpha \subset V_{\alpha+1}$.)*

Suppose that this were not true. Then there would exist a least ordinal α with $V_\alpha \not\subset V_{\alpha+1}$, where $\alpha \geqslant 2$. If α is not a limiting ordinal, $\alpha = \beta+1, Y \in X \in V_\alpha$, and $Y \notin V_\alpha$, then we obtain a contradiction as follows: $X \in V_{\beta+1} = \mathcal{P}(V_\beta) \Rightarrow X \subset V_\beta \Rightarrow Y \in V_\beta \Rightarrow Y \in V_{\beta+1} = V_\alpha$, since for β it is still true that $V_\beta \subset V_{\beta+1}$ by our choice of α. If α is a limit ordinal, the argument is analogous (find $\gamma < \alpha$ with $Y \in X \in V_\gamma$ and $Y \notin V_\alpha$). □

We define the *rank* of any set $X \in K$ as follows: rank $X = \alpha$ if α is the least ordinal such that $X \in V_{\alpha+1}$. If $Y \in X$, then rank $X \geqslant$ rank $Y + 1$.

14. *All ordinals belong to V, and rank $\alpha = \alpha$.*

We first show that $\alpha \in V_{\alpha+1}$ for all ordinals α. This is true for $\alpha = 0$. Suppose that α is the least ordinal with $\alpha \notin V_{\alpha+1}$. If $\alpha = \beta + 1$, then $\beta \in V_{\beta+1}$, so that β and $\{\beta\} \in V_{\beta+2} = \mathcal{P}(V_{\beta+1})$, and hence $\alpha = \beta+1 = \beta \cup \{\beta\} \in V_{\beta+2} = V_{\alpha+1}$, a contradiction. On the other hand, if α is a limit ordinal, then $\alpha = \cup_{\beta<\alpha}\beta$ and $\beta \in V_{\beta+1} \subset V_\alpha$ by the choice of α, so that $\alpha = \cup_{\beta<\alpha}\beta \subset \cup_{\beta<\alpha} V_\beta = V_\alpha$, and $\alpha \in \mathcal{P}(V_\alpha) = V_{\alpha+1}$, a contradiction. Therefore, rank $\alpha \leqslant \alpha$. We similarly prove strict equality. □

15. The universe V is closed with respect to the standard set operations: difference, union, intersection, forming $\mathcal{P}(X)$ and $\cup_{Y\in X} Y$, and "collecting" sets indexed by any set: $\{X_Y | Y \in Z\}$. In particular, if $X, Y \in V_\alpha$, then the pair $\{X, Y\}$ is in $V_{\alpha+1}$. We write $\{X\}$ in place of $\{X, X\}$.

16. Direct products, relations, and functions can also be defined as elements of V using a device of Kuratowski. The intuitive notion of an ordered pair of sets $X, Y \in V$ is realized by means of the set

$$\langle X, Y\rangle = \{\{X\}, \{X, Y\}\} \in V.$$

As elements of V, ordered pairs are characterized by the following properties: an ordered pair is a set of two elements X' and Y', one of which is a subset of the other (say $X' \subset Y'$); if $X' \subset Y'$, then $X' = \{X\}$ is a one-element set, and X is called the *first term of the pair*; Y' is a set of at most two elements, and its element Y that is different from X (if it exists) or X itself (otherwise) is called the *second term of the pair*. Thus, $\langle X, Y\rangle = \langle X'', Y''\rangle$ if and only if $X = X$" and $Y = Y$", which justifies the name "ordered pair."

We emphasize that this definition is introduced so that the direct product construction does not leave the universe V, and so that a set corresponding to a direct product can be described in terms of the relation $\in$, i.e., in the language L_1Set.

An ordered n-tuple of sets is defined as

$$\langle X_1, \ldots, X_n \rangle = \langle \cdots \langle \langle X_1, X_2 \rangle, X_3 \rangle \cdots \rangle.$$

We define the direct product of two sets as

$$X \times Y = \{\langle U, W \rangle | U \in X, W \in Y\}.$$

Similarly,

$$X_1 \times \cdots \times X_n = (\cdots((X_1 \times X_2) \times X_3) \times \cdots).$$

We note that in general, $(X \times Y) \times Z \neq X \times (Y \times Z)$; we have only a canonical one-to-one correspondence between these two sets. But it is usually harmless to take the liberty of identifying the two sets and writing $X \times Y \times Z$.

A *binary* relation (or correspondence) r is a set (or class) all of whose elements are ordered pairs. If $r \in V$ is a relation, then its *domain of definition* dom(r) is the class of all first terms in the elements of r, and the *range of values* rng(r) is the class of all second terms.

A function is a binary relation in which each element is uniquely determined by its first term. Thus, functions that are maps of sets in V are identified with their graphs. If f is a function, we often write $W = f(U)$ instead of $\langle U, W \rangle \in f$. In addition, we set

$$f^{-1}(X) = \{Y | f(Y) \in X\},$$
$$f|_X = \{\langle U, W \rangle \in f | U \in X\}.$$

A *family* $\{X_Y | Y \in Z\}$ as an element of V is defined to be a function consisting of pairs $\{\langle Y, X_Y \rangle | Y \in Z\}$, and so on.

We again emphasize that the most important feature of these definitions is that we do not introduce any new objects besides elements of V, or any new relations other than those expressible in terms of $\in$. It should also be noted that in accordance with the usual ("extensional") notion, a *property* of the elements of a set $X \in V$ is a subset $Y \subset X$ (consisting of all elements with this property). Thus, $Y \in V$, so that properties, properties of properties, properties of sets of properties, . . . (with transfinite iteration) are elements of V.

The "universe" V has earned its name.

17. Finally, we show that a chain $X_1 \in X_2 \in \cdots$ of elements of V must terminate (of course, with the empty set).

We prove that if X is nonempty, then there exists a $Y \in X$ with $Y \cap X = \varnothing$ (the desired result is obtained if we apply this to the set X of terms in the chain). In fact, let Y be the element of least rank in X (which exists because the ranks, since they are ordinals, are well-ordered). If we had $X \cap Y = \varnothing$, then any element $Z \in X \cap Y$ would have lower rank than Y, a contradiction.

18. *Connection with the axioms of* $\mathrm{L_1Set}$. The point of view adopted in this book is as follows.

The intuitive notion of a set, to which we appealed when constructing the universe V, is the primary material. The language $\mathrm{L_1Set}$ was devised in order to write formal texts based on this material that are equivalent to our intuitive arguments concerning V. The axioms of $\mathrm{L_1Set}$ (including the logical axioms) are obtained as a result of analyzing intuitive proofs. Our criterion for the completeness of this list is that we can write a formal deduction that translates any intuitive proof. The fact that we are able to do this must be proved by a rather large compendium of formal texts, which can be found in other books on logic. In particular, in $\mathrm{L_1Set}$ we can write the formula "$\forall x\ \exists \text{ordinal}\ \alpha(x \in V_\alpha)$" and deduce it from the axioms. This formula is the formal expression of our restriction to sets in V.

The question of the formal consistency of the Zermelo–Fraenkel axioms must remain a matter of faith, unless and until a formal inconsistency is demonstrated. So far all the proofs that have been based on these axioms have never led to a contradiction; rather, they have opened up before us the rich world of classical and modern mathematics. This world has a certain reality and life of its own, which depends little on the formalisms called upon to describe it.

The discovery of a contradiction in any of various formalisms, even if it should occur, would merely serve to clarify, refine, and perhaps reconstruct certain of our ideas, as has happened several times in the past, but would not lead to their downfall.

The Last Digression. Truth as Value and Duty: Lessons of Mathematics.

1. **Introduction.** Imagine that you open your morning newspaper and read the following report:

Brownsville, AR. *A local object partially immersed in a liquid was buoyed upward Tuesday by a force equal to the weight of the liquid displaced by that object, witnesses at the scene reported. As of press time, the object is still maintaining positive buoyancy.*

In fact, I did read this report in the *Onion* [1]; I have abridged it only to add a Fénéonian touch.

If this book had been dedicated to the nature of the comical, one could have produced an interesting analysis of the clever silliness of this parody. But since we are preoccupied with mathematical truth, I will use it in order to illustrate the differences between the attitudes to truth among practitioners of social sciences and law on the one hand, and that of physicists, on the other.

[1] The Onion is a satirical newspaper, owned by an American "fake news" organization Onion, Inc., based in New York.

To put it crudely, in social sciences information comes from witnesses; but in what sense was Archimedes' role in his discovery that of a witness, and are the experimental observations generating/supporting a physical theory on an equal footing with the observations of witnesses to a crime scene, or respondents to a poll?

Now imagine another report, which could have been posted on the website of the Department of Physics of Cambridge University:

The Cavendish Laboratory News & Features bulletin announced yesterday that a Cavendish student has the won The Science, Engineering and Technology award. He managed to measure the constant π with unprecedented precision: $\pi = 31415925\ldots$ with an error ± 2 at the last digit.

I must confess right away that I did not read but simply fabricated this spoof in order to stress the further differences between the attitudes toward truth now held by physicists and by mathematicians respectively.

Literally speaking, such an announcement would make perfect sense: the *mathematical* constant π *can* be measured with some precision, in the same way that any *physical* constant such as the speed of light c or the mass of the electron can be measured. The maximum achievable precision, at least of a "naive" direct measurement of π, is determined by the degree to which we can approximate ideal Euclidean rigid bodies by real physical ones. The limits to this approximation are set by the atomic structure of matter, and in the final analysis, by quantum effects.

On the other hand, in order to get in principle as many digits of π as one wishes, measurements are not required at all. Instead, one can use one of the many existing formulas/algorithms/software codes and do it on a sheet of paper, a pocket calculator, or a supercomputer. This time the limits of precision are determined by the physical limitations of our calculator: the size of the sheet of paper, memory of computer, construction of the output device, available time ....

What I want to stress now is that π imagined as an infinite sequence of its digits is not amenable to a "finite" calculation: even the number of digits of π equal to the number of atoms in the observable universe would not exhaust π. As Wisława Szymborska beautifully put it:

heaven and earth shall pass away,
but not pi, that won't happen,
it still has an okay five,
and quite a fine eight,
and all but final seven,
prodding and prodding a plodding eternity
to last.

Nevertheless, mathematicians speak about π and work with π as if it were a completely well defined entity, graspable in its entirety not only by one exceptional supermind, but by the minds of all trained researchers, never doubting that when they speak of π, they speak about one and the same ideal object, as rigid as if it really existed in some Platonic world.

One facet of this rigidity can be expressed by a few theorems implying that whatever power series, integral, limit, and software code we might use to calculate π and whatever precision we choose, we will always get the same result. If we do not, either our formula was wrong, or the calculator made a mistake/there was a bug in the code/output device could not cope with the quantity of information ....

Contemplating this example, we may grasp the meaning of the succinct description of mathematics by Davis and Hersh: *"the study of mental objects with reproducible properties."*

However, I want to use this example in order to stress that most of the deep mathematical truths are about *infinity* and infinitary mental constructs rather than experimentally verifiable finitary—and finite—operations that can be modeled using actual objects of the physical world.

2. **Infinity, Georg Cantor, and truth.** Before Georg Cantor, infinity appeared in mathematical theorems mostly implicitly, through the quantifier "all" (which also could be only implicit as in most of Euclid's theorems).

Cantor proved the first theorem ever in which infinities themselves were objects of consideration and of a highly nontrivial discovery.

When Cantor first presented his diagonal argument in a letter to Dedekind in 1873, it was worded differently and used only to prove that the cardinality of the natural numbers is strictly less than that of the real numbers. The discovery of the proof itself was in a sense hardly less important than the discovery of the definition of *what it means for one infinity to be larger than another one.*

As soon as this was achieved, Cantor started thinking about the cardinality of the reals compared with that of the pairs of reals, or, geometrically, sets of points of a curve and of a surface respectively. They turned out to be equal! If we have a pair of numbers (α, β) in $(0, 1)$, Cantor suggested to produce from them the third number $\gamma \in (0, 1)$ by putting the decimal digits of α in the odd places and those of β in the even places. One sees that conversely, (α, β) can be reconstructed from γ. Dedekind, who was informed by Cantor's letter about this discovery as well, remarked that this does not quite work because some rational numbers have two decimal representations, such as $0499999\ldots = 05000000\ldots$. Cantor had to spend some time to amend the proof, but this was a minor embarrassment, in comparison with the fascinating novelty of the fact itself: *"Ce que je vous ai communiqué tout récemment est pour moisi inattendue, si nouveau, que je ne pourrai pour ainsi dire pas arriver à une certaine tranquillité d'esprit avant que je n'aie reçu, très honoré ami, votre jugement sur son exactitude. Tant que vous ne m'aurez pas approuvé, je ne puis que dire: je le vois, mais je ne le crois pas."*

"I see it but I do not believe it," Cantor famously wrote to Dedekind.

This returns us to the basic question on the nature of truth.

We are reminded that the notion of "truth" is a reification of a certain relationship between humans and *texts/utterances/statements*, the relationship that is called "belief," "conviction," or "faith," and which itself should be analyzed, together with other primary notions invoked in this definition.

S. Blackburn in his keynote talk "Truth and Ourselves: The Elusive Last Word" at The Balzan Symposium on Truth, 2008, extensively discussed other relationships of humans to texts, such as *scepticism, conservatism, relativism, deflationism.* However, in the long run all of them are secondary in the practice of a researcher in mathematics.

So I will return to truth.

I will skip analysis of the notion of "humans" :=) and will only sketch what must be said about texts, sources of conviction, and methods of conviction peculiar to mathematics.

(*i*) *Texts.* Alfred North Whitehead allegedly said that all of Western philosophy was but a footnote to Plato.

The underlying metaphor of such a statement is, "Philosophy is a text," the sum total of all philosophic utterances.

Mathematics decidedly is *not* a text, at least not in the same sense as philosophy. There are no authoritative books or articles to which subsequent generations turn again and again for wisdom. Except for historians, nobody reads Euclid, Newton, Leibniz, or Hilbert in order to study geometry, calculus, or mathematical logic. The life span of any mathematical paper or book can be years; in the best (and exceptional) case, decades. Mathematical wisdom, if not forgotten, lives as an invariant of all its (re)presentations in a permanently self-renewing discourse.

(*ii*) *Sources and methods of conviction.* Mathematical truth is not revealed, and its acceptance is not imposed by any authority.

Moreover, mathematical truth decidedly is not something that can be ascertained, as Justice Oliver Wendell Holmes put it, by "the majority vote of the nation that could lick all the others." Equally laughable is his idea that "the best test of truth is the power of the thought to get itself accepted in the competition of the market."

If this means that truth is not a democratic value, then something is wrong with our conception of democracy.

Ideally, the truth of a mathematical statement is ensured by a proof, and the ideal picture of a proof is a sequence of elementary arguments whose rules of formation are explicitly laid down before the proof even begins, and ideally are common for all proofs that have been devised and can be devised in the future.

This ideal picture is so rigid that it can itself become the subject of mathematical study, and the first two chapters of this book were dedicated to the presentation of the results of this soul-searching activity of our transgenerational community.

Of course, real-life proofs are rendered in a peculiar mixture of a natural language, formulas, motivations, and examples. They are much more condensed than imaginary formal proofs. The ways of condensing them are not systematic in any way. We are prone to mistakes, to taking on trust others' results that can be mistaken as well, and to relying upon authority and revelations from our teachers. (All of this should have been discussed together with the notion of "humans," which I have wisely avoided.)

Moreover, the discovery of truth may, and usually does, involve experimentation, nowadays vast and computer-assisted, false steps, sudden insights, and all that which makes mathematical creativity so fascinating to its adepts.

One metaphor of proof is a route, which might be a desert track, boring and unimpressive until one finally reaches the oasis of one's destination, or a footpath in green hills, exciting and energizing, opening great vistas of unexplored lands and seductive offshoots, leading far away even after the initial destination point has been reached.

3. Mathematics and Cognition

[...] "mismanagement and grief": here you have that enormous distance between cause and effect covered in one line. Just as math preaches how to do it.

J. Brodsky. On "September 1, 1939" by W. H. Auden.

Mathematics is most visible to the general public when it posits itself as an applied science, and in this role the notion of mathematical truth acquires distinctly new features. For example, our initial discussion of π as an essentialy nonfinitary ("irrational") real number becomes pointless; whenever π enters any practical calculation, the first few digits are all that matters.

In a wider context than just applied science, mathematics can be fruitfully conceived as a toolkit containing powerful cognitive devices. I have argued elsewhere that these devices can be roughly divided into three overlapping domains: models, theories, and metaphors. Quoting from my book *Mathematics as Metaphor*:

> A mathematical *model* describes a certain range of phenomena qualitatively or quantitatively but feels uneasy pretending to be something more.
> From Ptolemy's epicycles (describing planetary motions, ca 150) to the Standard Model (describing interactions of elementary particles, ca 1960), quantitative models cling to the observable reality by adjusting numerical values of sometimes dozens of free parameters (≥ 20 for the Standard Model). Such models can be remarkably precise.
> Qualitative models offer insights into *stability/instability, attractors* which are limiting states tending to occur independently of initial conditions, *critical phenomena* in complex systems which happen when the system crosses a boundary between two phase states, or two basins of different attractors. [...]
> What distinguishes a (mathematically formulated physical) *theory* from a model is primarily its higher aspirations. A modern physical theory generally purports that it would describe the world with absolute precision if only it (the world) consisted of some restricted variety of stuff: massive point particles obeying only the law of gravity; electromagnetic field in a vacuum; and the like. [...]
> A recurrent driving force generating theories is a concept of a reality beyond and above the material world, reality which may be grasped

> only by mathematical tools. From Plato's solids to Galileo's "language of nature" to quantum superstrings, this psychological attitude can be traced sometimes even if it conflicts with the explicit philosophical positions of the researchers.
>
> A (mathematical) *metaphor*, when it aspires to be a cognitive tool, postulates that some complex range of phenomena might be compared to a mathematical construction. The most recent mathematical metaphor I have in mind is Artificial Intelligence (AI). On the one hand, AI is a body of knowledge related to computers and a new, technologically created reality, consisting of hardware, software, Internet etc. On the other hand, it is a potential model of functioning of biological brains and minds. In its entirety, it has not reached the status of a model: we have no systematic, coherent and extensive list of correspondences between chips and neurons, computer algorithms and brain algorithms. But we can and do use our extensive knowledge of algorithms and computers (because they were created by us) to generate educated guesses about structure and function of the central neural system [...].
>
> A mathematical theory is an invitation to build applicable models. A mathematical metaphor is an invitation to ponder upon what we know.

As an aside, let us note that George Lakoff's *definition* of poetic metaphors such as "love is a journey" (in G. Lakoff. "The Contemporary Theory of Metaphor." In: A. Ortony (ed.), *Metaphor and Thought* (2nd ed.). Cambridge Univ. Press, 1993) is itself expressed as a mathematical metaphor using the characteristic Cantor–Bourbaki mental images and vocabulary: *"More technically, the metaphor can be understood as a mapping (in the mathematical sense) from a source domain (in this case, journeys) to a target domain (in this case, love). The mapping is tightly structured. There are ontological correspondences, according to which entities in the domain of love (e.g. the lovers, their common goals, their difficulties, the love relationship, etc.) correspond systematically to entities in the domain of a journey (the travellers, the vehicle, destinations, etc.)."*

When a mathematical construction is used as a cognitive tool, the discussion of truth becomes loaded with new meanings: a model, a theory, or a metaphor must be true to a certain reality, more tangible and real than the Platonic "reality" of pure mathematics. In fact, philosophers of science routinely discussed truth precisely in this context. Karl Popper's vision of scientific theories in terms of falsifiability (versus verifiability) is quite appropriate in the context of highly mathematicised theories as well.

What I want to stress here, however, is one aspect of contemporary mathematical models that is historically very recent. Namely, models are more and more widely used as "black boxes" with hidden computerized input procedures, and oracular outputs prescribing behavior of human users.

Mary Poovey, discussing financial markets from this viewpoint, remarks in her insightful essay "Can Numbers Ensure Honesty? Unrealistic Expectations and the US Accounting Scandal" (*Notices of the AMS*, vol. 50:1, Jan. 2003, pp. 27–35), that what she calls "representations" (computerized bookkeeping or the numbers a trader enters in a computer) tend to replace the actual exchange of cash or commodities. *"This conflation of representation and exchange has all kinds of material effects* [...] *for when representation can influence or take the place of exchanges, the values at stake become notional too: they can grow exponentially or collapse at the stroke of key."*

In fact, actions of traders, banks, hedge funds, and the like are to a considerable degree determined by the statistical models of financial markets encoded in the software of their computers. These models, now essentially defining financial markets, thus become a hidden and highly influential part of the actions, our computerized "collective unconscious." As such, they cannot even be judged according to the usual criteria of choosing models that better reflect the behavior of a process being modeled. They are *part* of any such process.

What becomes more essential than their empirical adequacy is, for example, their stabilizing or destabilizing potential. Risk management assuming mild variability and small risks can collapse when a disaster occurs, ruining many participants of the game; risk management based on models that use pessimistic "Lévy distributions" rather than omnipresent Gaussians paradoxically tends to flatten the shock waves and thus avoid major disasters (B. Mandelbrot).

4. **Truth as value.** When in the twentieth century mathematicians got involved in heated discussions about the so-called crisis in the foundations of mathematics, several issues were intermingled.

Philosophically minded logicians and professional philosophers were engaged with the nature and accessibility of mathematical truth (and reliability of our mental tools used in the process of acquiring it).

Logicists (finitists, formalists, intuitionists) were elaborating severe normative prescriptions trying to outlaw dangerous mental experiments with infinity, nonconstructivity, and *reductio ad absurdum.*

For a working mathematician, when he/she is concerned at all, "foundations" is simply a general term for the historically variable set of rules and principles of organization of the body of mathematical knowledge, both existing and being created. From this viewpoint, the most influential foundational achievement in the twentieth century was an ambitious project of the Bourbaki group, building all mathematics, including logic, around set-theoretic "structures" and making Cantor's language of sets a common vernacular of algebraists, geometers, probabilists, and all other practitioners of our trade. These days, this vernacular, with all its vocabulary and ingrained mental habits, is being slowly replaced by the languages of category theory and homotopy theory and their higher extensions. Respectively, the basic "left-brain" intuition of sets, composed of distinguishable elements, is giving way to a new, more "right-brain" basic intuition dealing with spacelike and continuous primary images, both deformable and deforming.

In Western ethnomathematics, truth is best understood as *a central value, ever to be pursued,* rather than anything achieved. Practical efficiency, authority, success in competition, faith, all other clashing values must recede in the mind of a mathematician when he or she gets down to work.

The most interesting intracultural interactions of mathematics are as well those that are not direct but rather proceed with the mediation of value systems.

Coda. Every four years, mathematicians from all over the world meet at the International Congress of Mathematicians (ICM), to discuss whatever interesting developments happened recently in their domains of expertise. One of the traditions of these congresses is a series of lectures for the general public.

In 1998, our congress met in Berlin, and Hans Magnus Enzensberger, the renowned poet and essayist, deeply interested in mathematics, spoke about "Zugbrücke außer Betrieb: Die Mathematik im Jenseits der Kultur": the drawbridge to the castle of mathematics is out of service. The main concern of his talk was a deplorable lack of mathematical culture and communication between the general public and mathematicians, leading to alienation and mutual mistrust.

At the end of his talk Enzensberger quotes an imaginary dialogue, where a mathematician is chatting with a fictitional layman "Seamus Android" (see I. Stewart. *The Problems of Mathematics.* Oxford Univ. Press, 1987).

"Mathematician: It's one of the most important discoveries of the last decade!

Android: Can you *explain* it in words ordinary mortals can understand?

Mathematician: Look, buster, if ordinary mortals could understand it, you wouldn't need mathematicians to do the job for you, right? You can't get a feeling for what's going on without understanding the technical details. How can I talk about manifolds without mentioning that the theorems only work if the manifolds are finite-dimensional paracompact Hausdorff with empty boundary?

Android: Lie a bit.

Mathematician: Oh, but I couldn't do that!

Android: Why not? Everybody *else* does."

And here I must play God and say to both Android and Mathematician: "Oh, no! Don't lie—because everybody else does."

III

The Continuum Problem and Forcing

1 The Problem: Results, Ideas

1.1. Cantor introduced two fundamental ideas in the theory of infinite sets: he discovered (or invented?) the scale of cardinalities of infinite sets, and gave a proof that this scale is unbounded. We recall that two sets M and N are said to *have the same cardinality* (card M = card N) if there exists a one-to-one correspondence between them. We write card $M \leqslant$ card N if M has the same cardinality as a subset of N. We say that M and N are *comparable* if either card $M \leqslant$ card N or card $N \leqslant$ card M. We write card $M >$ card N if card $M \geqslant$ card N but M and N do not have the same cardinality.

1.2. Theorem (Cantor, Schröder, Bernstein, Zermelo)

(a) *Any two sets are comparable. If both* card $M \leqslant$ card N *and* card $N \leqslant$ card M, *then* card M = card N. *In other words, the cardinalities are linearly ordered.*

(b) *Let $\mathcal{P}(M)$ be the set of all subsets of M. Then* card $\mathcal{P}(M) >$ card M. *In particular, there does not exist a largest cardinality.*

(c) *In any class of cardinalities there is a least cardinality. In other words, the cardinalities are well-ordered.*

PROOF.

(a) Suppose M has the same cardinality as the subset $M' \subset N$ and N has the same cardinality as the subset $N_1 \subset M \cong M'$. We identify M with M'. We then have three sets $N_1 \subset M \subset N$ and a one-to-one correspondence $f : N \to N_1$. We must construct a one-to-one correspondence $g : N \to M$. Here is an explicit definition of such a map:

$$g(x) = \begin{cases} f(x), & \text{if } x \in f^n(N)\backslash f^n(M) \text{ for some } n \geqslant 0, \\ x, & \text{otherwise.} \end{cases}$$

Here $f^n(y) = f(f(\cdots f(y)\cdots))$ (n times); $f^n(N) = \{f^n(y) | y \in N\}$, and $f^0(y) = y$. We leave the verification that g has the required properties to the reader.

Yu. I. Manin, *A Course in Mathematical Logic for Mathematicians, Second Edition*, Graduate Texts in Mathematics 53, DOI 10.1007/978-1-4419-0615-1_3,

To prove that any two sets are comparable, it is sufficient to show that any set can be well-ordered, since Lemma 5 of the appendix to Chapter II implies that well-ordered sets are comparable to each other. Let M be any set. For every nonempty subset $N \subset M$ choose an element $c(N) \in N$. We call a well-ordering $<$ of a subset $M' \subset M$ *admissible* (with respect to c) if $c(M \setminus \hat{X}) = X$ for all $X \in M'$, where $\hat{X} = \{Y | Y \in M', Y < X\}$.

We claim that if $M' \neq M''$ are two subsets of M having admissible well-orderings, then one set is an initial segment of the other, and the orderings are compatible. In fact, as in Section 7(a) of the appendix to Chapter II, we prove that the canonical isomorphism f of, say, M' with an initial segment of M'' is the identity embedding: if $f(X) \neq X$ and X is the least element with this property, then

$$f(\hat{X}) = \hat{X}, \qquad X = c(M \setminus \hat{X}) \Rightarrow X = c(M \setminus f(\hat{X})) = f(X),$$

which is a contradiction.

It is now easy to see that the union M' of all subsets of M that have a well-ordering admissible with respect to c itself has an admissible ordering; moreover, M' coincides with M, since otherwise we could embed M' in $M' \cup \{c(M \setminus M')\}$.

In particular, it follows that any set has the same cardinality as some ordinal, and hence the same cardinality as a unique cardinal. This justifies the use of the term "cardinality" and the use of cardinals as our standard scale of cardinalities (see Section 11 of the appendix to Chapter II).

(b) Since $\mathcal{P}(M)$ contains all the one-element subsets of M, we have card $\mathcal{P}(M) \geqslant$ card M. In addition, any map $f : M \dashrightarrow \mathcal{P}(M)$ cannot be one-to-one (or even onto). In fact, we set

$$N = \{z | z \notin f(z)\} \in \mathcal{P}(M),$$

and show that N is not contained in the image of f. If there existed an $n \in M$ such that $N = f(n)$, we would immediately obtain a contradiction by considering the relationship of n to N:

$$n \in N \Rightarrow n \in f(n) \Rightarrow n \notin N \quad \text{by the definition of } N;$$
$$n \notin N \Rightarrow n \notin f(n) \Rightarrow n \in N \quad \text{by the definition of } N.$$

This is Cantor's famous "diagonal process."

(c) The well-ordering of the cardinals is established at the same time as their comparability in the first stage of the theory of ordinals (see the Appendix to Chapter II). □

1.3. *Remark.* This proof of the lemma that any set can be well-ordered is essentially due to Zermelo. It was probably what prompted the most severe objections to the axiom of choice. The intuitive idea behind the proof reduces to a recipe for choosing one element after another from the set M until all of

M is exhausted. In this form it is immediately apparent that the prescription is "physically" unthinkable, and to many of Zermelo's contemporaries the whole proof seemed to be nothing but a trick. For example, the idea of "first" choosing an element $c(N)$ in each subset $N \subset M$ met with the following objection of Lebesgue. If the elements we choose are not characterized by any special properties, how do we know that we are always thinking about the same elements throughout the proof? But today, except for specialists in the foundations of mathematics, hardly any working mathematicians share these doubts.

We now formulate the basic problem that will concern us during the next two chapters. We shall write $\mathrm{card}\,\mathcal{P}(M) = 2^{\mathrm{card} M}$, in analogy to the finite case. The continuum is 2^{ω_0}.

1.4. *The continuum problem.* What place does the continuum occupy on the scale of cardinalities?

By Theorem 1.2(b), we have $2^{\omega_0} > \omega_0$. Hence, in any case, $2^{\omega_0} \geqslant \omega_1$. On the other hand, if $2^{\omega_0} > \omega_1, 2^{\omega_0} > \omega_2, \ldots, 2^{\omega_0} > \omega_n, \ldots$ for any n, then we would have $2^{\omega_0} > \omega_{\omega_0}$, since the continuum cannot be a union of countably many subsets of lower cardinality (König).

1.5. *The continuum hypothesis* (CH). $2^{\omega_0} = \omega_1$.

The generalized continuum hypothesis asserts that $2^{\mathrm{card} M}$ comes immediately after $\mathrm{card}\, M$ for any infinite M. Here is what we know about this question:

1.6. **Theorem**

(a) *The negation of the continuum hypothesis cannot be deduced from the other axioms of set theory if those axioms are consistent (Gödel).*

(b) *The continuum hypothesis cannot be deduced from the other axioms of set theory if those axioms are consistent (Cohen).*

The same holds true for the generalized continuum hypothesis.

If we grant that the axioms of set theory and the logical means of expression and deduction in $\mathrm{L}_1\mathrm{Set}$, which are implicit in the statement of Theorem 1.6, actually exhaust the apparatus for constructing proofs in modern mathematics, then we can say that the continuum problem is the first known example of an absolutely undecidable problem. Although Gödel's incompleteness theorem provides concrete examples of undecidable propositions in any formal system having reasonable properties, these examples can be decided in an "obvious" way in some higher system. The situation with the continuum problem seems much more difficult. If we agree that it is a meaningful question, then it can be decided only by introducing a new principle of proof. Various possibilities for doing this have been discussed, but none of the suggested new axioms for set theory seem sufficiently convincing or, more important, sufficiently useful in "real" mathematics. In the hundred years since the introduction of transfinite induction, not a single new method of constructing sets has come into common use (see, however, the end of IV.7 (added in the second edition)). Incidentally, the basic idea in Gödel's proof of Theorem 1.6(a) actually consists in verifying that all the old methods allow us to construct at most ω_1 subsets of ω_0 (or, equivalently, at most ω_1 real numbers).

1.7. *Gödel's idea.* Gödel considers the basic set-theoretic operations—forming pairs, products, complements, sums, and so on—and constructs the class of all sets that are obtained by transfinite iteration of these operations, starting from $\varnothing$. Such sets are called *constructible* sets. It is a priori completely unclear whether all subsets of $\{0, 1, 2, \ldots\}$ are constructible, or, more generally, whether all sets in the universe V are constructible. (It turns out that this problem is formally undecidable to the same extent as the continuum problem.) But we find that within the class of constructible sets, the number of subsets of $\{0, 1, 2, \ldots\}$ is equal to ω_1—most likely because we have omitted a vast number of nonconstructible sets. Meanwhile, all the axioms of set theory, restricted to this class, are true (in a reasonable meaning of "true"), as are all deductions from these axioms. Hence the negation of the CH is not deducible, since it is false in this model. The next chapter will be devoted to Gödel's theorem.

1.8. *Cohen's idea.* We shall present this idea in the version due to Scott and Solovay. First we give its application to a certain simplified problem, concerned with a language weaker than $\mathrm{L}_1\mathrm{Set}$; then in §§4–8 we present the application to $\mathrm{L}_1\mathrm{Set}$. For another version of Cohen's idea, see §9.

We shall discuss the CH in the following form: *there does not exist a subset of the real numbers* $\mathbf{R}$ *whose cardinality is strictly between that of* $\{0, 1, 2, \ldots\}$ *and that of* $\mathbf{R}$. In fact, if we had $2^{\omega_0} > \omega_1$, then any subset of $\mathbf{R}$ of cardinality ω_1 would have such an intermediate cardinality.

In order to show that this assertion is not deducible, which is equivalent to Cohen's theorem, it suffices to construct a model of the real numbers in which all the axioms and all propositions deducible from them are fulfilled and in which a set of intermediate cardinality exists. This model will be the set $\overline{R}$ of random variables on a very big probability space Ω. For a suitable choice of Ω, $\overline{R}$ will be so big that within the model there exists a set of intermediate cardinality, containing $\overline{N}$ (the integers of the model) and contained in $\overline{R}$ (the continuum of the model).

Of course, it cannot be quite this simple; there must be some obstacle to carrying out this program. The obstacle is that almost all the properties of $\mathbf{R}$, including most of the axioms, turn out to be false for $\overline{R}$, so that $\overline{R}$ cannot be a model for $\mathbf{R}$ in the usual sense of the word. Cohen's basic idea was to develop a method for overcoming this difficulty. He replaced the property of an assertion being true by another property, which we shall temporarily call "truth" in quotes, and which has the necessary formal properties. Namely, all the axioms of $\mathbf{R}$ are "true" in $\overline{R}$, all deductions from "true" assertions using the rules of logic again lead to "true" assertions, and the CH is not "true," and hence is not deducible from the axioms. We now show in greater detail how this is done.

1.9. Let I be a set of cardinality $> \omega_1$. We set

$$\Omega = [0,1]^I, \text{ with Lebesgue measure,}$$
$$\overline{R} = \text{the set of random variables on } \Omega$$
$$= \text{the set of measurable real-valued functions on } \Omega.$$

1.10. **Theorem**

(a) *All the axioms of the real numbers and all deductions from them are "true" for* $\overline{R}$.
(b) *The* CH *is not "true" for* $\overline{R}$.

Here we say that an assertion P about random variables $\bar{x}, \bar{y}, \ldots \in \overline{R}$ is "true" if the following condition is fulfilled:

> for each point $\omega \in \Omega$ we consider the values $\bar{x}(\omega), \bar{y}(\omega), \ldots$ of the random variables $\bar{x}, \bar{y}, \ldots$ and form the assertion P_ω about these ordinary real numbers; then for almost all $\omega \in \Omega$ (i.e., all but a set of measure 0) P_ω is true in the usual sense of the word.

Briefly, "truth" means experimental truth with probability one.

EXAMPLE. Let P be the assertion that "**R** has no zero divisors," i.e., "if $x, y \in \mathbf{R}$ are such that $xy = 0$, then either $x = 0$ or $y = 0$." Then the assertion "$\overline{R}$ has no zero divisors" is, of course, not true. However, it is "true" because: if $\bar{x}, \bar{y} \in \overline{R}$ are such that $\bar{x}\bar{y} = 0$, then for almost all $\omega \in \Omega$ either $\bar{x}(\omega) = 0$ or $\bar{y}(\omega) = 0$.

1.11. In order to give a precise meaning to the definition of "truth" and learn how to verify effectively the "truth" of rather complicated assertions, we must introduce a formal language, in this case the language of real numbers. This formal language is a mathematical object, and the precise formulation of Theorem 1.10 will concern this object, and not **R** or $\overline{R}$ at all.

The connection between this language and **R** is given by a system of informal recipes that tell how to translate the usual intuitive texts about **R** into this language, and by a system of theorems that tell us that the translation is always possible and that the recipes are faithful to the informal texts. The role of $\overline{R}$ is reduced to that of auxiliary construction that is used to define and compute a special "truth" function on the formulas of the language. Thus we see the role of logic in the program.

1.12. A detailed proof of Theorem 1.10 would be rather lengthy and nontrivial for several reasons. In the first place, a certain amount of space must be devoted to describing the formal language and the axioms of **R** in this language. We must then verify that all the axioms are "true" and that the CH is not "true"—this amounts to one or two dozen verifications, each of which involves an inductive argument with infinite sums and products in the Boolean algebra of measurable sets in Ω. However, the most serious difficulties arise because the meaning of every assertion changes considerably in going from **R** to $\overline{R}$, and not always in a convenient direction. We shall illustrate this qualitative aspect by attempting to explain why the CH is not "true," and why this is nontrivial.

As we have said, we want to construct a subset $\overline{M}$ of $\overline{R}$ having cardinality intermediate between the cardinality of $\overline{N}$ and the cardinality of $\overline{R}$. We do this as follows: For any $i \in I$, let the random variable $\bar{x}_i : [0,1]^I \to [0,I]$ be the ith projection. Choose a subset $\mathcal{J} \subset I$ such that $\omega_0 <$ card $\mathcal{J} <$card I (this is possible if I is large), and set

$$\overline{M} = \{x_j | j \in \mathcal{J}\} \subset \overline{R}.$$

Then card $\overline{N}$ < card $\overline{M}$ < card $\overline{R}$ is true in the usual meaning of the word. However, we must show that the corresponding assertion is "true" in our Pickwickian sense. But then the role of the integers is assumed by the "locally integral" random variables (whose values are integral with probability one), and these random variables can have cardinality much greater than ω_0. Thus, the required lower estimate for card $\overline{M}$ becomes much more serious. Similarly, if we formalize our naive description of $\overline{M}$ and then interpret it in $\overline{R}$, then $\overline{M}$ takes on a new meaning, and leads to a much larger set than the "real" $\overline{M}$. Thus, it is also unclear that the upper inequality for card $\overline{M}$ still holds. It seems almost miraculous that everything eventually falls into place.

The plan for the rest of the chapter is as follows. In §2 and §3 we give a (shortened) exposition for the second-order language of real numbers of this abbreviated version of the theorem that the CH is not deducible. If the reader is interested only in the complete proof for L_1Set, he may skip to §4, where we introduce the Boolean-valued "universe of random sets," which takes the place of V. In §§5–7 we verify that the Zermelo–Fraenkel axioms are "true," and in §8 we verify that the CH is "false." Finally, in §9 we discuss Cohen's original method, which is more syntactic and involves somewhat different intuitive ideas.

2 A Language of Real Analysis

2.1. In this section we describe a formal language based on the theory of real numbers. In particular, this means that the variables x, y, z will be considered as names of real numbers. However, if we try to use a first-order language to formulate the assertions we are interested in, such as the continuum hypothesis CH, or even the completeness axiom (which differentiates the real numbers from the rational numbers), we find that we are not able to do this. In fact, in these assertions we need to refer to arbitrary subsets (or relations of degree one) of the real numbers, whereas first-order languages do not have symbols for variable relations (compare with Section 3.17 of Chapter I).

This leads us to consider the second-order language L_2Real, which is the most economical language in which the axioms and the CH can be expressed. We shall give a brief description of this language, for the most part noting only those features that show the connections with the real numbers and those that are peculiar to second-order languages.

2.2. *The language* L_2Real. The alphabet consists of the variable symbols $x, y, z, \ldots$; the symbols for degree-1 functions $f, g, h, \ldots$; the constants 0 and 1; the degree-2 operations $+$ and $\cdot$; the degree-2 relations $=$ and $\leqslant$; and the same connectives, quantifiers, and parentheses as in languages of $\mathfrak{L}_1$. The *terms* are $x, y, z, \ldots$ and 0 and 1; and also $f(t), t_1 \cdot t_2$, and $t_1 + t_2$ if f is a function symbol and t, t_1, and t_2 are terms. The terms are names of real numbers.

The *atomic formulas* are $t_1 = t_2$ and $t_1 \leqslant t_2$, where t_1 and t_2 are terms. The set of *formulas* is defined inductively exactly as in languages of $\mathfrak{L}_1$, with one addition: $\forall f(Q)$ and $\exists f(Q)$ are formulas if Q is a formula and f is the symbol for a variable function. The notions of a free occurrence of a variable (x or f), of a closed formula, and so on carry over to L_2Real in the obvious way. We shall use the same type of abbreviated notation here as in Chapter I. The standard interpretation of formulas that is implicit in the language should be obvious from the definitions and from the following examples.

2.3. *The formula* $Z(y)$: "*y is an integer.*" It is perhaps not completely obvious how to write this formula. We can write, "y can be obtained from 0 by repeatedly adding or subtracting 1," or else "any function f that has period 1 and vanishes at 0 must also vanish at y," i.e.,

$$Z(y): \quad \forall f\Big(\big(f(0) = 0 \wedge \forall x (f(x) = f(x+1))\big) \Rightarrow f(y) = 0\Big).$$

2.4. *The formula* CH: "*Any subset of* $\mathbf{R}$ *either has the same cardinality as* $\mathbf{R}$, *or else is countable or finite.*"

We first restate the formula in different words: "Given a set of zeros of any function h, either there exists a function g mapping it onto all $\mathbf{R}$, or else there exists a function f mapping the integers onto all of this set." We then have

$$\begin{aligned}\text{CH: } \forall h\big(\exists g\ \forall y\ \exists x(h(x) = 0 \wedge y = g(x)) \vee \exists f\ \forall y(h(y)\\ = 0 \Rightarrow \exists x(Z(x) \wedge y = f(x)))\big).\end{aligned}$$

Notice that the formula $Z(x)$ occurs as part of the CH.

We further write the completeness axiom C:

2.5. *The formula* C: "*Any subset of* $\mathbf{R}$ (*the set of values of a function* f) *that is bounded from above has a least upper bound z.*" We write

$$\text{C}: \quad \forall f\,(\exists y \forall x (f(x) \leqslant y) \Rightarrow \exists z\ \forall y\,(\forall x (f(x) \leqslant y) \Leftrightarrow z \leqslant y)).$$

All the other formulas we are interested in are simpler and do not require any special comment.

We now give a precise definition of the property of "truth" for closed formulas in L_2Real; this property was described informally in §1. We emphasize that it is not an absolute property, but rather depends on the choice of the probability space Ω that is used to construct the "model" of the real numbers.

2.6. *The algebra of truth values.* As in §1, we set

$I =$ a set;
$\Omega = [0,1]^I$ with Lebesgue measure;
$B =$ the algebra of measurable sets in Ω modulo sets of measure zero;
$0 =$ the class of the empty set in B;
$1 =$ the class of Ω in B.

We have the following operations in B:

a', the "complement" of the element $a \in B$;
$a \wedge b$, the "intersection" of two elements $a, b \in B$;
$a \vee b$, the "union" of two elements $a, b \in B$.

These operations satisfy the usual identities and give a Boolean algebra structure on B. We write $a \leqslant b$ if $a \wedge b = a$.

Moreover, the operations of intersection and union extend uniquely to *infinite families* of elements, and continue to satisfy the usual identities that hold in the algebra of all subsets of any given set. We shall omit the verification of all this. We note only that sets here are identified "modulo sets of measure zero," and that identities of the type $(A \bmod 0) \wedge (B \bmod 0) = (A \cap B) \bmod 0$ do not carry over to infinite families.

Finally, B satisfies the following *countable chain condition*: if $a_\alpha \wedge a_\beta = 0$ for all distinct indices α and β then $a_\alpha \neq 0$ for at most countably many indices α. This follows because Lebesgue measure is positive and additive. Technically speaking, B is a *complete Boolean algebra with the countable chain condition.* The precise origin of B and the fact that it has a measure play a less important role.

2.7. *The interpretation set.* We now introduce a large set $\overline{M}$, each point ξ of which corresponds to the assignment of certain values to all the symbols in the alphabet of L_2Real. If ξ is fixed, each formula becomes a concrete statement about measurable functions (random variables) on Ω and about functionals on them (compare with §2 of Chapter II).

More precisely, we set

$\overline{R} =$ the set of measurable real-valued functions on Ω;
$\overline{R}^{(1)} =$ the set of all possible maps $\bar{f} : \overline{R} \Rightarrow \overline{R}$ that satisfy the condition

$\forall \bar{x}, \bar{y} \in \overline{R}$

$$\left(\text{the set } \{\omega \in \Omega | \bar{x}(\omega) = \bar{y}(\omega)\} \leqslant \{\omega \in \Omega | \bar{f}(\bar{x})(\omega) = \bar{f}(\bar{y})(\omega)\} \bmod 0\right).$$

The definition of $\overline{R}^{(1)}$ has the following intuitive meaning. If we ignore the "mod 0," the condition simply means that the value of the random variable

$\bar{f}(\bar{x})$ at each trial (each point in Ω) must be determined by the value of $\bar{x}$ *at this trial.* Of course, this is a very natural requirement if we want functions $\bar{f}$ to be adequate reflections of properties of ordinary real-valued functions in the sense of §1. The addition of "mod 0" weakens this requirement by saying "with conditional probability one."

We now return to the set $\overline{M}$. A point $\xi \in \overline{M}$ consists of a choice of

$x^\xi \in \overline{R}$, for each variable symbol x;

$f^\xi \in \overline{R}^{(1)}$, for each symbol f for a variable function.

Here is the interpretation of the expressions in the language that corresponds to a given choice of ξ:

(a) *Terms.* Let t be a term, and let $\xi \in \overline{M}$. Then $t^\xi \in \overline{R}$ is the random variable that is defined inductively in the obvious way.

(b) *The truth function* $\| \;\|$ *on atomic formulas.* Let P be the atomic formula $t_1 \leqslant t_2$ or $t_1 = t_2$. Its truth value at a point $\xi \in \overline{M}$ is the element of the algebra B that is defined as follows:

$$\|t_1 \leqslant t_2\|(\xi) = \left\{\omega \in \Omega | t_1^\xi(\omega) \leqslant t_2^\xi(\omega)\right\} \text{ mod } 0,$$

and similarly for $t_1 = t_2$.

(c) *The truth function* $\|P\|(\xi)$ *in the general case.* The general definition proceeds by induction. The rules when formulas are joined by connectives are the same as in Section 5.7 of Chapter II:

$$\begin{aligned}
\|\neg P\| &= \|P\|', \\
\|P \vee Q\| &= \|P\| \vee \|Q\|, \\
\|P \wedge Q\| &= \|P\| \wedge \|Q\|, \\
\|P \Rightarrow Q\| &= \|P\|' \vee \|Q\|, \\
\|P \Leftrightarrow Q\| &= (\|P\| \wedge \|Q\|) \vee (\|P\|' \wedge \|Q\|').
\end{aligned}$$

Here, for brevity, we have omitted the ξ. Finally,

$$\|\forall x P\|(\xi) = \bigwedge_{\xi'} \|P\|(\xi') \quad \text{(over all } \xi' \text{ that differ from } \xi \text{ only by a variation of } x\text{);}$$

$$\|\exists x P\|(\xi) = \bigvee_{\xi'} \|P\|(\xi') \quad \text{(over the same } \xi'\text{);}$$

and similarly when we quantify over variable functions. Intuitively, the value of the truth function of an assertion about random variables is the set of trials mod 0 for which this assertion becomes true as a fact about real numbers.

2.8. **Lemma.** *If P is a closed formula, then $\|P\|(\xi)$ does not depend on the choice of $\xi \in \overline{M}$ and takes only the value* 0 *or* 1.

This is proved by a simple induction on the length of P. It is just as easy to prove a more general fact: if P is any formula and ξ and ξ' do not differ on variables that occur freely in P, then $\|P\|(\xi) = \|P\|(\xi')$. Compare with Proposition 2.10 in Chapter II.

This value of $\|P\|(\xi)$ that is common for all ξ if P is closed can be denoted simply by $\|P\|$. We are now ready to formulate the basic definition of this section:

2.9. **Definition.** A formula P in $\mathrm{L_2Real}$ is said to be "true" if $\|P\|(\xi) = 1$ for all $\xi \in \overline{M}$.

3 The Continuum Hypothesis Is Not Deducible in $\mathrm{L_2}$ Real

3.1. **Fundamental Lemma**

(a) *"Truth" is preserved under the rules of deduction.*
(b) *The first-order logical axioms and the versions of them in* $\mathrm{L_2Real}$ *are "true."*
(c) *The special axioms of* $\mathrm{L_2Real}$ *are "true."*
(d) *The* CH *is not "true" if* card $I > \omega_1$.

This lemma implies the following theorem

3.2. **Theorem.** *The* CH *is not deducible from the axioms in* $\mathrm{L_2Real}$.

In this section we give those parts of the proof of the fundamental lemma that are also essential for the "real" Cohen theorem, as well as for our simplified problem. We note that Theorem 3.2 is weaker than Cohen's theorem because the language $\mathrm{L_2Real}$ contains fewer means of expression than the language of set theory. Although the continuum hypothesis can be stated in $\mathrm{L_2Real}$, because of Gödel's general results we have no basis for expecting, even if the CH were deducible, that the proof could also be given in this language. For example, the deduction could require us to introduce functionals of functions, functionals of functionals, and so on. The language of set theory, To which we shall return in §4, contains the means for considering all of these finite and even transfinite levels at once.

3.3. Proof of 3.1(a). If $\|P\| = 1$ and $\|P \Rightarrow Q\| = 1$, then $\|P\|' = 0$ and $\|P\|' \vee \|Q\| = 1$, so that $\|Q\| = 1$. Secondly, if $\|P\| = 1$, then $\|P\|(\xi) = 1$ for all $\xi \in \overline{M}$; but then (here ξ' runs through all variations of ξ along x)

$$\|\forall x P\|(\xi) = \bigwedge_{\xi'} \|P\|(\xi') = \bigwedge_{\xi'} 1 = 1. \qquad \square$$

We similarly prove this for Gen over functions.

3.4. Proof of 3.1(b) (sketch).

Tautologies. Their "truth" is proved in §5 of Chapter II.

Quantifier axioms. The proof proceeds by induction on the length of the formulas in the axiom schemes. Since it is completely straightforward, we shall omit it.

3.5. Proof of 3.1(c) (sketch). We shall list the axioms and make some brief comments.

The special axioms of set theory: The axioms of equality and the axiom (schema) of choice

$$\text{AC}: \quad \forall x\, \exists y P(x,y) \Rightarrow \exists f\, \forall x P(x, f(x)),$$

where P is any formula which does not have any free variables except x and y, and where f is free for y in P.

The special axioms of field theory: The axioms of the additive group, the axioms of the multiplicative group, and the distributivity of addition with respect to multiplication.

The special order axioms:

$$x \leqslant y \vee y \leqslant x,$$
$$(x \leqslant y \wedge y \leqslant x) \Leftrightarrow x = y,$$
$$x \leqslant y \Rightarrow (x + z \leqslant y + z),$$
$$(x \leqslant y \wedge 0 \leqslant z) \Rightarrow xz \leqslant yz.$$

The completeness axiom (see 2.5).

Among these axioms, the greatest effort is needed to verify that the axiom of choice and the completeness axiom are "true." But these computations resemble those in the proof that the CH is false, which will be given in detail below. Hence, the verification of these two axioms will be omitted.

The first axiom of equality is trivial. The second axiom is first verified for atomic formulas P, and then we use induction on the length of P. The argument is rather tedious, but simple.

The axioms of an ordered field are verified without difficulty. We shall limit ourselves to one example: "every nonzero number has an inverse," i.e.

$$\|\forall x(\neg(x=0) \Rightarrow \exists y(xy=1))\| = \bigwedge_{\bar{x}\in\overline{R}} \left(\|\bar{x}=0\| \vee \bigvee_{\bar{y}\in\overline{R}} \|\bar{x}\bar{y}=1\| \right).$$

To verify that this truth value equals 1, it suffices to prove this for each term on the right, i.e., for each fixed $\bar{x} \in \overline{R}$. Then, in turn, for that $\bar{x}$ it suffices to construct a random variable $\bar{y} \in \overline{R}$ such that $\|\bar{x} = 0\| \bigvee \|\bar{x}\bar{y} = 1\| = 1$. We set

$$\bar{y}(\omega) = \begin{cases} \bar{x}(\omega)^{-1}, & \text{if } \bar{x}(\omega) \neq 0, \\ 0, & \text{if } \bar{x}(\omega) \neq 0. \end{cases} \qquad \square$$

3.6. Proof of 3.1(d). We first recall the formula for the CH:

$$\forall h(\exists g\, \forall y\, \exists x(h(x)=0 \wedge y = g(x)) \vee$$
$$\exists f\, \forall y(h(y)=0 \Rightarrow \exists x(Z(x) \wedge y = f(x))).$$

We let P_1 and P_2 denote the first and the second alternatives in this formula. Thus, the CH has the form $\forall h(P_1 \vee P_2)$. We must prove that $\|\forall h(P_1 \vee P_2)\|(\xi) = 0$

for any point $\xi \in \overline{M}$. By the definition in 2.7,

$$\|\forall h\ (P_1 \vee P_2)\|(\xi) = \bigwedge_{\xi'} (\|P_1\|(\xi') \vee \|P_2\|(\xi')),$$

where ξ' runs through all variations of ξ along h. To show that this value is 0, it suffices to find a point ξ' such that $\|P_1\|(\xi') = \|P_2\|(\xi') = 0$. Since all the variables except h are bound in P_1 and P_2, choosing ξ' is equivalent to choosing $h^{\xi'} = \bar{h} \in \overline{R}^{(1)}$. We shall give $\bar{h}$ explicitly; this will be a function "whose set of zeros has intermediate cardinality."

To do this, as in §1 we fix a subset $\mathcal{J} \subset I$ having cardinality strictly between ω_0 and card I. Recall that for each $i \in I$, $\bar{x}_i \in \overline{R}$ is the "ith coordinate" function. Further, for each random variable $\bar{x} \in \overline{R}$, we choose a subset $\Omega(\bar{x}) \subset \Omega$ such that

$$\bigvee_{j \in \mathcal{J}} \|\bar{x} = \bar{x}_j\| = \Omega(\bar{x}) \mod 0$$

(here we use the completeness of B). Finally, we define $\bar{h} \in \overline{R}^{(1)}$ as follows for every $\bar{x} \in \overline{R}$ and $\omega \in \Omega$:

$$\bar{h}(\bar{x})(\omega) = \begin{cases} 0, & \text{if } \omega \in \Omega(\bar{x}), \\ 1, & \text{otherwise.} \end{cases}$$

3.7. Correctness Lemma

(a) *For fixed $\bar{x}$, $\bar{h}(\bar{x})$ is measurable as a function of ω, so that $\bar{h}$ maps $\overline{R}$ to $\overline{R}$.*
(b) *For every $\bar{x} \in \overline{R}$ we have*

$$\|\bar{h}(\bar{x}) = 0\| = \bigvee_{j \in \mathcal{J}} \|\bar{x} = \bar{x}_j\|.$$

(c) $\bar{h} \in \overline{R}^{(1)}$ *(see 2.7), so that there exists a point $\xi' \in M$ for which $h^{\xi'} = \bar{h}$.*

PROOF.

(a) $\bar{h}(\bar{x})$ takes only the values 0 and 1 on Ω, and the set where it takes each of these two values is measurable by the definition and by the completeness of B.

(b) is obvious from the definition.

(c) We must verify that for all $\bar{x}, \bar{y} \in \overline{R}$ we have

$$\{\omega \in \Omega | \bar{x}(\omega) = \bar{y}(\omega)\} \leqslant \{\omega \in \Omega | \bar{h}(\bar{x})(\omega) = \bar{h}(\bar{y}(\omega))\} \mod 0.$$

We shall show that the set of points $\omega \in \Omega$ for which both $\bar{x}(\omega) = \bar{y}(\omega)$ and $\bar{h}(\bar{x})(\omega) \neq \bar{h}(\bar{y})(\omega)$ has measure zero.

It suffices to consider the case $\bar{h}(\bar{x})(\omega) = 0, \bar{h}(\bar{y})(\omega) = 1$, i.e., to show that

$$\|\bar{x} = \bar{y}\| \wedge \|\bar{h}(\bar{x}) = 0\| \wedge \|\bar{h}(\bar{y}) = 1\| = 0.$$

We write the second term in the form $\bigvee_{j \in \mathcal{J}} \|\bar{x} = \bar{x}_j\|$ (by 3.7(b)) and apply the distributive axiom to the first and second terms (where we use the completeness of B). We further use the fact that $\|\bar{x} = \bar{y}\| \wedge \|\bar{x} = \bar{x}_j\| \leqslant \|\bar{y} = \bar{x}_j\|$. We then obtain

$$\|\bar{x} = \bar{y}\| \wedge \|\bar{h}(\bar{x}) = 0\| \leqslant \bigvee_{j \in \mathcal{J}} \|\bar{y} = \bar{x}_j\| = \|\bar{h}(\bar{y}) = 0\|,$$

which immediately gives us the required result. □

Explanation. Since the choice of $\bar{h}$ is the essential step in the proof, we would like to give some motivation for this choice. Recall that h is the name of the function the cardinality of whose set of zeros interests us. We choose a concrete $\bar{h}$ to "disprove" the CH in such a way that the "almost everywhere zeros" of $\bar{h}$ include the elements of the set $\{x_j | j \in \mathcal{J}\}$, which has intermediate cardinality in the naive sense of the word (compare with §1). However, $\bar{h}$ cannot be an arbitrary map from $\overline{R}$ to $\overline{R}$; it must satisfy the strong condition $\bar{h} \in \overline{R}^{(1)}$. Hence, along with all the $\bar{x}_j$, the almost everywhere zeros of $\bar{h}$ might also have to include various other $\bar{y} \in \overline{R}$, and might have to "partly include" still other $\bar{z} \in \overline{R}$. We say "partly include" to convey the possibility that $\|\bar{h}(\bar{z}) = 0\|$ is neither 0 nor 1, so that $\bar{z}$ has a "certain probability" of being a zero of $\bar{h}$.

Thus, the "set of zeros" of $\bar{h}$ might be bigger than we want, and we might expect to encounter difficulties in proving that this set cannot be mapped onto all of $\overline{R}$ (the alternative P_1). On the other hand, it would seem that this situation would make it trivial to disprove the alternative P_2 (mapping Z onto the entire set of zeros). But even this is wrong! As we noted before, we can have $\|Z(\bar{x})\| = 1$ for many $\bar{x}$ that are not constant integer functions on Ω. Moreover, for still other $\bar{x}$ we have $\|Z(\bar{x})\| \neq 0, 1$, so that the "set of integers" in our model has grown considerably.

A final remark: In this discussion we have been essentially dealing with the concept of a "B-random set," which will be a central idea in what follows (see §4). That is, the "set of zeros of $\bar{h}$" is random in the sense that for each $\bar{z} \in \overline{R}$, the assertion "$\bar{z} \in$ (zeros of $\bar{h}$)" is naturally assigned the Boolean truth value $\|\bar{h}(\bar{z}) = 0\|$.

We now return to the proof that $\|\mathrm{CH}\| = 0$.

3.8. Proof that $\|P_1\|(\xi') = 0$. By the rules for computing truth functions, we obtain

$$\|P_1\|(\xi') = \bigvee_{\bar{g}} \bigwedge_{\bar{y}} \bigvee_{\bar{x}} \left\{ \|\bar{h}(\bar{x}) = 0\| \wedge \|\bar{y} = \bar{g}(\bar{x})\| \right\},$$

where $\bar{h}$ was defined above, $\bar{g}$ runs through all elements of $\overline{R}^{(1)}$, and $\bar{x}$ and $\bar{y}$ run through all elements of $\overline{R}$. We suppose that $\|P_1\|(\xi') \neq 0$, and show that this leads to a contradiction. We write the above formula for $\|P_1\|(\xi')$ as $\bigvee_{\bar{g}} a(\bar{g})$.

If $\|P_1\|(\xi') \neq 0$, then $a(\bar{g}) \neq 0$ for some concrete function $\bar{g} \in \overline{R}^{(1)}$. We take this function $\bar{g}$ and set

$$a = \bigwedge_{\bar{y}} \bigvee_{\bar{x}} \Big(\bigvee_{j \in \mathcal{J}} \|\bar{x} = \bar{x}_j\| \wedge \|\bar{y} = \bar{g}(\bar{x})\| \Big).$$

Here we have substituted $\bigvee_{j\in\mathcal{J}} \|\bar{x} = \bar{x}_j\|$ for $\|\bar{h}(\bar{x}) = 0\|$ using 3.7(b). Furthermore, we obtain $\|\bar{x} = \bar{x}_j\| \wedge \|\bar{y} = \bar{g}(\bar{x})\| \leqslant \|\bar{y} = \bar{g}(\bar{x}_j)\|$. Using this and distributivity, we obtain

$$a \leqslant \bigwedge_{\bar{y}} \bigvee_{j\in\mathcal{J}} \|\bar{y} = \bar{g}(\bar{x}_j)\|.$$

In particular, for each $\bar{x}_i$ in place of $\bar{y}$, we have

$$a \leqslant \bigvee_{j\in\mathcal{J}} \|\bar{x}_i = \bar{g}\left(\bar{x}_j\right)\|.$$

If, as we have supposed, $a \neq 0$, then for each i there exists a $j(i) \in \mathcal{J}$ such that

$$\|\bar{x}_i = \bar{g}\left(\bar{x}_{j(i)}\right)\| \neq 0.$$

Since I is uncountable and card $\mathcal{J}$ < card I, it follows that there exists a $j_0 \in \mathcal{J}$ such that $j_0 = j(i)$ for all i in an uncountable subset $I_0 \subset I$. But this contradicts the countable chain condition on B, because the terms in the family $\|\bar{x}_i = \bar{g}(\bar{x}_{j_0})\| (i \in I_0)$ are pairwise disjoint. In fact,

$$\|\bar{x}_{i_1} = \bar{g}\left(\bar{x}_{j_0}\right)\| \wedge \|\bar{x}_{i_2} = \bar{g}\left(\bar{x}_{j_0}\right)\| \leqslant \|\bar{x}_{i_1} = \bar{x}_{i_2}\| = 0$$

if $i_1 \neq i_2$. □

Notice to what extent this proof parallels the "naive" argument in §1. By assumption, the function $\bar{y}$ maps the zeros of $\bar{h}$ onto $\overline{R}$ "with nonzero probability." But the exact meaning of the computations cannot readily be stated in words.

Computation of $\|Z(y)\|$. The formula for $Z(y)$, "y is an integer," was given in 2.3. Since this formula occurs in P_2, we must compute $\|Z(y)\|$ in order to compute $\|P_2\|$.

3.9. Lemma. *Let* $\eta \in \overline{M}$ *and* $y^\eta = y \in \overline{R}$. *Then*

$$\|Z(y)\|(\eta) = \bigvee_{n\in Z} \|\bar{y} = n\| = \{\omega \in \Omega | \bar{y}(\omega) \in Z\} \bmod 0.$$

PROOF. We must show that

$$\bigwedge_{\bar{f}} \left(\|\bar{f}(0) = 0\|' \vee \left(\bigvee_{\bar{x}} \|\bar{f}(\bar{x}) = \bar{f}(\bar{x}+1)\|' \right) \bigvee \|\bar{f}(\bar{y}) = 0\| \right) = \bigvee_{n\in Z} \|\bar{y} = n\|.$$

We prove this equality by proving inequality in both directions.

The inequality $\leqslant$. It suffices to find a concrete function $\bar{f} \in \overline{R}^{(1)}$ for which the corresponding term on the left is contained in the right-hand side. We define $\bar{f}$ by setting $\bar{f}(\bar{x})(\omega) = \sin^2 \pi\bar{x}(\omega)$ (here, instead of $\sin^2 \pi z$, we could take any measurable function with period 1 and zeros only at the integers). It is easy to see that $\bar{f}(\bar{x}) \in \overline{R}$ and $\bar{f} \in \overline{R}^{(1)}$. Then $\|\bar{f}(0) = 0\|' = 0$ and $\|\bar{f}(\bar{x}) = \bar{f}(\bar{x}+1)\|' = 0$. Hence we need only verify that

$$\|\sin^2 \pi\bar{y} = 0\| \leqslant \bigvee_{n\in Z} \|\bar{y} = n\|,$$

and this is obvious.

The inequality $\geqslant$. It suffices to show that for any fixed values of $n \in Z$, $\bar{f} \in \overline{R}^{(1)}$ and $\bar{y} \in \overline{R}$, we have

$$\|\bar{y} = n\| \leqslant b \vee c,$$

where

$$b = \|\bar{f}(0) = 0\|' \vee \left(\bigvee_{\bar{x}} \|\bar{f}(\bar{x}) = \bar{f}(\bar{x}+1)\|' \right); \quad c = \|\bar{f}(\bar{y}) = 0\|.$$

But the inclusion $a \leqslant b \vee c$ is equivalent to $a \wedge c' \leqslant b$. Furthermore, in our situation we have

$$a \wedge c' = \|\bar{y} = n\| \wedge \|\bar{f}(\bar{y}) = 0\|' \leqslant \|\bar{f}(n) = 0\|'.$$

(Here n in $\bar{f}(n)$ is the constant random variable that is everywhere equal to n.)

It is thus sufficient to see that

$$\|\bar{f}(n) = 0\|' \leqslant \|\bar{f}(0) = 0\|' \vee \left(\bigvee_{\bar{x}} \|\bar{f}(\bar{x}) = \bar{f}(\bar{x}+1)\|' \right),$$

or, taking complements, that

$$\|\bar{f}(n) = 0\| \geqslant \|\bar{f}(0) = 0\| \wedge \left(\bigwedge_{\bar{x}} \|\bar{f}(\bar{x}) = \bar{f}(\bar{x}+1)\| \right).$$

The right side can become larger only if we only take the intersection over the terms with $\bar{x} = 0, 1, 2, \ldots, n-1$. But this obviously gives

$$\|\bar{f}(0) = 0\| \wedge \|\bar{f}(0) = \bar{f}(1) = \cdots = \bar{f}(n)\| \leqslant \|\bar{f}(n) = 0\|. \qquad \square$$

3.10. Proof that $\|P_2\|(\xi') = 0$. Using Lemma 3.9 and the rules for computing truth functions, we find that

$$\|P_2\|(\xi') = \bigvee_{\bar{f}} \bigwedge_{\bar{y}} \left(\|\bar{h}(\bar{y}) = 0\|' \vee \bigvee_{\bar{x}} \left(\bigvee_{n} \|\bar{x} = n\| \wedge \|\bar{y} = \bar{f}(\bar{x})\| \right) \right).$$

Since $\bar{f} \in \overline{R}^{(1)}$ we have $\|\bar{x} = n\| \leqslant \|\bar{f}(\bar{x}) = \bar{f}(n)\|$, so that $\|\bar{x} = n\| \wedge \|\bar{y} = \bar{f}(\bar{x})\| \leqslant \|\bar{y} = \bar{f}(n)\|$.

Now it suffices to prove that the term corresponding to any concrete choice of $\bar{f}$ is equal to 0. We suppose that this is not the case, and show that we obtain a contradiction. Let $a \neq 0$ be the term corresponding to $\bar{f}$. By the previous paragraph, we have

$$a \leqslant \bigwedge_{\bar{y}} \left(\|\bar{h}(\bar{y}) = 0\|' \vee \bigvee_{n} \|\bar{y} = \bar{f}(n)\| \right).$$

In particular, for every $j \in \mathcal{J}$ we must have (with $\bar{x}_j$ in place of $\bar{y}$)

$$a \leqslant \bigvee_{n} \|\bar{x}_j = \bar{f}(n)\|$$

(where we have $\|\bar{h}(\bar{x}_j) = 0\|' = 0$ by 3.7(b)). Hence, for every j there exists an integer $n(j)$ such that $0 \neq \|\bar{x}_j = \bar{f}(n(j))\|$. Since $\mathcal{J}$ is uncountable, there exist an n_0 and an uncountable subset $\mathcal{J}_0 \subset \mathcal{J}$ such that $n(j_0) = n_0$ for all $j_0 \in \mathcal{J}_0$. Then the $\|\bar{x}_j = \bar{f}(n_0)\|$ for $j \in \mathcal{J}_0$ form an uncountable set of pairwise disjoint nonzero elements of B. This contradicts the countable chain condition on B.

4 Boolean-Valued Universes

4.1. In this section we fix a complete Boolean algebra B (see 2.6) and construct the universe V^B of "B-random sets." It will be a model for the Zermelo–Fraenkel axioms in the same generalized sense in which the random variables $\overline{R}$ were a model for the real numbers $\mathbf{R}$ in §3. In §§5–7 we verify that all the axioms of L$_1$Set are "true," and then in §8 we verify that the continuum hypothesis is "false" for a suitable choice of B.

The objects of V^B will be denoted by capital letters $X, Y, Z, \ldots$. Any two objects determine elements $\|X \in Y\| \in B$ and $\|X = Y\| \in B$. The intuitive meaning, say, of the first of these is as follows: if B is the algebra of measurable sets in a probability space, then $\|X \in Y\|$ is the maximal set on which "X is an element of Y with probability one." Since we do not deal with probability measures in the general case, we shall simply call the elements of B "probabilities," and then $\|X \in Y\|$ is simply the probability that X belongs to Y.

It is not trivial to construct precise definitions, because we want the axiom of extensionality to be "true." If a random set must be uniquely determined by its elements (which are also random), even in a generalized sense, then this random set cannot be "too" random (see 4.3).

We shall assume that as a set B is an element of the von Neumann universe V. Then all the objects of V^B will also be elements of V, and all our constructions can be expressed in L$_1$Set. In principle, this allows us to take a more formalistic point of view than we shall in fact take. The proof given below of the independence of the CH could then be used as a guide for constructing a much more syntactic version, based on an "internal interpretation" of the language L$_1$Set in itself. In this context the assumption that the Zermelo–Fraenkel axioms are consistent in the statement of Theorem 1.6 becomes a necessary precaution, since (by Gödel's result) this consistency cannot be established using only the language L$_1$Set itself. However, in our treatment this condition is pure hypocrisy, since by assuming the "existence" of the universe V, which is a model for the axioms, we automatically "prove" that those axioms are consistent (see Section 18 of the appendix to Chapter II).

4.2. *Construction of* V^B. For every ordinal α we construct the set V_α^B by transfinite recursion, and then set $V^B = \cup_\alpha V_\alpha^B$. The first step is $V_0^B = \varnothing$.

Inductive assumption. The set V_α^B is defined for the ordinal $\alpha \geqslant 0$; for every element $X \in V_\alpha^B$ the set $D(X) \subset V_\alpha^B$ is defined (its intuitive meaning will be explained below); for every pair of elements $X, Y \in V_\alpha^B$ the "Boolean truth functions"

$$\|X \in Y\| \in B, \qquad \|X = Y\| \in B$$

are defined (intuitively, they should be thought of as the "probability that X is an element of Y" and the "probability that X coincides with Y," respectively).

By assumption, this data satisfies the following conditions:

$$(a)\ \text{If } \beta_1 \leqslant \beta_2 \leqslant \alpha_1, \text{ then } V^B_{\beta_1} \leqslant V^B_{\beta_2}.$$

$$(b)\ \text{If } \beta < \alpha \text{ and } X \in V^B_{\beta+1} \backslash V^B_\beta, \text{ then } D(X) = V^B_\beta. \qquad (1)_\alpha$$

$$(c_1)\ \|X \in Y\| = \bigvee\nolimits_{Z\in D(Y)} (\|X = Z\| \wedge \|Z \in Y\|)$$

(the condition $(1)_\alpha$ expresses the requirement that the formula $x \in y \Leftrightarrow \exists z(x = z \wedge z \in y)$, which is easily deduced from the Zermelo–Fraenkel axioms, must be "true").

$$(c_2)\ \|X = Y\| = \left(\bigwedge_{Z\in D(X)} \|Z \in X\|' \vee \|Z \in Y\|\right) \wedge \left(\bigwedge_{Z\in D(Y)} \|Z \in Y\|' \vee \|Z \in X\|\right) \qquad (2)_\alpha$$

(this condition expresses the "truth" of the formula $x = y \Leftrightarrow (\forall z\ (z \in x \rightarrow z \in y) \wedge \forall z\ (z \in y \Rightarrow z \in x))$. We note that it is not completely clear at this point why, for example, in $(1)_\alpha$ we took the union only over Z in $D(Y)$; it would seem natural to take all Z. Later we shall see that the formula remains true if we take the Boolean union over all Z.

This completes the description of the data for V^B_α. We now give explicitly the recursive construction of $V^B_{\alpha+1}$ and the corresponding data.

Definition of $V^B_{\alpha+1}$ and D. We set $V^B_{\alpha+1} = V^B_\alpha \cup V^{B^*}_{\alpha+1}$, where $V^{B^*}_{\alpha+1}$ consists of all possible functions Z with domain of definition V^B_α and range of values $\subset B$ that satisfy the following "extensionality condition":

$$\|X = Y\| \wedge Z(X) = \|X = Y\| \wedge Z(Y), \quad \text{for all } X, Y \in V^B_\alpha. \qquad (3)$$

A little later we shall define $\|X \in Z\| = Z(X)$ for $X \in V^B_\alpha$ and $Z \in V^B_{\alpha+1} \backslash V^B_\alpha$. Thus, as before, (3) can be thought of as reflecting the formula

$$(x = y \wedge x \in z) \Leftrightarrow (x = y \wedge y \in z).$$

Compare also with the comment in 2.7 concerning the definition of $\overline{R}^{(1)}$.

We shall call the elements of $V^B_{\alpha+1} \backslash V^B_\alpha$ *new* elements (of rank $\alpha + 1$), and we shall call the elements of V^B_α *old* elements. We set $D(Z) = V^B_\alpha$ if Z is a new element.

Definition of the Boolean truth functions. These functions have already been defined for pairs of old elements. We further set

$$\|X \in Y\| = Y(X), \text{ if } X \text{ is old and } Y \text{ is new;} \qquad (4)$$

$$\|X = Y\| = \left(\bigwedge_{Z \in D(X)} \|Z \in X\|' \vee \|Z \in Y\| \right)$$
$$\wedge \left(\bigwedge_{Z \in D(Y)} \|Z \in Y\|' \vee \|Z \in X\| \right). \qquad (5)$$

Because of $(2)_\alpha$, (5) automatically holds if X and Y are both old elements; in the other cases, (5) uniquely determines $\|X = Y\|$ if we use (4) and the fact that Z runs only through old elements in (5). Finally, we set

$$\|X \in Y\| = \bigvee_{Z \in D(Y)} \|X = Z\| \wedge \|Z \in Y\| \qquad (6)$$

if X is a new element and Y is either new or old. The right side is uniquely determined using (4) and (5), since $D(Y) \subset V_\alpha^B$.

Formulas (4) and (6) show the following. As a first approximation we might say that a random set Y of rank α "consists" of sets Z of lower rank that occur in Y with probability $Y(Z)$; these probabilities can be chosen rather arbitrarily, subject only to the extensionality condition (3).

However, we then find (in formula (6) for new X and old Y) that we must automatically "include" more and more elements X in Y with probabilities already assigned by formula (6). It is conditions (3) and (6) that prevent our sets from being completely random.

Definition of V_α^B and other data for limiting ordinals α. We simply set $V_\alpha^B = \cup_{\beta<\alpha} V_\beta^B$, and then all the other data has already been determined.

4.3. *Verification that the definitions are correct.* Properties 4.2 (a) and (b) are obviously preserved in going from α to $\alpha+1$; we must verify $(1)_{a+1}$ and $(2)_{a+1}$. Now the only identity here that is not completely obvious is obtained by taking X old and Y new in $(1)_{a+1}$:

$$Y(X) = \bigvee_{Z \in V_\alpha^\beta} \|X = Z\| \wedge Y(Z).$$

This is verified as follows. We obtain $\geqslant$ by writing the right-hand side in the form $\bigvee_Z \|X = Z\| \wedge Y(X)$ using (3). We obtain $\leqslant$ by considering the term with $Z = X$ and taking into account that $\|X = X\| = 1$ for all X (as follows immediately from (5)).

This completes the construction of the Boolean-valued universe.

4.4. Examples and remarks. We examine some special cases of these constructions in order to clarify their structure.

(a) Obviously $V_1^B = \{\varnothing\}$, since there exists a unique "empty" function whose domain of definition is the subset $V_0^B = \varnothing$. We compute $V_2^B = V_1^B \cup V_2^{B^*}$.

We let $\{\varnothing\}_b \in V_2^{B^*}$ denote the function of the one-element set V_1^B that takes the value $b \in B$. All these functions are extensional, so that

$$V_2^B = \{\varnothing, \{\varnothing\}_b, \text{ for all } b \in B\}.$$

It follows from (4) that

$$\|\varnothing \in \{\varnothing\}_b\| = b.$$

It is clear from (5) that

$$\|\varnothing = \{\varnothing\}_b\| = b'.$$

Intuitively, these formulas mean that $\{\varnothing\}_b$ consists of one element $\varnothing$ "over b" and is empty away from b. Again applying (5), we obtain

$$\|\{\varnothing\}_a = \{\varnothing\}_b\| = (a' \vee b) \wedge (a \vee b') = (a \wedge b) \vee (a' \wedge b').$$

Thus, $\{\varnothing\}_a$ and $\{\varnothing\}_b$ coincide when either they are both empty or they both consist of one element $\varnothing$: this agrees with intuition. Now applying (6), we obtain

$$\|\{\varnothing\}_a \in \{\varnothing\}_b\| = \|\{\varnothing\}_a = \varnothing \,\| \wedge \|\varnothing \in \{\varnothing\}_b\| = a' \wedge b$$

(i.e., the only possible inclusion, which has the form $\varnothing \in \{\varnothing\}$, holds when $\{\varnothing\}_a$ is empty and $\{\varnothing\}_b$ is nonempty).

Finally, let $X \in V_3^{B^*}$ be an extensional function on the subset V_2^B with values in B. Then, by (6),

$$\|X \in \{\varnothing\}_b\| = \|X = \varnothing\| \wedge \|\varnothing \in \{\varnothing\}_b\| = \|X = \varnothing\| \wedge b,$$

and by (5),

$$\begin{aligned}\|X = \varnothing\| &= \left(\bigwedge_{a \in B} \|\{\varnothing\}_a \in X\|'\right) \wedge \|\varnothing \in X\|' \\ &= \left(\bigvee_{a \in B} \|\{\varnothing\}_a \in X\| \vee \|\varnothing \in X\|\right)'.\end{aligned}$$

Thus, intuitively, $\|X = \varnothing\|$ means the complement of the support of X in B, and $\|X \in \{\varnothing\}_b\|$ is the set where both X is empty and $\{\varnothing\}_b$ is nonempty, which again agrees with the usual formula $\varnothing \in \{\varnothing\}$. This shows how new objects X can be random elements of old objects with nonzero probabilities.

(b) We consider the case $B = \{0,1\}$. The corresponding probability space consists of one point, so our random sets become completely determined. What happens is this: the universe V^B maps naturally onto the von Neumann universe V in such a way that if $\tilde{X}$ denotes the image of $X \in V^B$, then all X and Y satisfy the conditions

$$\|X \in Y\| = 1 \Leftrightarrow \tilde{X} \in \tilde{Y},$$
$$\|X = Y\| = 1 \Leftrightarrow \tilde{X} = \tilde{Y}.$$

To construct this map we first set $\tilde{\varnothing} = \varnothing$. We now suppose that the map $V_\alpha^{\{0,1\}} \to V_\alpha$ has already been constructed with the required properties, and we extend the map to $\alpha + 1$. To do this, for any new element $X \in V_{\alpha+1}^{\{0,1\}}$ we first find the subset of $V_\alpha^{\{0,1\}}$ on which X takes the value 1, and we then take the image of this subset in V_α, which is an element $\tilde{X}$ of $\mathcal{P}(V_\alpha) = V_{\alpha+1}$; by definition, our map takes X to this $\tilde{X}$. We leave the verification of the properties of this map to the reader.

(c) *Boolean truth functions for the formulas in* $\mathrm{L_1Set}$.

We define these truth functions in an analogous manner to §2. We introduce the interpretation class $\overline{M}$: each point $\xi \in \overline{M}$ assigns to every variable symbol x in $\mathrm{L_1Set}$ some object $x^\xi = X$ of the universe V^B. We further assume that every point ξ maps the symbol $\varnothing$ in $\mathrm{L_1Set}$ to the empty set.

If P is the atomic formula $x \in y$ or $x = y$ in $\mathrm{L_1Set}$, then $\|P\|(\xi)$ is defined to be $\|x^\xi \in y^\xi\| \in B$ or $\|x^\xi = y^\xi\| \in B$, respectively. The value of $\|P\|(\xi)$ for all other P is defined inductively using exactly the same formulas as in Section 2.7. We need only note that although the expressions $\bigvee_\xi a_\xi$ and $\bigwedge_\xi a_\xi$ must be taken over families indexed by the *class* $\overline{M}$ when we compute with quantifiers, all the *different* elements of such a family form a *subset* of B, so that such an expression makes sense. We shall call a formula P "true" (in the model V^B) if $\|P\|(\xi) = 1$ for all ξ, and we shall call P "false" if $\|P\|(\xi) = 0$ for all ξ.

As in §3 of Chapter II, it can be verified that all the tautologies and logical quantifier axioms are "true" and that the rules of deduction preserve "truth." Hence, it remains for us to show that the Zermelo–Fraenkel axioms are "true" (for any B) and that the continuum hypothesis is "false" (for suitable B).

5 The Axiom of Extensionality Is "True"

We begin by proving some relations between the truth functions. First of all, it is clear from formula (5) in §4 that $\|X = Y\| = \|Y = X\|$ and $\|X = X\| = 1$. The following lemma is a less immediate consequence of the formulas.

5.1. **Lemma.** *For any* $X, Y, Z \in V^B$ *we have*

$$\|X = Y\| \wedge \|Y = Z\| \leqslant \|X = Z\|, \tag{I}$$

$$\|X = Y\| \wedge \|Y \in Z\| \leqslant \|X \in Z\|, \tag{II}$$

$$\|X \in Y\| \wedge \|Y = Z\| \leqslant \|X = Z\|. \tag{III}$$

PROOF.

(a) (III) *holds if* $X \in D(Y)$. In fact, then by formula (5) in §4,

$$\|Y = Z\| \leqslant \|X \in Y\|' \vee \|X \in Z\|,$$

so that if we intersect both sides with $\|X \in Y\|$, we obtain (III).

(b) (III) *holds if* $X, Y \in V_\alpha^B$ and Z is a new element of $V_{\alpha+1}^B$. In fact, we choose $U \in D(Y)$ and apply the special case of (III) proved in (a):

$$\|U \in Y\| \wedge \|Y = Z\| \leqslant \|U \in Z\|.$$

We take the Boolean intersection of both sides with $\|X = U\|$ and then the Boolean sum over all $U \in D(Y)$. Now applying formula (6) in §4 to the left-hand side and using distributivity, we obtain

$$\begin{aligned}\|X \in Y\| \wedge \|Y = Z\| &\leqslant \bigvee_{U \in D(Y)} \|X = U\| \wedge \|U \in Z\| \\ &\leqslant \bigvee_{U \in D(Z) = V_\alpha^B} \|X = U\| \wedge \|U \in Z\| = \|X \in Z\|.\end{aligned}$$

(c) (I) *holds in* $V_{\alpha+1}^B$ *if* (III) *holds in* V_α^B. We consider an element $U \in D(X) \in V_\alpha^B$. By (a), we have

$$\|U \in X\| \wedge \|X = Y\| \leqslant \|U \in Y\|.$$

We take the Boolean intersection with $\|Y = Z\|$:

$$\|U \in X\| \wedge \|X = Y\| \wedge \|Y = Z\| \leqslant \|U \in Y\| \wedge \|Y = Z\|.$$

Here the right side is always $\leqslant \|U \in Z\|$. In fact, if $Y \in V_\alpha^B$ this follows by part (b) or by the induction assumption, and if Y is a new element of $V_{\alpha+1}^B$ then it follows by part (a).

We have thus shown that for all $X, Y, Z \in V_{\alpha+1}^B$ and all $U \in D(X)$,

$$\|U \in X\| \wedge \|X = Y\| \wedge \|Y = Z\| \leqslant \|U \in Z\|$$

Because $a \wedge b \leqslant c$ implies $b \leqslant \|a' \vee c\|$ in any Boolean algebra, we then obtain

$$\|X = Y\| \wedge \|Y = Z\| \leqslant \|U \in X\|' \vee \|U \in Z\|,$$

and hence

$$\|X = Y\| \wedge \|Y = Z\| \leqslant \bigwedge_{U \in D(X)} \|U \in X\|' \vee \|U \in Z\|.$$

Interchanging X and Z, we find that for all $U \in D(Z)$,

$$\|Z = Y\| \wedge \|Y = X\| \leqslant \bigwedge_{U \in D(Z)} \|U \in Z\|' \vee \|U \in X\|.$$

These last two formulas, together with (5), clearly imply (I).

(d) (II) holds in $V_{\alpha+1}^B$ if (I) holds in $V_{\alpha+1}^B$. In fact, let $U \in D(Z)$. By (I), we have

$$\|X = Y\| \wedge \|Y = U\| \leqslant \|X = U\|.$$

We take the Boolean intersection with $\|U \in Z\|$ and then the Boolean sum over all $U \in D(Z)$:

$$\|X = Y\| \wedge \left(\bigvee_{U \in D(Z)} \|U \in Z\| \wedge \|Y = U\| \right) \leqslant \bigvee_{U \in D(Z)} \|Z = U\| \wedge \|U \in Z\|.$$

Applying $(1)_{\alpha+1}$ in §4, we obtain (II).

(e) (III) holds in $V^B_{\alpha+1}$ if (II) holds in $V^B_{\alpha+1}$. In fact, let $U \in D(Y)$. By part (a), we have

$$\|U \in Y\| \wedge \|Y = Z\| \leqslant \|U \in Z\|.$$

Intersecting with $\|X = U\|$ and applying (II) to the right-hand side, we obtain

$$\|X = U\| \wedge \|U \in Y\| \wedge \|Y = Z\| \leqslant \|X \in Z\|.$$

Finally, if we take the Boolean sum over all $U \in D(Y)$ and use formula (1) in §4, we obtain (III). □

Obviously, parts (a)–(e) prove the inductive step for α to $\alpha+1$. We are now in a position to establish the basic result of this section.

5.2. **Proposition.** *The axiom of extensionality*

$$x = y \Leftrightarrow \forall z (z \in x \Leftrightarrow z \in y)$$

is "true."

PROOF. The formula $\|P \Leftrightarrow Q\|(\xi) = 1$ is equivalent to $\|P\|(\xi) = \|Q\|(\xi)$. It is therefore sufficient to prove that for all $X, Y \in V^B$,

$$\|X = Y\| = \bigwedge_{Z \in V^B} (\|Z \in X\| \vee \|Z \in Y\|') \wedge (\|Z \in X\|' \vee \|Z \in Y\|).$$

The inequality $\geqslant$ follows immediately from formula (2) in §4. To obtain the opposite inequality, we write two obvious corollaries of formula (III) in Lemma 5.1:

$$\|X = Y\| \leqslant \|Z \in X\| \vee \|Z \in Y\|',$$
$$\|X = Y\| \leqslant \|Z \in X\|' \vee \|Z \in Y\|,$$

and we take the intersection over all Z. The proposition is proved. □

We note that formula (2) implies the following general extensionality property: for all $X, Y, Z \in V^B$,

$$\|X = Y\| \wedge \|Y \in Z\| = \|X = Y\| \wedge \|X \in Z\|.$$

5.3. **Corollary.** *The axioms of equality in* $\mathrm{L_1Set}$ *are "true."*

In fact (see Proposition 4.6 in Chapter II), the axioms of equality in our case consist of the "true" formula $x = x$, the axiom of extensionality (in the form $x = y \Rightarrow (P(x) \Rightarrow P(y))$ with $P(x) = z \in x$), and the "true" formula $x = y \Rightarrow (x \in z \Rightarrow y \in z)$ (in which $P(x) = x \in z$), since the only atomic formulas $P(x)$ in $\mathrm{L_1Set}$ are $z \in x$ and $x \in z$. □

5.4. *Remark.* In most computations, we shall need to know only the values of $\|X \in Y\|$ and $\|X = Y\|$, and not the precise definition of the objects X and Y. In this connection, we note that the following two binary relations on V^B coincide (as easily follows from (III) and the axiom of extensionality):

$$(a)\ \|X = Y\| = 1,$$

$$(b)\ \forall Z \in V^B, \quad \|Z \in X\| = \|Z \in Y\|.$$

We shall call such X and Y *equivalent* and write $X \sim Y$.

6 The Axioms of Pairing, Union, Power Set, and Regularity Are "True"

6.1. The computations in the previous section show that the basic work in ensuring that the axiom of extensionality is "true" was already incorporated into the definition of the universe V^B. The explicit formulas for recursively computing $\|X \in Y\|$ and $\|X = Y\|$ reflected so many special properties of inclusion and equality that together they guaranteed that the general axiom must hold.

In order to verify several of the other axioms, we must essentially define in V^B analogues of certain operations in V, such as forming the unordered pair and the set of subsets. These operations can be defined by means of formulas in $\mathrm{L_1Set}$. However, recall that if $P(x)$ is a formula with one free variable x, then the $x^\xi \in V$ for which $P(x)(\xi)$ is true generally form a class and not a set.

It will be convenient to introduce the auxiliary notion of a "random class" in V^B. Using this concept, we shall often construct the operations in V^B in two stages: the value of the operation will at first be a random class, which we then "identify" with a random set using a separate argument.

6.2. **Definition.**

(a) A random class is any function W on V^B with values in B that satisfies the following extensionality condition:

$$W(X) \wedge \|X = Y\| = W(Y) \wedge \|X = Y\|, \quad \text{for all } X, Y \in V^B.$$

(b) A random class W is said to be equivalent to a random set $Z \in V^B$ (written $W \sim Z$) if

$$W(X) = \|X \in Z\|, \quad \text{for all } X \in V^B.$$

6.3. EXAMPLES AND REMARKS

(a) For any random set Z the function $X \mapsto \|X \in Z\|$ is extensional by (II), §5, and so is a random class. By analogy, we often write $\|X \in W\|$ instead of $W(X)$ if W is any random class.

(b) There exist random classes that are not equivalent to random sets. One such example is the "universal" random class $W(X) = 1$ for all X. (If W were a set, we would have $\|W \in W\| = 1$, contradicting the regularity axiom, which will be shown to be "true" below.)

(c) Let W be a random class, and let α be any ordinal. We define the element $W_\alpha \in V^B_{\alpha+1}$ as follows: $D(W_\alpha) = V^B_\alpha, W_\alpha =$ the restriction of W to V^B_α (as a function; see 4.2). It is easy to see that for all $X \in V^B$ we have

$$\|X \in W_\alpha\| \leqslant \|X \in W\|. \tag{1}$$

In fact, let $U \in V^B_\alpha$ and $X \in V^B$. We then have

$$\|X = U\| \wedge W_\alpha(U) = \|X = U\| \wedge W(U) = \|X = U\| \wedge W(X) \leqslant W(X),$$

so that by (6), §4,

$$\|X \in W_\alpha\| = \bigvee_{U \in V^B_\alpha} \|X = U\| \wedge W_\alpha(U) \leqslant W(X) = \|X \in W\|.$$

We shall often show that some class W in which we are interested is equivalent to a set by finding an ordinal α such that $W \sim W_\alpha$. It is clear from (1) that this follows if $\|X \in W\| \leqslant \|X \in W_\alpha\|$ for all X.

(d) Let W, W_1, and W_2 be random classes. Then $W', W_1 \wedge W_2$, and $W_1 \vee W_2$ are also random classes, since the extensionality condition is trivially verified for these functions. We shall write $W_1 \cap W_2$ and $W_1 \cup W_2$ instead of $W_1 \wedge W_2$ and $W_1 \vee W_2$, respectively.

(e) Let W be a random class, and let X be a random set. We show that $W \cap X$ is equivalent to a random set. More precisely, if $D(X) = V^B_\alpha$, then $W \cap X \sim (W \cap X)_\alpha$. In fact, for any $Y \in V^B$ it follows by (6), §4, that

$$\begin{aligned}\|Y \in (W \cap X)_\alpha\| &= \bigvee_{U \in V^B_\alpha} \|U = Y\| \wedge \|U \in (W \cap X)_\alpha\| \\ &= \bigvee_{U \in V^B_\alpha} (\|U = Y\| \wedge \|U \in W\|) \wedge \|U \in X\| \\ &= \bigvee_{U \in V^B_\alpha} \|U = Y\| \wedge \|Y \in W\| \wedge \|U \in X\| \\ &= \|Y \in W\| \wedge \|Y \in X\| = \|Y \in W \cap X\|.\end{aligned}$$

This result implies that the separation axioms are "true" (see Section 4.9(b) of Chapter II).

The following proposition gives a general method for constructing random classes.

6.4. **Proposition.** *Let $P(x, y_1, \dots, y_n)$ be a formula that does not contain any free variables besides $x, y_1, \dots, y_n$. Let $Y_1, \dots, Y_n \in V^B$ be fixed. Then the function*

$$X \mapsto W(X) = \|P(X, Y_1, \dots, Y_n)\|$$

is a random class.

Intuitively, W contains every set X with probability equal to the probability that $P(X, \dots, Y_n)$ is true. $Y_1, \dots, Y_n$ play the role of "constants."

PROOF. We use the "truth" of the following axiom of equality:

$$\|\forall x\ \forall y_1 \cdots \forall y_n (x = y \Rightarrow (P(x, y_1, \dots, y_n) \Rightarrow P(y, y_1, \dots, y_n)))\| = 1.$$

If we take a point ξ in the interpretation class that assigns to $x, y, y_1, \dots, y_n$ the values $X, Y, Y_1, \dots, Y_n$, respectively, then we find that

$$\|X = Y\| \leqslant \|P(X, Y_1, \dots, Y_n)\|' \vee \|P(Y, Y_1, \dots, Y_n)\|,$$

or

$$\|X = Y\| \wedge W(X) \leqslant W(Y),$$

so that W is extensional. □

We are now ready to verify the axioms.

6.5. **Proposition.** *The axiom of pairing*

$$\forall u\ \forall w\ \exists x\ \forall z (z \in x \Leftrightarrow z = u \vee z = w)$$

is "true."

PROOF. By definition we have

$$\|\forall u\ \forall w\ \exists x\ \forall z (z \in x \Leftrightarrow z = u \vee z = w)\|$$
$$= \bigwedge_U \bigwedge_W \bigvee_X \bigwedge_Z \|Z \in X \Leftrightarrow Z = U \vee Z = W\|.$$

Hence to prove the theorem if suffices if for any $U, W \in V^B$, we find an $X \in V^B$ such that for all $Z \in V^B$,

$$\|Z \in X\| = \|Z = U\| \vee \|Z = W\|. \tag{2}$$

For fixed U and W we consider the right side of (2) as a function of Z. This function is a random class X by Proposition 6.4, since it corresponds to the formula $z = U \vee z = W$. We show that it is equivalent to a random set; more precisely, if $U, W \in V_\alpha^B$, then $X \sim X_\alpha$. By the remark at the end of 6.3(c), it suffices to verify that for all Z

$$\|Z \in X\| \leqslant \|Z \in X_\alpha\|.$$

But since $\|U \in X_\alpha\| = 1$, it follows by formula (II) in §5 that

$$\|Z = U\| \leqslant \|Z \in X_\alpha\|,$$

and similarly

$$\|Z = W\| \leqslant \|Z \in X_\alpha\|,$$

which gives the required inequality. □

6.6. **Proposition.** *The axiom of union*

$$\forall x\, \exists y\, \forall u (\exists z (u \in z \wedge z \in x) \Leftrightarrow u \in y)$$

is "true."

PROOF. We fix $X \in V^B$ and construct a random set Y such that for all $U \in V^B$,

$$\|U \in Y\| = \|\exists z (U \in z \wedge z \in X)\| = \bigvee_{Z \in V^B} \|U \in Z\| \wedge \|Z \in X\|.$$

By Proposition 6.4, there exists a random class Y with this property. We show that if $D(X) = V_\alpha^B$, then $Y \sim Y_\alpha$. Since $D(Y_\alpha) = D(X)$, we have

$$\begin{aligned} \|U \in Y_\alpha\| &= \bigvee_{Z \in D(X)} \|U = Z\| \wedge \|Z \in Y_\alpha\| \\ &= \bigvee_{Z \in D(X)} \|U = Z\| \wedge \Big(\bigvee_{Z_1 \in V^B} \|Z \in Z_1\| \wedge \|Z_1 \in X\| \Big). \end{aligned} \tag{3}$$

We show that the inner sum in (3) may be taken only over $Z_1 \in D(X)$. In fact, for any Z_1,

$$\|Z_1 \in X\| = \bigvee_{Z_2 \in D(X)} \|Z_1 = Z_2\| \wedge \|Z_2 \in X\|,$$

so that

$$\begin{aligned} \|Z \in Z_1\| \wedge \|Z_1 \in X\| &= \bigvee_{Z_2 \in D(X)} \|Z \in Z_1\| \wedge \|Z_1 = Z_2\| \wedge \|Z_2 \in X\| \\ &\leqslant \bigvee_{Z_2 \in D(X)} \|Z \in Z_2\| \wedge \|Z_2 \in X\|. \end{aligned} \tag{4}$$

Taking this into account, in (3) we first sum over Z for fixed $Z_1 \in D(X)$. Since $D(Z_1) \leqslant D(X)$, the sum over $Z \in D(X)$ coincides with the sum over $Z \in D(Z_1)$, and is equal to $\|U \in Z_1\|$. Thus,

$$\begin{aligned} \|U \in Y_\alpha\| &= \bigvee_{Z_1 \in D(X)} \|U \in Z_1\| \wedge \|Z_1 \in X\| \\ &\leqslant \bigvee_{Z_1 \in V^B} \|U \in Z_1\| \wedge \|Z_1 \in X\| = \|U \in Y\|, \end{aligned}$$

by (4). □

6.7. **Proposition.** *The power set axiom*

$$\forall x\ \exists y\ \forall z(z \subset x \Leftrightarrow z \in y)$$

is "true." (Recall that $z \subset x$ is abbreviated notation for $\forall u(u \in z \Rightarrow u \in x)$.)

PROOF. We fix $X \in V^B$ and construct a $Y \in V^B$ such that for all $Z \in V^B$,

$$\|Z \in Y\| = \|Z \subset X\| = \bigwedge_{U \in V^B} \|U \in Z\|' \vee \|U \in X\|.$$

By Proposition 6.4, the right side defines Y as a random class. We show that if $D(X) = V_\alpha^B$, then $Y \sim Y_{\alpha+1}$.

We first construct the element $Z_\alpha \in V_{\alpha+1}^B$ by considering Z as a random class. By (1) we have $\|U \in Z_\alpha\|' \geqslant \|U \in Z\|$, so that

$$\|Z \in Y\| \leqslant \|Z_\alpha \in Y\| = \|Z_\alpha \in Y_{\alpha+1}\|. \tag{5}$$

If we prove the inequality

$$\|Z \in Y\| \leqslant \|Z_\alpha = Z\|, \tag{6}$$

it will immediately follow from (5) and (6) that $Y \sim Y_{\alpha+1}$, since by (II), §5,

$$\|Z \in Y\| \leqslant \|Z_\alpha \in Y_{\alpha+1}\| \wedge \|Z_\alpha = Z\| \leqslant \|Z \in Y_{\alpha+1}\|.$$

It remains to verify (6).

First let $U \in D(X) = V_\alpha^B$. Then $\|U \in Z_\alpha\| = \|U \in Z\|$, so that $\|U \in Z_\alpha \Leftrightarrow U \in Z\|' = 0$, and a fortiori

$$\|U \in X\| \wedge \|U \in Z_\alpha \Leftrightarrow U \in Z\|' = 0. \tag{7}$$

As U varies, the left side of (7) determines a random class of the form $X \cap W$, where W corresponds to the formula $\neg(u \in Z_\alpha \Leftrightarrow u \in Z)$. Since $D(X) = V_\alpha^B$, it follows by 6.3(c) that $X \cap W \sim (X \cap W)_\alpha$. But according to (7), $(X \cap W)_\alpha$ is the zero function on V_α^B. Thus, $\|U \in X \cap W\| = 0$ for all $U \in V^B$. Consequently,

$$\|U \in X\| \leqslant \|U \in Z_\alpha \Leftrightarrow U \in Z\| \quad \text{for all } U. \tag{8}$$

To prove (6), we now write the left-and right-hand sides separately (using the "truth" of the formula $Z_\alpha = Z \Leftrightarrow \forall u(u \in Z_\alpha \Leftrightarrow u \in Z)$):

$$\|Z \in Y\| = \bigwedge_{U \in V^B} \|U \in Z\|' \vee \|U \in X\|,$$

$$\|Z_\alpha = Z\| = \bigwedge_{U \in V^B} \|U \in Z_\alpha \Leftrightarrow U \in Z\|.$$

It is now clear that the inequality in (6) holds term by term. In fact, for $\|U \in X\|$ this follows from (8), and for $\|U \in Z\|'$ it follows because

$$\|U \in Z_\alpha \Leftrightarrow U \in Z\| = (\|U \in Z_\alpha\|' \vee \|U \in Z\|) \wedge (\|U \in Z_\alpha\| \vee \|U \in Z\|')$$

and $\|U \in Z\|' \leqslant \|U \in Z_\alpha\|'$ for all U. □

6.8. **Proposition.** *The regularity axiom*

$$\forall x(\exists y\ (y \in x) \Rightarrow \exists y\ (y \in x \wedge y \cap x = \varnothing))$$

is "true."

Proof. We fix $X \in V^B$. The axiom with the "constant" X in place of x has the form $R \Rightarrow S$. We must show that $\|R \Rightarrow S\| = 1$. It suffices to prove that $\|R\| \wedge \|S\|' = 0$, where

$$\|R\| = \bigvee_{Y \in V^B} \|Y \in X\|, \tag{9}$$

$$\|S\|' = \bigwedge_{Y \in V^B} \|Y \in X\|' \vee \left(\bigvee_{Z \in V^B} \|Z \in Y\| \wedge \|Z \in X\| \right). \tag{10}$$

We suppose that $\|R\| \wedge \|S\|' = a \neq 0$, and show that this leads to a contradiction. It follows from (9) and (10) that there exists a $Y \in V^B$ such that $\|Y \in X\| \wedge a \neq 0$. We choose Y to have the least rank of any element with this property.

It is again clear from (9) and (10) that

$$\|Y \in X\| \wedge a \leqslant \bigvee_{Z \in V^B} \|Z \in Y\| \wedge \|Z \in X\|.$$

On the right we may sum only over $Z \in D(Y)$, without changing the value of the sum. Hence, there must exist a $Z \in D(Y)$ such that

$$\|Z \in X\| \wedge \|Y \in X\| \wedge\ a \neq 0,$$

so that $\|Z \in X\| \wedge a \neq 0$. But the rank of Z is less than the rank of Y, contradicting the choice of Y. □

7 The Axioms of Infinity, Replacement, and Choice Are "True"

7.1. We begin this section by describing two more methods for constructing random sets. The first of them, which is very widely used, solves the following problem. Suppose we are given a set of objects $X_i \in V^B, i \in I$, and a set of elements $a_i \in B$. We would like to construct a random set X that contains each

X_i with probability a_i, but such an X might not exist. However, it turns out that there always exists an X with $\|X_i \in X\| \geqslant a_i$ for all $i \in I$; moreover, there exists a least X with this property.

7.2. Lemma.

(a) *Under the conditions in* 7.1, *the function* X *of* Y

$$\|Y \in X\| = \bigvee_{i \in I} a_i \wedge \|Y \in X_i\| \tag{1}$$

is a random class X *that is equivalent to a random set. In addition,* $\|X_i \in X\| \geqslant a_i$, *and if* X' *is any random class such that* $\|X_i \in X'\| \geqslant a_i$ *for each* i, *then* $\|Y \in X'\| \geqslant \|Y \in X\|$ *for all* Y.

We shall say that X (or the equivalent random set) *collects the* X_i *with probabilities* a_i.

(b) *Under the same conditions, the function* Z *of* Y

$$\|Y \in Z\| = \bigvee_{i} a_i \wedge \|Y \in X_i\| \tag{2}$$

is a random class Z *that is equivalent to a random set. If we also have* $a_i \wedge a_j = 0$ *for all* $i \neq j$, *then* $\|Z = X_i\| \geqslant a_i$, *and for any random class* Z' *such that* $\|Z' = X_i\| \geqslant a_i$ *for each* i, *we have* $\|Y \in Z'\| \geqslant \|Y \in Z\|$ *for all* Y.

We shall say that Z *glues together the* X_i *with probabilities* a_i.

PROOF. It is easily verified that the functions Z and X defined by formulas (1) and (2) are extensional.

There exists an ordinal α such that $X_i \in V_\alpha^B$ for all i. We show that $X \sim X_\alpha$ and $Z \sim Z_\alpha$. For any $Y \in V^B$ we have

$$\begin{aligned}
\|Y \in X_\alpha\| &= \bigvee_{U \in V_\alpha^B} \|Y = U\| \wedge \|U \in X_\alpha\| \\
&= \bigvee_{U \in V_\alpha^B} \bigvee_{i} \|Y = U\| \wedge a_i \wedge \|U = X_i\| \\
&= \bigvee_{U \in V_\alpha^B} \bigvee_{i} a_i \wedge \|Y = X_i\| \wedge \|U = X_i\|.
\end{aligned}$$

If we consider the term with $U = X_i$ on the right, we obtain $a_i \wedge \|Y = X_i\| \leqslant \|Y \in X_\alpha\|$, so that $\|Y \in X\| \leqslant \|Y \in X_\alpha\|$ by (1), and the assertion follows by 6.3(c).

Similarly, for any $Y \in V^B$ we have

$$\begin{aligned}
\|Y \in Z_\alpha\| &= \bigvee_{U \in V_\alpha^B} \bigvee_{i} \|Y = U\| \wedge a_i \wedge \|U \in X_i\| \\
&= \bigvee_{U \in V_\alpha^B} \bigvee_{i} a_i \wedge \|Y \in X_i\| \wedge \|Y = U\|.
\end{aligned}$$

Since $\|Y \in X_i\| = \bigvee_{U \in V_\alpha^B} \|Y = U\| \wedge \|Y \in X_i\|$, it follows that $a_i \wedge \|Y \in X_i\| \leqslant \|Y \in Z_\alpha\|$, and $\|Y \in Z\| \leqslant \|Y \in Z_\alpha\|$ by (2).

Now let X' and Z' be any random sets with the properties in (a) and (b). It is clear from (1) that $\|X_i \in X\| \geqslant a_i$. If $\|X_i \in X'\| \geqslant a_i$, for each i, then $\|Y \in X'\| = \bigvee_U \|Y = U\| \wedge \|U \in X'\| \geqslant \bigvee_i \|Y = X_i\| \wedge \|X_i \in X'\| \geqslant \|Y \in X\|$ by (1).

Similarly, if $a_i \wedge a_j = 0$ for $i \neq j$ then it is clear from (2) that $a_i \wedge \|Y \in Z\| = a_i \wedge \|Y \in X_i\|$, so that

$$a_i \wedge \|X_i = Z\| = \bigvee_Y a_i \wedge \|Y \in X_i \Leftrightarrow Y \in Z\| = a_i$$

and $\|X_i = Z\| \geqslant a_i$. Now if $\|X_i = Z'\| \geqslant a_i$ for each i, then

$$\begin{aligned}\|Y \in Z'\| &\geqslant \|Y \in Z'\| \wedge \|Z' = X_i\| \\ &= \|Y \in X_i\| \wedge \|Z' = X_i\| \geqslant a_i \wedge \|Y \in X_i\|,\end{aligned}$$

so that $\|Y \in Z'\| \geqslant \|Y \in Z\|$. □

Here is our first application of Lemma 7.2(a):

7.3. **Proposition.** *The axiom of infinity*

$$\exists x(\varnothing \in x \wedge \forall u(u \in x \Rightarrow \{u\} \in x))$$

is "true."

PROOF. When we proved that the axiom of pairing is "true," we constructed for any $U, W \in V^B$ an element $Z \in V^B$ (unique up to equivalence) with the property that $\|Y \in Z\| = \|Y = U \vee Y = W\|$ for all Y. It is natural to let $\{U, W\}^B$ denote this element Z, and let $\{U\}^B = \{U, U\}^B$.

We now verify the axiom of infinity. We set $X_0 = \varnothing, X_1 = \{\varnothing\}^B, \ldots, X_n = \{X_{n-1}\}^B, \ldots$. Further, we let $X \in V^B$ be the element that collects all the X_i with probabilities 1. We show that

$$\|\varnothing \in X \wedge \forall u(u \in X \Rightarrow \{u\} \in X)\| = 1.$$

It is obviously sufficient to prove that for all $U \in V^B$ we have $\|U \in X\| \leqslant \|\{U\}^B \in X\|$, that is, by (1);

$$\bigvee\nolimits_{i=0}^{\infty} \|U = X_i\| \leqslant \bigvee\nolimits_{i=0}^{\infty} \|\{u\}^B = X_i\|.$$

In fact, since the formula $u = x \Leftrightarrow \{u\} = \{x\}$ is "true," and since $X_{i+1} = \{X_i\}^B$, it immediately follows that

$$\|U = X_i\| = \|\{U\}^B = X_{i+1}\|.$$ □

7.4. **Lemma.** *Let W be a random class. Then there exists an element $X \in V^B$ such that*

$$\bigvee_{U \in V^B} W(U) = W(X).$$

The left-hand side may be represented in the form $\|\exists x(x \in W)\| = \|W \neq \varnothing\|$. Hence, intuitively, the lemma says that the probability that a given class is nonempty coincides with the probability that a suitable element occurs in it.

PROOF. We first show that there exists an ordinal β such that $\bigvee_{U \in V^B} W(U) = \bigvee_{U \in V^B_\beta} W(U)$. In fact, let $a_\gamma = \bigvee_{U \in V^B_\gamma} W(U)$, and for any $a \in B$ set $\gamma(a) = \min(\gamma | a_\gamma > a)$ (or $\gamma(a) = 0$ if $a_\gamma \not> a$ for all γ). Finally, set $\beta = \sup_{a \in B} \gamma(a)$. This is an ordinal, because B is a set. If $\gamma > \beta$, then $a_\gamma \geqslant a_\beta$ by monotonicity, but we cannot have $a_\gamma > a_\beta$ because of the choice of β.

Thus, let $\bigvee_U W(U) = \bigvee_{U \in V^B_\beta} W(U)$. We index all the elements in V^B_β by an initial segment of ordinals (by the axiom of choice!): $V^B_\beta = \{U_\alpha\}_{\alpha \in I}$. We set

$$a_\alpha = W(U_\alpha) \wedge \left(\bigvee_{\gamma < \alpha} W(U_\gamma) \right)', \quad \alpha \in I.$$

Obviously $a_\alpha \wedge a_\gamma = 0$ for $\alpha \neq \gamma$. Using Lemma 7.2(b), we glue together the sets U_α with probabilities $a_\alpha (\alpha \in I)$. We obtain a set X satisfying the conditions $\|X = U_\alpha\| \geqslant a_\alpha \geqslant W(U_\alpha)$. Using the extensionality of W, we obtain

$$W(X) \geqslant \bigvee_{\alpha \in I} \|X = U_\alpha\| \wedge W(U_\alpha) = \bigvee_{\alpha \in I} W(U_\alpha) = \bigvee_{U \in V^B} W(U). \qquad \square$$

7.5. **Proposition.** *The replacement axiom*

$$\forall \bar{z}\, \forall u \Big(\forall x (x \in u \Rightarrow \exists ! y\, P(x, y, \bar{z})) \Rightarrow \exists w\, \forall y (y \in w \Leftrightarrow \exists x (x \in u \wedge P(z, y, \bar{z}))) \Big)$$

is "true" (here $\bar{z} = \langle z_1, \ldots, z_n \rangle$).

PROOF. We fix a "vector" $\overline{Z} = \langle Z_1, \ldots, Z_n \rangle$ with $Z_i \in V^B$ and an element $U \in V^B$. We shall write $P(x, y)$ instead of $P(x, y, \overline{Z})$. If we write the axiom with the "constants" Z_i and U in the form $R \Rightarrow S$, then we must prove that $\|R \Rightarrow S\| = 1$.

7.6. *The special case:* If $\|R\| = 1$, then $\|S\| = 1$.

We first show how the general case follows from this special case. Let $a \in B$, and let B_a denote the set $\{b \in B | b \leqslant a\}$. The operations on B induce a Boolean algebra structure on B_a with unit element $1_a = a$. The natural mapping $B \to B_a : b \mapsto b \wedge a$ is a homomorphism. An easy induction on a allows us to construct a surjective map of universes $V^B \to V^{B_a} : X \mapsto X_a$ such that for all $X, Y \in V^B$ we have

$$\|X_a \in Y_a\| = \|X \in Y\| \wedge a,$$
$$\|X_a = Y_a\| = \|X = Y\| \wedge a.$$

Now, to prove Proposition 7.5 from the special case 7.6, we choose $a = \|R\|$. Then $\|R\|_a = 1_a$, so that 7.6 implies that $\|S\|_a = 1_a$. This means that $\|S\| \geqslant a$, and hence $\|R \Rightarrow S\| = 1$. (Here we have used 7.6 in V^{B_a}; clearly $\|R\|_a = \|R_a\|$, where R_a is the obvious image of R in V^{B_a}.)

7.7. PROOF OF 7.6. The condition $\|R\| = 1$ means that for any $X \in V^B$,

$$\|X \in U\| \leqslant \|\exists! y \; P(X, y)\|. \tag{3}$$

To show that $\|S\| = 1$, it is sufficient if given $U \in V^B$, we find a $W \in V^B$ such that for all $Y \in V^B$,

$$\|Y \in W\| = \bigvee_{X \in V^B} \|X \in U\| \wedge \|P(X, Y)\|. \tag{4}$$

It follows from 6.5 that the formula (4) defines W as a random class. We find an ordinal α such that $W \sim W_\alpha$.

To do this, we first note that in (4) we may take the sum only over

$$\|Y \in W\| = \bigvee_{X \in D(U)} \|X \in U\| \wedge \|P(X, Y)\| \tag{5}$$

(the argument here is the same as after formula (3) in §6). We now apply Lemma 7.4 to the class $W_X(Y) = \|P(X, Y)\|$. It follows that for every $X \in D(U)$ there exists an element $Y_X \in V^B$ such that

$$\|\exists y \; P(X, y)\| = \|P(X, Y_X)\|. \tag{6}$$

(Because $\|\exists! y \;\; P(X, y)\| \leqslant \|\exists y P(X, y)\|$, we can use these Y_X to estimate $\|X \in U\|$ with the help of (9) below.)

We set $\alpha_X = \min(\alpha | Y_X \in V_\alpha^B)$, and

$$\alpha = \sup(\alpha_X | X \in D(U)),$$

and then show that $W \sim W_\alpha$ for this α. We must verify that $\|Y \in W\| \leqslant \|Y \in W_\alpha\|$ for every Y. By (5) and by formula (II) in §5, this follows if for any $X \in D(U)$ we have

$$\|X \in U\| \wedge \|P(X, Y)\| \leqslant \|Y = Y_X\| \wedge \|Y_X \in W_\alpha\|. \tag{7}$$

In the first place, by (3), (6), (5), and the definition of α, we have

$$\begin{aligned} \|X \in U\| &\leqslant \|P(X, Y_X)\|, \\ \|X \in U\| &\leqslant \|Y_X \in W\| = \|Y_X \in W_\alpha\|. \end{aligned} \tag{8}$$

Further, we consider the following formula, which is "true" because it is deducible from the logical axioms and the axioms of equality:

$$\forall x (\exists ! y \; P(x, y) \wedge P(x, y_1) \wedge P(x, y_2) \Rightarrow y_1 = y_2).$$

We thereby obtain

$$\|\exists ! y \; P(X, y)\| \wedge \|P(X, Y)\| \wedge \|P(X, Y_X)\| \leqslant \|Y = Y_X\|. \tag{9}$$

Finally, it follows from (3), (8), and (9) that

$$\|X \in U\| \wedge \|P(X,Y)\| \leqslant \|Y = Y_X\| \wedge \|Y_X \in W_\alpha\|,$$

i.e., we have (7).

7.8. Proposition. *The axiom of choice is "true."*

PROOF. Recall that the axiom of choice has the form $\forall x\, \exists y(Q \wedge R \wedge S \wedge T)$, where

Q denotes $\forall z(z \in y \Rightarrow \exists\, u\, \exists\, w(z = \langle u, w\rangle))$("$y$ is a binary relation");

R denotes $\forall u \forall w_1 \forall w_2(\langle u, w_1\rangle \in y \wedge \langle u, w_2\rangle \in y \Rightarrow w_1 = w_2)$("$y$ is a function");

S denotes $\forall u(\exists w(\langle u, w\rangle \in y) \Rightarrow u \in x)$("the domain of definition of y is contained in x");

T denotes $\forall u(u \neq \varnothing \wedge u \in x \Rightarrow \exists w(w \in u \wedge \langle u, w\rangle \in y))$("the domain of definition of y coincides with x, and y chooses one element from each nonempty element of x").

We fix $X \in V^B$ and construct the corresponding "choosing function" Y. To do this:

(a) We index $D(X)$ by an initial segment of ordinals:

$$D(X) = \{U_0, U_1, \ldots, U_\alpha, \ldots\}, \quad \alpha \in I.$$

(b) For each $U_\alpha \in D(X)$ we use Lemma 7.4 to find an element $W_\alpha \in V^B$ such that

$$\|W_\alpha \in U_\alpha\| = \bigvee_{W \in V^B} \|W \in U_\alpha\|.$$

(c) For each $\alpha \in I$ we set

$$a_\alpha = \|U_\alpha \in X\| \wedge \left(\bigvee_{\beta<\alpha} \|U_\beta \in X\|' \vee \|U_\beta = U_\alpha\|' \right).$$

(d) Finally, we let Y denote the set that collects the "ordered pairs" $\langle U_\alpha, W_\alpha\rangle^B$ with probabilities $a_\alpha, \alpha \in I$. Here, of course, $\langle U, W\rangle^B = \{\{U\}^B, \{U, W\}^B\}$.

The idea of this construction is as follows. In each U_α we choose the element W_α that belongs to U_α "with the largest possible probability." We then put together the graph of the choice function Y from the "pairs" $\langle U_\alpha, W_\alpha\rangle^B$, where we take the pairs in the order they are indexed, but include a given $\langle U_\alpha, W_\alpha\rangle^B$ only to the extent that U_α "was not already considered earlier as belonging to X."

We now substitute X and Y in place of x and y in the axiom of choice, and, letting Q, R, S, and T now denote the corresponding formulas with these constants, we show that $\|Q\| = \|R\| = \|S\| = \|T\| = 1$. We shall constantly be using the following formula, which follows from (1) and the definition of Y;

$$\|Z \in Y\| = \bigvee_{a} \|Z = \langle U_\alpha, W_\alpha \rangle^B\| \wedge a_\alpha. \tag{10}$$

7.9. $\|Q\| = 1$. By the definition of Q, this means that for all $Z \in V^B$ we must have

$$\|Z \in Y\| = \bigvee_{U,W} \|Z = \langle U, W \rangle^B\|,$$

but this is obvious from (10).

7.10. $\|R\| = 1$. By the definition of R, for any $U, W^1, W^2 \in V^B$ we must prove the inequality

$$\|\langle U, W^1 \rangle^B \in Y\| \wedge \|\langle U, W^2 \rangle^B \in Y\| \leqslant \|W^1 = W^2\|.$$

Using (10), we rewrite the left-hand side in the form

$$\bigvee_{\alpha,\beta} \|U = U_\alpha\| \wedge \|W^1 = W_\alpha\| \wedge a_\alpha \wedge \|U = U_\beta\| \wedge \|W^2 = W_\beta\| \wedge a_\beta.$$

Since $\|U = U_\alpha\| \wedge \|U = U_\beta\| \leqslant \|U_\alpha = U_\beta\|$ and $\|U_\alpha = U_\beta\| \wedge a_\alpha \wedge a_\beta = 0$ for $\alpha \neq \beta$ (see the definition of a_α), it follows that in this sum we need only consider the terms with $\alpha = \beta$. But such a term is $\leqslant \|W^1 = W_\alpha\| \wedge \|W^2 = W_\alpha\| \leqslant \|W^1 = W^2\|$, as required.

7.11. $\|S\| = 1$. This is equivalent to the inequality

$$\|\langle U, W \rangle^B \in Y\| \leqslant \|U \in X\|.$$

But by (10), the left-hand side equals

$$\begin{aligned}\bigvee_{\alpha} \|U = U_\alpha\| \wedge \|W = W_\alpha\| \wedge a_\alpha &\leqslant \bigvee_{\alpha} \|U = U_\alpha\| \wedge \|W = W_\alpha\| \wedge \|U_\alpha \in X\| \\ &\leqslant \bigvee_{\alpha} \|U = U_\alpha\| \wedge \|U_\alpha \in X\| = \|U \in X\|.\end{aligned}$$

7.12. $\|T\| = 1$. We must prove that for any $U \in V^B$,

$$\|U \in X\| \wedge \|U \neq \varnothing\| \leqslant \bigvee_{W \in V^B} \|W \in U\| \wedge \|\langle U, W \rangle^B \in Y\|. \tag{11}$$

We first show that it suffices to prove (11) for $U \in D(X)$, i.e., for all U_α, $\alpha \in I$. In fact, suppose (11) holds for all U_α. Then for $U \in V^B$ we have

$$\|U \in X\| = \bigvee_{\alpha} \|U = U_\alpha\| \wedge \|U_\alpha \in X\|,$$

$$\|U \neq \varnothing\| = \bigvee_{U_1 \in V^B} \|U_1 \in U\|,$$

and hence

$$\begin{aligned}
\|U \in X\| \wedge \|U \neq \varnothing\| &= \bigvee_{\alpha, U_1} \|U_1 \in U\| \wedge \|U = U_\alpha\| \wedge \|U_\alpha \in X\| \\
&\leqslant \bigvee_{\alpha, U_1} \|U_1 \in U_\alpha\| \wedge \|U = U_\alpha\| \wedge \|U_\alpha \in X\| \\
&\qquad\qquad \text{(by (III) in §5)} \\
&= \bigvee_{\alpha} \|U_\alpha \neq \varnothing\| \wedge \|U = U_\alpha\| \wedge \|U_\alpha \in X\| \\
&\leqslant \bigvee_{\alpha, W \in V^B} \|W \in U_\alpha\| \wedge \|\langle U_\alpha, W\rangle^B \in Y\| \wedge \|U = U_\alpha\| \\
&\qquad\qquad \text{(by (11) for } U_\alpha) \\
&\leqslant \bigvee_{W} \|W \in U\| \wedge \|\langle U, W\rangle^B \in Y\|.
\end{aligned}$$

(Here we used the fact that

$$\begin{aligned}
&\|\langle U_\alpha, W\rangle^B \in Y\| \wedge \|U = U_\alpha\| \\
&\quad = \bigvee_{\beta} \|U_\alpha = U_\beta\| \wedge \|W = W_\beta\| \wedge a_\beta \wedge \|U = U_\alpha\| \\
&\quad \leqslant \bigvee_{\beta} \|U = U_\beta\| \wedge \|W = W_\beta\| \wedge a_\beta \\
&\quad = \|\langle U, W\rangle^B \in Y\|.)
\end{aligned}$$

Thus, it remains to prove (11) for $U_\alpha, \alpha \in I$. Now

$$\|U_\alpha \neq \varnothing\| = \|\exists w(w \in U_\alpha)\| = \bigvee_{W} \|W \in U_\alpha\| = \|W_\alpha \in U_\alpha\|.$$

Hence (11) can be rewritten

$$\|U_\alpha \in X\| \wedge \|W_\alpha \in U_\alpha\| \leqslant \bigvee_{W} \|W \in U_\alpha\| \wedge \|\langle U_\alpha, W\rangle^B \in Y\|. \tag{12}$$

We prove this by induction on α. (12) is obvious for $\alpha = 0$, since the term on the right with $W = W_0$ coincides with the left-hand side. Suppose (12) holds for $\beta < \alpha$.

By the definition of a_α, we have

$$\|U_\alpha \in X\| = a_\alpha \vee \left(\bigvee_{\beta < \alpha} \|U_\beta \in X\| \wedge \|U_\beta = U_\alpha\| \right).$$

If we substitute this formula in the left-hand side of (12), we find that we must prove two inequalities:

$$a_\alpha \wedge \|W_\alpha \in U_\alpha\| \leqslant \bigvee_{W} \|W \in U_\alpha\| \wedge \|\langle U_\alpha, W\rangle^B \in Y\|, \tag{13}$$

$$\begin{aligned}
&\|U_\beta \in X\| \wedge \|U_\beta = U_\alpha\| \wedge \|W_\alpha \in U_\alpha\| \\
&\qquad \leqslant \bigvee_{W} \|\langle U_\alpha, W\rangle^B \in Y\| \wedge \|W \in U_\alpha\|, \quad \textit{for all } \beta < \alpha.
\end{aligned} \tag{14}$$

The inequality (13) is obvious if we look at the term on the right with $W = W_\alpha$. The inequality (14) reduces to the induction assumption as follows. The left-hand side of (14) is

$$\begin{aligned}&\leqslant \|U_\beta \in X\| \wedge \|U_\beta = U_\alpha\| \wedge \|W_\alpha \in U_\beta\| \\ &\leqslant \|U_\beta \in X\| \wedge \|U_\beta = U_\alpha\| \wedge \|W_\beta \in U_\beta\|\end{aligned}$$

by the definition of W_β. Further, using the induction assumption and extensionality, we have

$$\begin{aligned}&\|U_\beta = U_\alpha\| \wedge \|U_\beta \in X\| \wedge \|W_\beta \in U_\beta\| \\ &\qquad \leqslant \bigvee_W \|W \in U_\beta\| \wedge \|\langle U_\beta, W\rangle^B \in Y\| \wedge \|U_\beta = U_\alpha\| \\ &\qquad \leqslant \bigvee_W \|W \in U_\alpha\| \wedge \|\langle U_\alpha, W\rangle^B \in Y\|,\end{aligned}$$

which completes the verification of the axiom of choice. □

8 The Continuum Hypothesis Is "False" for Suitable B

8.1. We recall (Lemma 7.2(a)) that the set $X \in V^B$ collects the sets $\{X_i\}$ with probabilities $a_i \in B(i \in I)$ if $\|Y \in X\| = \bigvee_i \|Y = X_i\| \wedge a_i$ for all Y. Using this definition, we can introduce a useful canonical mapping $t \mapsto \hat{t}$ from the von Neumann universe V to the universe V^B. Let $\hat{\varnothing} = \varnothing$ (recall that $\|Y \in \varnothing\| = 0$ for all Y), and if $\hat{s}$ has already been defined for all $s \in V_\alpha$, then for $t \in V_{\alpha+1}$, we *let $\hat{t}$ collect all the $\hat{s}$ for $s \in t$ with probability* 1. In other words, for any $Y \in V^B$,

$$\|Y \in \hat{t}\| = \bigvee_{s \in t} \|Y = \hat{s}\|. \tag{1}$$

(Here the collecting set $\hat{t}$ is not uniquely defined, i.e., it is defined only modulo equivalence, so that, strictly speaking, we should also specify the rank of $\hat{t}$, for example by saying that it equals the rank of t. This is not essential for us, however, since we shall be interested only in the truth functions, which do not change if we replace an object by an equivalent object.)

We now formulate some additional conditions (besides completeness) that must be imposed on the Boolean algebra B for the purposes of this section. Recall that ω_0 is the first infinite ordinal, ω_1 is the first ordinal having cardinality $> \omega_0$, and ω_2 is the first ordinal having cardinality $> \omega_1$.

8.2. *Conditions on B.*

(a) The countable chain condition, which, we recall, says that if we have a family of elements $\{a_i\}, i \in I$, such that at $a_i \neq 0$ and $a_i \wedge a_j = 0$ for $i \neq j$, then I is at most a countable set.
(b) There exists a family of elements $b(n, \alpha) \in B$, indexed by the set $\omega_0 \times \omega_2$, with the following property: if $Z(\alpha)$ collects the elements $\hat{n}, n \in \omega_0$, with probabilities $b(n, \alpha)$, then $\|Z(\alpha) = Z(\beta)\| = 0$ for $\alpha \neq \beta, \alpha, \beta \in \omega_2$.

The second condition has the following intuitive meaning. It is easy to see that $\|Z(\alpha) \subset \hat{\omega}_0\| = 1$. In fact, this equality is equivalent to $\|\forall x(x \in Z(\alpha) \Rightarrow x \in \hat{\omega}_0)\| = 1$, i.e., to

$$\forall X \in V^B, \qquad \|X \in Z(\alpha)\| \leqslant \|X \in \hat{\omega}_0\|,$$

and this is obvious from (1), since $\hat{\omega}_0$ collects the $\hat{n}$ with probability 1, and $Z(\alpha)$ collects the $\hat{n}$ with probabilities $b(n,\alpha) \leqslant 1$.

Thus, condition (b) means that we can find ω_2 distinct subsets $Z(\alpha) \subset \hat{\omega}_0$, so that, in the naive sense, we have card $\mathcal{P}(\hat{\omega}_0) > \omega_1$. This is precisely the negation of the continuum hypothesis. Of course, it is still necessary to show that this intuitive idea can be made into a proof.

8.3. *The existence of B with the required properties.* We could use measurable sets, as in §3. However, in order to vary our approach, and to prepare for §9, we give another construction. Let $\{0,1\}$ be the discrete two-point space, let $I = \omega_0 \times \omega_2$, and let $S = \{0,1\}^I$ be the space of vectors whose coordinates are indexed by I and take the values 0 or 1. We introduce the direct product topology on S. It has a standard basis of open sets consisting of all vectors whose coordinates indexed by a finite subset $\mathcal{J} \subset I$ are fixed.

If $a \subset S$, we set

$$a^{'} = \text{the complement of the closure of } a \text{ in } S,$$

and we set $a'' = (a^{'})^{'}$. Sets $a \subset S$ with $a'' = a$ are called regular open sets in S.

8.4. **Theorem.** Let

$$B = \{a \subset S | a'' = a\},$$
$$a \wedge b = a \cap b,$$
$$a \vee b = (a \cup b)''.$$

Then B with the operations $\wedge, \vee$, and $^{'}$ is a complete Boolean algebra with the countable chain condition, and $\bigvee_i a_i = (\cup_i a_i)''$ for any family of $a_i \in B$.

We omit the proof (see J. B. Rosser, *Simplified Independence Proofs*, Academic Press, New York, 1969, Chapter 2).

8.5. **Lemma.** *Under the conditions in* 8.4, *let*

$$b(n,\alpha) = \textit{the set of vectors with } 1 \textit{ in the } (n,\alpha) \textit{ place},$$

and let $Z(\alpha)$ be defined as in 8.2(b). *Then*

$$\|Z(\alpha) = Z(\beta)\| = 0, \quad \textit{for } \alpha \neq \beta.$$

PROOF. By formula (5) in §4, we have

$$\|Z(\alpha) = Z(\beta)\| = \bigwedge_{n \in \omega_0} (b(n,\alpha) \vee b(n,\beta)) \wedge (b(n,\alpha)^{'} \wedge b(n,\beta)^{'}).$$

The right side can become larger only if we replace $\wedge$ by $\cap$ and $\vee$ by $\cup$; here the primes $'$ coincide with the ordinary complements. If we had $\|Z(\alpha) = Z(\beta)\| \neq 0$, then there would exist an element X in the standard basis of the the topology (see the beginning of 8.3) that is contained in

$$\bigcap_{n\in\omega_0} (b(n,\alpha)\cap b(n,\beta)) \cup (b(n,\alpha)' \cap b(n,\beta)').$$

But this intersection consists of all vectors having the same (n,α)-coordinate and (n,β)-coordinate for all n, while all coordinates except for a finite number range freely in any element X of the standard basis of the topology. □

8.6. *Formulation of the negation of the continuum hypothesis.* We shall prove that the following is "true":

$$\neg\mathrm{CH}:\quad \begin{array}{l}\forall x((\text{“}x \text{ is an ordinal”} \wedge \text{“}x \text{ is not finite”} \wedge \forall y(y \in x \Rightarrow \text{“}y\\ \text{is finite”})) \Rightarrow \exists w(\text{“there is no function from } x \text{ onto all of}\\ w\text{”} \wedge \text{“there is no function from } w \text{ onto } \mathcal{P}(x)\text{”})).\end{array}$$

Here:

x is finite: $\forall y(y \subset x \wedge y \neq x \Rightarrow$ "there is no function from y onto all of x").

We leave the translation of the other abbreviated notation to the reader.

The premise in $\neg$ CH says that "x is the first infinite ordinal," and the conclusion says that "w is a set having cardinality intermediate between that of x and that of $\mathcal{P}(x)$." We shall abbreviate $\neg$ CH as follows:

$$\forall x(P(x) \Rightarrow \exists w(Q_1(x,w) \wedge Q_2(x,w))). \tag{2}$$

8.7. **Reduction Lemma.** *Let $P(x)$ and $Q(x)$ be two formulas in the Zermelo–Fraenkel language having one free variable x and satisfying the properties*

The formula $\exists! x\ P(x)$ is deducible from the axioms, and

$X_0 \in V^B$ is an element such that $\|P(X_0)\| = 1$.

Then $\|P(X)\| = \|X = X_0\|$ for all X, and if $\|Q(X_0)\| = 1$, it follows that $\|\forall x P(X) \Rightarrow Q(X))\| = 1$.

PROOF. We first note that $|\exists x\ P(x)\| \geqslant \|\exists! x\ P(X)\| = 1$, since all the axioms are "true" in V^B, and the rules of deduction preserve "truth." It hence follows from Lemma 7.4 that there exists an object $X_0 \in V^B$ with $\|P(X_0)\| = 1$.

Further, $P(x) \wedge P(y) \Rightarrow x = y$ is also deducible, so that if we apply this with X in place of x and X_0 in place of y, we find that

$$\|P(X)\| \leqslant \|X = X_0\|. \tag{3}$$

But $\|P(X)\| \wedge \|X = X_0\| = P(X_0) \wedge \|X = X_0\| = \|X = X_0\|$. Hence the inequality in (3) may be replaced by equality.

Finally, we suppose that $\|Q(X_0)\| = 1$. Then, by what was just proved,

$$\begin{aligned}\|P(X)\| &= \|Q(X_0)\| \wedge \|X = X_0\| = \|Q(X)\| \wedge \|X = X_0\| \\ &= \|Q(X)\| \wedge \|P(X)\|,\end{aligned}$$

so that $\|P(X)\| \leqslant \|Q(X)\|$, and $\forall x(P(x) \Rightarrow \|Q(X))\| = 1$. □

This lemma can be applied to $\neg$CH in the form (2), since the formula $\exists!\, x\, P(X)$, where $P(x)$ is the premise "x is the first infinite ordinal," is deducible from the axioms. We shall not give this formal deduction, and shall consider the uniqueness of ω_0 to be common knowledge. Now, by Lemma 8.7, to verify $\neg$CH it suffices to prove the following facts:

8.8. $\|P(\hat{\omega}_0)\| = 1$. (In other words, $\hat{\omega}_0$ plays the role of X_0 in our situation.)

8.9. $\|Q_1(\hat{\omega}_0, \hat{\omega}_1\| = 1$.

8.10. $\|Q_2(\hat{\omega}_0, \hat{\omega}_1\| = 1$. (This then implies that $\|\exists\omega(Q_1(\hat{\omega}_0, \omega) \wedge Q_2(\hat{\omega}_0, \omega))\| = 1$, and completes the verification of the conditions of the lemma.) 8.8. is verified almost mechanically, and we leave it as an exercise.

8.11. Verification Of 8.9. We must show that if B satisfies the countable chain condition, then

$$\|\exists \text{ a function from } \hat{\omega}_0 \text{ onto all of } \hat{\omega}_1\| = 0.$$

The proof that follows carries over word for word to the more general case, when instead of ω_0 and ω_1, we take any pair $s, t \in V$ such that card $s <$ card t and card s is infinite.

We suppose that

$$0 \neq a = \|\exists f(f \text{ is a function } \wedge \forall y(y \in \hat{\omega}_1 \Rightarrow \exists x(x \in \hat{\omega}_0 \wedge \langle x, y\rangle \in f)))\|,$$

and we show that this leads to a contradiction. There must exist an $F \in V^B$ such that

$$a \leqslant \|F \text{ is a function }\| \wedge \left(\bigwedge_Y \cdots\right).$$

For every $\alpha \in \omega_1$, we consider the term in $\bigwedge_Y \cdots$ corresponding to $Y = \hat{\alpha}$ and use the fact that $\|\hat{\alpha} \in \hat{\omega}_1\| = 1$. We obtain

$$a \leqslant \|F \text{ is a function}\| \wedge \left(\bigvee_X \|X \in \hat{\omega}_0\| \wedge \|\langle X, \hat{\alpha}\rangle^B \in F\|\right). \tag{4}$$

By (1), we have

$$\begin{aligned}\|X \in \hat{\omega}_0\| \wedge \|\langle X, \hat{\alpha}\rangle^B \in F\| &= \bigvee_{n<\omega_0} \|X = \hat{n}\| \wedge \|\langle X, \hat{\alpha}\rangle^B \in F\| \\ &= \bigvee_{n<\omega_0} \|X = \hat{n}\| \wedge \|\langle \hat{n}, \hat{\alpha}\rangle^B \in F\|,\end{aligned}$$

so that if we sum first over X and then over n, we may write (4) in the form

$$a \leqslant \|F \text{ is a function}\| \wedge \Big(\bigvee_{n<\omega_0} \|\langle \hat{n}, \hat{\alpha}\rangle^B \in F\| \Big).$$

Hence, for every $\alpha < w_1$ there is an $n(\alpha) < \omega_0$ such that

$$\|F \text{ is a function}\| \wedge \|\langle n(\hat{\alpha}), \hat{\alpha}\rangle^B \in F\| \neq 0.$$

Then there exist an n_0 and a subset $\mathcal{J} \subseteq \omega_1$ of cardinality ω_1 such that

$$0 \neq a_\alpha = \|F \text{ is a function}\| \wedge \|\langle \hat{n}_0, \hat{\alpha}\rangle^B \in F\|, \quad \text{for all } \alpha \in \mathcal{J}.$$

It remains to show that $a_\alpha \wedge a_\beta = 0$ for $\alpha \neq \beta$, which contradicts the countable chain condition on B. Now by the definition of a function

$$a_\alpha \wedge a_\beta = \|F \text{ is a function}\| \wedge \|\langle \hat{n}_0, \hat{\alpha}\rangle^B \in F\| \wedge \|\langle \hat{n}_0, \hat{\beta}\rangle^B \in F\| \leqslant \|\hat{\alpha} = \hat{\beta}\|,$$

so that it suffices to show that $\alpha \neq \beta$ implies $\|\hat{\alpha} = \hat{\beta}\| = 0$.

In fact, if, say, $\gamma \in \alpha$ but $\gamma \notin \beta$, then the formula (5) in §4 for $\|\hat{\alpha} = \hat{\beta}\|$ has a zero term, namely $\|\hat{\gamma} \in \hat{\alpha}\|' \vee \|\hat{\gamma} = \hat{\beta}\|$. (To check that $\|\hat{\gamma} = \hat{\beta}\| = 0$ if $\gamma \notin \beta$ we have to know that $\|\hat{\gamma} = \hat{\delta}\| = 0$ if $\gamma \neq \delta$, but we have to know this only for γ and δ of lower rank than α and β, so that the detailed proof uses induction on the rank.) □

8.12. Verification of 8.10. We must show that

$$\|\exists \text{ a function from } \hat{\omega}_1, \text{onto } \mathcal{P}(\hat{\omega}_0)\| = 0,$$

that is, that

$$\|\exists g\,(g \text{ is function } \wedge \forall z (z \subset \hat{\omega}_0 \Rightarrow \exists y (y \in \hat{\omega}_1 \wedge \langle y, z\rangle \in g)))\| = 0.$$

Suppose that for some $G \in V^B$ we have

$$0 \neq a = \|G \text{ is a function }\| \wedge \Big(\bigwedge_Z \cdots \Big).$$

For every $\alpha < \omega_2$ we consider the term corresponding to $Z = Z(\alpha)$ (see the definition in 8.2 and 8.5), and we use the fact that

$$0 \neq a \leqslant \|G \text{ is a function}\| \wedge \Big(\bigvee_Y \|Y \in \hat{\omega}_1\| \wedge \|\langle Y, Z(\alpha)\rangle^B \in G\| \Big). \tag{5}$$

By (1), we have

$$\begin{aligned}\|Y \in \hat{\omega}_1\| \wedge \|\langle Y, Z(\alpha)\rangle^B \in G\| &= \bigvee_{\beta<\omega_1} \|Y = \hat{\beta}\| \wedge \|\langle Y, Z(\alpha)\rangle^B \in G\| \\ &= \bigvee_{\beta<\omega_1} \|Y = \hat{\beta}\| \wedge \|\langle \hat{\beta}, Z(\alpha)\rangle^B \in G\|.\end{aligned}$$

Summing first over Y, we rewrite (5) in the form

$$0 \neq a \leqslant \|G \text{ is a function}\| \wedge \bigvee_{\beta<\omega_1} \|\langle\hat{\beta}, Z(\alpha)\rangle^B \in G\|.$$

Hence, for every $\alpha < \omega_2$ there is a $\beta(\alpha) < \omega_1$ such that

$$0 \neq a_\alpha = \|G \text{ is a function}\| \wedge \|\langle\beta(\alpha), Z(\alpha)\rangle^B \in G\|.$$

Then there exist a $\beta_0 < \omega_1$ and a subset $\mathcal{J} \subset \omega_2$ of cardinality ω_2 such that

$$0 \neq a_\alpha = \|G \text{ is a function}\| \wedge \|\langle\hat{\beta}_0, Z(\alpha)\rangle^B \in G\|, \quad \text{for all } \alpha \in \mathcal{J}.$$

As in 8.11, we obtain a contradiction to the countable chain condition if we show that $a_\alpha \wedge a_\beta = 0$ for $\alpha \neq \beta$. But this follows from

$$a_\alpha \wedge a_\beta \leqslant \|Z(\alpha) = Z(\beta)\| = 0$$

by Lemma 8.5. □

9 Forcing

9.1. By choosing the Boolean algebra B in various ways, one can use the corresponding models V^B to show that many different assertions P are consistent with the Zermelo–Fraenkel axioms. But each choice of B for a given P such that $\|P\| = 1$ in V^B presents a separate problem.

There is another interpretation of this method that is closer to Cohen's original idea. From this point of view we start not with a universe V and a Boolean algebra B, but with an (often countable) transitive model M and an ordered set C of "forcing conditions." It is usually more obvious how to choose a suitable C than how to choose a suitable B for proving that a given proposition P is consistent. One might say that B embodies the "physical meaning" of the problem, while C expresses its "logical meaning." Anyway, it is not difficult to go from one version to the other, and in either case it takes about the same amount of work to verify the "truth" of the axioms.

In this section we discuss the second version, using forcing, with most of the proofs omitted. The details can be found in Cohen's original article, and also in Jech's book *Lectures in Set Theory with Particular Emphasis on the Method of Forcing*, Springer–Verlag Lecture Notes in Mathematics 217, 1971, and in J. R. Shoenfield's article "Unramified forcing," *Proc. Symp. in Pure Math.*, vol. 13, 1, 357–381 (American Math. Soc, Providence, 1972).

9.2. Before introducing the general concept of forcing, we consider a special case that arises in a typical problem.

Let X and Y be two sets, for example $\mathcal{P}(\omega_0)$ and ω_2. We consider the proposition P : "card $X \geqslant$ card Y," which in this special case is the negation of the CH. One possible approach to constructing a model (in the usual rather than Boolean sense) of L_1Set in which P is true is as follows.

We take our original countable transitive model M of set theory (i.e., of the special axioms of L_1Set), which was shown to exist in §7 of Chapter II. Let X_M and Y_M be the "representatives" of X and Y in M. (This means that if, say, X is defined by the formula $\exists!\, x\, P(x)$, then $X_M = x^\xi$, where ξ is a point of the interpretation class for which $|P(x)|_M(\xi) = 1$; see §7 of Chapter II.) We assume that X_M is infinite and Y_M is nonempty. Then "from an external point of view" X_M is countable and Y_M is at most countable, so there automatically exists a function F that maps X_M onto all of Y_M. A natural idea would be to add (the graph of) F to M, i.e., to consider the least countable model N of the axioms that contains M and F. Then N has a map from X_M onto Y_M, but it is very likely that $X_N \neq X_M$ and $Y_N \neq Y_M$. What we need in N is a map from X_N onto Y_N.

As we have shown when discussing Skolem's paradox in Chapter II, at least for certain pairs (such as $X = \omega_0, Y = \mathcal{P}(\omega_0)$), we cannot obtain a map from X onto Y in this way. In those cases in which we can construct such a map, we must choose F very carefully. Cohen's idea was that F, rather than being chosen so as to satisfy some conditions, should be chosen so as to avoid reflecting any specific properties of M, i.e., F should be "generic." We shall formulate this more precisely.

It turns out to be important to start not by choosing F directly, but by choosing the set

$$G = \{\text{restrictions of } F \text{ to finite subsets of } X_M\}.$$

Clearly, F is uniquely determined from $G : F = \cup_{g \in G} g$ (recall that a function is the same as its graph). Hence F is contained in any model that contains G. But now we must give an axiomatic characterization of the suitable G without using F explicitly. Here are the properties that G must satisfy:

9.3.

(a) $G \subseteq C$, where C is the set of maps from finite subsets of X_M to Y_M. It is important that $C \in M$, because the formula in L_1Set that defines C is (M, V)-absolute. We need this remark in order to motivate the general definitions later.

(b) $\varnothing \in G$; if $p \in G$ and $q \in C$, where $q \subseteq p$, then $q \in G$; for any $p_1, p_2 \in G$ there is $p \in G$ such that $p \supseteq p_1 \cup p_2$.

Suppose we have chosen such a set G of maps from finite subsets of X_M to Y_M. Then $\cup_{g \in G} g$ is also a map from some subset of X_M to Y_M. In order for this map to be defined on all of X_M and to be surjective, it is necessary and sufficient for the following additional conditions to hold:

$$\forall Z \in X_M, \qquad G \cap \{p \in C | p \text{ is defined at } Z\} \neq \varnothing,$$
$$\forall Z \in Y_M, \qquad G \cap \{q \in C | q \text{ takes the value Z}\} \neq \varnothing.$$

We call a subset $D \subseteq C$ *dense in* C if for all $p \in C$ there is a $q \in D$ with $p \subseteq q$. The set of maps p defined at Z and the set of maps q taking the value

Z are dense, and, moreover, are elements of M by the same consideration of (M, V)-absoluteness. Hence the two requirements at the end of the last paragraph are included in the last condition, that G be *generic*:

(c) $G \cap W \neq \varnothing$ for all dense subsets $D \subseteq C$ that are elements of M.

Although it is not yet evident, it is precisely the condition that G be generic that ensures that the properties of the sets X_M and Y_M will be preserved as much as possible after we add G to the model.

We now define the general concept of "forcing conditions."

9.4. *Forcing conditions.* These are the elements in any partially ordered set $(C, <)$ that has a maximal element 1. Usually C and $<$ lie in the original model M.

A set G is called *generic over* M (relative to C) if the following conditions hold:

(a) $G \subseteq C$;
(b) $1 \in G$; if $p \in G$ and $q \in C$, where $q \geqslant p$, then $q \in G$; for any $p_1, p_2 \in G$ there is a $p \in G$ such that $p \leqslant p_1$ and $p \leqslant p_2$;
(c) $G \cap D \neq \varnothing$ for all dense subsets $D \subsetneq C$ with $D \in M$ (D is dense if for all $p \in C$, there is a $q \in D$ with $q \leqslant p$).

If the reader compares this definition with the special case in 9.3, he or she will notice that we have replaced $\subseteq$ by $\geqslant$ and $\varnothing$ by 1. This is in keeping with Cohen's original point of view, according to which $p \geqslant q$ if, when p is considered as a "condition" imposed, say, on F, more F's satisfy p than q. (Each p fixes the restriction of F to some finite subset of X_M.)

9.5. *The existence of generic sets. Let M and C be fixed. If $M \cap \mathcal{P}(C)$ is countable, then for every $p \in C$ there exists a generic set G containing p.*

In fact, we index the elements of $M \cap \mathcal{P}(X)$ as $X_1, X_2, X_3, \ldots$ and then set

$$p_1 = p, \quad p_{n+1} = \begin{cases} p_n, & \text{if } p_n \leqslant q \text{ for all } q \in X_n; \\ \text{any } q \in X_n \text{ such that } q < p_n, & \text{otherwise.} \end{cases}$$

Finally, we set $G = \{q \in C | \exists n (p_n \leqslant q)\}$.

Conditions (a) and (b) for G to be generic are trivial to verify. Condition (c) follows because if $D \in M$ and D is dense, then there exist n and q for which $D = X_n, q \in X_n$, and $q \leqslant p_n$, so that $p_{n+1} \in D \cap G$.

9.6. *The connection with Boolean models.* As mentioned before, we have considerable freedom in our choice of the set C of forcing conditions and the generic subset $G \subseteq C$. Exactly how one "forces" a given proposition P was explained briefly in 9.2. We now show how to construct an axiom model $M[G]$ that contains M and G, once C and G have already been chosen.

The article by Shoenfield gives a direct construction, but we shall make use of an analogy with V^B, as in Jech's presentation. In this approach $M[G]$ is constructed in three basic steps:

(a) Corresponding to the set C we construct a canonical complete Boolean algebra B.
(b) We construct a Boolean universe M^B over B that is "relativized" by means of M.
(c) We construct a canonical maximal ideal $I_G \subseteq B$ determined by G and the "fiber" of the universe M^B over the quotient algebra $B/I_G \cong \{0,1\}$. It is this fiber that will be the model $M[G]$.

We now discuss these steps separately and in more detail.

9.7. *Ordered sets and Boolean algebras.* Every Boolean algebra B has a canonical partial ordering: $a \leqslant b$ if $a \wedge b = a$. All elements of the structure of B are uniquely determined by this partial ordering. The induced ordering on $B - \{0\}$ is *separable.* By definition, this means that if $a, b \neq 0$ and $a \leqslant b$, then there exists $c \leqslant a, c \neq 0$, such that there is no $d \neq 0$ for which $d \leqslant b$ and $d \leqslant c$. (It suffices to take $c = a \wedge b^{'}$.) Such b and c are called *disjoint.*

Now let C be a fixed partially ordered set. We consider the class of (non-strictly) order-preserving maps of C into different *complete* Boolean algebras B such that 0 is not contained in the image.

9.8. **Proposition.** *In this class of maps there exists a unique universal map* $e : C \to B$ *with the following properties:*

(a) $e(c)$ *is the maximal separable ordered quotient set of* C *such that* $c_1, c_2 \in C$ *are disjoint* $\Leftrightarrow e(c_1), e(c_2) \in B$ *are disjoint;*
(b) $e(c)$ *is dense in* $B - \{0\}$.

B can be realized as the algebra of regular open sets in the space C with the topology defined by the basis $U_C = \{x \in C | x \leqslant c\}, c \in C$.

Now we can indicate how I_G is constructed from the generic subset $G \subseteq C$:

$$G_1 = \{b \in B | \exists p \in G, e(p) \leqslant b\},$$
$$I_G = B \setminus G_1.$$

It is not hard to prove that I_G is a maximal ideal in B, i.e., the kernel of a Boolean homomorphism $B \to \{0,1\}$. The set G_1 is precisely the preimage of 1 under this homomorphism. Since G is generic in C, we have the following property of G_1: *for any subset* $A \subseteq B$ *such that* $\bigvee_{a \in A} a = 1$ *and* $a_1 \wedge a_2 = 0$ *whenever* $a_1 \neq a_2 \in A$, *there exists a unique element* $a \in A \cap G_1$.

9.9. *The universe* M^B. This universe is constructed from M and B in exactly the same way as V^B was constructed from V and B, with one essential difference: *all constructions are relativized with respect to* M. This means that instead of B, we take the algebra B_M that "represents" B in M (see 9.2); only ordinals $\alpha \in M$ are used in the construction of M_α^B, and so on. A rigorous presentation of these constructions would require much more formalization using the expressive means in $\mathrm{L_1Set}$ than seems desirable in this section. In such a presentation both the general plan and the details of the work would remain essentially the same as before.

The basic result of these constructions is that to every closed formula P in L_1Set with constants in M corresponds a Boolean truth value $\|P\| \in B_M$. Here the value 1 corresponds to the axioms, and deductions preserve "truth."

The next step cuts down the size of M^B, again giving a transitive standard submodel.

9.10. *Construction* of $M[G]$. For brevity, we shall write B instead of B_M, and so on. The construction essentially consists in going from "random" sets $X, Y \in M^B$ to "determined" sets $\overline{X}, \overline{Y}$, where we say that $\overline{X} \in \overline{Y}$ if the truth value $\|X \in Y\|$ goes to 1 under the homomorphism $B \to B/I_G = \{0,1\}$, i.e., if $\|X \in Y\| \in G_1$ (see 9.8). More precisely, we inductively define

$$i(\varnothing) = \varnothing,$$

and let $M[G]$ denote the image of the map $i : M^B \to V$. This notation is justified by the following result. Suppose that C and $<$ belong to M and that the subset $G \subseteq C$ is generic.

9.11. **Proposition.** *$M[G]$ is a model for the Zermelo–Fraenkel axioms that contains M and G. If M is countable, then $M[G]$ is the least such model.*

$M[G]$ contains M for the following reason. If we let $X \mapsto \hat{X}$ denote the map $M \to M^B$ that is constructed as in 8.1, then it is easy to show that $\hat{X} = X$.

$M[G]$ contains G because $G = \overline{G'}$, where G' is the object in M^B that collects all the $\hat{b}, b \in B$, with probability 1.

$M[G]$ is an axiom model basically because M^B is a Boolean axiom model. However, here we use in an essential way the assumption that G is generic. (Shoenfield verifies this result directly, without using M^B.)

9.12. Example. We return to the assertion "card $\mathcal{P}(\omega_0) \geqslant (\omega)_2$" in 9.2. By the above discussion, to prove that it is consistent with the axioms we choose a countable model M and then set

$$C = \{\text{maps of finite subsets of } \mathcal{P}(\omega_0) \text{ to } \omega_2\},$$
$$G \subseteq C = \text{a generic subset of } C.$$

If we consider a map from a subset of $\mathcal{P}(\omega_0)$ to ω_2 as a function from $\omega_0 \times \omega_2$ to $\{0,1\}$, and if, instead of "relative" constructions in M, we consider "absolute" constructions in V, then the Boolean algebra B that we obtain from C turns out to be the same algebra that was constructed in 8.3 and 8.4. This explains the appearance of B. The ideal I_G did not play any role in §8 because we were not trying to construct a standard model.

9.13. We conclude with a very general theorem of Easton, which shows how little we understand the behavior of the function 2^k (k a cardinal).

Let α be a limit ordinal. Its *cofinality* cf (α) is the least ordinal β such that α is the union of β ordinals less than α. An infinite cardinal k is called *regular* if $\mathrm{cf}(k) = k$ and is called *singular* if $\mathrm{cf}(k) < k$. König (1905) proved that $\mathrm{cf}(2^k) > k$.

9.14. **Theorem** (Easton, 1965). *Let F be any (nonstrictly) monotonic function on a subclass of the regular cardinals that takes values in the class of cardinals and that satisfies:* cf $(\aleph_{F(k)}) > \aleph_k$. *Then the assertion "$\forall$ regular $k \in$ dom $F, 2^{\aleph_k} = \aleph_{F(k)}$" does not contradict the Zermelo–Fraenkel axioms.*

If the domain of F is a set, Easton's theorem can be obtained using a model of the form $M[G]$, where M is a model in which the generalized continuum hypothesis holds (Gödel proved that such an M exists; see the next(chapter). If the domain of F is a class (for example, the class of all regular cardinals), the concept of forcing must be generalized to the case that C is a class.

For singular cardinals κ, the following result is known (Silver's theorem).

Let κ be singular, cf(κ) uncountable. Denote by κ^+ the successor cardinal to κ. If $2^{\mathrm{cf}(\lambda)} = \lambda^+$ for all infinite cardinals $\lambda < \kappa$, then $2^{\mathrm{cf}(\kappa)} = \kappa^+$.

IV

The Continuum Problem and Constructible Sets

1 Gödel's Constructible Universe

1.1. In this section we introduce the subclass $L \subset V$—"Gödel's constructible universe"—and establish its fundamental properties. Perhaps the shortest description of L is that it is the smallest transitive model of the axioms of $\mathrm{L_1Set}$ that contains all the ordinals. But the working definition of L, from which the name "constructible universe" is derived, is rather different.

We consider the following operations $F_1, \ldots, F_8$ on sets:

$$
\begin{aligned}
F_1(X,Y) &= \{X,Y\},\\
F_2(X,Y) &= X \backslash Y,\\
F_3(X,Y) &= X \times Y,\\
F_4(X) &= \{U \mid \exists W(\langle U,W\rangle \in X)\} = \mathrm{dom}\, X,\\
F_5(X) &= \{\langle U,W\rangle \mid U,W \in X;\ U \in W\},\\
F_6(X) &= \{\langle U_1,U_2,U_3\rangle \mid \langle U_2,U_3,U_1\rangle \in X\},\\
F_7(X) &= \{\langle U_1,U_2,U_3\rangle \mid \langle U_3,U_2,U_1\rangle \in X\},\\
F_8(X) &= \{\langle U_1,U_2,U_3\rangle \mid \langle U_1,U_3,U_2\rangle \in X\}.
\end{aligned}
$$

We say that a set (or class) Y is closed with respect to an operation F of degree r if we have $F(Z_1, \ldots, Z_r) \in Y$ for all $Z_1, \ldots, Z_r \in Y$ such that $F(Z_1, \ldots, Z_r)$ is defined. For every $X \in V$ we let $\mathcal{J}(X)$ denote the smallest set $Y \supset X$ that is closed with respect to the operations $F_1, \ldots, F_8$. It will later be shown (Section 1.4) that $\mathcal{J}(X)$ actually is a set. The following construction is analogous to the definition of V.

1.2. **Definition.**

$$
\begin{aligned}
L_0 &= \varnothing;\\
L_{\alpha+1} &= \mathcal{P}(L_\alpha) \cap \mathcal{J}(L_\alpha \cup \{L_\alpha\});\\
L_\alpha &= \bigcup_{\beta<\alpha} L_\beta, \ \textit{if } \alpha \textit{ is a limit ordinal};\\
L &= \cup\, L_\alpha.
\end{aligned}
$$

The elements of L are called *constructible sets.*

Yu. I. Manin, *A Course in Mathematical Logic for Mathematicians, Second Edition*, Graduate Texts in Mathematics 53, DOI 10.1007/978-1-4419-0615-1_4,

The operations $F_1, \dots, F_8$ and simple combinations of them, together with the transfinite recursion in the definition of L, exhaust the arsenal of primitive set-theoretic constructions used in mathematics. This can be seen by looking at Bourbaki's "compendium of the results of set theory," upon which all subsequent material in their voluminous treatise on the foundations of mathematics is based. The only way we could possibly (but not necessarily) leave L would be to apply the axiom of choice. This could happen provided that L is strictly less than V; but, as mentioned before, this question is undecidable in the Zermelo–Fraenkel axiom system (see also 5.16 below). Gödel was of the opinion that L does not exhaust V, as are most specialists who accept the semantics of $\mathrm{L_1Set}$.

Of course, the constructibility of the elements of L should not be understood in a finitistic sense. The sets we construct at the $(\alpha + 1)$th stage are only the subsets of L_α that are obtained from the elements of the sets L_α and $\{L_\alpha\}$ using the explicit constructions F_i. But when we consider all the ordinals indexing the stages, we see that L is hopelessly infinite. Nevertheless, in many respects the construction of L is simpler than that of V, and L seems to provide a convenient framework for mathematics.

We now list some properties of L that follow easily from the definitions. The specific nature of the operations F_i plays a very secondary role in these properties.

1.3. $L_n = V_n$ *for all* $n \leqslant \omega_0$. This is true for L_0. Suppose it is true for L_n. It is clear from the definition that $L_n \in L_{n+1}$ and $\{X\} \in L_{n+1}$ for all $X \in L_n$. Moreover, any subset of L_n can be represented as a finite difference $(\cdots(L_n \backslash \{X_1\}) \backslash \{X_2\}) \backslash \cdots \backslash \{X_k\}$, where the $X_i \in L_n$ are the elements not in the given subset.

1.4. card $L_\alpha =$ card α *for all infinite ordinals* α. In fact, for $X \in V$ let

$$\Phi(X) = X \cup \bigcup_{i=1}^{3} F_i''(X \times X) \cup \bigcup_{j=4}^{8} F_i''(X),$$

where $F''(X) = \{F(Y) | Y \in X\}$ is the image of F restricted to the elements of X. Then $\mathcal{J}(X) = \bigcup_{n=0}^{\infty} \Phi^n(X)$. It is hence clear that card $\mathcal{J}(X) =$ card X if X is infinite. We now prove the assertion 1.4 by induction on α.

Obviously card $L_\alpha \geqslant$ card α. Suppose that α is the least infinite ordinal for which card $L_\alpha >$ card α. By 1.3, we have $\alpha > \omega_0$. α cannot be a limit ordinal, or we would have card $L_\alpha = \Sigma_{\beta<\alpha}$ card $\beta =$ card α. But the case $\alpha = \beta + 1$ is also impossible, since in that case card $L_\alpha \leqslant$ card $\mathcal{J}(L_\beta \cup \{L_\beta\}) =$ card $(L_\beta \cup \{L_\beta\}) =$ card $\beta =$ card α. □

In particular, the result 1.4 shows that beginning with $w_0 + 1$, the inclusion $L_\alpha \subset V_\alpha$ becomes a strict inequality, since card $V_{\omega_0+1} = 2^{\omega_0}$. Of course, this does not in principle exclude the possibility that $\forall_\alpha\ \exists\beta > \alpha,\ L_\beta \supset V_\alpha$, but it seems that there is no such β even for $\alpha = \omega_0 + 1$.

1.5. *L is transitive:* $Y \in X \in L_\alpha \Rightarrow Y \in L_\alpha$, *i.e.*, $L_\alpha \subset L_{\alpha+1}$. See Section 13 of the appendix to Chapter II; the proof is no different for L.

1.6. *L is a big class*: by definition, this means that *for any $X \in V$ with $X \subset L$ there exists a $Y \in L$ such that $X \subset Y$.*

On L we consider the function $\phi(x)$ that is equal to the least α for which $x \in L_\alpha$. Let $X \in V$, $X \subset L$. We consider the map ϕ restricted to X. By the replacement axiom, the values of ϕ form some set Y. The elements of Y are ordinals. Let $\beta = \cup Y$. Then for each $x \in X$ we have $\beta \geqslant \phi(x)$, so that $X \subset L_\beta$.

Effective numbering of L by ordinals.

We order pairs of ordinals $\langle \alpha, \beta \rangle$ by the relation

$$\begin{aligned}\langle \alpha_1, \beta_1 \rangle < \langle \alpha_2, \beta_2 \rangle \Leftrightarrow &\text{ either } \max(\alpha_1, \beta_1) < \max(\alpha_2, \beta_2),\\ &\text{ or else these maxima are equal and } \alpha_1 < \alpha_2,\\ &\text{ or else these maxima are equal and } \alpha_1 = \alpha_2\\ &\text{ and } \beta_1 < \beta_2.\end{aligned}$$

Further, we order triples $\langle i, \alpha, \beta \rangle$, where $i = 0, \dots, 8$, by the relation

$$\begin{aligned}\langle i_1, \alpha_1, \beta_1 \rangle < \langle i_2, \alpha_2, \beta_2 \rangle \Leftrightarrow &\text{ either } \langle \alpha_1, \beta_1 \rangle < \langle \alpha_2, \beta_2 \rangle,\\ &\text{ or else } \langle \alpha_1, \beta_1 \rangle = \langle \alpha_2, \beta_2 \rangle \text{ and } i_1 < i_2.\end{aligned}$$

We call these triples *important.*

1.7. **Lemma.** *The class of important triples is well-ordered by the relation $<$. In addition, the following assertions hold:*

(a) *The next triple after $\langle i, \alpha, \beta \rangle$ has the form*

$$\begin{aligned}&\langle i+1, \alpha, \beta \rangle, && \textit{if } i \leqslant 7;\\ &\langle 0, \alpha+1, \beta \rangle, && \textit{if } i = 8 \textit{ and } \alpha+1 < \beta;\\ &\langle 0, \alpha+1, 0 \rangle, && \textit{if } i = 8 \textit{ and } \alpha+1 = \beta;\\ &\langle 0, \alpha, \beta+1 \rangle, && \textit{if } i = 8 \textit{ and } \alpha > \beta;\\ &\langle 0, 0, \beta+1 \rangle, && \textit{if } i = 8 \textit{ and } \alpha = \beta.\end{aligned}$$

(b) *Limit triples have the form*

$$\begin{aligned}&\langle 0, \alpha, \beta \rangle, && \textit{if } \alpha+1 \leqslant \beta \textit{ and } \alpha \textit{ is a limit ordinal:}\\ & && \textit{this is the limit of } \langle i, \gamma, \beta \rangle, \gamma < \alpha;\\ &\langle 0, \alpha, 0 \rangle, && \textit{if } a \textit{ is a limit ordinal: this is the limit of } \langle i, \gamma, \alpha \rangle, \gamma < \alpha;\\ &\langle 0, \alpha, \beta \rangle, && \textit{if } \alpha \geqslant \beta \textit{ and } \beta \textit{ is a limit ordinal:}\\ & && \textit{this is the limit of } \langle i, \alpha, \gamma \rangle, \gamma < \beta;\\ &\langle 0, 0, \beta \rangle, && \textit{if } \beta \textit{ is a limit ordinal: this is the limit of } \langle i, \alpha, \gamma \rangle,\ \alpha < \beta,\\ & && \gamma < \beta.\end{aligned}$$

PROOF. The proof follows immediately from the definitions. We shall illustrate this by showing explicitly how to find the least triple in any nonempty class C of triples. We set

$$\gamma = \min\{\max(\alpha,\ \beta) | \langle i,\ \alpha,\ \beta\rangle \in C\};$$
$$C_\gamma = \{\langle i,\ \alpha,\ \beta\rangle \in C | \max(\alpha,\ \beta) = \gamma\}.$$

If C_γ does not contain any triples of the form $\langle i,\ \alpha,\ \gamma\rangle$, then let β_0 be the minimum of the third coordinates of triples in C_γ, and let i_0 be the least i such that $\langle i,\ \gamma,\ \beta_0\rangle \in C_\gamma$. Then $\langle i_0,\ \gamma,\ \beta_0\rangle$ is the least triple in C. Otherwise, let C_γ' consist of triples of the form $\langle i,\ \alpha,\ \gamma\rangle \in C_\gamma$, let α_0 be the minimum of the second coordinates in C_γ', and let i_0 be the least i such that $\langle i,\ \alpha_0,\ \gamma\rangle \in C_\gamma$. Then $\langle i_0,\ \alpha_0,\ \gamma\rangle$ is the least triple in C.

The exact form of assertions (a) and (b) will be needed only in §5. The lemma implies that there exists a unique order-preserving isomorphism

$$K : \{\text{ordinals}\} \Rightarrow \{\text{important triples}\}.$$

Using this isomorphism, we recursively define a numbering mapping

$$N : \{\text{ordinals}\} \Rightarrow L.$$

Since we have $\alpha < \gamma$ and $\beta < \gamma$ if $\gamma > 0,\ i > 0$, and $K(\gamma) = \langle i,\ \alpha,\ \beta\rangle$, we may set

$$N(\gamma) = \begin{cases} L_\alpha, & \text{for } i = 0; \\ F_i(N(\alpha),\ N(\beta)), & \text{for } i = 1,2,3; \\ F_i(N(\alpha)), & \text{for } i = 4,5,6,7,8. \end{cases}$$

1.8. **Lemma.**

(a) *The mapping N is correctly defined.*
(b) *The image of N coincides with all of L.*

PROOF.

(a) To verify correctness, it suffices to show that $\{L_\alpha\} \in L$ and that the class L is closed with respect to the operations F_i. In fact, then induction on γ shows that $N(\gamma) \in L$ if $N(\alpha) \in L$ for all $\alpha < \gamma$.

Let $X, Y \in L_\alpha$. Since L is transitive (see 1.5), we easily find that $F_1(X, Y)$, $F_2(X, Y)$, and $F_4(X)$ belong to $\mathcal{P}(L_\alpha)$, and hence to $L_{\alpha+1}$. For example,

$$U \in F_4(X) \Rightarrow \exists W \langle U, W\rangle \in L_\alpha \Rightarrow \{U\} \in L_\alpha \Rightarrow U \in L_\alpha.$$

Further, $X \times Y$ is a subset of the ordered pairs of elements in L_α. We showed that the unordered pairs lie in $L_{\alpha+1}$, so that the ordered pairs lie in $L_{\alpha+2}$, and finally $X \times Y \in L_{\alpha+3}$ and $F_5(X) \in L_{\alpha+4}$. Analogously, the elements of $F_i(X)$ for $i = 6, 7, 8$ are ordered triples of elements in L_α, so that $F_i(X) \in L_{\alpha+6}$.

(b) Let Z be the image of N. We show by induction on α that $L_\alpha \subset Z$. If α is a limit ordinal and $L_\gamma \subset Z$ for each $\gamma < \alpha$ then also $L_\alpha = \bigcup_{\gamma<\alpha} L_\gamma \subset Z$. Suppose $\alpha = \beta + 1$ and $L_\beta \subset Z$, and let $X \in L_\alpha$. Then $X \in \Phi^n(L_\beta \cup \{L_\beta\})$ and we show that $X \in Z$ by induction on n.

(b_1) $n = 0$. Then either $X \in L_\beta$ so $X \in Z$ by the induction hypothesis, or else $X = L_\beta$, in which case $X = N(\gamma)$ for γ such that $K(\gamma) = \langle 0, \beta, 0\rangle$.

(b_2) $n > 0$. Let $X = F_i(Y, Z)$, $i = 1, 2, 3$; $Y, Z \in \Phi^{n-1}(L_\beta \cup \{L_\beta\})$. By the induction hypothesis, $Y = N(\gamma_1)$ and $Z = N(\gamma_2)$ for some ordinals γ_1, γ_2. Therefore $X = N(\gamma)$, where $K(\gamma) = \langle i, \gamma_1, \gamma_2\rangle$.

Let $X = F_i(Y)$, $i = 4, \ldots, 8$; $Y \in \Phi^{n-1}(L_\beta \cup \{L_\beta\})$. The verification is analogous.

The lemma is proved. □

In §3 the numbering N will allow us to prove that a strong form of the axiom of choice is L-true. The fundamental step in the proof is to choose the element with the least N-number in each constructible set.

2 Definability and Absoluteness

2.1. Let $M \subset V$ be a nonempty class, and let P be a formula in $\mathrm{L_1Set}$. As in §7 of Chapter II, we shall consider the truth values $|P|_M(\xi)$ for $\xi \in \overline{M}$, where we take the standard interpetation of $\mathrm{L_1Set}$ in V restricted to M. We then say that the formula P is M-true if $|P|_M = 1$ for all ξ.

We shall also consider formulas "with constants in M," where we assume that the language $\mathrm{L_1Set}$ has been extended so that its alphabet includes names for all the elements of M. We shall designate these elements by the same letters as in the metalanguage ($X, Y, \ldots$ for sets; $\alpha, \beta, \ldots$ for ordinals, etc.), which we hope will not lead to confusion. We extend the definition of $|P|_M(\xi)$ to formulas with constants in M in the obvious way: we take $X^\xi = X$ for any constant X and any point ξ.

2.2. **Definition.** Let $X_i \in M$, $i = 1, \ldots, n$. Sets of the form

$$\{\langle y_1^\xi, \ldots, y_n^\xi\rangle | \xi \in \overline{M},\ y_i^\xi \in X_i \text{ for } i = 1, \ldots, n;\ |P|_M(\xi) = 1\} \subset X_1 \times \cdots \times X_n$$

are called M-definable sets. Here P runs through all formulas with constants in M and free variables in the set $\{y_1, \ldots, y_n\}$.

If $P(y_1, \ldots, y_n,\ Z_1, \ldots, Z_m)$ is such a formula (where the notation shows the constants and free variables) and if $y_i^\xi = Y_i$, we shall often write "$P(Y_1, \ldots, Y_n, Z_1, \ldots, Z_m)$ is M-true" instead of $|P|_M(\xi) = 1$.

The next proposition, which, in particular, is applicable to L, is a basic instrument for proving many assertions about L.

2.3. **Proposition.** *Let $M \subset V$ be a transitive big class (see 1.6) that is closed with respect to the operations $F_1, \ldots, F_8$. Then all M-definable sets are elements of M.*

PROOF. The proof is by induction on the number of connectives and quantifiers in the defining formula P.

(a) $P(y_1, \ldots, y_n; Z_1, \ldots, Z_m)$ is an atomic formula. It can have one of eight possible forms: the predicate can be either $\in$ or $=$, and on each side of $\in$ or $=$ we can have either a constant or a variable. But all of these cases reduce to two: $y_i \in y_j$ and $y_i \in Z_j$, if we are willing to make the formula a little more complicated. For example, since M is transitive, we have

$$\text{“}y = Z\text{”} \quad \text{defines the same set as } \forall z(z \in Z \Leftrightarrow z \in y),$$
$$\text{“}Z \in y\text{”} \quad \text{defines the same set as } \exists z(z = Z \wedge z \in y),$$

and so on. We therefore analyze these two basic cases.

(a_1) $y_i \in Z$. We have $Z \cap X_i = Z \backslash (Z \backslash X_i) \in M$, since Z and $X_i \in M$, and M is F_2-closed; and we have $X_1 \times \cdots \times X_{i-1} \times Z \cap X_i \times \cdots \times X_n \in M$, since M is F_3-closed. This last set is M-definable by the formula $y_i \in Z$, because M is transitive.

(a_2) $y_i \in y_j$. We use induction on $n \geqslant 3$. Let

$$Y = \{\langle Y_1, \ldots, Y_n\rangle | Y_k \in X_k \quad \text{for } k = 1, \ldots, n;\ Y_i \in Y_j\}.$$

The case $\langle i, j\rangle = \langle n-1, n\rangle$. Let $X_{n-1} \cup X_n \subset X \in M$. Then

$$Y = \\ \times F_6(F_5(X) \times (X_1 \times \cdots \times X_{n-2}) \cap (X_{n-1} \times X_n) \times (X_1 \times \cdots \times X_{n-2})).$$

The case $\langle i, j\rangle = \langle n, n-1\rangle$. Again let $X_{n-1} \cup X_n \subset X \in M$. Then

$$Y = \\ \times F_7(F_5(X) \times (X_1 \times \cdots \times X_{n-2}) \cap (X_{n-1} \times X_n) \times (X_1 \times \cdots \times X_{n-2})).$$

The case $n \notin \{i, j\}$. By the induction assumption, the set Y', which is M-defined by the formula $y_i \in y_j$ in $X_1 \times \cdots \times X_{n-1}$, lies in M. But $Y = Y' \times X_n$.

The case $n-1 \notin \{i, j\}$. Let Y' be M-defined by the formula $y_i \in y_j$ in $X_1 \times \cdots \times X_{n-2} \times X_n$. Then $Y = F_8(Y' \times X_{n-1})$.

The case $n = 2$ reduces to the case $n = 3$ by taking the direct product with $\{\varnothing\}$ and projecting. The projection of $X_1 \times \cdots \times X_n$ onto X_1 is $F_4 \circ \cdots \circ F_4$ ($n-1$ times).

(b) *Connectives.* $\wedge$ corresponds to intersection, and $\neg$ corresponds to taking the complement (relative to $X_1 \times \cdots \times X_n$). M is closed with respect to these operations, and the other connectives can be expressed in terms of these two.

(c) *Quantifiers.* It suffices to verify $\exists$. This corresponds to projecting, because M is a big class. More precisely, let Y be M-defined by the formula $\exists y_{n+1} P(y_1, \ldots, y_n, y_{n+1})$ in $X_1 \times \cdots \times X_n$. We have

$$\langle Y_1, \ldots, Y_n\rangle \in Y \Leftrightarrow \\ \text{there exists a } Y_{n+1} \in M \text{ such that } P(Y_1, \ldots, Y_{n+1}) \text{ is } M\text{-true}.$$

To each $\langle Y_1,\ldots,Y_n\rangle \in X_1\times\cdots\times X_n$ we associate the least ordinal a for which there exists $Y_{n+1}\in M\cap V_\alpha$ such that $P(Y_1,\ldots,Y_{n+1})$ is M-true, if there is such a Y_{n+1}. This gives rise to a function on $Y\subset X_1\times\cdots\times X_n$. Let A be the set of its values, and let $\beta=\cup A$. Then $X=M\cap V_\beta$ is a set, and $X\subset M$. Since M is a big class, there exists $X_{n+1}\in M$ such that $X\subset X_{n+1}$. By the induction assumption, the M-definable subset $Y'\subset X_1\times\cdots\times X_n\times X_{n+1}$ consisting of those points $\langle Y_1,\ldots,Y_{n+1}\rangle$ for which $P(Y_1,\ldots,Y_{n+1})$ is M-true belongs to M. But $Y=F_4(Y')$, and M is closed under F_4.

The proposition is proved. □

In order to be able to use Proposition 2.3, we need criteria for verifying M-truth. As remarked in §7 of Chapter II, the basic technical tool for this is the notion of *absoluteness.* A formula P is called M-absolute ((M,V)-absolute in the terminology of Chapter II) if $|P|_M(\xi)=|P|_V(\xi)$ for all $\xi\in\overline{M}\subset\overline{V}$. The standard method of proving that a formula is M-true is to prove that it is V-true and M-absolute.

The following lemma provides us with a large class of M-absolute formulas.

2.4. Lemma.

(a) *Atomic formulas are M-absolute for all M.*

(b) *If the formulas P, P_1, and P_2 are M-absolute, then so are the formulas $\neg P$ and $P_1 * P_2$ (where $*$ is any connective).*

(c) *Suppose that the class M is transitive, and is closed with respect to an operation f of degree r. If the formula P is M-absolute, then the "restricted quantifier" formulas*

$$\forall x(x\in f(y_1,\ldots,y_r)\Rightarrow P),$$
$$\exists x(x\in f(y_1,\ldots,y_r)\wedge P)$$

are also M-absolute.

PROOF. Part (c) is the only assertion that might not be completely obvious. Before proving it, we make one remark. The formula $x\in f(y_1,\ldots,y_r)$ is written in a suitable extension of L$_1$Set, and may be assumed to be V-equivalent to some formula $P(x,y_1,\ldots,y_r)$ in L$_1$Set (with constants in M) for which $\forall y_1,\ldots,\forall y_r\ \exists!x\,P$ or a restricted version of this formula is deducible from the Zermelo–Fraenkel axioms. This P determines the operation f. We also allow the case $r=0$; then f is simply a constant in M. We shall identify f with its standard interpretation, i.e., we shall denote terms by $f(Y_1,\ldots,Y_r)\in M$ for $Y_1,\ldots,Y_r\in M$.

Now let $\xi\in\overline{M}$, $y_i^\xi=Y_i\in M$, $Q=\exists x(x\in f(y_1,\ldots,y_r)\wedge P)$, $Y=f(Y_1,\ldots,Y_r)\in M$. Then

$$|Q|_M(\xi)=\sup_{X\in M}\ (|X\in Y|_M\cdot|P|_M(\xi')),$$

where the $\xi'\in\overline{M}$ are variations of ξ along x such that $x^{\xi'}=X$. Since P is absolute, it follows that $|P|_M(\xi')=|P|_V(\xi')$, and since M is transitive, it follows that if $X \notin M$, then $|X \in Y|_M = |X \in Y|_V = 0$. Hence, on the right we can write V everywhere in place of M and can let ξ' run through all variations of ξ along x in V with $x^{\xi'} = X$. The resulting expression equals $|Q|_V(\xi)$.

The quantifier $\forall$ can be handled analogously, or else can be reduced to $\exists$. The lemma is proved. □

We shall abbreviate the restricted quantifier formulas in 2.4(c) as

$$(\forall x \in f(y_1, \dots, y_r))P, \qquad (\exists x \in f(y_1, \dots, y_r))P,$$

respectively.

If all the quantifiers in a formula Q are restricted in this way, we say that Q is a Σ_0-formula.

As a first application of the results in 2.3 and 2.4, we prove the following fact.

2.5. **Proposition.** *All ordinals are constructible.*

PROOF. Suppose that this is not the case, and that β is the least nonconstructible ordinal. All of the elements in β are contained in L_α. Since L is transitive, it follows that all $\gamma \geqslant \beta$ are nonconstructible. Hence,

$$\beta = \{x | (x \text{ is an ordinal} \wedge x \in L_\alpha) \text{ is } V\text{-true}\}.$$

If we show that "V-true" may be replaced by "L-true" here, we immediately have a contradiction, since then $\beta \in L$ by Proposition 2.3.

To do this, it suffices to verify that the formula "x is an ordinal" is L-absolute. Using the regularity axiom, from which $\neg(y \in y)$ is deducible, we can write this formula in the following Σ_0-form:

$$(\forall y \in x)(\forall z \in y)(z \in x) \wedge (\forall y_1 \in x)(\forall y_2 \in x)(y_1 \in y_2 \vee y_2 \in y_1 \vee y_1 = y_2)$$

and then apply Lemma 2.4. □

3 The Constructible Universe as a Model for Set Theory

3.1. **Theorem.** *The Zermelo–Fraenkel axioms are L-true.*

PROOF. The general principle for verifying the axioms is to note that every set whose existence is stipulated in a given axiom can be represented as a set defined by a Σ_0-formula with constants in L. We only occasionally have to perform a direct verification that a subformula is L-absolute.

(a) *Empty set.* This axiom is equivalent to the Σ_0-formula $\neg\exists x(x \in \varnothing)$, which is V-true.

(b) *Extensionality.* This axiom can be represented in Σ_0-form. In addition, in Section 4.8 of Chapter II we verified this axiom for any transitive class.

(c) *Pairing.* A direct computation of the L-truth function gives 1, since L is closed with respect to forming pairs.

(d) *Regularity.* This follows by a direct computation using the transitivity of L.

(e) *Union.* Here it is somewhat more complicated to reduce the axiom to a Σ_0-formula. The axiom is written in the form

$$\forall x\, \exists y\, \forall u (\exists z (u \in z \wedge z \in x) \Leftrightarrow u \in y).$$

Let $\xi \in \overline{L}$, let ξ' be any variation of ξ along x, and let $X = x^{\xi'} \in L$. We must show that

$$|\exists y\, \forall u (\exists z (u \in z \wedge z \in X) \Leftrightarrow u \in y)|_L(\xi') = 1.$$

It suffices to find a $Y \in L$ such that

$$|\forall u (\exists z (u \in z \wedge z \in X) \Leftrightarrow u \in Y)|_L = 1,$$

i.e., such that for all $U \in L$,

$$|(\exists z \in X)(U \in z)|_L = |U \in Y|_L.$$

We can clearly take $Y = \bigcup_{Z \in X} Z$ if we show that Y is constructible. Since L is transitive, we know that all the elements of Y are constructible. Hence, there exists a constructible set Y' such that $Y' \supset Y$. Then Y can be represented as follows (where we replace V-truth by L-truth using Lemma 2.4):

$$Y = \{U | U \in Y';\ (\exists z \in X)(U \in z) \text{ is } L\text{-true}\}.$$

Now the required assertion follows by Proposition 2.3.

In what follows we shall usually omit explicit mention of the points $\xi \in \overline{L}$.

(f) *Power set axiom* $\forall x\, \exists y\, \forall z (z \subset x \Leftrightarrow z \in y)$. We fix $X \in L$, form the set $Y = \mathcal{P}(X) \cap L$ of constructible subsets of X, and show that Y is constructible. In fact, let $Y' \supset Y$, where Y' is constructible. Then by Lemma 2.4,

$$Y = \{Z | Z \in Y';\ (Z \subset X) \text{ is } L\text{-true}\},$$

because $Z \subset X$ has the Σ_0-form $(\forall z \in Z)(z \in X)$. Now a direct computation gives

$$|\forall z (z \subset X \Leftrightarrow z \in Y)|_L = 1.$$

(g) *Infinity.* This axiom is L-true because of the constructibility of the set $\{\varnothing, \{\varnothing\}, \{\{\varnothing\}\}, \dots\}$, which can be represented in the form

$$\{Y | Y \in L_{\omega_0};\ [Y = \varnothing \vee (\exists y \in L_{\omega_0})(Y = \{y\})] \text{ is } L\text{-true}\}.$$

(h) *Replacement.* Let $\overline{z} = \langle z_1, \dots, z_n \rangle$. This axiom is written in the form

$$\forall \overline{z}\, \forall u \Big(\forall x (x \in u \Rightarrow \exists! y P(x, y, \overline{z}))$$

$$\Rightarrow \exists w\, \forall y (y \in w \Leftrightarrow \exists x (x \in u \wedge P(x, y, \overline{z}))) \Big).$$

We fix $Z_1, \ldots, Z_n \in L, \overline{Z} = \langle Z_1, \ldots, Z_n \rangle$, and $U \in L$. It is sufficient to consider the case that the premise is L-true, i.e., for all $X \in L$,

$$|X \in U \Rightarrow \exists! y P(X, y, \overline{Z})|_L = 1.$$

We must find a value $W \in L$ of w for which the conclusion is L-true. We set $W^{'}$ = a constructive set containing as elements all constructive Y for which

$$(\exists x \in U) P(x, Y, \overline{Z}) \text{ is } L\text{-true.}$$

This set exists because since the premise of the axiom is L-true, it follows that each $X \in U$ corresponds to at most one constructible Y. We then set

$$W = \left\{Y | Y \in W^{'}; (\exists x \in U)\Big(P(x, Y, \overline{Z})\Big) \text{ is } L\text{-true}\right\}.$$

This set is constructible by Proposition 2.3, and it follows from the way it is defined that

$$\left|\forall y \Big(y \in W \Leftrightarrow \exists x (x \in U \wedge P(x, y, \overline{Z}))\Big)\right|_L = 1.$$

(i) *Axiom of choice.* The main intuitive point in the verification is the numbering N of the universe L that was constructed in 1.8. But the formal verification is much more complicated here than in the previous cases. A fair amount of work is needed to give a formalization of the construction in 1.7–1.8 that is sufficiently detailed to prove the following fact:

3.2. **Proposition.** *There exists a formula $N(x, y)$ in L with two free variables such that*

(a) *For any $X, Y \in V$, the formula $N(X, Y)$ is V-true if and only if X is an ordinal and $Y = N(X)$.*
(b) *$N(x, y)$ is L-absolute.*

We shall postpone the proof until §5, and shall make use of this proposition to verify the axiom of choice. We divide this verification into two steps.

3.3. Universal choice function. Let $X \in L$ be a nonempty set. We construct the function Y that for every nonempty $Z \in X$ chooses the element U in Z with the least N-number (see 1.8):

$$Y = \Big\{\langle Z, U \rangle | Z \in X,\ U \in \bigcup_{X' \in X} X^{'};\ U \in Z \wedge \exists w \Big(N(w, U) \wedge \forall z \big(z \in Z$$
$$\Rightarrow (z = U \ \vee \ \forall w^{'} (N(w^{'}, z) \Rightarrow w \in w^{'}))\big)\Big) \text{ is } V\text{-true}\Big\}.$$

We want to prove that $Y \in L$. By Proposition 2.3, this holds if we can define Y by means of the L-truth of a formula. We are not allowed mechanically to replace V by L, since it is not immediately obvious from its external form that this formula is L-absolute. We proceed as follows: taking into account the constructibility of the ordinals, we take all ordinals that occur as the least

N-numbers of the elements of the constructible set $\cup_{X' \in X} X' = \cup(X)$, and we find a constructible set W that contains these ordinals. Then we replace $\exists w$ by $\exists w \in W$ and $\forall w'$ by $\forall w' \in W$ in the formula. The set Y does not change, and now V-truth may be replaced by L-truth, as can be seen using Proposition 3.2 and Lemma 2.4.

3.4. We now compute the L-truth value of the axiom of choice:

$$\forall x(x \neq \varnothing \Rightarrow \exists y(y \text{ is a function} \wedge \text{dom } y = x \wedge (\forall z \in x)(z \neq \varnothing \Rightarrow y(z) \in z))).$$

It suffices to show that if we take a nonempty $X \in L$ and the constructible choice function $Y \in L$ in 3.3, then

$$|Y \text{ is a function}|_L = |\text{dom } Y = X|_L = |(\forall z \in X)(z \neq \varnothing \Rightarrow Y(z) \in z)|_L = 1.$$

The third formula here is V-true, and is written in Σ_0-form except for the subformula $Y(z) \in z$, which can be replaced by $(\forall u \in U(Y))(\langle z, u\rangle \in Y \Rightarrow u \in z)$. Thus, the third formula is L-absolute and hence L-true.

We verify that the first two formulas are absolute in §5. They are V-true by construction. This completes the proof of Proposition 3.1. □

We note that the same argument shows the following: *all the axioms*, with the possible exception of the axiom of choice, *are M-true for any transitive big class M that is closed with respect to the operations* $F_1, \ldots, F_8$.

4 The Generalized Continuum Hypothesis Is L-True

4.1. We wish to show that the assertion "card $\mathcal{P}(\omega_\alpha) = \omega_{\alpha+1}$" is L-true. A certain amount of caution is essential here, because cardinality is not an L-absolute notion. If Y is a constructible set, let $\text{card}_L(Y)$ be the least ordinal β for which there exists in L a one-to-one onto function $f : Y \to \beta$. Hence "card (Y) = card (Z)" is L-true iff $\text{card}_L(Y) = \text{card}_L(Z)$. Note that although $\text{card}_L(Y) \geqslant \text{card}\,(Y)$, equality fails if there are one-to-one onto functions $Y \to \beta$ in V, but no such function lies in L. The cardinal ω_α in L is the αth ordinal $\beta > \omega_0$ such that $\text{card}_L(\beta) = \beta$. Thus ω_α in L may not coincide with the "real" ω_α, that is, with ω_α in V.

We shall show that for each ordinal β and each constructible $X \subset \beta$ there is an ordinal γ with $X \in L_\gamma$ and $\text{card}_L(\gamma) = \text{card}_L(\beta)$. Hence $\mathcal{P}(\beta) \cap L \subset L_{\beta^+}$, where β^+ is the least ordinal greater than β such that $\text{card}_L(\beta^+) \neq \text{card}_L(\beta)$. The L-truth of the generalized continuum hypothesis will then follow if we show the L-truth of "card $(\beta^+) = \beta^+$."

Our proof exploits throughout a proposition that requires a good deal of work formalizing the construction of L within L_1Set.

4.2. **Proposition.** *There exists a formula $L(x, y)$ of* L_1Set *with two independent variables* *such that*

(a) *for any* X *and* Y *in* V, $L(X,Y)$ *is* V*-true* $\Leftrightarrow$ Y *is an ordinal and* $X \in L_\gamma$;

(b) *for any transitive model* $M \subset V$ *of the axioms (without the axiom of choice), the formula* $L(x,y)$ *is* M*-absolute. In particular, it is* L*-absolute.*

We again postpone the proof until §5.

4.3. **Lemma.** *Let* $X \subseteq \beta$ *be constructible. Then* $X \in L_\gamma$ *for some ordinal* γ *such that* $\operatorname{card}_L(\gamma) = \operatorname{card}_L(\beta)$.

PROOF. In this deduction, in addition to Proposition 4.2 we use versions of Propositions 7.3 and 7.6 of Chapter II that apply to the constructible universe. They are formulated precisely and proved below, in Sections 4.5 and 4.6.

Suppose that $X \subset \beta$ is constructible. Let δ be an ordinal such that $X \in L_\delta$. We enlarge the alphabet of $\mathrm{L_1Set}$ by adding names $\bar{\delta}$ and $\overline{X}$ for δ and X. Let $\mathcal{E}$ be the set of formulas

$$\{\text{axioms of } \mathrm{L_1Set}\} \cup \{L(\overline{X}, \bar{\delta})\}.$$

Let $N_0 \subset L$ be the set $\beta \cup \{X\} \cap \{\delta\}$. By Proposition 4.5 there is a constructible set N such that $N_0 \subset N$, all formulas in $\mathcal{E}$ are (N, L)-absolute, and $\operatorname{card}_L(N) = \operatorname{card}_L(\beta)$. Thus $(N, \in)$ is a model for the axioms and, by Proposition 4.2 (a), for $L(\overline{X}, \bar{\delta})$. Now N might not be transitive, but then by Proposition 4.6 there are a transitive axiom model (M, ε) and a constructible isomorphism $f : (N, \in) \xrightarrow{\sim} (M, \varepsilon)$. Hence $L(\overline{X}, \bar{\delta})$ is M-true and $\operatorname{card}_L(M) = \operatorname{card}_L(N)$. What are the interpretations of the constants $\overline{X}$ and $\bar{\delta}$ in M?

Since the set $\beta \subset N$ is transitive, it goes to itself under the isomorphism f; hence so does the set $X \subset \beta$. Let δ_M be the image of δ under f. Since by Proposition 4.2(b) the formula $L(x, y)$ is M-absolute, and $L(\overline{X}, \bar{\delta})$ is M-true, it follows that $L(X, \delta_M)$ is V-true, so that δ_M is an actual ordinal and $X \in L_{\delta_M}$. Moreover, since $\delta_M \in M$ and M is transitive, $\delta_M \subset M$; hence $\operatorname{card}_L(\delta_M) \leqslant \operatorname{card}_L(M)$. Letting γ be the larger of δ_M and β, we have $\operatorname{card}_L(\gamma) = \operatorname{card}_L(\beta)$ and $X \in L_\gamma$. The lemma is proved. □

4.4. DEDUCTION THAT THE GCH IS L-TRUE FROM THE LEMMA. Let β^+ be the smallest ordinal greater than β such that $\operatorname{card}_L(\beta^+) \neq \operatorname{card}_L(\beta)$. Then Lemma 4.3 implies the V-truth of the formula

$$\forall z (z \in L \Rightarrow (z \subset \beta \Rightarrow z \in L_{\beta^+})).$$

Since "$z \in L_{\beta^+}$" (i.e., the formula $L(z, \beta^+)$) is L-absolute, it follows that

$$\forall z (z \subset \beta \Rightarrow z \in L_{\beta^+})$$

is L-true. Now if β is the cardinal ω_α in L then β^+ is the cardinal $\omega_{\alpha+1}$ in L. Hence for each α we have shown the L-truth of

$$\mathcal{P}(\omega_\alpha) \subset L_{\omega_{\alpha+1}}.$$

We claim that the following formula is also L-true:

$$\mathrm{card}(L_{\omega_{\alpha+1}}) = \omega_{\alpha+1}.$$

Since "$\mathrm{card}(\mathcal{P}\omega_\alpha)) \leqslant \omega_{\alpha+1}$" is formally deducible in $\mathrm{L_1Set}$ from the preceding two formulas, and since all the axioms are L-true, this will show that the GCH is L-true.

Our claim is verified thus: In Section 1.4 we proved that $\mathrm{card}(L_\gamma) = \mathrm{card}(\gamma)$ for each ordinal γ. Indeed, that proof can be formalized in $\mathrm{L_1Set}$, using the formula $L(x, y)$ of Proposition 4.2. That is, the assertion "$\forall\gamma(\mathrm{card}(L_\gamma) = \mathrm{card}(\gamma))$" is deducible from the axioms (see 5.17). Since the axioms are L-true, this assertion is then L-true. But since "$\mathrm{card}(w_{\alpha+1}) = w_{\alpha+1}$" is trivially L-true, the claim follows. This completes the proof. □

4.5. **Proposition.** *Let $\mathcal{E}$ be a constructible countable set of L-true formulas in the language* $\mathrm{L_1Set}$, *and let M_0 be a constructible set. Then there exists a constructible set $M \supset M_0, \mathrm{card}_L(M) \leqslant \mathrm{card}_L(M_0) + \omega_0$, such that all of the formulas in $\mathcal{E}$ are (M, L)-absolute.*

PROOF. The general scheme is the same as in Section 7.3 of Chapter II, but some additional precautions are required. The main point is to prove that if $P(x, \overline{y}), \overline{y} = (y_1, \ldots, y_n)$, is a formula in $\mathcal{E}$, then there exists a constructible set $M \supset M_0$ with $\mathrm{card}_L(M) \leqslant \mathrm{card}_L(M_0) + \omega_0$ that can be constructed constructibly from P and has the property that $\exists x(P(x, \overline{y}))$ is (M, L)-absolute. After this we must verify constructible closure over all $P \in \mathcal{E}$.

We reproduce the construction in Section 7.3 of Chapter II. We construct the set M_i by induction. Let $\overline{Y} = \langle Y_1, \ldots, Y_n\rangle \in M_i \times \cdots \times M_i$. We let $\hat{M}_i(\overline{Y})$ denote the class $\{X | P(X, Y_1, \ldots, Y_n)$ is L-true$\}$. We let $\tilde{M}_i(\overline{Y})$ denote $\varnothing$ if $\hat{M}_i(\overline{Y})$ is empty, and $\hat{M}_i(\overline{Y}) \cap L_\alpha$ for the least α for which this intersection is nonempty otherwise. Since $L(x, y)$ is absolute (see §5), it is not hard to see that the function $\tilde{M}_i$, dom $\tilde{M}_i = M_i \times \cdots \times M_i$, is constructible. Because the constructible axiom of choice holds in L, we can obtain a constructible function F_i by choosing one element from each nonempty $\tilde{M}_i(\overline{Y})$. Let N_i be the set of values of $\tilde{M}_i$. This set is constructible, since all of our constructions are absolute; and if M_i is infinite, then $\mathrm{card}_L(N_i) = \mathrm{card}_L(M_i)$. We set $M_{i+1} = M_i \cup N_i$ and $M = \cup M_i$. The set M has the required properties; obviously, $\mathrm{card}_L(M) + \omega_0 = \mathrm{card}_L(M_0) + \omega_0$ in L. The formal transition from $\{M_i\}$ to M is realized by considering a function that "closes" M_0, as in Section 5.11 below. □

4.6. **Proposition.** *For every constructible set N such that the extensionality axiom is N-true there exist a unique constructible transitive set M and isomorphism $f : (N, \in) \tilde{\to} (M, \varepsilon)$.*

PROOF. The plan of proof is the same as in Section 7.6 of Chapter II. First let "f is a continuous $(\alpha + 1)$-sequence" be the formula "α is an ordinal"$\wedge$"f is a function"$\wedge$ dom$f = \alpha + 1 \wedge (\forall\beta \in \alpha + 1)(\beta$ a limit ordinal $\Rightarrow f(\beta) = \bigcup_{\gamma\in\beta} f(\gamma))$. This formula is shown to be L-absolute as in Section 5.14 below.

Now consider the L-absolute operation $\phi(Z) = \{X | X \in N \wedge X \cap N \subset Z\}$, and let $\varnothing_N$ be the unique member of N such that $\varnothing_N \cap N = \varnothing$. Finally, let $\psi(x, y)$ be the formula

$$(\exists f)(\text{“}f \text{ is a continuous } (x+1)\text{-sequence”} \wedge f(0) = \varnothing_N \wedge \\ \times (\forall \beta \in x)(f(\beta+1) = \phi(f(\beta))) \wedge y = f(x)).$$

Then ψ is L-absolute, as can be shown as in Sections 5.14 and 5.15 below, and $\psi(x, y)$ is L-true if and only if $y = N_x$ in the sense of Chapter II, Section 7.6.

We now set $\hat{N} = \cup_\alpha N_\alpha = \{z | (\exists \alpha)(\exists y \subset N)(\psi(\alpha, y) \wedge z \in y)\}$. We show that $\hat{N} = N$. Clearly $\hat{N} \subset N$, and if $N \backslash \hat{N} = Y$ were nonempty, it would follow by the regularity axiom, which holds in L, that $\exists Z(Z \in Y \wedge Z \cap Y = \varnothing)$. For this Z we would have $Z \subset \hat{N}$, hence $Z \subset N_\alpha$ for a suitable α, so that $Z \in N_{\alpha+1}$, which is a contradiction.

The implication $Z \subset \hat{N} \Rightarrow \exists \alpha(Z \subset N_\alpha)$, which we have used here, follows because there exists an absolute function on $\hat{N}$ that associates to each X the least α for which $X \in N_\alpha$. The replacement axiom shows that there exists an ordinal α_0, namely, the least upper bound of the values of this function, for which $\hat{N} = N = N_{\alpha_0}$. This ordinal, which is fixed for N, occurs in our subsequent construction, which is verified to be absolute as in §5.

Let “h is a constructing $(\alpha+1)$-sequence for N, M” be the formula “h is a continuous $(\alpha+1)$-sequence” $\wedge\ h(0) = \{\langle \varnothing_N, \varnothing \rangle\} \wedge$ “$(\forall \beta \in \alpha)(h(\beta+1)$ is a function $\wedge$ dom $h(\beta+1) = N_{\beta+1} \wedge$ the value of $h(\beta+1))$ on any $X \in N_{\beta+1}$ is the set of $h(\beta)$-images of elements of $X \cap N)$.” Then for each α there is a unique such h; let M_α be the image of $h(\alpha)$. For $\alpha = \alpha_0$ we obtain a function $h : N \to M = M_{\alpha_0}$, where M is our desired constructible set and h is a constructible $\in$-isomorphism.

The proposition is proved. □

5 Constructibility Formula

5.1. The purpose of this section is to prove Propositions 4.2 and 3.2. Both proofs are extremely straightforward, and simply consist in writing out explicitly the formulas $L(x, y)$ and $N(x, y)$ and verifying that the conditions in Lemma 2.4 apply. But since these formulas are very long, we perform the verifications in a series of “blocks,” in order to improve their appearance and to make the interpretation and verification of the conditions in 2.4 easier. As soon as a block (subformula) is constructed and its absoluteness is verified, we replace it by an abbreviated notation in the next formula.

The material within each subsection is arranged in the following order: first the abbreviated notation for the formula that is being constructed and shown to be absolute in the subsection; then the complete form of the formula; and finally any remarks that may be needed regarding absoluteness. The “complete form” of the formula may contain abbreviated notation for subformulas. If such a subformula has not yet been interpreted in detail and shown to be absolute, this is done right after the complete form.

By absoluteness we mean "M-absoluteness for any transitive model M for the axioms without the axiom of choice."

Sections 5.2–5.15 are devoted to the formula $L(x, y)$, and Sections 5.18–5.20 are devoted to the formula $N(x, y)$. As the material we are dealing with accumulates, we shall allow ourselves to omit more and more details and to rely on the reader's experience.

The formulas

$$z = \begin{cases} F_i(x, y), & i = 1, 2, 3; \\ F_j(y), & j = 4, 5, 6, 7, 8. \end{cases}$$

5.2. $z = \{x, y\}$: $(\forall u \in z)(u = x \vee u = y) \wedge x \in z \wedge y \in z$. This whole formula is clearly absolute by Lemma 2.4. From now on we shall not even comment on such simple cases.

5.3. $z = x \backslash y$: $(\forall u \in z)(u \in x \wedge u \notin y) \wedge (\forall u \in x)(u \notin y \Rightarrow u \in z)$.

5.4. $z = x \times y$: $(\forall u_1 \in x)(\forall u_2 \in y)(\langle u_1, u_2 \rangle \in z)$

$\wedge (\forall u \in z)(\exists u_1 \in x)(\exists u_2 \in y)(u = \langle u_1, u_2 \rangle)$;

$\langle u_1, u_2 \rangle \in z$: $(\exists v \in z)(v = \langle u_1, u_2 \rangle)$;

$u = \langle u_1, u_2 \rangle$: $(\forall v \in u)(v = \{u_1\} \vee v = \{u_1, u_2\})$

$\wedge \{u_1\} \in u \wedge \{u_1, u_2\} \in u$;

$\{u_1, u_2\} \in u$: $(\exists v \in u)(v = \{u_1, u_2\})$.

5.5. $Z = F_4(y) = \operatorname{dom} y$: $(\forall u \in z)(\exists v \in \bigcup\bigcup (y))(\langle u, v \rangle \in y)$

$\wedge (\forall u \in \bigcup\bigcup (y))(\forall v \in \bigcup\bigcup (y))(\langle u, v \rangle \in y \Rightarrow u \in z)$.

Here $\bigcup\bigcup$ appears because $\langle u, v \rangle = \{\{u\}, \{u, v\}\} \in y \Rightarrow u, v \in \bigcup\bigcup (y)$. This formula is absolute, since a transitive model is closed with respect to the operation $\bigcup$ (see 3.1(e)). We shall write $\bigcup^2 = \bigcup\bigcup$, and so on.

5.6. $z = F_5(y)$: $(\forall u \in z)(\exists v \in y)(\exists w \in y)(v \in w \wedge u = \langle v, w \rangle) \wedge (\forall v \in y)$ $(\forall w \in y)(v \in w \Rightarrow \langle v, w \rangle \in z)$.

5.7. $z = F_6(y)$: $(\forall u \in z)(\exists u_1 \in \bigcup^4(y))(\exists u_2 \in \bigcup^4(y))(\exists u_3 \in \bigcup^2(y))(\langle u_1, u_2, u_3 \rangle \in y \wedge u = \langle u_3, u_1, u_2 \rangle) \wedge (\forall u_1 \in \bigcup^4(y))(\forall u_2 \in \bigcup^4(y))(\forall u_3 \in \bigcup^2(y))(\langle u_1, u_2, u_3 \rangle \in y \Rightarrow \langle u_3, u_1, u_2 \rangle \in z)$. Here $\bigcup^4$ appears for the same reason as $\bigcup^2$ in 5.5. The formulas $\langle u_1, u_2, u_3 \rangle \in y$, etc., are shown to be absolute in the same way as in 5.4.

The operations F_7 and F_8 are treated analogously to F_6.

The formulas

$$y = \begin{cases} F_i''(x \times x), & \text{for } i = 1, 2, 3; \\ F_j''(x), & \text{for } j = 4, 5, 6, 7, 8. \end{cases}$$

5.8. $y = F_i''(x \times x)$, $i = 1, 2, 3$:

$$(\forall u \in y)(\exists u_1 \in x)(\exists u_2 \in x)(u = F_i(u_1, u_2))$$
$$\wedge(\forall u_1 \in x)(\forall u_2 \in x)(F_i(u_1, u_2) \in y),$$

where $F_i(u_1, u_2) \in y: \ (\exists v \in y)(v = F_i(u_1, u_2))$.

5.9. $y = F_j''(x), j = 4, \ldots, 8$: $(\forall u \in y)(\exists v \in x)(u = F_j''(v)) \wedge (\forall v \in x)$ $(F_j''(v) \in y)$.

5.10. $y = \Phi(x)$ (see 1.4):

$$(\forall z \in y)(z \in x \vee z \in F_1''(x \times x) \vee \cdots \vee z \in F_8''(x)) \wedge (\forall z \in x)(z \in y)$$
$$\wedge (\forall z \in F_1''(x \times x))(z \in y) \wedge \cdots \wedge (\forall z \in F_8'' y(x))(z \in y).$$

The class L is closed with respect to the operations F_i''. In fact, suppose, for example, that $i \geqslant 4$, and let $X \in L$. Let $U \in L$ be a set containing all $F_i(Y)$ for $Y \in X$. Then

$$F_1''(X) = \{Z | Z \in U, \ (\exists y \in X)(Z = F_i(y)) \text{ is } V\text{-true}\}.$$

Since the formula $Z = F_i(y)$ has been shown to be absolute, we may replace "V-true" by "L-true" here, and then apply Proposition 2.3. Thus, the formula $y = \Phi(x)$ is L-absolute by Lemma 2.4.

If M is an arbitrary transitive model, then the verification that M is closed with respect to F_i'' is somewhat different. Namely, the formula $\forall x \ \exists! y(y = F_i''(x))$ is obviously V-true. The formal deduction of this formula does not use the axiom of choice. Hence, the formula is M-true for any transitive model M. We therefore have $Y \in M$ if $X \in M$, where $Y = F_i''(X)$. We shall use this device many times in what follows.

5.11. "*g closes x*," which is short for "*g is a function on* ω_0, *and* $g(n) = \Phi^n(x)$ *for all* $n \in \omega_0$." We write the formula with the constant ω_0 and the free variables g and x:

$$\text{"}g \text{ is a function"} \wedge F_4(g) = \omega_0 \wedge g(0) = x$$
$$\wedge (\forall n \in \omega_0)(g(n+1) = \Phi(g(n))).$$

Here:

(a) "g is a function":

$$(\forall u \in g)(\exists u_1 \in \cup^2(g))(\exists u_2 \in \cup^2(g))(u = \langle u_1, u_2 \rangle)$$
$$\wedge (\forall u_1 \in \cup^2(g))(\forall u_2 \in \cup^2(g))(\forall u_3 \in \cup^2(g))$$
$$(\langle u_1, u_2 \rangle \in g \wedge \langle u_1, u_3 \rangle \in g \Rightarrow u_2 = u_3).$$

(b) $g(0) = x: \ \langle \varnothing, x \rangle \in g$.
(c) $g(n+1) = \Phi(g(n))$:

$$(\exists y \in \cup^2(g))(\langle n, y \rangle \in g \wedge \langle n \cup \{n\}, \ \Phi(y) \rangle \in g),$$

where

$$\langle n \cup \{n\},\ \Phi(y)\rangle \in g: \quad (\exists u \in \bigcup^2(g))(\exists v \in \bigcup^2(g))$$
$$(u = n \cup \{n\} \wedge v = \Phi(y) \wedge \langle u, v\rangle \in g).$$

Since $\omega_0 \in M$, the formula 5.11 is now easily seen to be absolute by the previous results.

In 5.11 we took the liberty of using g and n for variables of $\mathrm{L_1Set}$ in order to make the formulas intuitively clearer. In what follows we shall also use α, β, K, and N as variables, thereby temporarily ignoring our convention of using only lowercase letters at the end of the Latin alphabet.

5.12. $y \in \mathcal{J}x : \exists g($"$g$ closes x"$\wedge(\exists n \in \omega_0)(\langle n, y\rangle \in g))$. Here the quantifier over g is not restricted. Since the formula under the $\exists g$ sign is absolute, we may conclude directly from the definition $\|_L(\xi) = \|_V(\xi), \xi \in \overline{M}$, that $y \in \mathcal{J}x$ is also absolute, provided we show that for any $X \in M$, the function $G \in V$ that closes X lies in M. The formula $\forall x\ \exists!\ g$ ("g closes x") is obviously V-true. If we formalize the verification of this fact, we see that this formula is deducible from the axioms without the axiom of choice. Hence it is M-true. This implies that for any $X \in M$ we have $G \in M$.

5.13. $y \in \mathcal{P}(x) \cap \mathcal{J}(x \cup \{x\}) : \ (\forall z \in y)(\forall v \in z)(v \in x) \wedge y \in \mathcal{J}(x \cup \{x\})$.

5.14. "f is the constructing $(\alpha + 1)$-sequence," which is short for "α is an ordinal"$\wedge$ "f is a function"$\wedge$ $\mathrm{dom}\, f = \alpha + 1 \wedge (\forall \beta \in \alpha + 1)(f(\beta) = L_\beta)$.

Here:

(a) $(\forall \beta \in \alpha + 1)(f(\beta) = L_\beta)$:

$$(\forall \beta \in \alpha + 1)((\beta \text{ is a limit ordinal } \Rightarrow f(\beta) = \bigcup_{\gamma\in\beta} f(\gamma))$$
$$\wedge(f(\beta + 1) = \mathcal{P}(f(\beta)) \cap \mathcal{J}(f(\beta) \cup \{f(\beta)\}))).$$

(b) "β is a limit ordinal": "β is an ordinal"$\wedge(\forall \alpha \in \beta)(\beta \neq \alpha \cup \{\alpha\})$.

(c) $f(\beta) = \bigcup_{\gamma\in\beta} f(\gamma) : \quad (\exists v \in \bigcup^2(f))(v = \bigcup_{\gamma\in\beta} f(\gamma) \wedge \langle \beta, v\rangle \in f)$;

$$v = \bigcup_{\gamma\in\beta} f(\gamma): \quad (\forall u \in v)(\exists \gamma \in \beta)(u \in f(\gamma))$$
$$\wedge\, (\forall u \in \bigcup^3(f))(u \in f(\gamma) \Rightarrow u \in v);$$
$$u \in f(\gamma): \quad (\exists w \in \bigcup^2(f))(\langle \gamma, w\rangle \in f \wedge u \in w).$$

(d) $f(\beta + 1) = \mathcal{P}(f(\beta)) \cap \mathcal{J}(f(\beta) \cup \{f(\beta)\})$:

$$(\exists u \in \bigcup^2(f))(\langle \beta + 1, u\rangle \in f \wedge (\forall v \in u)$$
$$(v \in \mathcal{P}(f(\beta)) \cap \mathcal{J}(f(\beta) \cup \{f(\beta)\}))$$
$$\wedge\, \forall v(v \in \mathcal{P}(f(\beta)) \cap \mathcal{J}(f(\beta) \cup \{f(\beta)\}) \Rightarrow v \in u));$$
$$v \in \mathcal{P}(f(\beta)) \cap \mathcal{J}(f(\beta) \cup \{f(\beta)\}):$$
$$(\exists u \in \bigcup^2(f))(\langle \beta, u\rangle \in f \wedge v \in \mathcal{P}(u) \cap \mathcal{J}(u \cup \{u\})).$$

Finally, in order to verify directly that the subformula

$$\forall v(v \in \mathcal{P}(f(\beta)) \cap \mathcal{J}(f(\beta) \cup \{f(\beta)\}) \Rightarrow v \in u)$$

is M-absolute, it suffices to show that M is closed with respect to the operation $X \mapsto \mathcal{P}(X) \cap \mathcal{J}(X \cup \{X\})$. But M is closed with respect to both $\mathcal{J}$ and $X \mapsto \mathcal{P}(X) \cap M$, so the verification is complete.

5.15. $L(x, y)$: "y is an ordinal and $x \in L_y$": "y is an ordinal"$\wedge\exists f$("f is the constructing $(y+1)$-sequence"$\wedge(\exists z \in \cup^2(f))(\langle y, z\rangle \in f \wedge x \in z)$). Since the quantifier $\exists f$ is not bounded, in order to verify this last absoluteness statement we must show that the constructing $(Y+1)$-sequence F is an element of M for any ordinal Y in M. We use the same argument as in 5.12: the formula $\forall y$(y is an ordinal $\Rightarrow \exists! f$(f is the constructing $(y+1)$-sequence)) not only is V-true, but also is deducible from the axioms without the axiom of choice; therefore it is M-true.

This completes the proof of Proposition 4.2. □

5.16. *Remark.* The formula $\forall x\ \exists y\ L(x, y)$ is often written in the form $V = L$, and is called the *axiom of constructibility.* The absoluteness of $L(x, y)$ implies that the following formula is L-true:

$$|\forall x\ \exists y\ L(x, y)|_L = \inf_{X \in L} \sup_{Y \in L} |L(X, Y)|_L = \inf_{X \in L} \sup_{Y \in L} |L(X, Y)|_V = 1.$$

Hence, this formula is consistent with the Zermelo–Fraenkel axioms. On the other hand, $V = L$ implies the generalized continuum hypothesis (GCH), and since the negation of the GCH is also consistent with the Zermelo–Fraenkel axioms, it follows that $\neg(V = L)$ is consistent with the axioms.

We now proceed to the proof of Proposition 3.2. This proof follows the same plan as the proof of Proposition 4.2. We return to the conventions and constructions in 1.7–1.8.

5.17. *Remark.* In Section 4.4 we exploited the fact that the assertion "$\alpha \geqslant \omega_0 \Rightarrow \text{card}(L_\alpha) = \text{card}(\alpha)$" is formally deducible from the axioms of L_1Set (without the axiom of choice). We may now see that such a formal deduction can be obtained by exactly mimicking the proof in Section 1.4. Indeed, from the definition of $L(x, y)$ we have the formal deducibility of "$L_{\alpha+1} = \mathcal{P}(L_\alpha) \cap \mathcal{J}(L_\alpha \cup \{L_\alpha\})$" and "$\beta$ a limit ordinal $\Rightarrow L_\beta = \cup_{\gamma\in\beta} L_\gamma$". Moreover, the following are deducible: "$\text{card}(X) < \omega_0 \Rightarrow \text{card}(X) < \text{card}(\mathcal{P}(X)) < \omega_0$" and "$\text{card}(X) \geqslant \omega_0 \Rightarrow \text{card}(\mathcal{J}(X)) = \text{card}(X)$." As a result, the assertions "$\text{card}(L_{\omega_0}) = \omega_0$," "$\text{card}(L_a) \geqslant \omega_0 \Rightarrow \text{card}(L_{\alpha+1}) = \text{card}(L_\alpha)$," and "$\beta$ a limit ordinal $\Rightarrow \text{card}(L_\beta) = \text{card}(\cup_{\gamma\in\beta} L_\gamma)$" are all deducible. And from these and the axioms of L_1Set the desired assertion may be deduced (using, in particular, the deducibility of "$\text{card}(\omega_0) = \omega_0$," "$\alpha \geqslant \omega_0 \Rightarrow \text{card}(\alpha+1) = \text{card}(\alpha)$," "$\beta$ is a limit ordinal $\Rightarrow \beta = \cup_{\gamma\in\beta}\gamma$," and in addition an instance of transfinite induction on the ordinals, which is of course also formally deducible in L_1Set).

5.18. *The formula* $H(K, x)$: K is a function $\wedge$ x is an ordinal $\wedge$ dom $K = x+1 \wedge K(0) = \langle 0,0,0\rangle \wedge (\forall y \in x+1)(K(y)$ is an important triple $\wedge K(y+1)$ is the next important triple after $K(y)) \wedge (y$ is a limit ordinal $\Rightarrow K(y) = \lim_{z\in y} K(z))$ *is absolute.*

We shall not analyze the subformulas that have been considered before. The following subformulas remain:

(a) "$K(y)$ is an important triple $\wedge K(y+1)$ is the next important triple after $K(y)$";
(b) $K(y) = \lim_{z\in y} K(z)$.

We shall have to use the absoluteness of the auxiliary formula "$y = x_{(i)}$," which is short for "x is an important triple (i.t.) and y is the ith coordinate of x," where $i = 1, 2$, or 3. That is;

$$(\exists u_1 \in \mathsf{U}^3(x))(\exists u_2 \in \mathsf{U}^3(x))(\exists u_3 \in \mathsf{U}(x)) \\ \times (x = \langle u_1, u_2, u_3\rangle \wedge u_1 \text{ is an ordinal} \wedge u_1 \leqslant 8 \\ \wedge u_2 \text{ is an ordinal } \wedge u_3 \text{ is an ordinal} \wedge y = u_i).$$

The *complete form of* (a) is

$$(\exists u \in \mathsf{U}(K))(\exists v \in \mathsf{U}(K))(\langle y, u\rangle \in K \wedge \langle y+1, v\rangle \in K \\ \wedge u \text{ is an i.t.} \wedge v \text{ is the i.t. after } u).$$

According to Lemma 1.7(a), "u is an i.t. $\wedge v$ is the i.t. after u" can be written in the form $\bigvee_{i=1}^5 C_i(u, v)$, where $C_i(u, v)$ is the formalization of the ith alternative in 1.7(a). For example,

$$C_1: \quad u \text{ is an i.t.} \wedge v \text{ is an i.t.} \wedge u_{(1)} \leqslant 7 \wedge v_{(1)} = u_{(1)} + 1 \\ \wedge v_{(2)} = u_{(2)} \wedge v_{(3)} = u_{(3)};$$

$$C_2: \quad u \text{ is an i.t.} \wedge v \text{ is an i.t.} \wedge u_{(1)} = 8 \wedge u_{(2)} + 1 < u_{(3)} \\ \wedge v_{(1)} = 0 \wedge v_{(2)} = u_{(2)} + 1 \wedge v_{(3)} = 0.$$

The other C_i are analogous, and are absolute for the same reasons.

The *complete form of* (b). Here we need to know that the following auxiliary formulas are absolute:

$$u = \bigcup_{z\in y} K(z)_{(i)},\ i = 2 \text{ or } 3: \quad (\forall v \in u)(\exists z \in y)(v = K(z)_{(i)}) \\ \wedge (\forall z \in y)(\exists v \in u)(v = K(z)_{(i)});$$

$$v = K(z)_{(i)}: \quad (\exists w \in \mathsf{U}(K))(\langle z, w\rangle \in K \wedge w \text{ is an i.t.} \wedge v = w_{(i)}).$$

Then, using Lemma 1.7(b), we explain the formula $K(y) = \lim_{z\in y} K(z)$ as follows:

$$K(y)_{(1)} =$$

$$0 \wedge \exists u_2\, \exists u_3 \left(u_2 = \bigcup_{z\in y} K(z)_{(2)} \wedge u_3 = \bigcup_{z\in y} K(z)_{(3)} \wedge \bigvee_{i=1}^{4} D_i(u_2, u_3, y) \right),$$

where the alternatives D_i have the following structure, depending on "how $K(z)$ approaches $K(y)$":

$$\begin{aligned}
D_1:\quad & u_2 \in u_3 \wedge u_2 \text{ is a limit ordinal} \wedge ((\exists z \in y)(K(z)_{(3)} = u_3) \\
& \quad \rightarrow k(y)_{(2)} = u_2 \wedge K(y)_{(3)} = u_3); \\
D_2:\quad & u_2 = u_3 \wedge u_2 \text{ is a limit ordinal} \wedge ((\exists z \in y)(K(z)_{(3)} = u_3) \\
& \quad \wedge (\forall z \in y)(K(z)_{(2)} \in u_2) \rightarrow k(y)_{(2)} = u_2 \wedge K(y)_{(3)} = 0); \\
D_3:\quad & u_2 \geqslant u_3 \wedge u_3 \text{ is a limit ordinal} \wedge ((\forall z \in y)(K(z)_{(3)} \in u_3) \\
& \quad \rightarrow K(y)_{(2)} = u_2 \in K(y)_{(3)} = u_3); \\
D_4:\quad & u_2 = u_3 \wedge u_2 \text{ is a limit ordinal} \wedge ((\forall z \in y)(K(z)_{(2)} \in u_2 \\
& \quad \wedge K(z)_{(3)} \in u_3) \rightarrow K(y)_{(2)} = 0 \wedge K(y)_{(3)} = u_3).
\end{aligned}$$

It is therefore obvious that the D_i are absolute. Even though the quantifiers $\exists u_2$ and $\exists u_3$ are not restricted, there is no problem, since when $K^\xi, y^\xi \in L$, this formula can be V-true only if $u_2^{\xi'}$ and $u_3^{\xi'}$ are uniquely determined ordinals and lie in L, which gives us L-truth.

5.19. *The formula* $S(N, x)$: "x is an ordinal $\wedge$ N is a function $\wedge$ dom $N = x+1 \wedge (\forall y \leqslant x+1)(N(y)$ is a constructible set with N-number $y)$" *is absolute.*

We shall need to know that the following auxiliary formula is absolute:

$$y = (x)_i, \quad i = 1, 2, 3, \quad \text{where } K(x) = \langle (x)_1, (x_2), (x)_3 \rangle$$

(not to be confused with the formula $y = x_{(i)}$ in 5.16, which occurs here as a subformula): x is an ordinal $\wedge \exists K(H(K, x) \wedge (\exists u \in \cup(K))(\langle x, u\rangle \in K \wedge y = u_{(i)}))$. Even though $\exists K$ is not restricted, this does not cause any problem, because for every ordinal $x^\xi \in L$, the value of K^ξ making $H(K^\xi, x^\xi)$ V-true lies in L. In fact, the V-true formula

$$\forall x (x \text{ is an ordinal} \Rightarrow \exists! K(H(K, x)))$$

is deducible from the axioms without the axiom of choice, and hence is L-true.

We now return to $S(N, x)$. We need only show that the subformula "$N(y)$ is a constructible set with N-number y" is absolute. By definition, this subformula can be written as $\bigvee_{i=0}^{8} Q_i(y, N)$, where the alternatives have the form

$$\begin{aligned}
& Q_0\text{: } (y)_1 = 0 \wedge \langle y, L_{(y)_2}\rangle \in N; \\
& Q_i,\ 1 \leqslant i \leqslant 3\text{: } (y)_1 = i \wedge \langle y, F_i(N((y)_2), N((y)_3))\rangle \in N; \\
& Q_i,\ 4 \leqslant i \leqslant 8\text{: } (y)_1 = i \wedge \langle y, F_i(N((y)_2))\rangle \in N.
\end{aligned}$$

The absoluteness of the subformulas that have not been analyzed is clear from the following complete forms of these formulas:

(a) $\langle y, L_{(y)_2}\rangle \in N$: $(\exists z \in \cup(N))(\langle y, z\rangle \in N \wedge z \in L_{(y)_2})$;
$z = L_{(y)_2}$: $(\exists u \in y+1)(u = (y)_2 \wedge z = L_u)$;
$z = L_u$: $(\forall v \in z)(v \in L_u) \wedge \forall v(v \in L_u \Rightarrow v \in z)$.

We can verify directly that the last subformula, with the unrestricted quantifier $\forall v$*, is absolute, since* $L_U \in L$ *for any ordinal* U*, and* L *is transitive.*

(b) $\langle y, F_i(N((y)_2))\rangle \in N, \qquad i = 4, \dots, 8$:
$(\exists u, v, w \in \cup(N))(u = (y)_2 \wedge \langle u, v\rangle \in N \wedge w = F_i(v) \wedge \langle y, w\rangle \in N)$.

(c) $\langle y, F_i(N((y)_2), N((y)_3))\rangle \in N, \qquad i = 1, 2, 3$:
$(\exists u_2, u_3, v_2, v_3, w \in \cup\, (N))(u_2 = (y)_2 \wedge u_3 = (y)_3 \wedge \langle u_2, v_2\rangle \in N$
$\wedge \langle u_3, v_3\rangle \in N \wedge w = F_i(u_2, v_3) \wedge \langle y, w\rangle \in N)$.

5.20. *The formula* $N(x, y)$: "x is an ordinal $\wedge y = N(x)$" *is absolute.*

In fact, this formula is written in the form

$$\exists N(S(N, x+1) \wedge \langle x, y\rangle \in N).$$

There is no problem with $\exists N$ being unrestricted, since we can apply the same type of argument as we have used many times before: for any ordinal x^ξ there is a unique $N^{\xi'}$ making this formula V-true, and then $N^{\xi'} \in L$, since the formula $\forall x$ (x is an ordinal $\Rightarrow \exists! N(S(N, x+1))$) is deducible from the axioms without the axiom of choice, and hence is L-true.

This completes the proof of Proposition 3.2. □

6 Remarks on Formalization

Gödel's theory, to which this chapter is devoted, is usually presented in a more syntactic version. We shall now briefly describe the system of basic ideas and the most important changes in the proofs in this version, in which the least possible appeal is made to the semantics.

6.1. Let $Q(x)$ be a formula in $\mathrm{L_1Set}$ with one free variable x. Let ZF be the set of all the (logical, special, and equality) axioms of $\mathrm{L_1Set}$ except for the axiom of choice. $Q(x)$ is said to be *transitive* if

$$\mathrm{ZF} \vdash (Q(x) \wedge y \in x) \Rightarrow Q(y).$$

6.2. The *relativization* P_Q of a formula P in $\mathrm{L_1Set}$ relative to Q is defined by induction on the number of connectives and quantifiers in P:

$$\begin{aligned}
(x \in y)_Q &\quad \text{is} \quad Q(x) \wedge Q(y) \Rightarrow x \in y;\\
(x = y)_Q &\quad \text{is} \quad Q(x) \wedge Q(y) \Rightarrow x = y;\\
(\neg P)_Q &\quad \text{is} \quad \neg(P_Q);\\
(P_1 * P_2)_Q &\quad \text{is} \quad (P_1)_Q * (P_2)_Q, \quad \text{for any connective } *;\\
(\forall x P)_Q &\quad \text{is} \quad \forall x (Q(x) \Rightarrow P);\\
(\exists x P)_Q &\quad \text{is} \quad \exists x (Q(x) \wedge P).
\end{aligned}$$

6.3. $Q(x)$ is called an (*internal*) *model* of $\mathrm{L_1Set}$ if for any axiom $P \in \mathrm{ZF}$ we have

$$\mathrm{ZF} \vdash P_Q.$$

This model is *transitive* if Q is transitive.

A formula $P(y_1, \dots, y_n)$ is called *Q-absolute* if

$$\mathrm{ZF} \vdash (Q(y_1) \wedge \cdots \wedge Q(y_n)) \Rightarrow (P \Leftrightarrow P_Q).$$

6.4. The connection between these concepts and our earlier ones is as follows. Every formula $Q(x)$ determines a class $M = \{X \in V | Q(X) \text{ is } V\text{-true}\}$. This class M has the property that

$$|P|_M(\xi) = |P_Q|_V(\xi), \quad \forall \xi \in \overline{M},$$

for any formula P (as can easily be proved by induction on the number of connectives and quantifiers in P). Thus, to give a syntactic reformulation of our proofs we must make the following changes throughout;

(a) We consider only classes M that are defined by formulas Q, and all references to M are replaced by references to Q.
(b) We everywhere replace "P is V-true" by "P is deducible from ZF."
(c) We everywhere replace "P is M-true" by "P_Q is deducible from ZF."
(d) We everywhere replace "P is M-absolute" by "P is Q-absolute."

In order for the new assertions on deducibility from ZF to become sufficiently obvious, we must either do some additional work formalizing the proofs or else give more careful intuitive proofs. In particular, we must find finite subsets of ZF from which the various facts are deducible. The basic results are stated as follows in the new syntactic language:

6.5. $\exists y\ L(x, y)$ *"is" a transitive internal model of* $\mathrm{L_1Set}$.

6.6. $\mathrm{ZF} \vdash (\text{axiom of choice})_{\exists y\ L(x,y)}$.

6.7. $\mathrm{ZF} \vdash (\text{generalized continuum hypothesis})_{\exists y\ L(x,y)}$.

6.8. Thus, a completely syntactic version of Gödel's theory would consist of all the deductions implicit in 6.5–6.7, without any commentary. Of course, such a treatment has never been written. The formula $\exists y L(x, y)$ alone takes up several pages; without appealing to semantics, it would be impossible either to think up, or to explain, or even to copy down all this without making mistakes. The deductions of all the required relativized formulas $P_{\exists y\ L(x,y)}$ would also be extremely long. This situation gives us an instructive example of what was discussed in "Digression: Proof" in Chapter II.

7 What Is the Cardinality of the Continuum?

After all we have learned about the Zermelo–Fraenkel language and axiom system, it might seem naive to return to this question. But we must do so if we consider mathematical meaning to be our primary concern.

Some specialists in the foundations of mathematics espouse a different point of view. Namely, they answer that the question itself is meaningless. It seems that Paul Cohen himself tends toward this viewpoint, at the same time admitting that "this is a hard decision" (P. Cohen, Comments on the foundations of set theory, *Proc. Symp. Pure Math.*, vol. XIII, part I, American Math. Soc., Providence 1971, p. 12).

From this point of view it is natural to reject almost the entire semantics of $\mathrm{L_1Set}$, including all the V_α starting with $\alpha = \omega_0 + 1$ in the von Neumann universe. No halfway solutions can help matters, especially since questions concerning higher axioms of infinity or the so-called measurable cardinals are in an even worse position than the CH.

It thus becomes necessary to try to find alternative languages and semantics. Here the differences of opinion are wide and irreconcilable. The most clear-cut position is that of the constructivists, although even among them there are different shades of opinion. The constructivists do not recognize infinity as a usable concept, and reject ineffective existence proofs. (It turns out that in practice they often replace these ineffective proofs by a more carefully differentiated word usage—"there cannot not exist," or "there quasi-exists"—which is nearly synonymous with certain linguistic precautions adopted in classical texts.) In our opinion, the shortcoming in their point of view is that constructivism is in no sense "another mathematics." It is, rather, a sophisticated subsystem of classical mathematics, which rejects the extremes in classical mathematics and carefully nourishes its effective computational apparatus.

Unfortunately, it seems that it is these "extremes"—bold extrapolations, abstractions that are infinite and do not lend themselves to a constructivist interpretation—that make classical mathematics effective. One should try to imagine how much help mathematics could have provided twentieth-century quantum physics if for the past hundred years it had developed using only abstractions from "constructive objects." Most likely, the standard calculations with infinite-dimensional representations of Lie groups that today play an important role in understanding the microworld would simply never have occurred to anyone.

It is not impossible that a new (or a completely forgotten old) conception of the continuum, in which the continuum has no "cardinality," could be found in the course of a deep investigation of the external world. The notion of a set consisting of elements may actually be adequate only for finite or countable sets, and "higher infinities" may turn out to be abstractions from objects of a completely different type.

Physics seems to point up a difference in principle between "counting" and the Eudoxus–Dedekind idealization of measurement. The counting procedure applies to regions of attraction—"attractors" (R. Thom)—that are units not having sharp boundaries. The parts of a unit, even if they have physical meaning, are nevertheless attractors of a different sort. But even these ideas apparently stop making sense in the microworld.

If nature has a fundamentally statistical aspect, it might be fruitful to consider mathematical models in which the statistical aspect appears as an undefined concept. The unexpected richness of the nonstandard interpretations of classical mathematics in Boolean-valued models agrees with the suggestion that all the words we say should be understood in a new way.

7.2. We now discuss a less radical point of view on the continuum problem, according to which this question of its cardinality is meaningful. Then the main problem once again becomes how to determine the place of the continuum on the scale of alephs.

Cohen concludes his book with the following opinion: "A point of view which the author feels may eventually come to be accepted is that CH is *obviously* false.... C is greater than $\aleph_n, \aleph_w, \aleph_\alpha$ where $\alpha = \aleph_w$ etc. This point of view regards C as an incredibly rich set given to us by one bold new axiom, which can never be approached by any piecemeal process of construction."

We thus have a conjectural estimate from below for C, and nothing more—not even a conjecture as to whether the cardinal C is regular or singular.

Of course, the real problem consists not only in guessing a plausible conjecture, but in supporting it with sufficiently convincing indirect evidence for it to become widely accepted, even if not proved. What sort of evidence could this be? In discussing new axioms for set theory, Gödel writes:

> there may exist ... other (hitherto unknown) axioms of set theory which a more profound understanding of the concepts underlying logic and mathematics would enable us to recognize as implied by these concepts.
>
> Furthermore, however, even disregarding the intrinsic necessity of some new axiom, and even in case it had no intrinsic necessity at all, a decision about its truth is possible also in another way, namely, inductively by studying its "success," that is, its fruitfulness in consequences and in particular in "verifiable" consequences, i.e., consequences demonstrable without the new axiom, whose proofs by means of the new axiom, however, are considerably simpler and easier to discover, and make it possible to condense into one proof many different proofs. The axioms for the system of real numbers, rejected by the intuitionists, have in this sense been verified to some extent owing to the fact that analytic number theory frequently allows us to prove number theoretic theorems which can subsequently be verified by elementary methods. A much higher degree of verification than that, however, is conceivable. There might exist axioms so abundant in their verifiable consequences, shedding so much light upon a whole discipline, and furnishing such powerful methods for solving given problems (and even solving them, as far as that is possible, in a constructivistic way) that quite irrespective of their intrinsic necessity they would have to be assumed at least in the same sense as any well established physical theory (K. Gödel, What is Cantor's continuum problem? *Amer. Math. Monthly*, vol. 54, no. 9, 1947).

There is little to add here to this ardently expressed hope. But see §8 of Chapter VII, where it is shown using an idea of Gödel's own that *any* new independent axiom can shorten to an arbitrary extent the proofs of suitable assertions that are provable without the axiom. This result somewhat weakens our confidence in pragmatic criteria for truth.

7.3. More than two decades after the publication of the first edition of this book, Hugh W. Woodin introduced interesting new ideas about the continuum hypothesis.

His constructions enrich both our set-theoretic intuition and its formal language, in an intuitively consistent way.

We will very briefly explain Woodin's approach, following his notes "The continuum hypothesis. I, II," *Notices AMS*, 48 (2001), no. 6, 567–576, and no. 3, 681–690. We will work in the constructible universe of Section IV.1.

Call a set X *transitive* if each element of an element of X belongs to X. The transitive closure of X is the minimal transitive set containing X.

Let k be an infinite cardinal, and $H(k)$ the set of all sets X whose transitive closure is of cardinality $\leq k$. Accepting the axiom of choice, one sees that any constructible set belongs to some $H(k)$. Let $k_0, k_1, k_2, \dots$ be the increasing sequence of the first infinite cardinals. Woodin easily reinterprets $H(k_0)$ as the semiring of natural numbers **N** with addition and multiplication, and, with some effort, $H(k_1)$ as a particular structure on the set of subsets of this semiring. These efforts are justified by providing a list of axioms for these structures that are intuitive and provide a basis for generalization to $H(k_2)$.

Having thus set the stage, Woodin takes up $H(k_2)$ and introduces an extension of first-order logic and a new axiom modestly called $(*)$.

Here the *grand finale* arrives: in this context Woodin can prove that $2^{\aleph_0} = \aleph_2$.

The following quotation from his second paper nicely concludes the discussion of this whole section:

"So, is the continuum hypothesis solvable? Perhaps, I am not completely confident the 'solution' I have sketched *is* the solution, but it is for me a convincing evidence that there is a solution. Thus, I now believe the continuum hypothesis is solvable, which is a fundamental change in my view of set theory."

Part II
COMPUTABILITY

V

Recursive Functions and Church's Thesis

1 Introduction. Intuitive Computability

1.1. The first part of this book was primarily concerned with *mathematical proof*; we showed that the analogous concept in formal languages is that of formal deduction, after which the most interesting results were that certain intuitive mathematical assertions (such as the continuum hypothesis and its negation) are not deducible.

Our primary concern in the second part of the book is the notion of a *determinate computational process*, that is, the processing of information, or, briefly, the notion of an algorithm. In §2 we give a precise and presumably complete characterization of everything that can be obtained using computational algorithms. Then the most interesting results turn out to be assertions that certain intuitively defined functions cannot be computed by an algorithm (Chapter VI).

Both the theory of proof and the theory of computation can be presented in large part independently of one another. This is the approach we have adopted, even though it does not correspond to the historical development. But when the machinery of both theories has been developed to a certain point, it becomes possible to apply each theory to investigate the other. The third part of the book is devoted to such applications.

In this section we describe informally the main focal points of the theory of computability. We appeal to the reader's intuitive notion of algorithms, which can be conveniently used to illuminate the structure and interrelations of the basic concepts.

When we make these concepts precise in the next section, we shall not give a description of the algorithms themselves, but rather of their results, i.e., *computable functions.* The concept of an algorithm seems to lose too much in any formalization, while the notion of algorithmic computability seems not to lose anything essential.

1.2. We now introduce several simple basic concepts. Let X and Y be two sets. A *partial function* (or mapping) from X to Y is any pair $\langle D(f), f\rangle$ consisting of a subset $D(f) \subset X$ and a mapping $f : D(f) \to Y$. Here $D(f)$ (instead of

Yu. I. Manin, *A Course in Mathematical Logic for Mathematicians, Second Edition*, Graduate Texts in Mathematics 53, DOI 10.1007/978-1-4419-0615-1_5,

the earlier dom f) is called the domain of definition of f; f is defined at a point $x \in X$ if $x \in D(f)$; f is nowhere defined if $D(f)$ is empty; and there exists a unique nowhere defined partial function.

We let $\mathbf{Z}^+ = \{1, 2, 3, \ldots\}$ denote the set of natural numbers, *excluding zero.* (It is not necessary, only convenient, to exclude zero.) If $n \geqslant 1$, we let $(\mathbf{Z}^+)^n$ denote the n-fold direct product of $\mathbf{Z}^+$ with itself, i.e., the set of ordered n-tuples $\langle x_1, \ldots, x_n \rangle$, $x_i \in \mathbf{Z}^+$. It is convenient to let $(\mathbf{Z}^+)^0$ denote the set consisting of an arbitrary element, denoted by "$\cdot$". The basic objects of our concern will be partial functions from $(\mathbf{Z}^+)^m$ to $(\mathbf{Z}^+)^n$ for various m and n. When we classify these functions according to their computability, the reader can think of the word "program" as referring to a program for a universal computer that is written without regard to time or memory limitations. Here every program for computing a function has a special "blank space" in which to insert the value of the argument.

1.3. *The basic informal definitions.* (a) A partial function f from $(\mathbf{Z}^+)^m$ to $(\mathbf{Z}^+)^n$ is called *computable* if there exists a "program" that, whenever a vector $x \in (\mathbf{Z}^+)^m$ is entered in the input, gives as output

$$\begin{array}{ll} f(x), & \text{if } x \in D(f); \\ 0, & \text{if } x \notin D(f). \end{array}$$

Here 0 merely indicates that f is not defined at x; we could allow the output in this case to be anything *not in* $(\mathbf{Z}^+)^n$.

(b) A partial function f from $(\mathbf{Z}^+)^m$ to $(\mathbf{Z}^+)^n$ is called *semicomputable* if there exists a "program" that, whenever a vector $x \in (\mathbf{Z}^+)^m$ is entered in the input, gives $f(x)$ as output if $x \in D(f)$, and either gives 0 as output or else works infinitely long without stopping if $x \notin D(f)$.

In particular, *computable functions are semicomputable,* and *everywhere defined semicomputable functions are computable.*

(c) A partial function f is called *noncomputable* if it does not satisfy condition (b) (and a fortiori (a)).

1.4. *Comments*

(a) The most basic of these three concepts is semicomputability, since computability reduces to this property. In fact, to determine whether a semi-computable function is computable, we proceed as follows.

Let $X \subset Y$ be two sets. By the *characteristic function* of X in Y we mean the function $\chi_X : Y \to \mathbf{Z}^+$ such that

$$\chi_X(x) = \begin{cases} 1, & \text{if } x \in X; \\ 2, & \text{if } x \notin X. \end{cases}$$

Note that χ_X is everywhere defined on Y.

Now let f be a semicomputable function from $(\mathbf{Z}^+)^m$ to $(\mathbf{Z}^+)^n$. If f were computable as well, then the characteristic function of $D(f)$ would also be computable: simply add to the program that computes f the instructions "send 0 to 2, and anything not 0 to 1, and print as output." Conversely, if $\chi_{D(f)}$ is computable, then so is f: in front of the program that semicomputes f, put the program that computes $\chi_{D(f)}$ and then the instruction to give 0 as output immediately if $\chi_{D(f)}(x) = 2$ and to continue with the program for f with x as the argument if $\chi_{D(f)}(x) = 1$. Thus, since the everywhere defined function $\chi_{D(f)}$ is computable if and only if it is semicomputable, we have f is computable $\Leftrightarrow$ f is semicomputable and $\chi_{D(f)}$ semicomputable. Later, we shall first formalize the concept of semicomputability, and then take the right side of this equivalence as the formalization of computability.

(b) *There exist noncomputable functions.* In fact, any program is a finite text in a finite alphabet, so that the set of programs is countable, while the set of all functions $\mathbf{Z}^+ \to \mathbf{Z}^+$ is uncountable. (For a critical discussion of this argument, see 1.5 below.)

AN EXAMPLE OF A NONCOMPUTABLE FUNCTION. We consider the language of arithmetic SAr, which was described in §10 of Chapter II, and number the formulas of this language as explained in §11 of Chapter II. We define a function f by stipulating that

$$f(x)\begin{cases} = 1, & \text{if the } x\text{th formula is true in the standard interpretation;} \\ \text{is not defined,} & \text{if the } x\text{th formula is false.} \end{cases}$$

The function f is noncomputable. In Chapter VII we shall see that this follows because the set $D(f)$ is not definable in arithmetic, by Tarski's theorem.

In other words, it is impossible (even in principle) to distinguish the set of all number-theoretic truths by writing a single program (even a very long and complicated one) that could tell from a statement's formulation whether it is true. Of course, to prove this result requires a much deeper analysis of the concept of computability.

(c) *There exist functions that are semicomputable but not computable.* We first give a typical example of a program that semicomputes a function. We consider the following function f from $\mathbf{Z}^+$ to $\mathbf{Z}^+$, which is defined in terms of Fermat's problem:

$$f(n)\begin{cases} = 1, & \text{if there exist } x, y, z \in \mathbf{Z}^+ \text{ for which } x^{n+2} + y^{n+2} = z^{n+2}; \\ \text{is not defined,} & \text{otherwise.} \end{cases}$$

Here is a program that semicomputes f: after entering n in the input, run through all vectors $\langle x, y, z\rangle$ in a suitable order. (For example, according to increasing $x + y + z$, and for given $x + y + z$, in lexicographic order.) For each such vector verify whether $x^{n+2} + y^{n+2} = z^{n+2}$. If this equation holds, give 1 as output; otherwise, go on to the next $\langle x, y, z\rangle$.

Hence, f is semicomputable. But it is not known whether f is computable. According to Fermat's conjecture, f is nowhere defined (and hence computable!). The strongest theoretical results known concerning f—the so-called criteria of Kummer, Wieferich, Vandiver, and others—may be regarded as a sort of approximation to proving that f is *computable*, not that f is nowhere defined. That is, in order to verify the Fermat conjecture successively for various values of n, we must perform a (machine) computation (whose size grows rapidly with n) to determine $\chi_{D(f)}$ at the point n, when this determination is possible.[1]

There is an analogous example of a semicomputable function that we actually know is not computable. In Chapter VI we prove that there exists a polynomial $P(t, x_1, \ldots, x_n)$ with integer coefficients such that the function

$$g(t) \begin{cases} = 1, & \text{if the equation } P(t, x_1, \ldots, x_n) = 0 \text{ is solvable} \\ & \text{with } x_1, \ldots, x_n \in \mathbf{Z}^q; \\ \text{is not defined}, & \text{otherwise}, \end{cases}$$

is not computable. This function is semicomputable by the same argument as in the case of the function connected with Fermat's equation.

1.5. *Critical discussion of the above proofs.* Before proceeding further, we consider from a more critical point of view, for example, the argument in 1.4(b). The first weak point that catches our attention is that we did not say precisely what a program is. But this is not essential; for any fixed definition we choose, a program must in any case be a text in a finite alphabet if it at all corresponds to our intuitive notions, and there are countably many such texts. A much stronger objection to the argument goes roughly as follows: what justification do we have for working with just *one* definition of what a program is? Could there perhaps exist an increasing hierarchy of precisely describable "methods of computation," so that for every function from $\mathbf{Z}^+$ to $\mathbf{Z}^+$ we could choose a corresponding program that could compute this function?

A fundamental discovery in the theory of computability was that this last question has a negative answer. We now have a unique and final formal notion that corresponds to the intuitive idea of semicomputability. It can be stated as follows:

1.6. **Church's Thesis** (weakest form). *It is possible to give explicitly:*

(a) *a family of basic semicomputable functions;*
(b) *a family of elementary operations that, starting from any semicomputable functions, allow new semicomputable functions to be constructed;*

with the property that any semicomputable function can be obtained in a finite number of steps, where each step consists in applying one of the elementary operations to the functions constructed before and those in the family (a).

[1] Since the publication of the first Edition, Fermat's conjecture was proved by Wiles, so now we know that f is computable and empty. The reader may wish to replace in our discussion f by another function, say characteristic function of the set of numbers n of such primes p_n that $p_{n+1} = p_{n+2}$. This is another old number-theoretic problem. It remains unsolved.

1.7. *Comment.* Church's thesis will be given a precise formulation in the next section: the basic functions and the elementary operations will be given explicitly. The exact mathematical theory of computability begins at that point. But it seemed important to indicate first the general significance of the discovery that such families of functions and operations exist at all and can even be given explicitly, a result that is far from obvious.

This is an experimental fact, one of the most important discovered by logic. In the next section we discuss evidence of its value and usefulness. Now we merely note that this fact is related to the finiteness of the basic logical and set-theoretic principles of mathematics (implicit, for example, in $\mathrm{L_1Set}$), but is not identical to this finiteness.

2 Partial Recursive Functions

2.1. In this section we give the precise definition and the basic properties of a class of partial functions from $(\mathbf{Z}^+)^m$ to $(\mathbf{Z}^+)^n$, which we take as an adequate formalization of the class of semicomputable functions. We give the definition in a way parallel to the statement of Church's thesis in 1.6.

2.2. *The basic functions*

$$\mathrm{suc}: \mathbf{Z}^+ \to \mathbf{Z}^+, \quad \mathrm{suc}(x) = x+1;$$
$$1^{(n)}: (\mathbf{Z}^+)^n \to \mathbf{Z}^+, \quad 1^{(n)}(x_1,\dots,x_n) = 1, \quad n \geqslant 0;$$
$$\mathrm{pr}_i^n: (\mathbf{Z}^+)^n \to \mathbf{Z}^+, \quad \mathrm{pr}_i^n(x_1,\dots,x_n) = x_i, \quad n \geqslant 1.$$

2.3. *The elementary operations on partial functions*

(a) *Composition* (or *substitution*). This operation associates to every pair of partial functions f from $(\mathbf{Z}^+)^m$ to $(\mathbf{Z}^+)^n$ and g from $(\mathbf{Z}^+)^n$ to $(\mathbf{Z}^+)^p$ the function $h = g \circ f$ from $(\mathbf{Z}^+)^m$ to $(\mathbf{Z}^+)^p$ that is defined as follows:

$$D(g\circ f) = f^{-1}(D(g)) = \{x \in (\mathbf{Z}^+)^m | x \in D(f), f(x) \in D(g)\};$$
$$(g\circ f)(x) = g(f(x)).$$

(b) *Juxtaposition.* This operation associates to partial functions f_i from $(\mathbf{Z}^+)^m$ to $(\mathbf{Z}^+)^{n_i}$, $i = 1,\dots,k$, the function $(f_1,\dots,f_k)$ from $(\mathbf{Z}^+)^m$ to $(\mathbf{Z}^+)^{n_1} \times \dots \times (\mathbf{Z}^+)^{n_k}$ that is defined as follows:

$$D((f_1,\dots,f_k)) = D(f_1) \cap \dots \cap D(f_k);$$
$$(f_1,\dots,f_k)(x_1,\dots,x_m) = \langle f_1(x_1,\dots,x_m),\dots,f_k(x_1,\dots,x_m)\rangle.$$

(c) *Recursion.* This operation associates to a pair of partial functions f from $(\mathbf{Z}^+)^n$ to $\mathbf{Z}^+$ and g from $(\mathbf{Z}^+)^{n+2}$ to $\mathbf{Z}^+$ the partial function h from $(\mathbf{Z}^+)^{n+1}$ to $\mathbf{Z}^+$ that is defined by recursion on the last argument:

$$\begin{cases} h(x_1,\dots,x_n,1) = f(x_1,\dots,x_n) & \text{(initial condition);} \\ h(x_1,\dots,x_n,k+1) = g(x_1,\dots,x_n,k,h(x_1,\dots,x_n,k)), & \text{for } k \geqslant 1 \\ & \text{(recursive step).} \end{cases}$$

The domain of definition $D(h)$ is also defined by recursion:

$$\langle x_1,\ldots,x_n,1\rangle \in D(h) \Leftrightarrow \langle x_1,\ldots,x_n\rangle \in D(f),$$
$$\langle x_1,\ldots,x_n,k+1\rangle \in D(h) \Leftrightarrow \langle x_1,\ldots,x_n,k\rangle \in D(h), \quad \text{and}$$
$$\langle x_1,\ldots,x_n,k,h(x_1,\ldots,x_n,k)\rangle \in D(g) \quad \text{for } k \geqslant 1.$$

(d) *The μ-operator.* This operation associates to a partial function f from $(\mathbf{Z}^+)^{n+1}$ to $\mathbf{Z}^+$ the partial function h from $(\mathbf{Z}^+)^n$ to $\mathbf{Z}^+$ that is defined as follows:

$$D(h) = \{\langle x_1,\ldots,x_n\rangle | \exists x_{n+1} \geqslant 1, f(x_1,\ldots,x_n,x_{n+1}) = 1 \text{ and}$$
$$\langle x_1,\ldots,x_n,k\rangle \in D(f) \text{ for all } k \leqslant x_{n+1}\};$$
$$h(x_1,\ldots,x_n) = \min\{x_{n+1} | f(x_1,\ldots,x_n,x_{n+1}) = 1\}.$$

The general role of μ is to introduce "implicitly defined" functions, as is often done in many areas of mathematics. Three remarks about the definition of μ should be made at this point. First, we obviously chose the minimal y with $f(x_1,\ldots,x_n,y) = 1$ in order to ensure that the function h is single-valued. The second observation is that at first glance, it might seem that the domain of definition of h is artificially narrow. If, for example, we have $f(x_1,\ldots,x_n,2) = 1$ and $f(x_1,\ldots,x_n,1)$ is not defined, then we have taken $h(x_1,\ldots,x_n)$ to be undefined, rather than equal to 2. This is done because we want to preserve intuitive semicomputability in going from f to h, as will be discussed in somewhat greater detail below (see 2.7(a)).

Finally, we note that all the operations before μ, if applied to everywhere defined functions, give an everywhere defined function. This is obviously not the case for μ. Thus, μ is the only one of the operations that causes partial functions to arise unavoidably.

2.4. **Definition.**

(a) A sequence of partial functions $f_1,\ldots,f_N$ is called a partial recursive (respectively primitive recursive) description of the function $f_N = f$ if

f_1	belongs to the family of basic functions;
$f_i,\ i \geqslant 2,$	either belongs to the family of basic functions, or else is obtained by applying one of the elementary operations (respectively one of the elementary operations other than μ) to certain of the functions $f_1,\ldots,f_{i-1}$.

(b) A function f is called partial recursive (respectively primitive recursive) if it admits a partial recursive (respectively primitive recursive) description.

(The analogy with the definition of a deduction in a formal language immediately catches our attention, and can sometimes be of use.)

2.5. Church's Thesis (usual form)

(a) *A function f is semicomputable if and only if it is partial recursive.*
(b) *A function f is computable if and only if both f and $\chi_{D(f)}$ are partial recursive.*

Remark on terminology. Everywhere defined partial recursive functions are also called *general recursive* functions. If the domain of definition is either clear or not essential in a given context, we simply use the term "recursive." (Note that every primitive recursive function is general recursive.)

2.6. *Use of Church's thesis.* Before discussing in detail the arguments supporting Church's thesis, we indicate how it is used in practice in mathematics. Two basic applications are especially evident in the literature.

(a) *Church's thesis used for a definition of algorithmic undecidability.* Suppose we have a countable sequence of mathematical "problems" $P_1, P_2, \ldots$. Further, suppose that each problem has a "yes" or "no" answer, and that the conditions in P_n are written out "effectively" as a function of n. Such a sequence $P = (P_n)$ is called a "mass problem." We associate to such a problem a function f from $\mathbf{Z}^+$ to $\mathbf{Z}^+$

$$D(f) = \{i \in \mathbf{Z}^+ | P_i \text{ has "yes" for an answer}\};$$
$$f(i) = 1, \quad \text{if } i \in D(f).$$

A mass problem P is called *algorithmically decidable* if the functions f and $\chi_{D(f)}$ are partial recursive. Otherwise, P is called *algorithmically undecidable.* We also distinguish the case in which only $\chi_{D(f)}$ is not partial recursive from the case in which even f is not partial recursive. The second type of undecidability is worse than the first; we saw examples of this in §1. Finally, a whole hierarchy of "degrees of undecidability" can be rigorously defined and investigated.

A well-known example of a mass problem is the *problem of word identities in groups.* Let G be a finitely defined group, and let $a_1, \ldots, a_r \in G$ be elements. A "reduced word" in $a_1, \ldots, a_r$ is an expression of the form $a_{i_1}^{\varepsilon_i} \cdots a_{i_k}^{\varepsilon_k}$, where $k \geqslant 1$, $\varepsilon_j = \pm 1$, and $\varepsilon_j = \varepsilon_{j+1}$ whenever $i_j = i_{j+1}$. We number all the reduced words and ask the question P_n: "Does the nth word represent the unit element of the group G?" The "mass problem" (P_n) turns out to be algorithmically decidable for certain groups G and elements $a_1, \ldots, a_n$ and algorithmically undecidable for others (Novikov, Boone, Higman). The function f in this case is always partial recursive, but $\chi_{D(f)}$ is not always (see Chapter VIII).

For another example of an undecidable problem, this one connected with Diophantine equations, see Chapter VI.

(b) *Church's thesis as a heuristic principle.* The intuitive notion of "semicomputability" at first seems broader than the notion of "partial recursiveness," and many problems concerning partial recursive functions become much easier if we replace the conditions in the problems by informal ideas and allow such ideas to be used to solve the problems. For example, the formula

$e = \lim(1 + 1/n)^n$ and the Euclidean algorithm make it intuitively clear that the functions $f, g : \mathbf{Z}^+ \to \mathbf{Z}^+$ given by

$$f(n) = \text{the } n\text{th digit in the decimal expansion of } e,$$
$$g(n) = \text{the } n\text{th prime number}$$

are computable, but the verification that they are recursive requires rather painstaking constructions.

Church's thesis allows us to solve such problems in two stages: (1) finding an informal solution using any intuitive algorithms we need, and (2) formalizing the solution. The second stage presupposes a certain proficiency in finding a partial recursive description for a wide variety of semicomputable functions, and Church's thesis assures us that such a description exists.

As proofs of recursiveness become more and more numerous in the literature, it becomes increasingly common to go through only the first stage of the solution; a striking example of this is Hartley Rogers' book *Theory of Recursive Functions and Effective Computability* (McGraw-Hill, New York, 1967). We shall also take such liberties toward the end of this book. All the same, there is a certain danger in this practice. It is possible that the habit of increasingly using informal arguments delayed the discovery of such a fundamental fact as the result that recursively enumerable sets and Diophantine sets coincide.

2.7. *Arguments in support of Church's thesis*

(a) First of all, the basic functions clearly must be computable, no matter how we precisely define the notion of computability. Furthermore, when the elementary operations are applied to semicomputable functions, they again give a semicomputable function. A program to semicompute the latter function can easily be put together from the programs that semicompute the original functions. We shall consider only the case of the μ-operator in detail, leaving the simple construction of the other three programs to the reader.

In the notation of 2.3(d), let f be a semicomputable function from $(\mathbf{Z}^+)^{n+1}$ to $\mathbf{Z}^+$. In order to compute $h(x_1, \ldots, x_n)$, we go through the vectors $\langle x_1, x_2, \ldots, x_n, 1\rangle$, $\langle x_1, \ldots, x_n, 2\rangle, \ldots$ in the order of increasing last coordinate, and compute the values of f at these vectors. If $\langle x_1, \ldots, x_n\rangle \in D(h)$, where h is obtained from f by applying the μ-operator, then the program for f successively computes

$$f(x_1, \ldots, x_n, 1), \ldots, f(x_1, \ldots, x_n, y-1),$$

and finally $f(x_1, \ldots, x_n, y) = 1$. The least such y, if it exists, must be given as output; it will be the value of h at the point $\langle x_1, \ldots, x_n\rangle$. On the other hand, if it turns out that one of the values $f(x_1, \ldots, x_n, k)$ (before we reach $f = 1$) is not defined, then either the program that semicomputes f will work infinitely long, or else it will give an answer not in $\mathbf{Z}^+$, which must then be given as output. But then, by definition, h is not defined at the point $\langle x_1, \ldots, x_n\rangle$, and the behavior of the program for h still agrees with the definition of h being semicomputable.

From all this we conclude that *partial recursive functions are semicomputable.* However, the stronger part of Church's thesis is the converse: *semicomputable functions are partial recursive.* (The definition of computability in terms of semicomputability is simply taken from §1 without any changes.) As has been said, this result is an experimental fact. The experimental evidence for it is divided into several classes, which we consider in (b)–(d) below.

(b) In the literature we find a huge collection of recursive descriptions of various computable and semicomputable functions. See, for example, Rózsa Péter, *Recursive Functions* (Academic Press, New York, 1967). We shall give part of this list in the next section. We also find certain techniques for composing recursive descriptions that are applicable to entire classes of (semi)computable functions. Every time an author has tried to find a partial recursive description of a (semi)computable function, he has met with success.

(c) Turing proposed a mathematical characterization of an abstract computer, and gave strong arguments to the effect that this computer is universal, i.e., it can (semi)compute any (semi)computable function. His arguments came from a detailed analysis of the characteristic features of determinate computational processes. (We again recall that we have not at all concerned ourselves with formalizing computational processes, but only with the results of such processes.) It turned out that the class of functions that are semicomputable by Turing machines exactly coincides with the class of partial recursive functions.

(d) Church, Post, Markov, Kolmogorov, Uspenskiĭ, and others have proposed other deterministic schemes for processing information of a general (not necessarily number-theoretic) character. In all cases it has turned out that if the sets of input and output are numbered in a suitable "effective" way, these methods lead to a class of maps from $\mathbf{Z}^+$ to $\mathbf{Z}^+$ that coincides with some subclass of the partial recursive functions.

For further discussion of Church's thesis, we refer the reader to the literature; see, in particular, S. Kleene, *Introduction to Metamathematics* (Van Nostrand, New York–Toronto, 1952).

3 Basic Examples of Recursiveness

3.1. In this section we give a short list of recursive functions and a selection of basic techniques for proving recursiveness. Both these lists will subsequently be enlarged when needed (in particular, see Chapter VII).

3.2. (a) $\mathrm{sum}_2 : (\mathbf{Z}^+)^2 \to \mathbf{Z}^+, \quad \langle x_1, x_2\rangle \mapsto x_1 + x_2.$

Use recursion on x_2, starting from the initial condition

$$x_1 + 1 = \mathrm{sum}_2(x_1, 1) = \mathrm{suc}(x_1)$$

and applying the recursive step

$$x_1 + k + 1 = \mathrm{sum}_2(x_1, k + 1) = \mathrm{suc}(\mathrm{sum}_2(x_1, k)).$$

(b) $\mathrm{sum}_n : (\mathbf{Z}^+)^n \to \mathbf{Z}^+, \quad \langle x_1, \dots, x_n \rangle \mapsto \sum_{i=1}^n x_i, \quad n \geqslant 3.$

Suppose that we already know that sum_{n-1} is recursive. We can obtain sum_n by juxtaposition and composition as follows:

$$\mathrm{sum}_n = \mathrm{sum}_2 \circ (\mathrm{sum}_{n-1} \circ (\mathrm{pr}_1^n, \dots, \mathrm{pr}_{n-1}^n), \mathrm{pr}_n^n).$$

Another version is to use recursion on x_n, starting from the initial condition $\mathrm{suc} \circ \mathrm{sum}_{n-1}$ and applying the recursive step

$$\sum_{i=1}^{n-1} x_i + k + 1 = \mathrm{suc}(\mathrm{sum}_n(x_1, \dots, x_{n-1}, k)).$$

This choice of recursive descriptions, even of "natural" ones, will become even more numerous as the functions become more complicated.

3.3. (a) $\mathrm{prod}_2 : (\mathbf{Z}^+)^2 \to \mathbf{Z}^+, \quad \langle x_1, x_2 \rangle \mapsto x_1 x_2.$

Use recursion on x_2, starting from the initial condition x_1 and applying the recursive step

$$x_1(k+1) = x_1 k + x_1 = \mathrm{sum}_2(x_1 k, x_1).$$

(b) $\mathrm{prod}_n : (\mathbf{Z}^+)^n \to \mathbf{Z}^+, \langle x_1, \dots, x_n \rangle \mapsto x_1, \dots, x_n, \quad n \geqslant 3.$
$\mathrm{prod}_n = \mathrm{prod}_2 \circ (\mathrm{prod}_{n-1} \circ (\mathrm{pr}_1^n, \dots, \mathrm{pr}_{n-1}^n), \mathrm{pr}_n^n).$

3.4. (a) $\mathbf{Z}^+ \to \mathbf{Z}^+, \quad x \mapsto x \dot{-} 1 = \begin{cases} x-1, & \text{if } x \geqslant 2; \\ 1, & \text{if } x = 1. \end{cases}$

Use recursion with the functions

$$f : (\mathbf{Z}^+)^0 \to \mathbf{Z}^+, \cdot \mapsto 1;$$
$$g = \mathrm{pr}_1^2 : (\mathbf{Z}^+)^2 \to \mathbf{Z}^+, \quad \langle x_1, x_2 \rangle \mapsto x_1.$$

(b) $(\mathbf{Z}^+)^2 \to \mathbf{Z}^+$:

$$\langle x_1, x_2 \rangle \mapsto x_1 \dot{-} x_2 = \begin{cases} x_1 - x_2, & \text{if } x_1 > x_2; \\ 1, & \text{if } x_1 \leqslant x_2. \end{cases}$$

This "truncated difference" is obtained by applying recursion to the functions

$$f(x_1) = x_1 \dot{-} 1;$$
$$g(x_1, x_2, x_3) = x_3 \dot{-} 1.$$

3.5. $F : (\mathbf{Z}^+)^n \to \mathbf{Z}^+$, *where F is any polynomial in $x_1, \dots, x_n$ with integer coefficients that takes values only in $\mathbf{Z}^+$.*

If all the coefficients in F are nonnegative, then F is a sum of products of the functions $\mathrm{pr}_i^n : \langle x_1, \dots, x_n \rangle \mapsto x_i$. Otherwise, we write $F = F^+ - F^-$, where F^+ and F^- have nonnegative coefficients, and at all points of $(\mathbf{Z}^+)^n$ the nontruncated difference coincides with the truncated difference $F^+ \dot{-} F^-$ because of the assumption concerning F.

We shall often use the recursiveness of the function $(x_1 - x_2)^2 + 1$, or $h = (f - g)^2 + 1$, where f and g are recursive. This technique allows us to identify the set on which $f = g$ with the "level set of h at 1," i.e., the set on which $h = 1$.

3.6. "Step functions": for each $a, b, x_0 \in \mathbf{Z}^+$, the function defined by

$$s_{x_0}^{a,b}(x) = \begin{cases} a, & \text{for } x \leqslant x_0, \\ b, & \text{for } x > x_0. \end{cases}$$

If $x_0 = 1$, we obtain this function by recursion with initial value a and all the succeeding values b. In the general case we set

$$s_{x_0}^{a,b}(x) = s_1^{a,b}(x + 1 \dot{-} x_0).$$

3.7. $\text{rem}(x, y) =$ *the remainder in* $[1, x]$ (since we cannot use zero!) *when* y *is divided by* x.

We have

$$\text{rem}(x, 1) = 1,$$
$$\text{rem}(x, y + 1) = \begin{cases} 1, & \text{if } \text{rem}(x, y) = x; \\ \text{suc} \circ \text{rem}(x, y), & \text{if } \text{rem}(x, y) \neq x. \end{cases}$$

We now apply a somewhat artificial technique. We consider the step function $s = s_1^{2,1}$, i.e., $s(1) = 2$ and $s(x) = 1$ if $x \geqslant 2$, and we set

$$\phi(x, y) = s\big((\text{rem}(x, y) - x)^2 + 1\big).$$

Obviously,

$$\text{rem}(x, y) \neq x \Leftrightarrow \phi(x, y) = 1,$$
$$\text{rem}(x, y) = x \Leftrightarrow \phi(x, y) = 2,$$

so that

$$\text{rem}(x, y + 1) = 2\, \text{suc}(\text{rem}(x, y)) \dot{-} \phi(x, y)\, \text{suc}(\text{rem}(x, y)).$$

This gives a recursive definition of rem.

We next describe this technique in a more general form.

3.8. Suppose h is defined by "recursion with conditions," i.e.,

$$h(x_1, \ldots, x_n, 1) = f(x_1, \ldots, x_n); h(x_1, \ldots, x_n, k + 1)$$
$$= g_i(x_1, \ldots, x_n, k, h(x_1, \ldots, x_n, k)),$$

if the condition $C_i(x_1, \ldots, x_n, k, h)$ holds, $i = 1, \ldots, m$, where the exhaustive and mutually exclusive conditions C_i are given in the form

$$C_i \text{ is fulfilled} \Leftrightarrow \phi_i\big(x_1, \ldots, x_n, k, h(x_1, \ldots, x_n, k)\big) = 1,$$

with ϕ_i an everywhere defined recursive function that takes only the values 1 and 2. Then we can write the recursive step as follows:

$$h(x_1,\ldots,x_n,k+1) = 2\sum_{i=1}^{m} g_i(x_1,\ldots,x_n,k,h(x_1,\ldots,x_n k))$$
$$\dot{-}\sum_{i=1}^{m}(g_i\phi_i)(x_1,\ldots,x_n,k,h(x_1,\ldots,x_n,k)).$$

This device allows us to show that the following functions, which will be needed later, are primitive recursive:

3.9. $$\mathrm{qt}(x,y) = \begin{cases} \text{the integral part of } y/x, & \text{if } y/x \geqslant 1;\\ 1, & \text{if } y/x < 1. \end{cases}$$

We have

$$\mathrm{qt}(x,1) = 1;$$
$$\mathrm{qt}(x,y+1) = \begin{cases} \mathrm{qt}(x,y), & \text{if } \mathrm{rem}(x,y+1) \neq x;\\ \mathrm{qt}(x,y)+1, & \text{if } \mathrm{rem}(x,y+1) = x \text{ and } y+1 \neq x;\\ 1, & \text{if } y+1 = x. \end{cases}$$

We reduce the conditions to the standard form 3.8 using the functions

$$\tilde{s}\big((\mathrm{rem}(x,y+1)-x)^2+1\big),$$
$$s\big((\mathrm{rem}(x,y+1)-x)^2+1\big)\cdot\tilde{s}\big((x-y-1)^2+1\big),$$
$$s\big((x-y-1)^2+1\big),$$

where $s = s_1^{1,2}$ and $\tilde{s} = s_1^{2,1}$.

3.10. $\mathrm{rad}(x)$ = the integral part of $\sqrt{x}$.
We have

$$\mathrm{rad}(1) = 1,$$
$$\mathrm{rad}(x+1) = \begin{cases} \mathrm{rad}(x), & \text{if } \mathrm{qt}(\mathrm{rad}(x)+1,x+1) < \mathrm{rad}(x)+1;\\ \mathrm{rad}(x)+1, & \text{if } \mathrm{qt}(\mathrm{rad}(x)+1,x+1) = \mathrm{rad}(x)+1. \end{cases}$$

The reduction of these conditions to the standard form 3.8 will be left to the reader.

3.11. (a) $\min(x,y)$:

$$\min(x,1) = 1,$$
$$\min(x,y+1) = \begin{cases} \min(x,y), & \text{if } x \leqslant y;\\ \min(x,y)+1, & \text{if } x > y. \end{cases}$$

(b) $\max(x,y)$: analogous.

3.12. *If $f(x_1,\dots,x_n)$ is recursive, then*

$$Sf = \sum_{k=1}^{x_n} f(x_1,\dots,x_{n-1},k) \quad \text{and} \quad Pf = \prod_{k=1}^{x_n} f(x_1,\dots,x_{n-1},k)$$

are recursive

In fact,

$$Sf(x_1,\dots,x_{n-1},x_n+1) = Sf(x_1,\dots,x_n) + f(x_1,\dots,x_n+1),$$
$$Pf(x_1,\dots,x_{n-1},x_n+1) = Pf(x_1,\dots,x_n)\cdot f(x_1,\dots,x_n+1).$$

3.13. *If $f(x_1,\dots,x_n)$ is recursive, then so are the functions obtained from f by:*

(a) *any permutation of the arguments;*
(b) *adding any number of "dummy" arguments;*
(c) *identifying the elements of any subset of the arguments ($f(x,x)$ instead of $f(x,y)$, and so on).*

In fact, all of these functions can be obtained from f and the various pr_i^m using composition and juxtaposition.

3.14. *A map $f : (\mathbf{Z}^+)^m \to (\mathbf{Z}^+)^n$ is recursive if and only if all of its components $\mathrm{pr}_i^n \circ f$ are recursive.*

This is obvious.

In conclusion, we note that all the specific functions described above are primitive recursive, and that all the above general operations, when applied to primitive recursive functions, yield primitive recursive functions. Starting in the next section, we shall make essential use of the μ-operator, which was defined in 2.3(d).

4 Enumerable and Decidable Sets

4.1. **Definition.** A set $E \subset (\mathbf{Z}^+)^n$ is called *recursively enumerable* if there exists a partial recursive function f such that $E = D(f)$ (the domain of definition of f).

The discussion in §1 and §2 showed that recursive enumerability has the following intuitive meaning: there exists a program that identifies the elements x in E but that might not identify the elements not in E. Later, in 4.12 and 4.18, we shall give another intuitive description of recursively enumerable sets that is more closely related to the etymology of the name: these are *sets all of whose elements can be obtained using a suitable "generating" program (perhaps with repetitions and with no indication of the order in which the elements occur)*.

The concept of a recursively enumerable set occupies a central place in the theory of computability, alongside the concept of a partial recursive function. It will later be clear, in particular from Proposition 4.15, that either of these concepts can be reduced to the other one. However, only by using both ideas together do we obtain the flexibility necessary for efficient proofs.

We begin with the following simple fact.

Recall that the *level set* at m (or simply the m-level) of a function f from $(\mathbf{Z}^+)^n$ to $\mathbf{Z}^+$ is the set $E \subset D(f)$ such that

$$x \in E \Leftrightarrow f(x) = m.$$

4.2. **Proposition.** *The following three classes of sets coincide:*

(a) *Recursively enumerable sets.*
(b) *Level sets of partial recursive functions.*
(c) *Level sets at* 1 *of partial recursive functions.*

(a) $\subset$ (*c*). Suppose that E is recursively enumerable, so that $E = D(f)$, where f is partial recursive. Then $E =$ the 1-level of the function $1^{(1)} \circ f$.

(b) $=$ (*c*). The m-level of f coincides with the 1-level of $(f - m)^2 + 1$. The function $(f-m)^2+1$ is partial recursive whenever f is, by Proposition 3.5.

(c) $\subset$ (*a*). Suppose that E is the 1-level of a partial recursive function $f(x_1, \dots, x_n)$. Set

$$g(x_1, \dots, x_n) = \min\left\{y|(f(x_1, \dots, x_n) - 1)^2 + y = 1\right\}.$$

Obviously, g is partial recursive and $E = D(g)$. □

The following much more difficult assertion, along with its corollaries, constitutes the central result of this section.

4.3. **Theorem.** *The following two classes of sets coincide:*

(a) *Recursively enumerable sets.*
(b) *Projections of level sets of primitive recursive functions with values in* $\mathbf{Z}^+$.

4.4. First part of the proof. We first recall that if we are given a set $E \subset (\mathbf{Z}^+)^{n+m}$, then its projection ("onto the space of the first n coordinates") is the set $F \subset (\mathbf{Z}^+)^n$ that is defined as follows:

$$\langle x_1, \dots, x_n \rangle \in F$$
$$\Leftrightarrow \exists \langle y_1, \dots, y_m \rangle \in (\mathbf{Z}^+)^m, \quad \langle x_1, \dots, x_n, y_1, \dots, y_m \rangle \in E.$$

(From this point on, we shall not adhere to the practice in Part I of using different notation for "variable coordinates" and for particular values of the coordinates.) We similarly define the projection "onto the coordinates with indices $(i_1, \dots, i_n) \subset (1, \dots, n+m)$." The number m is called the *codimension* of the projection. The canonical map $E \to F$ (as well as its image) is also customarily called a projection, but this is not likely to cause any confusion.

For the time being we shall call projections of level sets of primitive recursive functions *primitive enumerable* sets. The first part of the proof consists in showing that primitive enumerable sets are recursively enumerable; the second part consists in verifying the converse implication.

Thus, let $f(x_1, \dots, x_n, x_{n+1}, \dots, x_{n+m})$ be a primitive recursive function, and let E be the projection of its 1-level onto the first n coordinates. (We need only consider 1-levels because of the consideration used once before: the k-level of f coincides with the 1-level of $f' = (f-k)^2 + 1$.) We explicitly construct a partial recursive function g such that $E = D(g)$.

We distinguish three cases, depending on the codimension of the projection: $m = 0, m = 1$, and $m \geqslant 2$.

Case (a): $m = 0$. Then $E =$ the 1-level of $f \Leftrightarrow E$ is recursively enumerable, by Proposition 4.2 (where g is constructed explicitly).

Case (b): $m = 1$. Let

$$g(x_1, \ldots, x_n) = \min\{x_{n+1} | f(x_1, \ldots, x_n, x_{n+1}) = 1\}.$$

Obviously, g is partial recursive, and $D(g) = E$. (Notice that we have used here the fact that $D(f) = (\mathbf{Z}^+)^{n+1}$.)

Case (c): $m \geqslant 2$. We reduce this case to the previous one using the following lemma, which is also important in many other situations and is of interest in its own right (as a statement that there is no notion of dimension in "recursive geometry").

4.5. **Lemma.** *For each $m \geqslant 1$ there exists a one-to-one mapping $t^{(m)}$: $\mathbf{Z}^+ \to (\mathbf{Z}^+)^m$ such that*

(a) *The function $t_i^{(m)} = \mathrm{pr}_i^m \circ t^{(m)}$ is primitive recursive for all $1 \leqslant i \leqslant m$.*
(b) *The inverse function $\tau^{(m)} : (\mathbf{Z}^+)^m \to \mathbf{Z}^+$ is primitive recursive.*

4.6. *How the lemma is used.* Suppose that the lemma is true. We apply it to the situation in case (c) in 4.4 as follows. For $m \geqslant 2$ we set

$$g(x_1, \ldots, x_n, y) = f(x_1, \ldots, x_n, t_1^{(m)}(y), \ldots, t_m^{(m)}(y)).$$

Obviously, g is primitive recursive if f is. It is easy to see that E coincides with the projection of the 1-level of g onto the first n coordinates. Since this is a projection of codimension 1, we have reduced this case to the previous one.

4.7. Proof of the lemma. The case $m = 1$ is trivial. We use induction on m, starting with $m = 2$.

Construction of $t^{(2)}$. We first construct $\tau^{(2)} : (\mathbf{Z}^+)^2 \to \mathbf{Z}^+$ explicitly by setting

$$\tau^{(2)}(x_1, x_2) = \frac{1}{2}\left((x_1 + x_2)^2 - x_1 - 3x_2 + 2\right).$$

It is easy to see that if we list the pairs $\langle x_1, x_2 \rangle \in (\mathbf{Z}^+)^2$ in "Cantor order," i.e., according to increasing $x_1 + x_2$ and, among those with given $x_1 + x_2$, according to increasing x_1 then $\tau^{(2)}(x_1, x_2)$ will be precisely the index of the pair $\langle x_1, x_2 \rangle$ in this list. Thus, $\tau^{(2)}$ is a one-to-one correspondence and, moreover, is primitive recursive (where we use Proposition 3.5 and then the recursiveness of qt in 3.9 to take care of the 1/2).

The calculation of the pair $\langle x_1, x_2 \rangle$ as a function of its index y is an elementary problem, and results in the following formulas for the inverse function $t^{(2)}$:

$$t_1^{(2)}(y) = y - \frac{1}{2}\left[\sqrt{2y - \frac{7}{4}} - \frac{1}{2}\right]\left(\left[\sqrt{2y - \frac{7}{4}} - \frac{1}{2}\right] + 1\right),$$

$$t_2^{(2)}(y) = \left[\sqrt{2y - \frac{7}{4}} - \frac{1}{2}\right] - t_1^{(2)}(y) + 2.$$

Here $[z]$ denotes the integral part of z. The verification that these functions are primitive recursive using the results (and techniques) of §3 is left to the reader as an exercise.

Construction of $t^{(m)}, m \geqslant 3$. Suppose that $t^{(m-1)}$ and $\tau^{(m-1)}$ have already been constructed with the required properties. We first set

$$\tau^{(m)}(x_1, \ldots, x_m) = \tau^{(2)}\left(\tau^{(m-1)}(x_1, \ldots, x_{m-1}), x_m\right).$$

It is clear that $\tau^{(m)}$ is one-to-one and primitive recursive. Solving the equation $\tau^{(2)}\left(\tau^{(m-1)}(x_1, \ldots, x_{m-1}), x_m\right) = y$ in two steps, we obtain the following formulas for the inverse function $t^{(m)}$:

$$t_m^{(m)}(y) = t_2^{(2)}(y),$$
$$t_i^{(m)}(y) = t_i^{(m-1)}(t_1^{(2)}(y)), \quad 1 \leqslant i \leqslant m-1.$$

The $t_i^{(m)}$ are primitive recursive by the induction assumption. This completes the proof of the lemma, and by the same token the first part of the proof of Theorem 4.3. □

SECOND PART OF THE PROOF. We must now show that every recursively enumerable set is primitive enumerable. We begin with the following property of the class of primitive enumerable sets.

4.8. **Lemma.** *The class of primitive enumerable sets is closed with respect to the following operations: finite direct product, finite intersection, finite union, and projection.*

PROOF. Let $E, E' \subset (\mathbf{Z}^+)^n$ and $E_1 \subset (\mathbf{Z}^+)^m$ be three primitive enumerable sets that are projections of the 1-levels of the primitive recursive functions f, f', and f_1, respectively:

$$x = \langle x_1, \ldots, x_n \rangle \in E \Leftrightarrow \exists y = \langle y_1, \ldots, y_r \rangle, \qquad f(x, y) = 1,$$
$$x = \langle x_1, \ldots, x_n \rangle \in E^1 \Leftrightarrow \exists z = \langle z_1, \ldots, z_q \rangle, \qquad f'(x, z) = 1,$$
$$u = \langle u_1, \ldots, u_m \rangle \in E_1 \Leftrightarrow \exists v = \langle v_1, \ldots, v_s \rangle, \qquad f_1(u, v) = 1.$$

We then have

$$E \times E_1 = \text{a projection of the 1-level of the function}$$
$$g(x, u; y, v) = f(x, y) \cdot f_1(u, v);$$

$$E \cup E' = \text{a projection of the 1-level of the function}$$
$$g(x; y, z) = (f(x, y) - 1)(f'(x, z) - 1) + 1;$$
$$E \cap E' = \text{a projection of the 1-level of the function}$$
$$g(x; y, z) = f(x, y) \cdot f'(x, z).$$

Closure with respect to the projection operation is clear from the definition. Lemma 4.8 is proved. □

Now let E be a recursively enumerable set. We realize E as the 1-level of a partial recursive function f from $(\mathbf{Z}^+)^n$ to $\mathbf{Z}^+$ using Proposition 4.2, and we note that to prove that E is primitive enumerable, it suffices to show that the *graph* $\Gamma_f \subset (\mathbf{Z}^+)^n \times \mathbf{Z}^+$ of f is primitive enumerable. In fact, it is clear that $E =$ the 1-level of $f =$ the projection of the set $\Gamma_f \cap [(\mathbf{Z}^+)^n \times \{1\}]$ onto the first n coordinates. Here the set $\{1\} \subset \mathbf{Z}^+$ is primitive enumerable (for example, by 3.6), so that if we prove that Γ_f is primitive enumerable, it will follow from Lemma 4.8 that the same is true for E. Thus, our problem is finally reduced to the following form: we must prove that *the graph of a partial recursive function f is primitive enumerable.* To do this we verify that first, the graphs of the simplest functions are primitive enumerable, and second, if we apply any of the elementary operations to functions having primitive enumerable graphs, then the resulting function also has a primitive enumerable graph.

Graphs of the basic functions

$$\Gamma_{\text{suc}} \subset (\mathbf{Z}^+)^2 = \text{the 1-level of } (x_1 + 1 - x_2)^2 + 1,$$
$$\Gamma_{1^{(n)}} \subset (\mathbf{Z}^+)^{n+1} = \text{the 1-level of } x_{n+1},$$
$$\Gamma_{\text{pr}_i^n} \subset (\mathbf{Z}^+)^{n+1} = \text{the 1-level of } (x_i - x_{n+1})^2 + 1.$$

Stability under juxtaposition. Let f and g be partial functions from $(\mathbf{Z}^+)^m$ to $(\mathbf{Z}^+)^p$ and $(\mathbf{Z}^+)^q$, respectively. Suppose that Γ_f and Γ_g are primitive enumerable. Then $\Gamma_{(f,g)} \subset (\mathbf{Z}^+)^m \times (\mathbf{Z}^+)^p \times (\mathbf{Z}^+)^q$ coincides with the intersection

$$(\Gamma_f \times (\mathbf{Z}^+)^q) \cap \text{perm}(\Gamma_g \times (\mathbf{Z}^+)^p),$$

where perm : $(\mathbf{Z}^+)^m \times (\mathbf{Z}^+)^q \times (\mathbf{Z}^+)^p \to (\mathbf{Z}^+)^m \times (\mathbf{Z}^+)^p \times (\mathbf{Z}^+)^q$ is the operation of permuting the last two factors:

$$\langle x^{(m)}, y^{(q)}, z^{(p)} \rangle \mapsto \langle x^{(m)}, z^{(p)}, y^{(q)} \rangle.$$

It is clear from Lemma 4.8 that $\Gamma_{(f,g)}$ is primitive enumerable.

Stability under composition. Let g be a partial function from $(\mathbf{Z}^+)^n$ to $(\mathbf{Z}^+)^m$, let f be a partial function from $(\mathbf{Z}^+)^m$ to $(\mathbf{Z}^+)^1$, and let $h = f \circ g$. Then $\Gamma_h =$ the projection of the set $(\Gamma_g \times (\mathbf{Z}^+)^p) \cap ((\mathbf{Z}^+)^n \times \Gamma_f)$ onto $(\mathbf{Z}^+)^n \times (\mathbf{Z}^+)^p$. As before, if Γ_f and Γ_g are primitive enumerable, then so is Γ_h by Lemma 4.8.

The stability relative to recursion and the μ-operator is much subtler. We shall need the following elegant and useful lemma.

4.9. **Lemma.** *There exists a primitive recursive function* $\text{Gd}(k, t)$ (*Gödel's function*) *with the following property: for any $N \in \mathbf{Z}^+$ and any finite sequence*

$a_1, \ldots, a_N \in \mathbf{Z}^+$ *of length* N, *there exists* $t \in \mathbf{Z}^+$ *such that* $\mathrm{Gd}(k,t) = a_k$ *for all* $1 \leqslant k \leqslant N$. (In other words, the function Gd allows us to consider integers as encoding arbitrarily long sequences of integers: $\mathrm{Gd}(k,t)$ is the kth member of the sequence encoded by t, and the existence assertion ensures that each sequence has an encoding.)

PROOF. We first set

$$\mathrm{gd}(u,k,t) = \mathrm{rem}(1+kt, u)$$

and show that gd has the same property as Gd if we are allowed to choose $\langle u,t \rangle \in (\mathbf{Z}^+)^2$. Once we show this, we can set $\mathrm{Gd}(k,y) = \mathrm{gd}(t_1^{(2)}(y), k, t_2^{(2)}(y))$, where $t^{(2)} : \mathbf{Z}^+ \to (\mathbf{Z}^+)^2$ is the isomorphism in Lemma 4.5. (It is not really essential to remove the extra parameter in $\mathrm{gd}(u,k,t)$, but working with $\mathrm{Gd}(k,t)$ will make some of the formulas shorter.)

Thus, suppose we are given $a_1, \ldots, a_N \in \mathbf{Z}^+$. We first choose $X \in \mathbf{Z}^+$ so as to satisfy $X \geqslant N$ and $1 + kX! > a_k$ for all $1 \leqslant k \leqslant N$. We then set $t = X!$. It is easy to see that if $k_1 \neq k_2$ and $k_1, k_2 \leqslant N$, then $1 + k_1X!$ and $1 + k_2X!$ are relatively prime, since any common divisor would have to divide $(k_1 - k_2)X!$, i.e., would have to consist of primes $\leqslant X$, but no such prime divides $1 + k_1X!$.

By the Chinese remainder theorem, there exists a solution $u \in \mathbf{Z}^+$ of the system of equations

$$u \equiv a_k \mod(1 + kX!), \qquad 1 \leqslant k \leqslant N.$$

It is then obvious that

$$\mathrm{gd}(u,k,t) = \mathrm{rem}(1 + kt, u) = a_k, \qquad 1 \leqslant k \leqslant N. \qquad \square$$

We now continue with the proof of Theorem 4.3.

4.10. *Stability relative to the* μ*-operator.* Let f be a partial function from $(\mathbf{Z}^+)^{n+1}$ to $\mathbf{Z}^+$ and let

$$g(x_1, \ldots, x_n) = \min\{y | f(x_1, \ldots, x_n, y) = 1\}.$$

Recall that the domain of definition of g consists of those $\langle x_1, \ldots, x_n \rangle$ for which such a y exists and $\langle x_1, \ldots, x_n, k \rangle \in D(f)$ for all k less than the least such y. We want to prove that if Γ_f is primitive enumerable, then so is Γ_g.

Suppose that Γ_f is the projection onto the first $n+1$ coordinates of the 1-level of a primitive recursive function F:

$$\begin{aligned}\phi =& f(x_1, \ldots, x_{n+1}) \\ &\Leftrightarrow \exists \langle y_1, \ldots, y_m \rangle,\ F(x_1, \ldots, x_{n+1}, \phi, y_1, \ldots, y_m) = 1\end{aligned}$$

(where ϕ has been used to denote the argument of F that becomes the value of f). As in 4.4, it suffices to consider the case $m = 1$, since if $m \geqslant 2$, then we can use Lemma 4.5 to replace the vector $\langle y_1, \ldots, y_m \rangle$ by a single y, and if $m = 0$, then we can introduce a "dummy argument" y on which F does not actually depend.

Thus, let $m = 1$. We introduce a function G of the arguments $x_1, \ldots, x_n, \gamma, y, t, t_1$ by setting $s(1) = 2$, $s(x) = 1$ for $x \geqslant 2$, and

$$F_k = F\big(x_1, \ldots, x_n, k, \mathrm{Gd}(k,t), \mathrm{Gd}(k,t_1)\big), \qquad k \geqslant 1,$$

$$G = F(x_1, \ldots, x_n, \gamma, 1, y) \prod_{k=1}^{\gamma-1} s(\mathrm{Gd}(k,t)) \cdot F_k.$$

Here $\prod_{k=1}^{0} = 1$ by definition. It is easy to see that G is primitive recursive, since it is obtained by recursion on γ from two other functions that are obviously primitive recursive. We shall show that Γ_g is the projection of the 1-level of G onto the coordinates $(x_1, \ldots, x_n, \gamma)$.

The inclusion $\mathrm{pr}(G = 1) \subset \Gamma_g$. Let $\langle x_1, \ldots, x_n, \gamma, y, t, t_1 \rangle$ be a point in the 1-level of G. We must verify that $\langle x_1, \ldots, x_n \rangle \in D(g)$ and that $\gamma = g(x_1, \ldots, x_n)$. In other words, we must show that

$$f(x_1, \ldots, x_n, \gamma) = 1;$$

$$f(x_1, \ldots, x_n, k) \quad \text{is defined and} > 1 \text{ for all } k \leqslant \gamma - 1.$$

Since $G = 1$ at the given point, it follows that all the factors in G equal 1 there. In particular, $F(x_1, \ldots, x_n, \gamma, 1, y) = 1$, which implies that $f(x_1, \ldots, x_n, \gamma) = 1$, because Γ_f is the projection of the 1-level of F. If $\gamma = 1$, there is nothing more to be proved.

Suppose $\gamma > 1$. Since the kth factor in the product $\prod_{k=1}^{\gamma-1}$ equals 1, we obtain

$$s(\mathrm{Gd}(k,t)) = 1 \Rightarrow \mathrm{Gd}(k,t) \geqslant 2,$$

$$F_k = 1 \Rightarrow \mathrm{Gd}(k,t) = f(x_1, \ldots, x_n, k) \geqslant 2,$$

as required.

The inclusion $\Gamma_g \subset \mathrm{pr}(G = 1)$. Let $\langle x_1, \ldots, x_n, \gamma \rangle \in \Gamma_g$. We must choose values for the remaining coordinates y, t, and t_1 in such a way as to make all the factors in G equal to 1.

First of all, $\langle x_1, \ldots, x_n, \gamma, 1 \rangle \in \Gamma_f$ by the definition of g. We find the necessary value of y by lifting this point from Γ_f to the 1-level of F. If $\gamma = 1$, we may choose arbitrary values of t and t_1.

Suppose $\gamma > 1$. We then find t from the system of equations

$$\mathrm{Gd}(k,t) = f(x_1, \ldots, x_n, k), \qquad \text{for all } 1 < k \leqslant \gamma - 1.$$

(Here the right side exists by the definition of $D(g)$.)

Finally, for each $k \leqslant y - 1$ we lift the point

$$\langle x_1, \ldots, x_n, k, \mathrm{Gd}(k,t) \rangle \in \Gamma_f$$

to a point on $F = 1$ having additional coordinate $y^{(k)}$, and then we find t_1 from the system of equations

$$\mathrm{Gd}(k, t_1) = y^{(k)}, \qquad 1 \leqslant k \leqslant \gamma - 1.$$

This makes all the factors in $\prod_{k=1}^{\gamma-1}$ equal to 1. In fact, $s(\mathrm{Gd}(k,t)) = 1$, since $\mathrm{Gd}(k,t) = f(x_1,\dots,x_n,k) \geqslant 2$ for $k \leqslant \gamma-1$, and, finally, $F_k = F(x_1,\dots,x_n,k, \mathrm{Gd}(k,t), \mathrm{Gd}(k,t_1)) = 1$ by the definition of t and t_1. □

4.11. *Stability relative to recursion.* We now carry out the last step in the proof of Theorem 4.3.

Let f and g be partial functions of n and $n+2$ variables, respectively, and let h be the function of $n+1$ variables that is obtained from f and g using recursion:

$$h(x_1,\dots,x_n,1) = f(x_1,\dots,x_n),$$
$$h(x_1,\dots,x_n,k+1) = g\big(x_1,\dots,x_n,k,h(x_1,\dots,x_n,k)\big).$$

We must show that if Γ_f and Γ_g are primitive enumerable, then so is Γ_h.

Let F and G be primitive recursive functions whose 1-levels project onto Γ_f and Γ_g, respectively:

$$\phi = f(x_1,\dots,x_n) \Leftrightarrow \exists y, F(x_1,\dots,x_n,\phi,y) = 1,$$
$$\gamma = g(x_1,\dots,x_{n+2}) \Leftrightarrow \exists z, G(x_1,\dots,x_{n+2},\gamma,z) = 1,$$

where, as in 4.10, it suffices to consider the case in which the projection codimension is 1.

We shall explicitly construct a function H whose 1-level projects onto Γ_h. H will be a function of the arguments $x_1,\dots,x_{n+1},\eta,y,t,t_1$ (where η is the argument that becomes the value of h). We set

$$\tilde{s}(1) = 1, \qquad \tilde{s}(x) = 2, \quad \text{for } x \geqslant 2;$$
$$G_k = G\big(x_1,\dots,x_n,k-1,\mathrm{Gd}(k-1,t),\mathrm{Gd}(k,t),\mathrm{Gd}(k,t_1)\big);$$
$$H = F\big(x_1,\dots,x_n,\mathrm{Gd}(1,t),y\big)\cdot\tilde{s}\Big[\big(\eta-\mathrm{Gd}(x_{n+1},t)\big)^2+1\Big]\prod_{k=2}^{x_{n+1}} G_k.$$

(We take $\prod_{k=2}^{x_{n+1}} = 1$ if $x_{n+1} = 1$.) As in 4.10, we easily verify that H is primitive recursive.

The inclusion $\mathrm{pr}(H=1) \subset \Gamma_h$. Let $\langle x_1,\dots,x_{n+1},\eta,y,t,t_1\rangle$ be a point on $H=1$. We must show that $h(x_1,\dots,x_{n+1}) = \eta$. Since the second factor in H equals 1, we first obtain $\eta = \mathrm{Gd}(x_{n+1},t)$. If we also have $x_{n+1} = 1$, then setting the first factor in H equal to 1 gives

$$\eta = \mathrm{Gd}(1,t) = f(x_1,\dots,x_n) = h(x_1,\dots,x_n,1).$$

Now suppose $x_{n+1} > 1$. In this case, using the equation $G_k = 1$ we find that for all $2 \leqslant k \leqslant x_{n+1}$,

$$\mathrm{Gd}(k,t) = g\big(x_1,\dots,x_n,k-1,\mathrm{Gd}(k-1,t)\big),$$

and using the equation $F=1$ and the definition of h we find that

$$\mathrm{Gd}(1,t) = f(x_1,\dots,x_n) = h(x_1,\dots,x_n,1).$$

If we increase k from $k = 1$ to $k = x_{n+1}$ and use the recursive definition of h, we see by induction on k that $\mathrm{Gd}(k,t) = h(x_1,\ldots,x_n,k)$ and, in particular,

$$\eta = \mathrm{Gd}(x_{n+1},t) = h(x_1,\ldots,x_n,x_{n+1}).$$

The inclusion $\Gamma_h \subset \mathrm{pr}(H=1)$. We are given a point

$$\langle x_1,\ldots,x_{n+1},h(x_1,\ldots,x_{n+1})\rangle \in \Gamma_h.$$

We let $\eta = h(x_1,\ldots,x_{n+1})$. We must also choose values of y, t, and t_1 so as to make H equal to 1.

If $x_{n+1} = 1$, we choose t such that $\mathrm{Gd}(1,t) = h(x_1,\ldots,x_n,1) = f(x_1,\ldots,x_n)$. We then lift the point $\langle x_1,\ldots,x_n,\mathrm{Gd}(1,t)\rangle \in \Gamma_f$ to a point on $F=1$. This gives us the value of y; t_1 may be chosen arbitrarily.

Now let $x_{n+1} > 1$. We first find t from the system of equations

$$\begin{aligned} \mathrm{Gd}(1,t) &= f(x_1,\ldots,x_n) = h(x_1,\ldots,x_n,1); \\ \mathrm{Gd}(k,t) &= h(x_1,\ldots,x_n,k) = g(x_1,\ldots,x_n,k-1,\mathrm{Gd}(k-1)), \\ &\times 2 \leqslant k \leqslant x_{n+1}. \end{aligned}$$

We then find y by lifting the point $\langle x_1,\ldots,x_n,\mathrm{Gd}(1,t)\rangle \in \Gamma_f$ to the 1-level of F. This makes the first two factors in H equal to 1.

We next lift the points

$$\langle x_1,\ldots,x_n,k-1,\mathrm{Gd}(k-1,t),\mathrm{Gd}(k,t)\rangle \in \Gamma_g, \qquad 2 \leqslant k \leqslant x_{n+1},$$

to the 1-level of G by adding coordinates $z^{(k)}$, and then solve the following system of equations for t_1:

$$\mathrm{Gd}(k,t_1) = z^{(k)}, \qquad 2 \leqslant k \leqslant x_{n+1}.$$

This makes the G_k factors in H equal to 1.

The proof of Theorem 4.3 is complete. □

4.12. *Explanation of the term "recursively enumerable set."* Theorem 4.3 shows that if E is recursively enumerable, then there exists a program that "generates" E (see 4.1). In fact, suppose E is the projection onto the first n coordinates of the 1-level of the primitive recursive function $f(x_1,\ldots,x_n,y)$. The program that generates E must run through the vectors $\langle x_1,\ldots,x_n,y\rangle$, say in Cantor order, compute f at each vector, and give $\langle x_1,\ldots,x_n\rangle$ as output if and only if f equals 1 (compare with Corollary 4.18 below). Unlike programs of the type described in §1, which can become stuck forever on an element not in E, a generating program sooner or later gives us any given element of E, and nothing other than such elements. However, if E is empty, we might never find this out.

We conclude this section by discussing the properties of the so-called decidable sets. Intuitively, $E \subset (\mathbf{Z}^+)^n$ is decidable if there exists a program that for every element of $(\mathbf{Z}^+)^n$ tells whether it belongs to E.

4.13. **Definition.** A set $E \subset (\mathbf{Z}^+)^n$ is called *decidable* if both it and its complement are recursively enumerable.

In §5 and in the next chapter we show that there exist sets that are recursively enumerable but not decidable. This result is closely connected with Gödel's incompleteness theorem, which is the subject of Chapter VII.

4.14. **Theorem.** *The following three classes of sets coincide:*

(a) *sets whose characteristic function is recursive;*
(b) *level sets of general recursive (i.e., everywhere defined partial recursive) functions;*
(c) *decidable sets.*

PROOF. The relations (a) = (b) and (b) $\subset$ (c) are obvious from what has already been proved. It thus remains to show that (c) $\subset$ (a).

Let $E \subset (\mathbf{Z}^+)^n$ be a decidable set, and let E' be its complement. By definition, $E = D(f)$ and $E' = D(f')$ for certain partial recursive functions f and f'. We may even assume that $f \equiv 1$ and $f' \equiv 2$ (where they are defined). We consider $\Gamma_f \cup \Gamma_{f'} \subset (\mathbf{Z}^+)^n \times \mathbf{Z}^+$. This union is obviously the graph Γ_g of the characteristic function g of the set E. It is clear from the proof of Lemma 4.8 that Γ_g is recursively enumerable whenever Γ_f and $\Gamma_{f'}$ are. Hence, the partial recursiveness of g is implied by the following result, which is also of independent interest.

4.15. **Proposition.** *In order for a partial function g from $(\mathbf{Z}^+)^n$ to $\mathbf{Z}^+$ to be partial recursive, it is necessary and sufficient that its graph Γ_g be recursively enumerable.*

PROOF. Necessity has already been proved.

We verify sufficiency. Since Γ_g is recursively enumerable, there exists a primitive recursive function $G(x_1, \ldots, x_n, \gamma, z)$ (see 4.10) such that $\Gamma_g =$ the projection of the 1-level of G onto $(x_1, \ldots, x_n, \gamma)$. We set

$$H(x_1, \ldots, x_n, u) = G(x_1, \ldots, x_n, t_1^{(2)}(u), t_2^{(2)}(u)),$$

where $u \mapsto \langle t_1^{(2)}(u), t_2^{(2)}(u) \rangle$ is the primitive recursive isomorphism $\mathbf{Z}^+ \to (\mathbf{Z}^+)^2$ described in 4.5 and 4.7. H is obviously primitive recursive. Finally, we set

$$h(x_1, \ldots, x_n) = \min\{u | H(x_1, \ldots, x_n, u) = 1\}.$$

This is a partial recursive function whose domain of definition coincides with $D(g)$ and that easily allows us to compute g:

$$g(x_1, \ldots, x_n) = t_1^{(2)}(h(x_1, \ldots, x_n)).$$

Thus, g is partial recursive, and the proof of Proposition 4.15 and Theorem 4.14 is complete. □

4.16. **Corollary.** *Every partial recursive function g has a description in which the μ-operator is applied only once.*

4.17. **Corollary.** *Every partial recursive function g that is everywhere defined has a description $g_1, \ldots, g_N = g$ in which all the functions g_i are everywhere defined.*

In fact, the description whose last part (starting with G) was constructed in 4.15 has this property.

4.18. **Corollary.** *The class of nonempty recursively enumerable sets coincides with the class of sets of values of primitive recursive functions.*

In fact, the set of values of a function f is a projection of the graph of f. Conversely, let $E \subset (\mathbf{Z}^+)^n$ be a nonempty enumerable set that is the projection onto the $(x_1, \ldots, x_n)$-space of the 1-level of a primitive recursive function $f(x_1, \ldots, x_n, y)$. Let $\langle e_1, \ldots, e_n \rangle$ be an arbitrary member of E. Then E coincides with the set of values of the primitive recursive function

$$g(z) = \begin{cases} \langle t_1^{(n+1)}(z), \ldots, t_n^{(n+1)}(z) \rangle, & \text{if } f(t_1^{(n+1)}(z), \ldots, t_n^{(n+1)}(z), t_{n+1}^{(n+1)}(z)) = 1; \\ \langle e_1, \ldots, e_n \rangle, & \text{if not.} \end{cases}$$

□

4.19. **Corollary.**

(a) *Finite sets and their complements in $(\mathbf{Z}^+)^n$ are decidable.*

(b) *Every partial function from $(\mathbf{Z}^+)^m$ to $(\mathbf{Z}^+)^n$ with a finite domain of definition is recursive and computable.*

In fact, the one-point set $\{a\} \subset \mathbf{Z}^+$ is a level for a suitable sum of two step functions, and its complement is a level for another such sum. Decidability is preserved under finite union and intersection, so we have (a) for $n = 1$. Then the isomorphism $\tau^{(n)}$ allows us to infer this result for all n.

This also implies (b), since the graphs of the mappings in (b) are finite, and therefore enumerable.

5 Elements of Recursive Geometry

5.1. Let $E \subset (\mathbf{Z}^+)^m$ be an enumerable set. We consider the structure on E given by the following data:

(a) $\mathcal{E} = \{E' | E' \subset E, E' \text{ is enumerable}\}$.

(b) For every $E' \in \mathcal{E}, \mathbf{R}(E') = \{f | D(f) = E', f : E' \to \mathbf{Z}^+ \text{ is recursive}\}$.

We let $\mathcal{R} =$ the set of pairs $\langle E', \mathbf{R}(E') \rangle, E' \in \mathcal{E}$.

We shall show that the structure $\{\mathcal{E}, \mathcal{R}\}$ has much in common with the structure "a topological space together with a sheaf." This allows us to find natural interpretations for certain well-known results about enumerable sets, and to ask new questions suggested by analogies with other geometrical theories.

We begin with some simple observations.

5.2. $\mathcal{E}$ *is a lattice, i.e., it is closed with respect to finite unions and intersections.*

Since $\mathcal{E}$ is not closed with respect to arbitrary infinite unions, we cannot consider $\mathcal{E}$ as the system of open subsets of E in some topology. Nevertheless, in Section 5.9 below we show that $\mathcal{E}$ is stable with respect to an important class of infinite unions. We shall say that $\mathcal{E}$ determines a *quasitopology* on E (which has properties similar to those of Grothendieck topologies, but does not satisfy all the axioms of the latter).

5.3. *Let* $E^{'}, E'' \in \mathcal{E}$ *and* $E^{'} \subset E''$. *Then the restriction of functions to* $E^{'}$ *gives a mapping* $\mathbf{R}(E'') \to \mathbf{R}(E^{'}) : f \mapsto f|_{E'}$.

In fact, let $c_{E'} \in \mathbf{R}(E^{'})$ and $c_{E'} = 1$ on $E^{'}$. Then $f|_{E'} = fc_{E'}$ is recursive whenever f and $c_{E'}$ are.

5.4. *Let* $E^{'} = \bigcup_{k=1}^{n} E_k$, *where* $E^{'}, E_k \in \mathcal{E}$. *Suppose that the* f_k *are in* $\mathbf{R}(E_k)$ *and are compatible on intersections:*

$$\forall i, j \leqslant n, \qquad f_i|_{E_i \cap E_j} = f_j|_{E_i \cap E_j}.$$

Then there exists an (obviously unique) function $f \in \mathbf{R}(E^{'})$ *such that* $\forall k \leqslant n, f|_{E_k} = f_k$.

We need only verify that $f \in \mathbf{R}(E^{'})$, since there obviously exists a function $f : E^{'} \to \mathbf{Z}^+$ that is "glued together" from the f_k. But the graph of f is the union of the finitely many enumerable graphs $\Gamma_{f_i} \subset E^{'} \times \mathbf{Z}^+$, and so is itself enumerable. We then use Proposition 4.15.

The results 5.3 and 5.4 allow us to consider $\mathcal{R}$ as a *sheaf* on the quasitopology $\mathcal{E}$.

5.5. *Let* E_1 *and* E_2 *be enumerable sets, and let* $f : E_1 \to E_2$ *be a recursive function. Then* f *induces a morphism of the corresponding quasitopologies with sheaves in the following sense:*

(a) *If* $E^{'} \subset E_2$ *is enumerable, then* $f^{-1}(E^{'}) \subset E_1$ *is enumerable.*
(b) *For every* $E^{'} \subset E_2$, *composition with* f *determines a mapping*

$$f^*_{E'} : \mathbf{R}(E^{'}) \to \mathbf{R}(f^{-1}(E^{'})).$$

The first part follows because $c_{f^{-1}(E')} = c_{E'} \circ f$ is recursive whenever $c_{E'}$. and f are; the second part is obvious.

One might get the impression that the pair $\langle \mathcal{E}, \mathcal{R} \rangle$ completely characterizes E independently of the embedding $E \subset (\mathbf{Z}^+)^m$. However, this is not the case.

5.6. **Proposition.** *Let* E_1 *and* E_2 *be enumerable infinite sets. Then there exists a bijection* $f : E_1 \tilde{\to} E_2$ *such that* f *and* f^{-1} *are (partial) recursive.* f *induces an isomorphism* $\langle \mathcal{E}_1, \mathcal{R}_1 \rangle \tilde{\to} \langle \mathcal{E}_2, \mathcal{R}_2 \rangle$.

PROOF. We establish the following more precise facts:

(a) If $E \subset \mathbf{Z}^+$ is infinite and decidable, then there exists a general recursive bijection $f : \mathbf{Z}^+ \tilde{\to} E$ for which f^{-1} is (partial) recursive and is an increasing function. The converse is also true.

(b) If $E \subset \mathbf{Z}^+$ is infinite and enumerable, then there exists a general recursive bijection $f : \mathbf{Z}^+ \tilde{\to} E$ with f^{-1} (partial) recursive.

First suppose E is decidable, and let $g(x) = 2$ for $x \in E$, $g(x) = 1$ for $x \notin E$, and $h = c_{E'}$. We set

$$f(z) = \min\left\{ y \,\middle|\, \left(\sum_{x=1}^{y} g(x) - y - z\right)^2 + 1 = 1 \right\} = \text{the } z\text{th element of } E.$$

It is easy to see that

$$f^{-1}(x) = \left(\sum_{y=1}^{x} g(y) - x\right) h(x) \begin{cases} \text{is equal to the index of } x \\ \text{as an element of } E, \text{ if } x \in E; \\ \text{is not defined, otherwise.} \end{cases}$$

Now suppose E is enumerable. By Corollary 4.18, there exists a primitive recursive function $g : \mathbf{Z}^+ \to E$ whose image coincides with E. We shall adjust g so that it becomes bijective. We set

$$F = \{k \in \mathbf{Z}^+ | \forall i < k, g(i) \neq g(k)\}.$$

This set is decidable, since it is the 1-level of the following primitive recursive function h:

$$h(1) = 1; \quad h(k) = \prod_{i=1}^{k-1} s((g(i) - g(k))^2 + 1), \quad \text{for } k \geqslant 2;$$

$$s(x) = \begin{cases} 1, & \text{for } x \geqslant 2, \\ 2, & \text{for } x = 1. \end{cases}$$

By the previous result, there exists a recursive bijection $g' : \mathbf{Z}^+ \tilde{\to} F$. Let $f = g \circ g'$. Since $g|_F : F \tilde{\to} E$ is a bijection, it follows that $f : \mathbf{Z}^+ \tilde{\to} E$ is also a bijection. The inverse function is partial recursive because

$$f^{-1}(x) = \min\{y | (f(y) - x)^2 + 1 = 1\}.$$

The proposition is proved. □

Because of this result we usually consider the embedding $E \subset (\mathbf{Z}^+)^m$ to be an essential element of the structure on E. In particular, we call E_1 and E_2 *isomorphic* if there exists a bijection between them that is induced by a recursive bijection of the ambient spaces.

The complete classification of enumerable sets up to isomorphism is not known, but many subtle results have been obtained in the theory of "reducibilities." We shall only go so far as to show, using a theorem that will be proved in the next chapter, that not all enumerable sets are decidable.

5.7. *Families.* Suppose that $m \geqslant 0$ and B is a set. By a *family of m-sets* (or an *m-family*) *over the base B* we mean any mapping $B \to \mathcal{P}((\mathbf{Z}^+)^m)$. If $E_k \subset (\mathbf{Z}^+)^m$ is the image of $k \in B$ under this mapping, we also denote this family by $\{E_k\}$. We call the set $E = \{\langle x, k\rangle | x \in E_k\} \subset (\mathbf{Z}^+)^m \times B$ the *total space* of the family.

Similarly, we call a mapping $B \to \{$partial functions from $(\mathbf{Z}^+)^m$ to $\mathbf{Z}^+\}$ a *family of m-functions over the base B.* We call the function $f : \langle x, k\rangle \mapsto f_k(x)$ for $x \in D(f_k)$ the *total function* of the family.

A family of m-sets (resp. m-functions) is said to be *enumerable* if $B \subset (\mathbf{Z}^+)^n$ for some n and if the total space is enumerable in $(\mathbf{Z}^+)^m \times (\mathbf{Z}^+)^n$ (resp. the total function is partial recursive on $(\mathbf{Z}^+)^m \times (\mathbf{Z}^+)^n$).

If $\{E_k\}$ is enumerable, then the set $\{k \in B | E_k$ is nonempty$\}$ is enumerable, since it is a projection of the total space E. Each of the E_k is enumerable, since it is the intersection $E \cap (\mathbf{Z}^+)^m \times \{k\}$.

Similarly, if $\{f_k\}$ is enumerable, then the set $\{k \in B | f_k$ is not the nowhere defined function$\}$ is enumerable, since it is a projection of the domain of definition of the total function f. Each of the f_k is partial recursive, since it is the restriction of f to the enumerable set $D(f) \cap (\mathbf{Z}^+)^m \times \{k\}$.

If $\{f_k\}$ is an enumerable family of m-functions, then $\{D(f_k)\}$ is an enumerable family of m-sets (with total space $D(f)$), and $\{\Gamma_{fk}\}$ is an enumerable family of $(m+1)$-sets (with total space Γ_f, or more precisely, Γ_f after a permutation of its factors).

An enumerable family $\{E_k\}$ (respectively $\{f_k\}$) is said to be *versal* if every enumerable m-set (resp. any partial recursive m-function) is among the elements of the family. (The word "versal" is borrowed from algebraic geometry, after removing the prefix "uni" which would indicate that each term in the family could occur only once.) In §8 of the next chapter we show that versal families *exist* for each m. This is one of the central results of the theory, since total spaces and total functions of versal families are the starting point

for practically all investigations of undecidability. Here we limit ourselves to the simplest and most fundamental application:

5.8. **Theorem.** *Let $\{E_k\}$ be a versal family of* 1*-sets over the base $B \subset \mathbf{Z}^+$. Then the set*

$$F = \{k | k \in E_k\}$$

is enumerable, but is not decidable.

PROOF. Let $E \subset \mathbf{Z}^+ \times \mathbf{Z}^+$ be the total space of the family. Then $F =$ the projection of $E \cap$ (diagonal in $\mathbf{Z}^+ \times \mathbf{Z}^+$) onto the first factor, and therefore is enumerable.

On the other hand, for every $k \in B$ we have $\overline{F} = \mathbf{Z}^+ \backslash F \neq E_k$, since k belongs to either $\overline{F}$ or E_k, but not to both. Since $\{E_k\}$ is a versal family, $\overline{F}$ cannot be enumerable. The theorem is proved. □

We now show how to use enumerable families to strengthen the results in 5.2 and 5.4. We return to the notation at the beginning of the section.

5.9. *$\mathcal{E}$ is closed with respect to taking the union of the elements of any enumerable family of subsets of E.*

In fact, suppose that $\{E_k^{'}\}$ is such a family and $E^{'}$ is its total space, where $E^{'} \subset (\mathbf{Z}^+)^m \times (\mathbf{Z}^+)^n$. Then

$$\bigcup_{k \in B} E_k^{'} = \text{the projection of } E^{'} \text{ on } (\mathbf{Z}^+)^m.$$

5.10. *Suppose that $\{f_k\}$ is an enumerable family of partial functions on E, $E_k^{'} = D(f_k), E^{'} = \bigcup_{k \in B} E_k^{'}$, and*

$$\forall i, j \in B, \qquad f_i|_{E_i^{'} \cap E_j^{'}} = f_j|_{E_i^{'} \cap E_j^{'}}.$$

Then there exists a unique function $f \in \mathbf{R}(E^{'})$ that is glued together from the f_k.

In fact, the graph Γ_f is enumerable, since it is the union of the enumerable family of enumerable sets Γ_{f_k}.

5.11. After these remarks it is natural to consider the following system of ideas by way of analogy with the theory of spaces with sheaves.

(a) Let $E = \bigcup_{k \in B} E_k$ be a covering of an enumerable set by an enumerable family. Then for any $n \geqslant 1$ the family $\{E_{k_1} \cap \cdots \cap E_{k_n} | \langle k_1, \ldots, k_n \rangle \in B^n \rangle\}$ is also an enumerable covering of E. In fact, let $E^{'}$ = the total space of $\{E_k\} \subset E \times B$, and let

$$\begin{aligned} E^{'n} &= \{\langle x_1, \ldots, x_n, k_1, \ldots, k_n \rangle | x_i \in E_{ki}, \ i = 1, \ldots, n\} \\ &\approx E^{'} \times \cdots \times E^{'} \ (n \text{ times}). \end{aligned}$$

Then the total space of the family $\{E_{k_1} \cap \cdots \cap E_{k_n}\}$ is isomorphic to (diagonal in E^n) $\times B^n \cap E^{'n}$.

(b) Using the same notation, we define the "recursive product" $\mathbf{R}\Pi_n \subset \Pi_{\langle k_1, \ldots, k_n \rangle \in B^n} (E_{k_1} \cap \cdots \cap E_{k_n})$ as follows: $\mathbf{R}\Pi_0 = \mathbf{R}(E), \mathbf{R}\Pi_n$ = the set of enumerable families $\{f_{\langle k_1, \ldots, k_n \rangle}\}$ over B^n such that

$$f_{\langle k_1, \ldots, k_n \rangle} \in \mathbf{R}(E_{k_1} \cap \cdots \cap E_{k_n}), \quad \text{for } n \geqslant 1.$$

(c) For every $n \geqslant 0$ we have the following boundary mappings:

$$\begin{aligned} \partial_i^n : \mathbf{R}\prod\nolimits_n &\to \mathbf{R}\prod\nolimits_{n+1} \qquad i = 1, \ldots, n+1 : \\ (\partial_i^n (\cdots f_{\langle k_1, \ldots, k_n \rangle} \cdots))_{\langle l_1, \ldots, l_{n+1} \rangle} & \\ = f_{\langle l_1, \ldots, l_{i-1}, \ l_{i+1}, \ldots, l_{n+1} \rangle | E_{l_1} \cap \cdots \cap E_{l_{n+1}}}. & \end{aligned}$$

(Note that we really do not have $\partial_i^n (\mathbf{R}\Pi_n) \subset \mathbf{R}\Pi_{n+1}$.)

It is possible to associate various types of "recursive Čech cohomology groups" of the covering $\bigcup_{k \in B} E_k$ to the object

$$\mathbf{R}\prod_0 = \mathbf{R}(E) \overset{\partial_0^0}{\rightarrow} \mathbf{R}\prod\nolimits_1 \underset{\partial_2^1}{\overset{\partial_1^1}{\rightrightarrows}} \mathbf{R}\prod\nolimits_2 \begin{smallmatrix}\rightarrow\\ \rightarrow\\ \rightarrow\end{smallmatrix} \cdots .$$

It would be interesting to study such cohomology groups. The result 5.10 shows that this complex is "exact at the first term."

The reader should not find it hard to imagine what other geometrical concepts would look like in this context. In particular, it would be worthwhile to study the quotients of enumerable sets by enumerable equivalence relations. Higman's theorem (see Chapter VIII) gives a characterization of groups in the category of such objects.

We conclude by giving several results on the structure of $\mathcal{E}$. Because of Proposition 5.6, we need only consider subsets of $\mathbf{Z}^+$; that is, we take $\mathcal{E} = \{E' | E' \subset \mathbf{Z}^+, E' \text{ enumerable}\}$.

5.12. **Proposition.** *There exist enumerable subsets $F \subset \mathbf{Z}^+$ having an infinite complement such that for any infinite $E \in \mathcal{E}$ we have $F \cap E \neq \emptyset$, so that $F \cap E$ is infinite.*

Such F are called *simple*. From a topological point of view they resemble dense open sets.

PROOF. Let $\{E_k\}$ be a versal family of 1-sets over $\mathbf{Z}^+$ with total space $E \subset \mathbf{Z}^+ \times \mathbf{Z}^+$. We set $E' = E \cap \{\langle x, k \rangle | x > 2k\}$. Since E' is enumerable, there exists a primitive recursive function with image E':

$$g = (g_1, g_2) : \mathbf{Z}^+ \to E' .$$

Let $h(k) = \min\{z | g_2(z) = k\}$, let $f(k) = g_1(h(k))$, and let F denote the set of values of f. F has an infinite complement, since $f(k) > 2k$. The intersection of F with an infinite E_k is nonempty, since any value of $g_1(z)$ when $g_2(z) = k$ lies in $E_k \cap E' = \emptyset$. The proposition is proved. □

5.13. **Proposition.**

(a) *The quotient lattice $\mathcal{E}/(\text{finite sets})$ has nontrivial maximal elements.*
(b) *Every nonsimple enumerable set with an infinite complement is contained in such a maximal element.*
(c) *There exist simple enumerable sets with an infinite complement that are not contained in any nontrivial maximal set.*

We refer the reader to Rogers' book for the proof of these and many other results.

VI

Diophantine Sets and Algorithmic Undecidability

1 The Basic Result

1.1. In §4 of Chapter V we showed that enumerable sets are the same thing as projections of level sets of primitive recursive functions. The projections of the level sets of a special kind of primitive recursive function—polynomials with coefficients in $\mathbf{Z}^+$—are called *Diophantine sets*. We note that this class does not become any larger if we allow the coefficients in the polynomial to lie in $\mathbf{Z}$. The basic purpose of this chapter is to prove the following deep result:

1.2. **Theorem** (M. Davis, H. Putnam, J. Robinson, Yu. Matiyasevič). *All enumerable sets are Diophantine.*

The plan of proof is described in §2. §§3–7 contain the intricate yet completely elementary constructions that make up the proof itself; these sections are not essential for understanding the subsequent material, and may be omitted if the reader so desires.

In §8 we use Theorem 1.2 to prove the existence of versal families of enumerable sets and functions. Recall that in §5 of Chapter V this result was shown to imply that enumerable sets exist that are undecidable, a fact we shall use in Section 1.3 below.

In §7, which stands somewhat apart from the rest of the chapter, we define the Kolmogorov complexity of recursive functions, establish the basic properties of this concept, and prove that the problem of computing the complexity is algorithmically undecidable.

In Chapter VII the following corollary of Theorem 1.2 will be used in an essential way: *enumerable sets are definable in* $\mathrm{L}_1\mathrm{Ar}$. In fact, by their very definition, Diophantine sets are defined by formulas of the form $\exists x_1 \cdots \exists x_n(p)$, where p is an atomic formula.

In the remainder of this section we describe the principal applications of Theorem 1.2: settling Hilbert's tenth problem, constructing polynomials that take only and all prime number values in $\mathbf{Z}^+$, and so on.

Yu. I. Manin, *A Course in Mathematical Logic for Mathematicians, Second Edition*, Graduate Texts in Mathematics 53, DOI 10.1007/978-1-4419-0615-1_6,

1.3. *Hilbert's tenth problem.* Hilbert stated it as follows:

> Suppose we are given a Diophantine equation with an arbitrary number of unknowns and with rational integer coefficients. Give a way in which it is possible to determine after a finite number of operations whether this equation is solvable in rational integers.

We show that the combination of Theorem 1.2, Theorem 5.8 of Chapter V (which follows from Theorem 1.2), and Church's thesis implies that this problem is undecidable.

First of all, any natural number is the sum of four integer squares (Lagrange). Hence $f(x_1, \ldots, x_n) = 0$ is solvable in $(\mathbf{Z}^+)^n$ if and only if the equation $f(1 + \Sigma_{i=1}^4 y_{i1}^2, \ldots, 1 + \Sigma_{i=1}^4 y_{in}^2) = 0$ is solvable in $(\mathbf{Z})^{4n}$. Consequently, it is sufficient to show that the mass problem "determining whether there are solutions in $(\mathbf{Z}^+)$" (see Section 2.6 of Chapter V) is algorithmically undecidable.

Let $E \subset \mathbf{Z}^+$ be an enumerable set that is not decidable. We represent E as the projection onto the t-coordinate of the 0-level of the polynomial $f_t = f(t; x_1, \ldots, x_n)$, where $f \in Z[t, x_1, \ldots, x_n]$. The equation $f_{t_0} = 0, t_0 \in \mathbf{Z}^+$, has a solution if and only if $t_0 \in E$. By the discussion in §2 of Chapter V, the corresponding mass problem for the family $\{f_t\}$ is algorithmically decidable if and only if the characteristic function of E is computable. But by our choice of E, this characteristic function is only semicomputable.

Thus, solvability in integers cannot be determined algorithmically even for a suitable one-parameter family of equations. The number of unknowns in the equation, and, in general, the codimension of the projection in Theorem 1.2, can be reduced to 13 (Matiyasevič, Robinson). The precise minimum is not known, although it is an interesting problem.

Finally, it should be noted that the construction of a Diophantine representation for any enumerable set E is completely effective in the sense that given a recursive description of f with $D(f) = E$ or of g with $g(\mathbf{Z}^+) = E$, we can write out the corresponding polynomial explicitly. The same holds for the construction of versal families, of an enumerable undecidable set, and so on. These are all constructive assertions, and not simple existence theorems.

1.4. *Polynomials that represent the prime numbers.* The search for "explicit formulas" for prime numbers was a traditional occupation of dedicated number theory enthusiasts for many centuries. Euler found the polynomial $x^2 + x + 41$, which takes a long series of only prime values. But it has long been known that the set of values at integer points of a polynomial f in $\mathbf{Z}[x_1, \ldots, x_n]$ cannot consist entirely of prime numbers: for example, if p and q are two sufficiently large primes, then the congruence $f \equiv 0 \bmod pq$ can be solved (in infinitely many ways). On the other hand, the problem becomes solvable in the class of primitive recursive functions: the function $\{i \mapsto \text{the } i\text{th prime}\}$ is itself primitive recursive (see §1 of Chapter VII), but for trivial reasons.

The nontrivial statement of the problem and the problem's solution involve Theorem 1.2: the set of prime numbers is the set of all *positive values at points in* $(\mathbf{Z}^+)^n$ *of a certain polynomial in* $\mathbf{Z}[x_1, \ldots, x_n]$ (or, if we prefer, n may be replaced by $4n$; see the reduction step in 1.3). Matiyasevič showed that there is a suitable polynomial of degree 37 in 24 variables.

This is actually a general result that has nothing to do with the specific properties of prime numbers:

1.5. **Proposition.** *Let $E \subset \mathbf{Z}^+$ be a Diophantine set. Then there exists a polynomial $g \in \mathbf{Z}[x_0, \ldots, x_n]$ such that E coincides with the set of positive values of g at points in $(\mathbf{Z}^+)^{n+1}$.*

PROOF. Let E be the projection of the 0-level of the polynomial $f(x_0, x_1, \ldots, x_n)$ onto the x_0-coordinate. We set

$$g = x_0[1 - f^2(x_0, x_1, \ldots, x_n)].$$

Clearly, the positive values of g are precisely the elements of E. □

It remains only to use the fact that the set of prime numbers is decidable, and hence Diophantine by Theorem 1.2.

The following sets are also sets of positive integer values of polynomials:

1.6. *The sequences* $\{1, 10, 100, \ldots, 10^k, \ldots\}$ and $\{1, 2^2, 3^{3^3}, \ldots, n^{n^{n^{\cdot^{\cdot^{\cdot^n}}}}}$ (n times)$, \ldots\}$.

It is amazing that the values of the corresponding polynomials can drop to zero and below in neighborhoods of points where these values are so large.

1.7. *The Fermat set $\{n | n > 2$ and $x^n + y^n + z^n = 0$ is solvable in $\mathbf{Z}\}$.* Thus, the variable n can be moved from the exponent to the coefficients of a Diophantine equation.

1.8. The set $\{10\varepsilon_1, 10^2\varepsilon_2, \ldots, 10^n\varepsilon_n, \ldots\}$, where ε_i is the ith digit after the decimal point in the decimal expansion of e (or π or $\sqrt[3]{2}$, or any other "computable" irrational number).

1.9. *The set of all partial fractions in the continued fraction expansion of e, or π, or $\sqrt[3]{2}$.*

We recall that in the case of $\sqrt[3]{2}$ it is not known whether this set is finite or infinite.

These examples show that many number-theoretic questions reduce to problems of the solvability of Diophantine equations. In Chapter VII we shall see that in a certain sense, "almost all of mathematics" reduces to such problems.

2 Plan of Proof

2.1. In this section we introduce some auxiliary notions and give the plan of proof for Theorem 1.2.

We shall temporarily introduce a class of sets that are intermediate between enumerable and Diophantine sets. In order to define this class, we consider the map that to every subset $E \subset (\mathbf{Z}^+)^n$ associates the set $F \subset (\mathbf{Z}^+)^n$ that is given by the following rule:

$$\langle x_1, \ldots, x_n \rangle \in F \Leftrightarrow \forall k \in [1, x_n], \langle x_1, \ldots, x_{n-1}; k \rangle \in E.$$

We shall say that F is obtained from E by applying the bounded universal quantifier to the nth coordinate. We define similarly the operation of applying the bounded universal quantifier to any coordinate.

2.2. **Definition-Lemma.** *Consider the following three classes of subsets of* $(\mathbf{Z}^+)^n$ *for each* n.

(I) *Projections of level sets of primitive recursive functions.*
(II) *The least class of sets that contains the level sets of polynomials with integer coefficients and that is closed with respect to taking finite direct products, finite unions, finite intersections, projections, and applying the bounded universal quantifier.*
(III) *Projections of level sets of polynomials with integer coefficients.*

The following assertions hold for these classes:

(a) *The class* (I) *coincides with the class of enumerable sets, and the class* (III) *coincides with the class of Diophantine sets. We shall call sets in the class* (II) *D-sets.*
(b) (I) $\supset$ (II) $\supset$ (III).

PROOF.

(a) In Theorem 4.3 of Chapter V we showed that the class of primitive enumerable sets coincides with the class of enumerable sets. The rest of (a) merely consists of definitions.

(b) Only the inclusion (II) $\subset$ (I) is not completely obvious. First of all, the m-level set of a polynomial f is the same as the 1-level set of the primitive recursive function $(f-m)^2+1$. Hence, to verify (II) $\subset$ (I) it suffices to show that the class (I) is closed with respect to (finite) direct product, union, intersection, and the bounded universal quantifier. All except for the last of these were established in Lemma 4.8 of Chapter V.

Finally, suppose F is the image of a primitive enumerable set E under the bounded universal quantifier:

$$\langle x_1, \ldots, x_{n-1}, x_n \rangle \in F \Leftrightarrow \forall k \leqslant x_n, \quad \langle x_1, \ldots, x_{n-1}, k \rangle \in E.$$

Starting with the function $f(x_1, \ldots, x_{n-1}, x_n; y_1, \ldots, y_m)$ whose 1-level projects onto E, we want to construct a function g whose 1-level projects onto F. A natural idea is to consider as an approximation to g the product

$$\prod_{k=1}^{x_n} f(x_1, \ldots, x_{n-1}, k; y_{1k}, \ldots, y_{mk}),$$

where the y_{ik} are "independent variables." The only problem is that the number of arguments of this "function" increases with x_n. To deal with this, we apply the Gödel function $\mathrm{Gd}(k,t)$, which was defined in Section 4.9 of Chapter V. The function g will now depend on $x_1,\dots,x_n$ and on m additional arguments $t_1,\dots,t_m$:

$$g(x_1,\dots,x_n;\ t_1,\dots,t_m)$$
$$=\prod_{k=1}^{x_n} f(x_1,\dots,x_{n-1},k;\mathrm{Gd}(k,t_1),\dots,\mathrm{Gd}(k,t_m)).$$

This function is primitive recursive, because the kth factor is obtained from f and Gd by substitution and identifying arguments, and then g is constructed from such factors by recursion.

We now verify that the set F is the projection of the 1-level of g onto the $\langle x_1,\dots,x_n\rangle$-coordinates. In fact, if $g(x_1,\dots,t_m)=1$, then for all $1\leqslant k\leqslant x_n$ we have $f(x_1,\dots,x_{n-1},k,\mathrm{Gd}(k,t_1),\dots,\mathrm{Gd}(k,t_m))=1$, i.e., for all $1\leqslant k\leqslant x_n$ the point $\langle x_1,\dots,x_{n-1},k\rangle$ belongs to E. This means that $\langle x_1,\dots,x_n\rangle\in F$.

Conversely, if $\langle x_1,\dots,x_n\rangle\in F$, then for $1\leqslant k\leqslant x_n$ we can lift the point $\langle x_1,\dots,x_{n-1},k\rangle$ to the 1-level of f. Let the y-coordinates of the resulting point be $y_{1,k},\dots,y_{m,k}$. We solve the following system of equations for the t_i:

$$\mathrm{Gd}(k,t_i)=y_{i,k},\quad \text{for all } 1\leqslant k\leqslant x_n.$$

This is possible by the fundamental property of Gd. The resulting values for the t_i, along with $x_1,\dots,x_n$, make g equal to one. This completes the proof of Lemma 2.2. □

2.3. The plan for the rest of the proof of Theorem 1.2 is as follows. In §3 we show that the classes (I) and (II) coincide, and in §§4–7 we show that (II) and (III) coincide.

2.4. *Remark.* In the course of proving Lemma 2.2, we obtained the following facts, which should always be kept in mind in what follows:

(a) In the definitions of the classes (I)–(III) we may always replace "level sets" by "1-level sets" (by going from f to $(f-m)^2+1$).
(b) All of the classes (I)–(III) are closed with respect to (finite) products, intersections, unions, and also projections. (The proof of this for the class (I) in Lemma 4.8 of Chapter V is also applicable to the class (III).)

We encounter much greater difficulty in treating the bounded universal quantifier. Indeed, the most technical part of the proof in §§4–7 is concerned with showing that the class of Diophantine sets is closed with respect to the bounded universal quantifier.

3 Enumerable Sets Are D-Sets

Let $f:(\mathbf{Z}^+)^n\to\mathbf{Z}^+$ be a primitive recursive function. Its 1-level can be represented as the projection onto the first n coordinates of the set $\Gamma_f\cap[(\mathbf{Z}^+)^n\times\{1\}]$, where Γ_f is the graph of f. Thus, an enumerable set can be obtained as a projection of the intersection of the graphs of two primitive recursive functions. Since, by definition, the class of D-sets is closed with respect to projections and intersections, the assertion in the title of this section follows from the following fact:

3.1. **Proposition.** *The graphs of primitive recursive functions are D-sets.*

PROOF. The graphs of the basic functions are Diophantine. The stability of the property of graphs "being D-sets" relative to the composition and juxtaposition of functions is verified by the same arguments as in the proof of Lemma 4.8 of Chapter V. It remains to prove the stability under recursion. We shall first of all need information about the graph of Gödel's function. Here it is more convenient to use gd instead of Gd.

3.2. **Lemma.** *The graph of the Gödel function* $\mathrm{gd}(u,k,t) = \mathrm{rem}(1 + kt,\ u)$ *is Diophantine, and a fortiori, a D-set.*

PROOF. The set

$$\Gamma_{\mathrm{gd}} = \{\langle u,k,t,\gamma\rangle \mid \gamma \text{ is the remainder when } u \text{ is divided by } 1 + kt\}$$

is the intersection of the following two sets in $(\mathbf{Z}^+)^4$:

$$E_1 : \gamma \leqslant 1 + kt;$$
$$E_2 : u - \gamma \geqslant 0 \quad \text{and is divisible by } 1 + kt.$$

Both E_1 and E_2 are Diophantine. In fact, E_1 is a projection of the 0-level of the polynomial $2 + kt - \gamma - y_1$, and E_2 is a projection of the 0-level of the polynomial $u - \gamma - (1 + kt)(y_2 - 1)$. The lemma is proved. □

3.3. **Corollary.** *Let f and g be functions of n and n+2 arguments, respectively, whose graphs are D-sets. Then the following equations determine D-sets in the* $(x_1,\ldots,x_{n+1},u,t,\ldots)$*-coordinate space (where any additional coordinates may follow the t):*

$$E : \mathrm{gd}(u,1,t) = f(x_1,\ldots,x_n);$$
$$F : \mathrm{gd}(u,x_{n+1}+1,t) = g(x_1,\ldots,x_{n+1},\mathrm{gd}(u,x_{n+1},t)).$$

PROOF. Introducing extra coordinates after the t amounts to taking the direct product with $(\mathbf{Z}^+)^p$, and this, of course, takes D-sets to D-sets.

E can be represented as a projection of the intersection of the sets $\mathrm{gd}(u,k,t) = w, f(x_1,\ldots,x_n) = w$, and $k = 1$ (where k and w are auxiliary coordinates). Since Γ_{gd} and Γ_f are D-sets, the same is true for E.

Similarly, F can be represented as a projection of the intersection of the sets

$$\mathrm{gd}(u,x_{n+1}+1,t) = w_1,$$
$$\mathrm{gd}(u,x_{n+1},t) = w_2,$$
$$g(x_1,\ldots,x_{n+1},w_2) = w_1.$$

These are D-sets, because Γ_g and Γ_{gd} are D-sets. □

3.4. PROOF OF PROPOSITION 3.1. Recall that it remains to verify the following assertion: Let h be the function defined recursively from functions f and g by the equations

$$h(x_1,\ldots,x_n,1) = f(x_1,\ldots,x_n),$$
$$h(x_1,\ldots,x_n,k+1) = g(x_1,\ldots,x_n,k,h(x_1,\ldots,x_n,k));$$

then the graph Γ_h,

$$\langle x_1,\ldots,x_{n+1},\eta\rangle \in \Gamma_h \Leftrightarrow \eta = h(x_1,\ldots,x_{n+1}),$$

is a D-set whenever the graphs Γ_f and Γ_g are D-sets.

First step. We set $\Gamma_h = \Gamma^1 \cup \Gamma^2$, where $x_{n+1} = 1$ on Γ^1 and $x_{n+1} \geqslant 2$ on Γ^2. Since

$$\langle x_1,\ldots,x_{n+1},\eta\rangle \in \Gamma^1 \Leftrightarrow x_{n+1} = 1 \quad \text{and} \quad \eta = f(x_1,\ldots,x_n),$$

it follows that Γ^1 is the intersection of $\Gamma_f \times \mathbf{Z}^+$ and a D-set, and therefore is a D-set. It remains to verify that Γ^2 is also a D-set.

Second step. In the $(x_1,\ldots,x_{n+1},\eta,u,t)$-coordinate space we consider the sets

$$\begin{aligned}
&E_1 : \eta = \mathrm{gd}(u,x_{n+1},t),\\
&E_2 : \mathrm{gd}(u,1,t) = f(x_1,\ldots,x_n),\\
&E_3 : x_{n+1} > 1, \quad \mathrm{gd}(u,k,t) = g(x_1,\ldots,x_n,k-1,\mathrm{gd}(u,k-1,t))\\
&\quad \text{for all } 2 \leqslant k \leqslant x_{n+1}.
\end{aligned}$$

It is easy to see that $\Gamma^2 = \mathrm{pr} \bigcap_{i=1}^3 E_i$. In fact, as in §4 of Chapter V, we obtain inclusion in one direction by comparing E_2 and E_3 with the inductive definition of h, and in the other direction by suitably choosing the parameters u and t in Gödel's function. Thus, it remains to show that the E_i are D-sets.

Third step. E_1 is the graph of gd with some additional coordinates. E_2 was shown to be a D-set in the proof of Corollary 3.3.

Finally, E_3 is "almost" obtained from the set F in Corollary 3.3 by applying the bounded universal quantifier to the x_{n+1}-coordinate. More precisely (for brevity, we ignore the η-coordinate);

$$\begin{aligned}
\langle x_1,\ldots,x_{n+1},u,t\rangle \in E_3 &\Leftrightarrow \forall k \in [2,x_{n+1}], \langle x_1,\ldots,x_n,k-1,u,t,\rangle \in F\\
&\Leftrightarrow \forall k \in [1,x_{n+1}-1], \langle x_1,\ldots,x_n,k,u,t,\rangle \in F.
\end{aligned}$$

Consequently, if we apply to F the bounded universal quantifier in the x_{n+1}-coordinate, we obtain a D-set that is the same as E_3 with the x_{n+1}-coordinates of all its points decreased by 1. So it remains to see that the operation of shifting back by 1 preserves the property of "being a D-set," and this follows easily from the definitions. The proof is complete. □

4 The Reduction

4.1. The next three sections are devoted to proving that the class of D-sets coincides with the class of Diophantine sets. As noted at the end of §2, it suffices to show that the class of Diophantine sets is closed with respect to the bounded universal quantifier.

Let $f(x_1, \ldots, x_n, k, y_1, \ldots, y_m)$ be any nonconstant polynomial with integer coefficients. f will be fixed for the duration of this section. Let d be the degree of f, and let c be the sum of the absolute values of its coefficients.

We define the set E by the condition

$$\langle x_1, \ldots, x_n, y\rangle \in E \Leftrightarrow \forall k \leqslant y\, \exists \langle y_1, \ldots, y_m\rangle, \quad f(x_1, \ldots, x_n, k, y_1, \ldots, y_m) = 0.$$

We want to show that E is Diophantine. In this section we prove the following reduction step, which is due to Davis, Putnam, and Robinson.

4.2. **Proposition.** *E is Diophantine if the following three sets are Diophantine:*

$$x_1 = x_2^{x_3};$$
$$x_1 = x_2!;$$
$$\frac{x_1}{x_2} = \binom{x_3/x_4}{x_5}, \qquad x_3 \geqslant x_4 x_5,$$

where $\binom{n}{k} = n(n-1)\cdots(n-k+1)/k!$ *is the "binomial coefficient."*

The proof of this and all subsequent propositions of this type follows a standard pattern. To show that E is Diophantine, we introduce auxiliary sets E_i with the following properties:

(a) $E = \bigcap_{i=1}^{N} E_i$;

(b) the E_i are Diophantine.

But usually we are not able to establish directly that all the E_i are Diophantine, so we apply the same procedure to certain of the E_i. Thus, the proof that E is Diophantine has a treelike pattern.

The exposition of each step will consist of the following stages: the introduction of auxiliary variables, which disappear when we project; explicit construction of the sets E_i; the proof of the inclusion $E \subset \operatorname{pr} \cap_{i=1}^{N} E_i$; and the proof of the inclusion $E \supset \operatorname{pr} \cap_{i=1}^{N} E_i$.

4.3. Proof of Proposition 4.2. We denote the auxiliary variables by the symbols Y, N, K, $Y_1, \ldots, Y_m$. We introduce the sets E_i in the $\langle x_1, \ldots, x_n, y, Y, N, K, Y_1, \ldots, Y_m\rangle$-space by the following relations:

$$E_1: \quad N \geqslant c \cdot (x_1 \cdots x_n y Y)^d, \qquad Y < Y_1, \ldots, Y < Y_m$$

(intuitively speaking, the right side of the first inequality gives a rough estimate for the value of the polynomial f at the point $\langle x_1,\ldots,x_n,y,y_1,\ldots,y_m\rangle$ if all $y_i \leqslant Y$).

$$E_2: \quad 1+KN! = \prod_{k=1}^{y}(1+kN!)$$

(this is a "large modulus"; $f = 0$ will be replaced by divisibility by this modulus).

$$E_3: f(x_1,\ldots,x_n,K,Y_1,\ldots,Y_m) \equiv 0 \bmod(1+KN!);$$
$$E_{3+i}: \prod_{j<Y}(Y_j - j) \equiv 0 \bmod(1+KN!), \quad i=1,\ldots,m.$$

We define the set E' as $\cap_{i=1}^{m+3} E_i$.

Proof of the inclusion $E \subset \operatorname{pr} E'$. Given a point $\langle x_1,\ldots,x_n,y\rangle \in E$, we must choose values for the other coordinates so that the relations $E_1,\ldots,E_{m+3}$ are fulfilled.

By the definition of E, each point $\langle x_1,\ldots,x_n,k\rangle, k \leqslant y$, can be lifted to the 0-level of f:

$$f(x_1,\ldots,x_n,k,y_{1k},\ldots,y_{mk}) = 0.$$

For Y we take the maximum of y and the y_{ik}. Then, as before, we find the Y_i and N by solving the system of Gödel equations

$$\mathrm{gd}(Y_i,k,N!) = y_{ik}, \quad \text{for all } 1 \leqslant k \leqslant y.$$

The proof of Gödel's lemma shows that the Y_i and N may be taken arbitrarily large, in particular, so as to satisfy E_1. The number K is uniquely determined by E_2.

All the choices have now been made. The relation E_{3+i} holds because by the definition of Y_i and gd, we can find a number $Y_i - j$ with $j \leqslant Y$, namely $j = y_{ik}$, such that $Y_i - j \equiv 0 \bmod(1+kN!)$, for every $k \leqslant y$. Hence, the product on the left in E_{3+i} is divisible by all the $1+kN!, 1 \leqslant k \leqslant y$, which are pairwise relatively prime, since $N \geqslant y$ by E_1. Therefore, this product is divisible by $1+KN!$.

Finally, to verify E_3 we note that E_2 implies the congruence $K \equiv k \bmod (1+kN!)$, $1 \leqslant k \leqslant y$, because $(1+KN!) - (1+kN!) \equiv 0 \bmod(1+kN!)$. But then, since $y_{ik} \equiv Y_i \bmod(1+kN!)$ by our choice of Y_i, we find that

$$\begin{aligned} f(x_1,\ldots,x_n,K,Y_1,\ldots,Y_m) &\equiv f(x_1,\ldots,x_n,k,y_{ik},\ldots,y_{mk}) \\ &\equiv 0 \bmod(1+kN!). \end{aligned}$$

Since the moduli $1+kN!$ are pairwise relatively prime, this congruence implies E_3.

Proof of the inclusion $\operatorname{pr} E' \subset E$. Given a point

$$\langle x_1,\ldots,x_n,y,Y,N,K,Y_1,\ldots,Y_m\rangle$$

whose coordinates satisfy the relations $E_1, \dots, E_{m+3}$, we must find a vector $\langle y_{1k}, \dots, y_{mk} \rangle$ for each $k \leqslant y$ such that

$$f(x_1, \dots, x_n, k, y_{1k}, \dots, y_{mk}) = 0.$$

To do this we let p_k denote any prime divisor of $1 + kN!$, and we set

$$y_{ik} = \textit{the remainder when } Y_i \textit{ is divided by } p_k.$$

We claim that these y_{ik} give us the required equality. In fact, E_3 implies that $f(x_1, \dots, x_n, k, y_{1k}, \dots, y_{mk}) \equiv 0 \bmod p_k$. It suffices to show that the number on the left is less than p_k. We have

$$\begin{aligned} p_k \text{ divides } & \prod_{j \leqslant Y} (Y_i - j) \text{ by } E_{3+i} \\ & \Rightarrow p_k \text{ divides } Y_i - j \text{ for some } j \leqslant Y \\ & \Rightarrow y_{ik} = \text{the remainder when } Y_i \text{ is divided by } p_k \leqslant Y \\ & \Rightarrow f(x_1, \dots, x_n, k, y_{1k}, \dots, y_{mk}) \leqslant c(x_1 \cdots x_n y Y)^d \leqslant N < p_k, \end{aligned}$$

where the second inequality in the last line follows from E_1, and the third inequality follows because p_k divides $1 + kN!$.

Conclusion of the proof. It remains to show that the sets $E_1, \dots, E_{m+3}$ are Diophantine if the sets in Proposition 6.1 are Diophantine. In fact, if we trivially introduce new variables and make substitutions, we can first reduce the verification that all the E_i are Diophantine to showing that the following sets are Diophantine:

$$\begin{aligned} x_1 &= x_2!; \\ x_1 &= \prod_{k \leqslant x_2} (1 + k x_3); \\ x_1 &= \prod_{j \leqslant x_3} (x_2 - j), \qquad x_2 > x_3. \end{aligned}$$

It then remains to notice that the second of these relations can be written in the form

$$x_1 = x_3^{x_2} \begin{bmatrix} \frac{1}{x_3} + x_2 \\ x_2 \end{bmatrix},$$

and the third relation can be written as

$$x_1 = x_3! \binom{x_2 - 1}{x_3}, \qquad x_2 > x_3.$$

This completes the proof of Proposition 4.2. □

5 Construction of a Special Diophantine Set

5.1. In this section we begin the proof that the three sets in Proposition 4.2 are Diophantine. In order that the reader may better appreciate this stage in the proof, we mention that the most troublesome obstacle here is the rapid growth of one of the coordinates in comparison to the others (for example, $x_1 = x_2!$). J. Robinson had the following key idea. She proved that if we know that *any* specific set in $(\mathbf{Z}^+)^2$ is Diophantine and has one coordinate that grows faster than any power of the other but slower than, say, x^x (for example, exponentially), we may then conclude that *all* enumerable sets are Diophantine. After this, Matiyasevič and Čudnovskiĭ were able to show that a certain set of that type (connected with Fibonacci numbers) is Diophantine. For a history of the question, see Matijasevič's article "Diophantine Sets" in *Uspehi Mat. Nauk*, vol. XXVII, No. 5 (1972) (translated in *Russian Math. Surveys*).

In this section we give a construction that is an improved version of the original construction. Its idea is based on the following observation. Let $x^2 - dy^2 = 1$ be Pell's equation (where $d \in \mathbf{Z}^+$ is not a perfect square). Its solutions $\langle x, y\rangle \in (\mathbf{Z}^+)^2$ form a semigroup with composition law

$$(x_1 + y_1\sqrt{d})(x_2 + y_2\sqrt{d}) = x_3 + y_3\sqrt{d}.$$

This is a cyclic semigroup. That is, let $\langle x_1, y_1\rangle$ be the solution with the least first coordinate. Then any other solution has the form $\langle x_n, y_n\rangle$, where $n \in \mathbf{Z}^+$, and

$$x_n + y_n\sqrt{d} = (x_1 + y_1\sqrt{d})^n.$$

We call n the *number* of the solution $\langle x_n, y_n\rangle$.

The coordinates x_n and y_n grow exponentially with n, so that the set of solutions of Pell's equation, and also the projections of this set on the x- and y-axes, are Diophantine sets having logarithmic density. This is not yet enough: we still have the problem of including the solution number n among the coordinates of a Diophantine set. Only then can we apply Robinson's technique. This is what will be done below.

5.2. *Notation.* We consider Pell's equation with variable d. Its first solution generally varies as a function of d in an uncontrollable fashion, so that it is convenient to choose only those d whose first solutions have the simple special form $\langle a, 1\rangle, a \in \mathbf{Z}^+$. Obviously, then $d = a^2 - 1$.

We shall call the equation $x^2 - (a^2 - 1)y^2 = 1$ the *a-equation.* We define the two sequences $x_n(a)$ and $y_n(a)$ as the coordinates of its nth solution:

$$x_n(a) + y_n(a)\sqrt{a^2 - 1} = \left(a + \sqrt{a^2 - 1}\right)^n.$$

For each n, a formal definition of $x_n(a)$ and $y_n(a)$ as polynomials in a can easily be given by induction on n. Then the expressions $x_n(a)$ and $y_n(a)$ will make sense for all $n \in \mathbf{Z}$ and $a \in \mathbf{C}$. In particular,

$$x_n(1) = 1, \quad y_n(1) = n;$$

and all the formulas given below remain true.

The basic result of this section is the following:

5.3. **Proposition.** *The set*

$$E: \quad y = y_n(a), \qquad a > 1;$$

in the $\langle y, n, a\rangle$-space is Diophantine.

The proof uses the elementary number-theoretic properties of the sequences $x_n(a)$ and $y_n(a)$, most of which will be verified at the end of the section (see 5.8). The idea for determining n in a Diophantine way from $\langle y, a\rangle$ is to observe that $y_n(a) \equiv n \bmod (a-1)$ (Lemma 5.4). This uniquely determines n as long as $n < a-1$. To pass to the general case, we introduce an auxiliary A-equation with A large, and find formulas for its nth solution (using y) in which n appears in only a Diophantine context.

Formally, the proof that E is Diophantine follows the pattern described in 4.2. In addition to the basic variables y, n, a, we introduce six auxiliary variables: x, x_1, y_1, A, x_2, y_2. We set

$$E_1: \; y \geqslant n, \qquad a > 1;$$

$$E_2: \; x^2 - (a^2-1)y^2 = 1;$$

$$E_3: \; y_1 \equiv 0 \bmod 2x^2y^2;$$

$$E_4: \; x_1^2 - (a^2-1)y_1^2 = 1;$$

$$E_5: \; A = a + x_1^2(x_1^2 - a);$$

$$E_6: \; x_2^2 - (A^2-1)y_2^2 = 1;$$

$$E_7: \; y_2 - y \equiv 0 \bmod x_1^2;$$

$$E_8: \; y_2 \equiv n \bmod 2y.$$

Let $E' = \cap_{i=1}^{8} E_i$. We show that $\mathrm{pr}\, E' = E$.

The *inclusion* $E \subset \mathrm{pr}\, E'$. Given $\langle y, n, a\rangle \in E$, we must find values for the other variables such that $E_1, \dots, E_8$ hold. As before, we shall not introduce any new symbols for these values; after we choose, say, a value for x, the letter x will become the name for this value.

E_1 is automatically satisfied: $y_n(a) \geqslant n$ for all $a \geqslant 1, n \geqslant 1$ (induction on n). We find x uniquely from $E_2 : x = x_n(a)$. We take $\langle x_1, y_1/2x^2y^2\rangle$ to be any solution of the Pell equation $X^2 - (a^2-1)(2x^2y^2)^2Y^2 = 1$; this gives E_4. A is found uniquely from E_5. We take $\langle x_2, y_2\rangle$ to be the nth solution of the A-equation. Now all choices have been made. To verify E_7 and E_8 we need two lemmas.

5.4. **Lemma.** $y_k(a) \equiv k \bmod (a-1)$.

5.5. **Lemma.** *If* $a \equiv b \bmod c$, *then* $y_n(a) \equiv y_n(b) \bmod c$.

These lemmas will be proved in 5.8.

We use these lemmas as follows. From E_5 we obtain

$$A = a + (1 + (a^2-1)y_1^2)(1 + (a^2-1)y_1^2 - a) \equiv 1 \bmod 2y,$$

because of E_3. Lemma 5.4 then gives $y_2 = y_n(A) \equiv n \bmod 2y$; this is E_8. Lemma 5.5 gives $y_n(A) \equiv y_n(a) \bmod x_1^2$ (because of E_5); this is E_7.

The inclusion pr $E' \subset E$. From the relations $E_1, \ldots, E_8$ we have only to prove that n is the number of the solution $\langle x, y\rangle$. Note that n occurs only in E_8.

For the time being we let N, N_1, and N_2 denote the numbers of the solutions $\langle x, y\rangle$, $\langle x_1, y_1\rangle$, and $\langle x_2, y_2\rangle$, respectively. We shall prove that

$$n \equiv N \quad \text{or} \quad n \equiv -N \bmod 2y.$$

Since we also have $y \geqslant n$ (by E_1) and $y \geqslant N$ (by the definition of N), it follows that $n = N$, as required. The number N_2 will be the "stepping stone" to get from n to N.

First of all, as before, it follows from E_5 that $A \equiv 1 \bmod 2y$, and then it follows from the definition of N_2 and Lemma 5.4 that $y_2 \equiv N_2 \bmod 2y$. But by E_8 we have $y_2 \equiv n \bmod 2y$; hence

$$N_2 \equiv n \bmod 2y.$$

Next, $A \equiv a \bmod x_1^2$ by E_5, and then $y_2 = y_{N_2}(A) \equiv y_{N_2}(a) \bmod x_1^2$ by Lemma 5.5. Using E_7, we have $y = y_N(a) \equiv y_2 \bmod x_1^2$. Hence

$$y_N(a) \equiv y_{N_2}(a) \bmod x_1^2.$$

We now need two more lemmas, which will be proved in 5.8.

5.6. **Lemma.** *If $y_i(a) \equiv y_j(a) \bmod x_n(a)$, where $a > 1$, then either $i \equiv j$ or $i \equiv -j \bmod 2n$.*

5.7. **Lemma.** *If $y_i(a)^2$ divides $y_j(a)$, then $y_i(a)$ divides j.*

If we apply Lemma 5.6 with N, N_2, and N_1 in place of i, j, and n, and use the last congruence proved, we obtain

$$N \equiv \pm N_2 \bmod 2N_1.$$

If we apply Lemma 5.7 with N and N_1 in place of i and j, and use E_3, we obtain $y | N_1$. Hence

$$N \equiv \pm N_2 \bmod 2y,$$

and since we have already shown that $N_2 \equiv n \bmod 2y$, this completes the proof. □

5.8. Proof of the Lemmas. We shall write x_n and y_n instead of $x_n(a)$ and $y_n(a)$. Using the formula

$$x_{nk} + y_{nk}\sqrt{a^2-1} = \left(x_n + y_n\sqrt{a^2-1}\right)^k,$$

we find that

$$y_{nk} = \sum_{\substack{j \leqslant k \\ j \equiv 1 (\text{mod } 2)}} \binom{k}{j} x_n^{k-j} y_n^j (a^2 - 1)^{(j-1)/2}.$$

In particular,

$$y_{nk} \equiv k x_n^{k-1} y_n \mod (a^2 - 1),$$

which gives Lemma 5.4 if we set $n = 1$. In addition, we have

$$y_{nk} \equiv k x_n^{k-1} y_n \mod y_n^3.$$

If we replace nk, k, and n by n, n/k, and k, respectively, we obtain

$$y_n \equiv \frac{n}{k} x_k^{n/k-1} y_k \mod y_k^3.$$

Since x_k and y_k are relatively prime, we have

$$y_n \equiv 0 \mod y_k^2 \Rightarrow \frac{n}{k} \equiv 0 \mod y_k \Rightarrow n \equiv 0 \mod y_k,$$

which gives Lemma 5.7.

If we write $y_n(a)$ as a polynomial in a with integer coefficients whose degree and coefficients depend only on n, we immediately obtain Lemma 5.5. It remains to prove Lemma 5.6.

First of all, the equation

$$x_{n \pm m} + \sqrt{a^2 - 1}\, y_{n \pm m} = \left(x_n + \sqrt{a^2 - 1}\, y_n\right)\left(x_m \pm \sqrt{a^2 - 1}\, y_m\right)$$

gives us

$$\begin{aligned} x_{n \pm m} &= x_n x_m \pm (a^2 - 1) y_n y_m, \\ y_{n \pm m} &= \pm x_n y_m + x_m y_n. \end{aligned}$$

Hence,

$$\begin{aligned} y_{2n \pm m} &= y_{n + (n \pm m)} \equiv x_{n \pm m} y_n \mod x_n \equiv \pm (a^2 - 1) y_n^2 y_m \mod x_n \\ &\equiv \mp y_m \mod x_n, \end{aligned}$$

and, similarly,

$$y_{4n \pm m} = y_{2n + (2n \pm m)} \equiv -y_{2n \pm m} \mod x_n \equiv y_{\pm m} \mod x_n.$$

This means that the class y_k mod x_n has period $4n$ as a function of k, and within $[1, 4n]$ its behavior is determined by its values on the first quarter-period $[1, n]$:

$$y_{2n \pm m} \equiv \mp y_m, \quad y_{\pm m} \equiv \pm y_m, \quad \text{for } 1 \leqslant m \leqslant n.$$

If $a \geqslant 3$, it is clear that Lemma 5.6 follows from these facts and from the inequality $y_m < \frac{1}{2} x_n$ for $1 \leqslant m \leqslant n$, which, in turn, follows because

$$4 y_m^2 < (a^2 - 1) y_n^2 + 1 = x_n^2.$$

If $a = 2$, then we only have $y_m < \frac{1}{2} x_n$ for $m \leqslant n - 1$, but this is still enough to complete the proof of the lemma in this case. □

6 The Graph of the Exponential Is Diophantine

6.1. **Proposition.** *The set*

$$E: \quad m = a^n$$

in the $\langle m, a, n\rangle$-space is Diophantine.

PROOF. It suffices to show that $E_0 = E \cap \{a | a > 1\}$ is Diophantine. If $a > 1$, we easily obtain by induction on n that

$$(2a-1)^n \leqslant y_{n+1}(a) \leqslant (2a)^n,$$

in the notation of §5. Hence, for any $N \geqslant 1$ we have

$$a^n\left(1-\frac{1}{2Na}\right)^n = \frac{(2Na-1)^n}{(2N)^n} \leqslant \frac{y_{n+1}(Na)}{y_{n+1}(N)} \leqslant \frac{(2Na)^n}{(2N-1)^n}$$
$$= a^n\left(1-\frac{1}{2N}\right)^{-n}.$$

Thus, if we choose N large enough so that both

$$\left(1-\frac{1}{2N}\right)^{-n} - 1 < \frac{1}{a^n} \quad \text{and} \quad 1-\left(1-\frac{1}{2Na}\right)^n < \frac{1}{a^n},$$

then we obtain $a^n = [y_{n+1}(Na)/y_{n+1}(N)]$ (where the brackets here and below denote the integral part of a number). E_0 is therefore a projection of the set E_1:

$$a > 1,$$
$$0 \leqslant y_{n+1}(Na) - y_{n+1}(N)m < y_{n+1}(N),$$
$$N > ?,$$

where a suitable lower bound for N must be inserted in place of ?, in such a way as to keep the last relation Diophantine. An elementary calculation shows that it suffices to set $N > 4n(y+1)$. The results in §5 then imply that E_1 is Diophantine if we trivially introduce the auxiliary relations

$$y' = y_{n+1}(N) \quad \text{and} \quad y'' = y_{n+1}(Na). \qquad \square$$

7 The Factorial and Binomial Coefficient Graphs Are Diophantine

In this section we carry out the last series of arguments.

7.1. **Proposition.** *The set*

$$E: \quad r = \binom{n}{k}, \quad n \geqslant k,$$

in the $\langle r, k, n\rangle$-space is Diophantine.

Here, by definition, $\binom{n}{k} = n(n-1)\cdots(n-k+1)/k!$. We shall need the following lemma.

7.2. **Lemma.** *If $u > n^k$, then $\binom{n}{k}$= the remainder when $[(u+1)^n/u^k]$ is divided by u.*

PROOF. We have

$$(u+1)^n/u^k = \sum_{i=k+1}^{n} \binom{n}{i} u^{i-k} + \binom{n}{k} + \sum_{i=0}^{k-1} \binom{n}{i} u^{i-k}.$$

The first sum is divisible by u, and the last sum is less than 1 if $u > n^k$. □

7.3. PROOF OF PROPOSITION 7.1. We introduce the auxiliary variables u and v, and take the relations

$$\begin{aligned} E_1&: \quad u > n^k; \\ E_2&: \quad v = [(u+1)^n/u^k]; \\ E_3&: \quad r \equiv v \mod u; \\ E_4&: \quad r < u; \\ E_5&: \quad n \geqslant k. \end{aligned}$$

Lemma 7.2 immediately implies that $E = \mathrm{pr} \cap_{i=1}^{5} E_i$. E_1 is Diophantine because of Proposition 6.1; E_3, E_4, and E_5 are obviously Diophantine. It also becomes obvious that E_2 is Diophantine if we write E_2 in the form

$$(u+1)^n \leqslant u^k v < (u+1)^n + u^k$$

and again use Proposition 6.1. This completes the proof. □

7.4. **Proposition.** *The set $E : m = k!$ is Diophantine.*

7.5. **Lemma.** *If $k > 0$ and $n > (2k)^{k+1}$, then $k! = \left[n^k \Big/ \binom{n}{k}\right]$. (This is proved by some simple estimates.)*

PROOF OF PROPOSITION 7.4. We take the auxiliary variable n and the relations

$$\begin{aligned} E_1&: \quad n > (2k)^{k+1}; \\ E_2&: \quad m = \left[n^k / \binom{n}{k}\right]. \end{aligned}$$

The rest is obvious (using Propositions 6.1 and 7.1). □

7.6. **Proposition.** *The set*

$$E: \quad \frac{x}{k} = \binom{p/q}{k}, \quad p > qk,$$

in the $\langle x, y, p, q, k\rangle$*-space is Diophantine.*

The proof that follows is a slightly more complicated version of the argument in 7.2 and 7.3.

7.7. **Lemma.** *Let* $a > 0$ *be an integer such that* $a \equiv 0 \mod q^k k!$ *and* $a > 2^{p-1}p^{k+1}$. *Then*

$$\binom{p/q}{k} = a^{-1}\left[a^{2k+1}(1+a^{-2})^{p/q}\right] - a\left[a^{2k-1}(1+a^{-2})^{p/q}\right].$$

This is proved using the binomial Taylor series for $(1 + a^{-2})^{p/q}$. The inequality $a > 2^{p-1}p^{k+1}$ allows us to throw away all the terms in the first sum starting with the $(k + 1)$th and all the terms in the second sum starting with the kth when we take the integral part. The congruence $a \equiv \mod q^k k!$ ensures that the partial sums are integers. □

7.8. Proof of Proposition 7.6. We use the auxiliary variables a, u_1, u_2, and v, and the following relations:

$$E_1\colon a \equiv 0 \mod q^k k!;$$
$$E_2\colon a > 2^{p-1}p^{k+1};$$
$$E_3\colon u_1/u_2 = a^{-1}\left[a^{2k+1}(1+a^{-2})^{p/q}\right];$$
$$E_4\colon v = a\left[a^{2k-1}(1+a^{-2})^{p/q}\right];$$
$$E_5\colon xu_2 = y(u_1 - vu_2).$$

It follows from Lemma 7.7 that $E = \mathrm{pr} \bigcap_{i=1}^{5} E_i$. E_1 and E_2 are immediately seen to be Diophantine from Propositions 6.1 and 7.1. E_3 and E_4 are shown to be Diophantine just as at the end of 7.3, except that this time we must raise the inequalities to the qth power after clearing denominators. E_5 is obviously Diophantine.

This concludes the proof of Theorem 1.2, that enumerable sets coincide with Diophantine sets. □

8 Versal Families

Versal families were defined and first used in Section 5.7 of Chapter V. The purpose of this section is to prove their existence, using the result that enumerable sets are Diophantine (Theorem 1.2).

8.1. **Theorem.** *For any* $m \geqslant 0$, *versal enumerable families of* m*-sets and* m*-functions over the base* $\mathbf{Z}^+$ *exist and can be effectively constructed.*

PROOF. We divide the proof into several steps. Recall that $\tau^{(2)} : (\mathbf{Z}^+)^2 \xrightarrow{\sim} \mathbf{Z}^+$ is the primitive recursive isomorphism constructed in §4 of Chapter V, and $\langle t_1^{(2)}, t_2^{(2)} \rangle$ is its inverse. We shall write t_1 and t_2 for brevity.

(a) *A versal family of polynomials in* $\mathbf{Z}^+[x_1, x_2, x_3, \ldots]$. We define polynomials $f[l] \in \mathbf{Z}^+[x_1, x_2, x_3, \ldots]$ by recursion on $l \in \mathbf{Z}^+, l \geqslant 4$:

$$\begin{aligned} f[1] = f[2] = f[3] &= 1; \\ f[4k] &= k; \\ f[4k+1] &= x_k; \\ f[4k+2] &= f[t_1(k)] + f[t_2(k)]; \\ f[4k+3] &= f[t_1(k)] f[t_2(k)]. \end{aligned}$$

The definition is correct, since $t_1(k), t_2(k) < 4k + 2$. The image of the map $k \mapsto f[k]$ coincides with all of $\mathbf{Z}^+[x_1, x_2, x_3, \ldots]$, since it contains $\mathbf{Z}^+$ (in the $4k$-places) and all the x_k (in the $4k+1$-places), and, whenever it contains two polynomials $f[k_1]$ and $f[k_2]$, it contains their sum (in the $4\tau^{(2)}(k_1, k_2)+2$-place) and their product (in the $4\tau^{(2)}(k_1, k_2)+3$-place). (Compare with the numbering of constructible sets by ordinals in Chapter V.)

(b) *Construction of a versal* 1*-family over* $\mathbf{Z}^+$. Let E_k be the projection onto the x_1-coordinate of the 0-level of the polynomial $f[t_1(k)] - f[t_2(k)]$. Since all the elements of $\mathbf{Z}[x_1, x_2, x_3, \ldots]$ can be represented as such a difference, it is clear that the family $\{E_k\}$ contains all enumerable sets.

(c) $\{E_k\}$ *is enumerable.* We must show that the total space $E = \{\langle i, j \rangle | i \in E_j\} \subset \mathbf{Z}^+ \times \mathbf{Z}^+$ is enumerable. We write the condition $i \in E_j$ in the form of an $\mathcal{L}_1$-type formula, in which all the quantified variables take values in $\mathbf{Z}^+$. We use the fact that $f[t_1(j)] - f[t_2(j)] \in \mathbf{Z}[x_1, \ldots, x_j]$. We have

$$\begin{aligned} \langle i, j \rangle \in E \Leftrightarrow & i \in E_j \Leftrightarrow \exists x_1 \cdots \exists x_j (x_1 = i \wedge f[t_1(j)] = f[t_2(j)]) \\ \Leftrightarrow & \exists t ((\exists x_1 \cdots \exists x_j \, \forall k \leqslant j (f[k] = \mathrm{Gd}(k, t))) \\ & \wedge \mathrm{Gd}(5, t) = i \wedge \mathrm{Gd}(t_1(j), t) = \mathrm{Gd}(t_2(j), t)), \end{aligned}$$

where $\mathrm{Gd}(k, t)$ is Gödel's function (see §4 of Chapter V). Furthermore, by the definition of $f[k]$,

$$\begin{aligned} & \exists x_1 \cdots \exists x_j \forall k \leqslant j (f[k] = \mathrm{Gd}(k, t)) \\ & \quad \Leftrightarrow \forall k \leqslant j ((k \leqslant 3 \wedge \mathrm{Gd}(k, t) = 1) \vee \exists l ((k = 4l \wedge \mathrm{Gd}(k, t) = l) \\ & \qquad \vee (k = 4l + 2 \wedge \mathrm{Gd}(k, t) = \mathrm{Gd}(t_1(l), t) + \mathrm{Gd}(t_2(l), t)) \\ & \qquad \vee (k = 4l + 3 \wedge \mathrm{Gd}(k, t) = \mathrm{Gd}(t_1(l), t) \mathrm{Gd}(t_2(l), t)))). \end{aligned}$$

Here the part of the formula after $\exists l$ defines a decidable set in $\langle k, t, l \rangle$-space. The quantifier $\exists l$ projects this set onto the $\langle k, t \rangle$-coordinates, thereby taking it to an enumerable set, and the bounded quantifier $\forall k \leqslant j$ preserves enumerability (see §2). Returning to the formula that defines E, we find that the set we have constructed so far must be intersected with two other decidable sets and then projected along the t-axis, so that the result is again enumerable.

(d) *Construction of a versal m-family over* $\mathbf{Z}^+$. The case $m = 0$ is trivial, and the case $m = 1$ has already been discussed. The case $m \geqslant 2$ reduces to the case $m = 1$ using the isomorphism $\tau^{(m)} : (\mathbf{Z}^+)^m \tilde{\Rightarrow} \mathbf{Z}^+$. In fact, let $E_k = E_k^{(1)}$ be a versal 1-family, and set $E_k^{(m)} = (\tau^{(m)})^{-1}(E_k^{(1)})$. The family $\{E_k^{(m)}\}$ is enumerable because

$$E^{(m)} = \{\langle x, k\rangle | x \in E_k^{(m)}\} = \left\{\langle (\tau^{(m)})^{-1}(x), k\rangle | x \in E_k^{(1)}\right\}$$
$$= (\tau^{(m)}, \mathrm{pr}_1^1)^{-1} E^{(1)}.$$

(e) *Construction of a versal family of* 1*-functions.* We take a versal 2-family $\{E_k^{(2)}\}$ with total space

$$E^{(2)} = \{\langle x, y, k\rangle | \langle x, y\rangle \in E_k^{(2)}\} \subset (\mathbf{Z}^+)^3.$$

Let $g(x, y, k, z)$ be a primitive recursive function such that the projection of its 1-level onto the $\langle x, y, k\rangle$-coordinates coincides with $E^{(2)}$. We set

$$f(x, k) = t_1^{(2)}\left(\min\left\{u | g(x, t_1^{(2)}(u), t_2^{(2)}(u)) = 1\right\}\right).$$

We claim that $\{f_k | f_k(x) = f(x, k)\}$ is a versal family of 1-functions. The total function is obviously partial recursive. We need only verify that every partial recursive 1-function f occurs in the family.

Let Γ_f be the graph of f, and let $\Gamma_f = E_{k_0}^{(2)}$, where $k_0 \in \mathbf{Z}^+$. We show that $f = f_{k_0}$. In fact,

$$\langle x, f(x)\rangle \in \Gamma_f = E_{k_0}^{(2)} \Leftrightarrow \langle x, f(x), k_0\rangle \in E^{(2)} \Leftrightarrow \exists z \in \mathbf{Z}^+,$$
$$g(x, f(x), k_0, z) = 1.$$

Among the $z \in \mathbf{Z}^+$ that make $g(x, f(x),\ k_0, z) = 1$, we choose the z for which the number u given by $\langle f(x), z\rangle = \langle t_1^{(2)}(u), t_2^{(2)}(u)\rangle$ is minimal. For this u we have $f_{k_0}(x) = t_1^{(2)}(u) = f(x)$, which proves the claim.

(f) *Construction of a versal family of m-functions.* The case $m = 0$ is trivial. If $\{f_k^{(1)}\}$ is a versal family of 1-functions, then for $m \geqslant 2$ we set

$$f_k^{(m)}(x_1, \ldots, x_m) = f_k^{(1)}(\tau^{(m)}(x_1, \ldots, x_m)),$$

thereby obtaining a versal family of m-functions.

The theorem is proved. □

8.2. The choice of versal families is far from unique. If $m > 1$, there does not exist a versal family that contains each function or each set exactly once (i.e., a universal family). Nevertheless, there are important methods of extracting invariant information from data about the position of a function or set in a versal family. The next section is devoted to this question.

9 Kolmogorov Complexity

9.1. Let $u = \{u_k\}$ be an enumerable family of m-functions over $\mathbf{Z}^+$, and let f be a partial recursive m-function. We define the *complexity of f relative to the family u* as

$$C_u(f) = \begin{cases} \min\{k|u_k = f\}, & \text{if such a } k \text{ exists;} \\ \infty, & \text{otherwise.} \end{cases}$$

We call the enumerable family u (asymptotically) *optimal* if for any other enumerable family v, there exists a constant $c_{u,v} > 0$ such that for every partial recursive m-function f we have

$$C_u(f) \leqslant c_{u,v} C_v(f).$$

If we take v to be any versal family, we see that an optimal family must be versal, i.e., $C_u(f)$ never takes the value ∞.

9.2. **Theorem** (Kolmogorov)

(a) *For any $m \geq 0$, optimal families exist and can be effectively constructed.*
(b) *If u and v are optimal families of m-functions, then for any m-function f,*

$$c_{v,u}^{-1} \leqslant C_u(f)/C_v(f) \leqslant c_{u,v}.$$

9.3. *Remarks*

(a) The measure of complexity $C_u(f)$ involves the following intuitive ideas. In order to define any enumerable family u, it is necessary to give only a finite amount of information, for example, a program that semicomputes the total function of u. Therefore, in order to define a specific function f that occurs in the family u, it suffices to give no more than

$$\log_2 C_u(f) + \text{const}$$

bits of information, namely, the program for u and the number of f in u.

(b) A family being optimal means that it can be used to compute *any* m-function, and that the loss in using it rather than any other family to compute a function is bounded by a constant that does not depend on the function.

(c) Finally, the inequality 9.2(b), which follows trivially from the definition of an optimal family, shows that to within an additive term that is bounded in absolute value, the logarithmic measure of complexity

$$K_u(f) = [\log_2 C_u(f)] + 1 \quad (\text{where``[ \ ]''} = \text{``integral part''})$$

does not depend on the choice of the optimal family u, and so is an asymptotic invariant of f.

9.4. Proof of Theorem 9.2. We first choose a recursive embedding $\theta : \mathbf{Z}^+ \times \mathbf{Z}^+ \to \mathbf{Z}^+$ that has a recursive inverse function and that satisfies the following linear growth condition in one of its arguments:

$$\theta(k,j) \leqslant k \cdot \phi(j), \quad \text{for all } k, j \in \mathbf{Z}^+ \text{ and some suitable } \phi : \mathbf{Z}^+ \to \mathbf{Z}^+.$$

For example, we could let $\theta_1(k,j) = (2k-1)2^j$ with $\phi_1(j) = 2^{j+1}$, or, following Kolmogorov, we could let

$$\theta_2(\overline{k_1 k_2 \cdots k_r}, \overline{j_1 j_2 \cdots j_s}) = \overline{j_1 j_1 \cdots j_s j_s 01 k_1 \cdots k_r},$$

where $k_\alpha, j_\beta \in \{0,1\}$ and the bar denotes the binary expansion of a number. Here $\phi_2(j) <$ const $\cdot\, j^2$, so that this function grows more slowly. (See also Section 9.8 below.)

Now let U be any versal family of $(m+1)$-functions. We define a family u of m-functions by setting

$$u(x_1, \ldots, x_m, k) = U(x_1, \ldots, x_m, \theta^{-1}(k)).$$

We show that the family u is optimal, with the following bound for the constants:

$$c_{u,v} \leqslant \phi(C_U(v)).$$

In fact, let f be a recursive m-function. It suffices to consider the case in which f occurs in the family v. Then

$$\begin{aligned} f(x_1, \ldots, x_m) &= v(x_1, \ldots, x_m; C_v(f)) \\ &= U(x_1, \ldots, x_m, C_v(f); C_U(v)) \\ &= u(x_1, \ldots, x_m, \theta(C_v(f), C_U(v))), \end{aligned}$$

so that

$$C_u(f) \leqslant \theta(C_v(f), C_U(v)) \leqslant C_v(f)\phi(C_U(v)).$$

The theorem is proved. □

9.5. Example. A 0-function f can be identified with the single value it takes, i.e., with a positive integer n. In this case, Theorem 9.2 gives us an almost invariant complexity $C_u(n)$ for the integers. We have:

(a) $C_u(n) \leqslant$ const$\cdot n$ for all n, since the function "n" appears in the nth place in the simplest versal family $u_n(\cdot) = n$.

(b) $C(n) \sim \min\{2^{j-1}(2k-1) | n$ is the kth value of the jth function in some versal family of 1-functions$\}$. (We write $f \sim g$ if f and g have the same domain of definition, and $f \leqslant$ const$\cdot g$ and $g \leqslant$ const$\cdot f$ for suitable constants. In relations of the type $C_u(f_k) \sim g(k)$, we often omit the designation of the optimal family u, which we take to be arbitrary, but fixed.)

It is clear from (b) that the complexity of the numbers p_n (the nth prime), n^2, or

$$n^{n^{n^{\cdot^{\cdot^{\cdot^{n}}}}}} \qquad (n \text{ times})$$

as $n \to \infty$ is asymptotically no greater than const $\cdot\, n$, since each of these is the nth value of a fixed recursive function. In 9.7(b) below, we shall lower this estimate to const $\cdot\, C(n)$.

Instead of integers, Kolmogorov and his collaborators considered finite binary sequences and constructed a theory that showed that the most complex binary sequences are those that approach random behavior. See the survey article by A. K. Zvonkin and L. A. Levin in *Uspehi Matem. Nauk*, vol. XXV, No. 6 (1970) (translated in *Russian Mathematical Surveys*), which contains a large bibliography.

9.6. **Proposition.**

(a) *Let*

$$F = f_0(f_1(x_1,\dots,x_m),\dots,f_n(x_1,\dots,x_m),x_{m+1},\dots,x_p),$$

where the f_i are recursive functions. Then

$$C(F) \leqslant \text{const}\cdot\prod_{i=1}^{n} C(f_i)\left(\log\prod_{i=1}^{n} C(f_i)\right)^{n-1}$$

if f_0 is fixed and f_i runs through all possible m-functions. Here const *depends on f_0 and on the families used to compute the complexity, but does not depend on $f_1,\dots,f_n$.*

(b) *If f_0 is also allowed to vary, then $\prod_{i=1}^{n}$ must be replaced by $\prod_{i=0}^{n}$ and $\log^{n-1}$ must be replaced by $\log^{n}$ on the right.*

9.7. *Special cases*

(a) If, for example, we set $f_0 = \text{sum}_2$ or prod_2, then we have

$$C(f_1+f_2), C(f_1f_2) \leqslant \text{ const } C(f_1)C(f_2)\log(C(f_1)C(f_2)).$$

(b) If we set $n=1$ and $m=0$, we find that for any enumerable family $\{f_k\}$,

$$C(f(k,x_1,\dots,x_p)) \leqslant \text{ const } C(k).$$

9.8. Proof of Proposition 9.6. First of all, for every $n \geqslant 1$ we define the following recursive bijection with a recursive inverse:

$$\theta^{(n)}(k_1,\dots,k_n) = \begin{cases} \text{the index of the } n\text{-tuple } \langle k_1,\dots,k_n\rangle \text{ if we order } n\text{-tuples} \\ \quad \text{according to increasing } \prod_{i=1}^{n} k_i, \text{ and in alphabetical order} \\ \quad \text{for fixed } \prod_{i=1}^{n} k_i. \end{cases}$$

It is easy to see (by induction on n) that

$$\theta^{(n)}(k_1, \ldots, k_n) \leqslant \text{ const } \prod_{i=1}^{n} k_i \left(\log \prod_{i=1}^{n} k_i \right)^{n-1} .$$

We define the function $\Theta : (\mathbf{Z}^+)^{n+1} \to \mathbf{Z}^+$ as follows:

$$\Theta(l_1, \ldots, l_{n+1}) = \theta(\theta^{(n)}(l_1, \ldots, l_n), l_{n+1}),$$

where θ is as described in 9.4.

We now consider two optimal families $v(x_1, \ldots, x_p, l)$ and $u(x_1, \ldots, x_m, k)$ of p-functions and m-functions, respectively. We use these two families to construct the families

$$\begin{aligned} &W(x_1, \ldots, x_p; k_1, \ldots, k_n, l) \\ &\qquad = v(u(x_1, \ldots, x_m, k_1), \ldots, u(x_1, \ldots, x_m, k_n), x_{m+1}, \ldots, x_p, l), \\ &\qquad\qquad w(x_1, \ldots, x_p, k) = W(x_1, \ldots, x_p, \Theta^{-1}(k)). \end{aligned}$$

The function F occurs in the

$$\theta\left(\theta^{(n)}(C_u(f_1), \ldots, C_u(f_n)), C_v(f_0)\right)$$

place in the family w. Then the estimate $\theta(k, j) \leqslant k \cdot \phi(j)$, along with the estimate for $\theta^{(n)}$, gives assertion (a).

We similarly obtain (b) if we replace Θ by $\theta^{(n+1)}$ in the definition of w. □

Remark. The function $\theta^{(n)}$ gives us the most economical estimate for $C(F)$ that is symmetrical in the $C(f_1), \ldots, C(f_n)$. In specific situations it might make sense to improve the estimate in certain of the $C(f_i)$ at the expense of worsening the estimate with respect to the others; this is done by suitably changing θ. For example, Kolmogorov's θ gives

$$C(f_1 + f_2) \leqslant \text{ const } C(f_1) C(f_2)^2,$$

which is better than

$$\text{const } C(f_1) C(f_2) \log(C(f_1) C(f_2))$$

if $C(f_2)$ grows much more slowly than $C(f_1)$.

9.9. **Theorem.** *The function $C(f)$ is not computable. More precisely, let $g(k)$ be any unbounded partial recursive function, and let $\{f_k\}$ be any enumerable family. Then it is false that $C(f_k)|_{D(g)} \sim g(k)$.*

Thus, $C(f_k)$ can be computable (even up to $\sim$) only on a set of indices k such that there are only finitely many different functions among the functions f_k; otherwise, $C(f_k)$ is not bounded on this set.

PROOF. Suppose that $C(f_k)|_{D(g)} \sim g(k)$. We show that there exists a general recursive function $h : \mathbf{Z}^+ \to \mathbf{Z}^+$ whose image is contained in $D(g)$ and such that $g \circ h$ is monotonically increasing. We then obtain a contradiction as follows. By 9.7(b), for all k we have

$$C(f_{h(k)}) \leqslant \text{ const } C(k),$$

and, by our assumption and by the fact that $g \circ h$ is increasing,

$$C(f_{h(k)}) \geqslant \text{ const } g(h(k)) \geqslant \text{ const} \cdot k.$$

But these two inequalities are incompatible, because $\liminf C(k)/k = 0$ (for example, $C(k^2)/k^2 \leqslant \text{const}/k$).

It remains to construct h. We choose a general recursive bijection $h_1 : \mathbf{Z}^+ \widetilde{\Rightarrow} D(g)$, using Proposition 5.6 of Chapter V, and we set

$$E = \{k | \forall i < k, g(h_1(i)) < g(h_1(k))\}.$$

This set is decidable and infinite, and $g \circ h_1$ is an increasing function on E.

Let $h_2 : \mathbf{Z}^+ \to E$ be an increasing general recursive bijection (again using Proposition 5.6 of Chapter V). Then $h = h_1 \circ h_2$ has the necessary properties. The theorem is proved. □

9.10. *Remarks*

(a) Theorem 9.9 shows that computing complexity is a problem demanding creativity: even if we find the number of a place where f occurs in an optimal family $\{u_k\}$, there is no algorithm that could tell us whether this function occurs even sooner.

(b) Since $C(k) \neq C(l) \Rightarrow k \neq l$, it follows that for all x and B,

$$\text{card } \{y | y \leqslant x, C(y) \leqslant x/B\} \leqslant x/B,$$

i.e., most numbers have a large complexity.

Nevertheless, it is not possible to give effectively a sequence of numbers that asymptotically have maximal complexity. More precisely, let $\{k_i\}$ be any increasing sequence with $C(k_i) \geqslant k_i/B$ for some constant B. Then the set $\{k_i\}$ does not contain a single infinite enumerable set E. Otherwise, we would be able to find an increasing general recursive function $h : \mathbf{Z}^+ \to E$, and would obtain a contradiction, as in Theorem 9.9.

(c) Let $u = \{u_k\}$ be any optimal family of m-functions. The "moments of first appearance" $\{k | \forall i < k, u_i \neq u_k\}$ actually form a sequence of asymptotically maximal complexity, since, by the definition and by 9.7(b), they satisfy

$$k = C_u(u_k) \leqslant \text{ const} \cdot C(k).$$

Thus, we might say that in an optimal family the functions first appear "at random moments."

The problem of computing $C(u_k)$ is complicated by the fact that, at least in the specific families in the proof of Theorem 9.2, any function appears infinitely often, so that if we are not lucky we might first notice the function arbitrarily far out from the place where it first appeared.

(d) Finally, we mention that at least one essential aspect of the complexity of computations has not been touched upon in our discussion of C_u. Namely, $\log_2 C(k)$ measures the *length of a program* that could compute k, but says nothing about the *time* it takes for such a program to work, let alone the possibilities for shortening the time by performing parallel computations, lengthening the program, and so on.

The concept of complexity is rather far removed from practical uses. But it seems to be such a fundamental idea that its role in theoretical mathematics is likely to grow.

Part III

PROVABILITY AND COMPUTABILITY

VII

Gödel's Incompleteness Theorem

1 Arithmetic of Syntax

1.1. In this section we show how the syntax of formal languages reduces in principle to arithmetic. We do this by identifying the symbols, expressions, and texts in a finite or countable alphabet A with certain natural numbers (i.e., by *numbering* them) in such a way that the syntactic operations (juxtaposition, substitution, etc.) are represented by recursive functions, and the syntactic relations (occurrence in an expression, "being a formula," etc.) are represented by decidable or enumerable sets.

In Chapter II we described how this technique works for Smullyan's language of arithmetic, but now we shall investigate it more systematically. Our first task is to show that the computability of syntactic operations and the decidability (enumerability) of syntactic relations on the sets of expressions and texts do not depend on how we number them, as long as we adhere to certain weak natural restrictions.

This independence of the method of numbering allows us to consider this numbering not only as a technical device, but also as a reflection of a deep equivalence between arithmetic and the combinatorial properties of formal texts. In modern computers, where a single store-location may serve consecutively as a number, a name (code), and a command, this equivalence between syntax and arithmetic is realized "in the flesh" and is accepted as a basic principle. This was not the case, however, in 1931, when Gödel first introduced the concept of numbering.

1.2. *Numbering.* Let S be a finite or countable set. By a *numbering* of S we mean any *injective* map $N : S \to \mathbf{Z}^+$ whose image is *decidable*. We call $N(s)$ the *N-number* of an element $s \in S$. We call two numberings N and M of a set S *equivalent* if the partial functions $N \circ M^{-1}$ and $M \circ N^{-1}$ from $\mathbf{Z}^+$ to $\mathbf{Z}^+$ are partial recursive. These functions are automatically *computable* (not only semicomputable), since their domains of definition are decidable (see §1–2 of Chapter V).

Yu. I. Manin, *A Course in Mathematical Logic for Mathematicians, Second Edition*,
Graduate Texts in Mathematics 53, DOI 10.1007/978-1-4419-0615-1_7,

The intuitive meaning of these definitions is clear: requiring the set of $N(s)$ to be decidable ensures that it is possible to determine whether a natural number has the property of "being the number of an element of S," and two numberings are equivalent when each of them can be effectively recovered from the other for any $s \in S$.

1.3. **Lemma.**

(a) *The relation of equivalence between numberings is reflexive, symmetric, and transitive.*
(b) *Any injective map from a finite set S to $\mathbf{Z}^+$ is a numbering, and any two numberings of a finite S are equivalent.*
(c) *Any numbering of an infinite set is equivalent to a numbering whose image is all of $\mathbf{Z}^+$.*

All this either is obvious or has already been proved. In particular, (c) follows from Proposition 5.2 in Chapter V.

1.4. Let S_1 and S_2 be two sets, and let $N_i : S_i \to \mathbf{Z}^+, i = 1, 2$, be numberings of them. We call a partial function $f : S_1 \to S_2$ *partial recursive relative to* $\langle N_1, N_2 \rangle$ if the map $N_2 \circ f \circ N_1^{-1}$ is partial recursive. A tautological example: any numbering function $N : S \to \mathbf{Z}^+$ is partial recursive relative to $\langle N, \text{identity} \rangle$.

A subset $T \subset S$ is said to be decidable, enumerable, arithmetical (i.e., definable in L_1Ar, see Chapter II, §2) relative to the numbering N_1 if the set $N_1(T)$ has the corresponding property.

1.5. **Lemma.** *If $\langle N_1, N_2 \rangle$ is replaced by a pair of equivalent numberings $\langle N_1', N_2' \rangle$ in* 1.4, *then the classes of recursive functions $f : S_1 \to S_2$ and of decidable, enumerable, and arithmetical subsets of S_1 do not change.*

PROOF. The composition of computable recursive functions is recursive and computable. The inverse image of a decidable (respectively enumerable) set with respect to a computable function is decidable (respectively enumerable). Finally, suppose that $f : \mathbf{Z}^+ \to \mathbf{Z}^+$ is a partial recursive function, and that $E \subset \mathbf{Z}^+$ is an arithmetical set. Then $f^{-1}(E) = \text{pr}_1((\mathbf{Z}^+ \times E) \cap \Gamma_f)(\text{in } \mathbf{Z}^+ \times \mathbf{Z}^+)$. Since $\mathbf{Z}^+ \times E$ is arithmetical and Γ_f is also arithmetical (even Diophantine), it follows that $f^{-1}(E)$ is arithmetical. □

1.6. Let S_i be sets with numberings N_i, $i = 1, \dots, r$. A numbering $N : S_1 \times \cdots \times S_r \to \mathbf{Z}^+$ is said to be *compatible* with $\langle N_1, \dots, N_r \rangle$ if the projection $\text{pr}_i : S_1 \times \cdots \times S_r \to S_i$ is recursive relative to $\langle N, N_i \rangle$ for all $i = 1, \dots, r$, and if the partial function

$$(\mathbf{Z}^+)^r \overset{(N_1^{-1}, \dots, N_r^{-1})}{\Rightarrow} S_1 \times \cdots \times S_r \overset{N}{\Rightarrow} \mathbf{Z}^+$$

is recursive. In other words, the N_i-numbers of the coordinates are computed from the N-number of the vector, and conversely.

1.7. **Lemma.**

(a) *In the notation of* 1.5, *for any* $\langle N_1, \dots, N_r\rangle$ *there exists a numbering* N *that is compatible with them. For example, for* $s_i \in S_i, i = 1, \dots, r$, *we may set*

$$N(s_1, \dots, s_r) = \tau^{(r)}(N_1(s_1), \dots, N_r(S_r))$$

(for the definition of $\tau^{(r)}$, see Section 4.5 in Chapter V).

(b) *If* N *is compatible with* $\langle N_1, \dots, N_r\rangle$, N *is equivalent to* M, *and* N_i *is equivalent to* M_i, *for* $i = 1, \dots, r$, *then* M *is compatible with* $\langle M_1, \dots, M_r\rangle$.

(c) *If* N *is compatible with* $\langle N_1, \dots, N_r\rangle$ *and* M *is compatible with* $\langle N_1, \dots, N_r\rangle$, *then* N *and* M *are equivalent. If* N *is compatible with* $\langle N_1, \dots, N_r\rangle$ *and also with* $\langle M_1, \dots M_r\rangle$, *then* N_i *and* M_i *are equivalent for all* $i = 1, \dots, r$.

What all this says is that the relationship of compatibility gives a one-to-one correspondence between families consisting of r equivalence classes of numberings of the sets $S_1, \dots, S_r$ and certain equivalence classes of numberings of $S_1 \times \cdots \times S_r$. This lemma is proved by mechanically checking the definitions.

1.8. Let $A^l = A \times \cdots \times A$ (l times), and let $S(A) = A^1 \cup A^2 \cup \cdots \cup A^1 \cup \cdots$. If A is an alphabet, then $S(A)$ is the set of expressions in the alphabet. Here $A^0 = \{\wedge\}$ consists of the empty expression. The function $S(A) \to \mathbf{Z}^+$ that takes the value p on each element of A^p is called the *length* of the expression. The "ith coordinate" partial function from $\mathbf{Z}^+ \times S(A)$ to A^1 given by $\langle i, \langle a_1, \dots, a_p\rangle\rangle \mapsto a_i$ is defined on the subset $\bigcup_{i=1}^{\infty} \{i\} \times (A^i \cup A^{i+1} \cup \cdots)$. The "juxtaposition" function from $S(A) \times S(A)$ to $S(A)$ takes

$$\langle\langle a_1, \dots, a_p\rangle, \langle b_1, \dots, b_q\rangle\rangle \text{ to } \langle a_1, \dots, a_p, \ b_1, \dots, b_q\rangle.$$

A numbering N of $S(A)$ is called *admissible* if the length function, the ith coordinate function, and the juxtaposition function are partial recursive relative to $\langle N, \mathrm{id}\rangle$, $\langle\langle \mathrm{id}, N\rangle, N\rangle$, and $\langle\langle N, N\rangle, N\rangle$, respectively. A numbering N of $S(A)$ is said to be *compatible* with a numbering N_0 of A if it is admissible and if the restriction of N to A^1 is equivalent to N_0 on A (where we identify A^1 with A).

Here is the basic result of this section:

1.9. **Proposition.**

(a) *If* N *is admissible, then any numbering equivalent to* N *is also admissible.*

(b) *If* N *if compatible with* N_0, N' *is equivalent to* N, *and* N_0' *is equivalent to* N_0, *then* N' *is compatible with* N_0'.

(c) *If* N *and* N' *are both compatible with* N_0 *then they are equivalent.*

(d) *For any numbering* N_0 *of* A, *there exists a compatible numbering* N *of* $S(A)$, *whose equivalence class is uniquely determined by the class of* N_0 *because of* (c).

PROOF. We obtain (a) and (b) formally from Lemma 1.6. To prove (c), we find the N-number of an expression from its N'-number as follows. Let $m * n = N(N^{-1}(m)N^{-1}(n))$ (where the argument of N is the *juxtaposition* of the two expressions $N^{-1}(m)$ and $N^{-1}(n)$). The partial function from $\mathbf{Z}^+ \times \mathbf{Z}^+$ to $\mathbf{Z}^+$ defined by $\langle m, n\rangle \mapsto m * n$ is recursive and associative, since N is admissible. Further, let $(k)_i = N$ (the ith coordinate of $N^{-1}(k)$). The partial function $\mathbf{Z}^+ \times \mathbf{Z}^+ \to \mathbf{Z}^+ : \langle k, i\rangle \mapsto (k)_i$ is recursive for the same reason. We similarly define $(k)'_i$ in terms of N'. Finally, let $l' : \mathbf{Z}^+ \to \mathbf{Z}^+$ be the partial function "the length of $N'^{-1}(k)$." It is also recursive.

Then we have

$$N \circ N'^{-1}(k) = N \circ N'^{-1}((k)'_1) * \cdots * N \circ N'^{-1}\left((k)'_{l'(k)}\right).$$

But the N'-numbers of the one-letter expressions $\{(k)'_i\}$ form a decidable subset of $\mathbf{Z}^+$ (namely, the 1-level of the computable function l'). The restriction of $N \circ N'^{-1}$ to this subset is a recursive function, since the restrictions of N and N' to A^1 are equivalent. We obtain (c) from this and from the recursiveness of $*$, $(k)'_i$, and l' (by applying induction on x to $*_{i=1}^{x} N \circ N'^{-1}((k)'_i)$ and then substituting $x = l'(k)$).

We prove (d) using an explicit construction of Gödel (the idea of which, incidentally, goes back to Leibniz).

(d$_1$) *Construction of N compatible with N_0:*

$$N(a_1, \dots, a_m) = p_1^{N_0(a_1)} \cdots p_m^{N_0(a_m)},$$

where $p_1 = 2, p_2 = 3, \dots$ are the prime numbers. Here $N(\Lambda) = 1$. We verify that N has the required properties.

(d$_2$) *N is a numbering.* First of all, $N : S(A) \to \mathbf{Z}^+$ is an embedding because $N_0 : A \to \mathbf{Z}^+$ is injective, and we have unique factorization in $\mathbf{Z}^+$.

We show that the image of N is decidable. In the first place, the set of prime numbers in $\mathbf{Z}^+$ is decidable, since it is the 2-level of the everywhere defined recursive function

$$n \mapsto \text{the number of divisors of } n = \sum_{k=1}^{n} d(k, n) \dot{-} n,$$

where (see §3 of Chapter V)

$$d(k, n) = s\left((\mathrm{rem}(k, n) - k)^2 + 1\right) = \begin{cases} 2, & \text{if } k | n, \\ 1, & \text{otherwise,} \end{cases}$$

$$s(1) = 2, \quad s(\geqslant 2) = 1.$$

Thus, the function $i \mapsto p_i$ is recursive (see the proof of Proposition 5.2 in Chapter V).

We now set

$$f(n, i, y) = s\left((\mathrm{rem}(p_i^y, n) - p_i^y)^2 + 1\right).$$

This function is recursive, and hence so is the function of (n, i)

$$v_i(n) = \min\{y | f(n, i, y) = 1\} = (\text{the power of } p_i \text{ which divides } n) + 1.$$

This implies that the "length" function is recursive:

$$l(n) = \text{the number of prime divisors of } n = \sum_{i=1}^{n} s(v_i(n)) \dot{-} n$$

(automatically $p_m \nmid n$ when $m > n$, since $p_m > m$).

Now let E be the image of N_0 in $\mathbf{Z}^+$. Then

$$\text{image of } N = \{n | \forall i \leqslant l(n),\ v_i(n) \in E + 1\}.$$

But the set $F = \{\langle i, n\rangle | v_i(n) \in E + 1\}$ is decidable, since it is the preimage of $E+1$ under v, and applying the bounded universal quantifier preserves the decidability. In fact, let $\chi_F(i,n) = 1$ if $\langle i, n\rangle \in F$ and $\chi_F(i,n) = 2$ if $\langle i, n\rangle \notin F$. Then the image of N is the 2-level of the following function of n: $s(\Pi_{i=1}^{l(n)} \chi_F(i,n))$.

(d$_3$) *N is admissible.* We have already shown that the length function is recursive. The ith coordinate function is represented by $[p_i^{v_i(n)}/p_i]$ (the integral part). Finally, juxtaposition is represented by the function

$$m * n = m \prod_{j=1}^{l(n)} p_{l(m)+j}^{v_j(n)-1},$$

which is recursive by what has already been proved.

We note that our number-theoretic functions are defined on all of $\mathbf{Z}^+$, not only on the Gödel numbers of any specific numbering. In what follows we shall point out when such an extension of the domains of definition is possible only if there is a special reason for mentioning this possibility.

(d$_4$) *N is compatible with N_0.* The functions $x \mapsto 2^x$ and $y \mapsto \log_2(y)$ ($y \in 2^{\mathbf{Z}^+}$) tell us how to go from one numbering to the other on one-letter expressions. These functions are obviously recursive.

This completes the proof of Proposition 1.9. □

1.10. *Concluding remarks.* Proposition 1.9 shows that if we are given an equivalence class of numberings of an alphabet A of a formal language, then this uniquely determines an equivalence class of numberings of the set of expressions $S(A)$, of the set of texts $S(S(A))$, and so on, all of which are compatible with the numberings of A in the given class. Hence, the set of recursive operations and the set of decidable or enumerable relations are invariantly defined on the expressions and texts. The only nonuniqueness that remains is the choice of the equivalence class of the numbering of A.

In all cases of which the author is aware, this choice is also determined canonically in the following way. Namely, *A is realized as a decidable subset of the expressions in some finite "protoalphabet" A_0*, where decidability is understood in the sense of any numbering of $S(A_0)$ that is compatible with any numbering of A_0. It follows from Lemmas 1.3 and 1.5 and Proposition 1.9 applied to A_0 that the resulting class of numberings of A will not depend on either the embedding of A in $S(A_0)$, the numbering of A_0, or even the choice of A_0 (where we recall that if $A_0 \subset A_1$ are finite, then $S(A_0) \subset S(A_1)$ is decidable).

From this point of view, it is natural to consider the nine-letter alphabet of SAr, which was described in §10 of Chapter II, to be a protoalphabet. Then $x, x', x'', x''', \ldots$ are elements of the "real alphabet." Smullyan's particular numbering system is very convenient for proving Tarski's theorem, but the "undefinability of truth" in SAr does not depend on the special form of this numbering, as should by now be completely clear.

More generally, any complete printed description of any alphabet A realizes A in the protoalphabet of available typographical symbols, which is of course finite, and thereby determines a canonical equivalence class of numberings of A.

2 Incompleteness Principles

2.1. Gödel's theorem on the incompleteness of formal theories can be given many precise formulations, none of which entirely exhausts its content. In this section, using the results obtained in §1, we shall try to separate the conceptual aspects of the theorem from the technical details needed to prove it for various languages.

2.2. Let A be a finite or countable alphabet with its canonical equivalence class of numberings, and let $S(A)$ be the set of expressions in A. We suppose that the following two subsets of $S(A)$ have somehow been defined:

(a) $T \subset S(A)$, the set of "true" expressions. For example, we might have been given a language with A as its alphabet, some sort of semantics for the language, and a truth function.
(b) $D \subset S(A)$, the set of "provable" or "deducible" expressions. This set might be described by giving "axioms" and "rules of deduction," or in some other way. We shall always assume that $D \subset T$, as the terminology suggests (it is possible to prove only what is true).

There is every reason to expect that if D and T have been constructed "in a natural way" in the process of formalizing some fragment of modern mathematics, then the following principles hold true.

2.3. *The set D is enumerable.* The intuitive arguments to support this assertion are as follows. Suppose that the "provable" expressions are those for which "proofs" exist. Here "proofs" are certain texts that, perhaps, are written in another alphabet B, i.e., they are elements of $S(S(B))$. (For example, theorems in $\mathrm{L_1Ar}$ may be proved in $\mathrm{L_1Set}$.) One minimal requirement for formal mathematical proofs is that it must be possible mechanically to determine that they are proofs, i.e., they must form a decidable subset of $S(S(B))$. (Here it would actually be sufficient to require that the set of "proofs" be enumerable.) Another unavoidable requirement is that from every proof we must be able to obtain mechanically the "expression proved" in $S(A)$. In other words, the partial function from $S(S(B))$ to $S(A)$ given by "proof" $\mapsto$ "expression proved" must be (semi)computable. But then the image of this function is enumerable. In §5 we show that the set of deducible formulas in $\mathfrak{L}_1$ is enumerable, in accordance with these informal considerations.

We note that a time aspect has implicitly entered into the discussion. A "proof" is understood to mean a "proof using the means accepted at the present time and (semi)identifiable as being accepted." If, for example, we introduce a new axiom of set theory and it becomes widely accepted, then the concept of a proof becomes broader, as happened with the axiom of choice (or, rather, the principle of transfinite induction, Zorn's lemma, ...). See the discussion in §7.

2.4. *The set T is not enumerable if the semantics of truth is rich enough to include elementary arithmetic.* We clearly have in mind some version of Tarski's theorem, which, in fact, even tells us that T is not an arithmetical set. In the next section we give several precise formulations of this principle. (See also Sections 7.3–7.4 below.)

2.5. **Gödel's incompleteness theorem** (General form). *All formal theories of mathematics satisfy the principles* 2.3 *and* 2.4. *Therefore, if a theory is sufficiently rich, it always contains true expressions that are not provable.*

3 Nonenumerability of True Formulas

The following criteria are all variations on a single theme, even if this is not obvious at first, namely, "self-reference, or the diagonal process."

3.1. *The language* SAr. We refer the reader to §10 of Chapter II for the description of this language and its standard interpretation. In §11 of Chapter II we showed that the set of numbers of true formulas in Smullyan's numbering system is nonarithmetical. This set is a fortiori nonenumerable, since enumerable sets are even Diophantine.

3.2. *The language* $\mathrm{L_1Ar}$. Here we give two versions of the argument, one of which gives the stronger result and the other of which gives the more concrete result. A third version, which is closer to Godel's original proof, will be described in §7.

(a) *Tarski's theorem for* $\mathrm{L_1Ar}$. The proof that the set of true formulas in $\mathrm{L_1Ar}$ is nonarithmetical can be reduced to Tarski's theorem for SAr in the following way. In the first place, the sets of formulas in $\mathrm{L_1Ar}$ and SAr are decidable in the set of all expressions (this will be shown for $\mathrm{L_1Ar}$ in §4).

In the second place, the translation map {formulas in SAr} $\overset{\mathrm{tr}}{\Rightarrow}$ {formulas in $\mathrm{L_1Ar}$}, which was described in §10 of Chapter II, is recursive (as is easily shown using the arguments in the next section). Since, the map tr preserves the truth function, we have $T_s = \mathrm{tr}^{-1}(T_{L_1})$ in the obvious notation. But then, if T_{L_1} were arithmetical, it would follow that T_s is also arithmetical (see the proof of

Lemma 1.5), which contradicts Tarski's theorem for SAr. It would be a useful exercise for the reader, after first reading §4, to carry out this proof in complete detail.

The following argument is simpler and more precise, but it only shows that T_{L_1} is nonenumerable, and not that it is nonarithmetical.

(b) Let $E \subset \mathbf{Z}^+$ be an enumerable but undecidable set (which exists by §5 of Chapter V). Let E be defined by the formula $P(x)$ in $\mathrm{L}_1\mathrm{Ar}$, which has one free variable x. For $n \geqslant 2$ we set $\bar{n} = +\Big(+ (\bar{1} + (\bar{1}, \bar{1})) \cdots \Big)$, which is the term-name for the integer n in the obvious canonical $\mathfrak{L}_1$-type notation. We consider the family of closed formulas $\{\neg(P(\bar{n}))|n \in \mathbf{Z}^+\}$ in $\mathrm{L}_1\mathrm{Ar}$.

3.3. Proposition.

(a) *The function* $\mathbf{Z}^+ \to$ *{formulas in* $\mathrm{L}_1\mathrm{Ar}$*} given by* $n \mapsto \neg P(\bar{n})$ *is recursive.*
(b) *The set* $\{n|\neg(P(\bar{n}))$ *is true}* *is nonenumerable.*

Corollary. T_{L_1} *is nonenumerable; more precisely, the set of true formulas in the family* $\{\neg(P(\bar{n}))\}$ *is nonenumerable.*

(If T_{L_1} were enumerable, its preimage in $\mathbf{Z}^+$ would also be enumerable.)

PROOF.

(a) Let the formula $\neg(P(x))$ have the form $R_1 \times R_2 \times \cdots \times R_{S+1}$, where x does not occur in the expressions R_i. Using the same notation as in the proof of Proposition 1.9, for a fixed numbering N of the set of expressions with juxtaposition function $*$ we have

$$N(\neg(P(\bar{n}))) = N(R_1) * N(\bar{n}) * N(R_2) * \cdots * N(R_{S+1}).$$

Hence, it suffices to show that the function $n \mapsto N(\bar{n})$ is recursive. But since $\overline{n+1} = +(\bar{1}, \bar{n})$, it follows that for $n \geqslant 1$,

$$N(\overline{n+1}) = N(+) * N(''('') * N(\bar{1}) * N(\bar{n}) * N('')''),$$

which expresses $N(\overline{n+1})$ recursively in terms of $N(\bar{n})$.

(b) $\{n|\neg(P(\bar{n})) \in T_{L_1}\} = \mathbf{Z}^+ \backslash E$ by the definition of the formula $P(x)$ defining E. But the complement of E is nonenumerable, since E is undecidable.

The proposition and the corollary are proved. □

3.4. *Languages at least as rich as* $\mathrm{L}_1\mathrm{Ar}$. Let L be an arbitrary language with a (finite or countable) alphabet A, in which we are given a set T of "true" expressions. We suppose that L is no poorer than a language of arithmetic in the following sense: *There exists a translation map*

$$\mathrm{tr} : \{\text{formulas in } \mathrm{L}_1\mathrm{Ar}\} \Rightarrow \{\text{expressions } \textit{in } A\}$$

that takes T_{L_1} to T, takes the complement of T_{L_1} to the complement of T, and is recursive.

Then T is nonenumerable.

Such a translation map can be constructed for L_1Set, for example. Proposition 3.3 shows that, actually, we need only know how to translate into L the formulas in the family $\neg(P(\bar{n}))$; this allows us to use a very modest language of arithmetic.

3.5. *Remarks*

(a) The series of Diophantine problems "Is $P(\bar{n})$ true?" i.e., "Does the Diophantine equation $F(n; x_1, \ldots, x_r) = 0$ have a solution in $\mathbf{Z}^+$?" (where F is a suitable polynomial with integer coefficients; see Chapter VI) has the property that no finitely describable collection of means of proof is adequate to answer this series of questions completely. One might say that even the theory of Diophantine equations is infinitely complicated.

(b) In some sense any problem in mathematics reduces to a Diophantine problem. In fact, after translating the problem into a suitable formal language, we may just ask, "Is the formula P or the formula $\neg P$ provable?" But this is precisely the same as asking whether the number of P (the number of $\neg P$) belongs to the enumerable set D of provable formulas, i.e., whether the Diophantine equation corresponding to D in the given series is decidable.

This gives somewhat unexpected support for Gauss's opinion regarding the queenly status of arithmetic. There even exists a "queen of the Diophantine equations" whose graph projects onto the set of numbers of formulas in L_1Set that are deducible from the Zermelo–Fraenkel axioms.

But of course, we normally ask "Is P true?" and not "Is P provable?" from this point of view, the most creative activity in mathematics is the discovery of new principles of proof that do not reduce to the "legacy of the past" and that again must be taken on faith. Set theory as a whole was the most recent such principle in the modern development of mathematics. The dramatic history of its creation and of the disputes surrounding its acceptance is worthy of a discovery of this magnitude.

It is amazing that within formal mathematics it is possible to say something about such informal things. See also §7 below.

4 Syntactic Analysis

4.1. This section contains the preliminary technical material that will be needed in §5, when we prove that the set of deducible formulas in a language of $\mathfrak{L}_1$ is enumerable.

Let L be a fixed language in $\mathfrak{L}_1$ having a finite or countable alphabet A. In order to shorten the technical work somewhat, we assume that we are working with a dialect that contains only the connectives $\neg$ and $\to$ and the quantifier $\forall$. This is not in any sense essential. As in §1, we have a canonical equivalence class of numberings of A, which determines numberings of $S(A)$, $S(S(A))$, and so on.

The terms "recursive," "decidable," etc. will be understood to refer to this class. Thus, we may omit explicit mention of the numbering in the statements of the basic results. But in the proofs it will be more convenient to work directly with a numbering. We therefore fix one of the numberings $N : S(A) \to \mathbf{Z}^+$ with juxtaposition function $*$, length function l, and ith coordinate function $(k)_i$, as in the proof of Proposition 1.9. We shall assume that $m * n > \max(m, n)$, i.e., the number of any part of an expression is strictly less than the number of the whole expression. Such an N is called a *Gödel numbering.*

In addition to the conditions given in §1, we require that N satisfy the following conditions regarding recognition of the syntactic characteristics of the symbols of the alphabet:

(a) *The sets of variables, of constants, of operations, and of relations in A are decidable.*
(b) *The "degree" function on the set of operations and relations is recursive.*

We are now ready to begin. But before reading further the reader is advised to review §1 of Chapter II.

4.2. *The partial function from $S(A) \times \mathbf{Z}^+$ to $\mathbf{Z}^+$ given by*

$$\langle \textit{an expression } P, i\rangle \mapsto \begin{cases} \textit{the place in } P \textit{ containing the right parenthesis} \\ \textit{that corresponds to the left parenthesis in the } i\text{th} \\ \textit{place in } P \end{cases}$$

is computable, i.e., it is recursive and has a decidable domain of definition.

PROOF. It will be convenient to use the following notation: if Q is a statement about integers in $\mathbf{Z}^+$, then

$$\|Q\| = \begin{cases} 1, & \text{if } Q \text{ is true}, \\ 2, & \text{if } Q \text{ is false}. \end{cases}$$

This is a truth function that has been adjusted so as to take values in $\mathbf{Z}^+$, which does not contain zero.

We construct a function $\mathrm{Par}(k, i) : \mathbf{Z}^+ \times \mathbf{Z}^+ \to \mathbf{Z}^+$ as follows: if $(k)_i$ is not defined, or if $(k)_i \neq N(\text{“(”})$, or if $(k)_i = N(\text{“(”})$ but $\forall j \in [i, l(k)], \sum_{m=i}^{j} \|(k)_m = N(\text{“(”})\| \neq \sum_{m=i}^{j} \|(k)_m = N(\text{“)”})\|$, let $\mathrm{Par}(k, i) = 1$; otherwise, let $\mathrm{Par}(k, i) = \min\{j | j \leqslant l(k) \text{ and } \sum_{m=i}^{j} \|(k)_m = N(\text{“(”})\| = \sum_{m=i}^{j} \|(k)_m = N(\text{“)”})\|\}$. Obviously, when restricted to $N^{-1}(S(A)) \times \mathbf{Z}^+$, the function $\mathrm{Par}(k, i)$ gives the place in the expression $N^{-1}(k)$ containing the ")" that corresponds to the "(" in the ith place if this is possible, and gives 1 when this is not possible. (Compare with Lemma 1.2 in §1 of Chapter II.) Hence, it suffices to show that $\mathrm{Par}(k, i)$ is recursive. But $\mathrm{Par}(k, i)$ has been defined by gluing together a finite number (four) of recursive functions having decidable domains of definition (by the properties of N). Thus, $\mathrm{Par}(k, i)$ is recursive. □

4.3. *The partial function $S(A) \to \mathbf{Z}^+$ given by (an expression P) $\mapsto$ (the number of terms in L that are juxtaposed to get P) is computable.*

We recall that this number is uniquely defined (§1 of Chapter II).

PROOF. We first construct a formula that defines the function

$$LT(k) = \begin{cases} l(k)+1, & \text{if } N^{-1}(k) \text{ is not a juxtaposition of terms;} \\ \text{the number of terms whose} & \text{otherwise} \\ \quad \text{juxtaposition is } N^{-1}(k), & \end{cases}$$

from $\mathbf{Z}^+$ to $\mathbf{Z}^+$ recursively in terms of its values on smaller values of the argument. The way to carry out this syntactic analysis of $N^{-1}(k)$ can be described verbally as follows: first see whether $N^{-1}((k)_1)$ is a variable or constant and, if it is, whether $N^{-1}((k)_2 * \cdots * (k)_{l(k)})$ is a juxtaposition of terms; if $N^{-1}((k)_1)$ is not a variable or constant, check whether it is an operation, and if it is, whether it is followed by "(", whether there is a corresponding ")", whether a juxtaposition of the required number of terms lies between the "(" and the ")", and whether ")" is followed by a juxtaposition of terms.

To describe this procedure systematically, we set

$$f_1(k) = \begin{cases} (k)_2 * \cdots * (k)_{l(k)}, & \text{if } l(k) \geqslant 2; \\ 1, & \text{otherwise;} \end{cases}$$

$$f_2(k) = \begin{cases} (k)_3 * \cdots * (k)_{\mathrm{Par}(k,2)-1}, & \text{if } 4 \leqslant \mathrm{Par}(k,2); \\ 1, & \text{otherwise;} \end{cases}$$

$$f_3(k) = \begin{cases} (k)_{\mathrm{Par}(k,2)+1} * \cdots * (k)_{l(k)}, & \text{if } 1 < \mathrm{Par}(k,2) < l(k); \\ 1, & \text{otherwise.} \end{cases}$$

All of these functions are recursive.

We now write the following recipe for computing $LT(k)$ recursively:

$$l(k) = 1 \quad \text{and} \quad \begin{cases} N^{-1}(k) \text{ is a variable} \Rightarrow LT(k) = 1, \\ N^{-1}(k) \text{ is a constant} \Rightarrow LT(k) = 1, \\ N^{-1}(k) \text{ is neither a variable nor a constant} \Rightarrow LT(k) = 2; \end{cases}$$

$$l(k) > 1 \quad \text{and} \quad N^{-1}((k)_1) \text{ is a variable} \Rightarrow LT(k) = 1 + LT(f_1(k));$$

$$l(k) > 1 \quad \text{and} \quad N^{-1}((k)_1) \text{ is a constant} \Rightarrow LT(k) = 1 + LT(f_1(k));$$

$$\begin{aligned} l(k) > 1, \quad & N^{-1}((k)_1) \text{ is an operation}, \quad (k)_2 = N(\text{“(”}), \\ & 4 \leqslant \mathrm{Par}(k,2) = l(k), \\ & \text{degree } N^{-1}((k)_1) = LT(f_2(k)) \leqslant l(f_2(k)) \Rightarrow LT(k) = 1; \end{aligned}$$

$$\begin{aligned} l(k) > 1, \quad & N^{-1}((k)_1) \text{ is an operation}, \quad (k)_2 = N(\text{“(”}), \\ & 4 \leqslant \mathrm{Par}(k,2) < l(k), \\ & \text{degree } N^{-1}((k)_1) = LT(f_2(k)) \leqslant l(f_2(k)), \\ & LT(f_3(k)) \leqslant l(f_3(k)) \Rightarrow LT(k) = 1 + LT(f_3(k)); \end{aligned}$$

$$\begin{aligned} l(k) > 1, \quad & \text{and none of the previous additional conditions hold} \\ & \Rightarrow LT(K) = 1 + l(k). \end{aligned}$$

To show that LT is recursive, we first note that for each of the above eight alternatives, we can easily construct a recursive function $h_i(k, x, y, z)$ with the following property:

$$\|k \text{ satisfies the } i\text{th alternative}\| = h_i(k, LT(f_1(k)), LT(f_2(k)), LT(f_3(k))),$$

and we can also construct a recursive function $v_i(k, x, y, z)$ with the property that k satisfies the ith alternative $\Rightarrow$

$$LT(k) = v_i(k, LT(f_1(k)), LT(f_2(k)), LT(f_3(k))).$$

We therefore have the equation

$$\begin{aligned} LT(k) &= 2\sum_{i=1}^{8} v_i(k, LT(f_1(k)), LT(f_2(k)), LT(f_3(k))) \\ &\quad \dot{-} \sum_{i=1}^{8} (h_i v_i)(k, LT(f_1(k)), LT(f_2(k)), LT(f_3(k))). \end{aligned}$$

Since $f_i(k) < k$ for $k > 1$, this formula allows us successively to compute the values of $LT(k)$, starting with $LT(1)$. But the recursion here computes the value at k not in terms of the value at $k-1$, but in terms of several earlier values. It is this that presents the basic difficulty in showing that the syntactic functions are recursive. We now describe the device for overcoming this difficulty here and in all future cases.

In general, let $\phi_1(k), \ldots, \phi_s(k)$ be recursive functions having the property that $\phi_i(k) < k$ for all $i \leqslant s$ and $k \geqslant 2$. Further, let $h(x_1, \ldots, x_m, k, y_1, \ldots, y_s)$ be a recursive function, and let $g(x_1, \ldots, x_n, k)$ be defined by the relations

$$\begin{aligned} g(x_1, \ldots, x_n, 1) &= \text{some known recursive function}, \\ g(x_1, \ldots, x_n, k+1) &= h(x_1, \ldots, x_n, k, g(x_1, \ldots, x_n, \phi_1(k)), \\ &\qquad \ldots, g(x_1, \ldots, x_n, \phi_S(k))). \end{aligned}$$

Using the juxtaposition function $*$, we let

$$G(x_1, \ldots, x_n, k) = \mathop{*}_{i=1}^{k} g(x_1, \ldots, x_n, i).$$

Since

$$g(x_1, \ldots, x_n, i) = G(x_1, \ldots, x_n, k))_i$$

for all $i \leqslant l(G(x_1, \ldots, x_n, k)) = k$, and in particular for the greatest such i, it follows that to verify that g is recursive, it suffices to show that G is recursive. But for $k \geqslant 2$ we have

$$\begin{aligned} &G(x_1, \ldots, x_n, k+1) \\ &\quad = G(x_1, \ldots, x_n, k) * g(x_1, \ldots, x_n, k+1) \\ &\quad = G(x_1, \ldots, x_n, k) \\ &\qquad * h(x_1, \ldots, x_n, k, (G(x_1, \ldots, x_n, k))_{\phi_1(k)}, \ldots, (G(x_1, \ldots, x_n, k))_{\phi_s(k)}), \end{aligned}$$

which is in the standard form for a recursive equation.

If we apply this device to LT, setting $n = 0, s = 3$, and $\phi_i(k) = f_i(k+1)$, we obtain the recursiveness of LT. □

Corollary. *The set of terms is decidable.*

In fact, this set is the 1-level of the computable function LT.

4.4. *The set of atomic formulas is decidable.*

In fact,

$$N^{-1}(k) \text{ is an atomic formula} \Leftrightarrow (k)_1 \text{ is a relation, } (k)_2 = N(\text{“(”}),$$
$$\text{Par}(k,2) = l(k) \geqslant 4,$$
$$\text{and degree } N^{-1}((k)_1) = LT(f_2(k)) \leqslant l(f_2(k)),$$

where $f_2(k)$ was defined in 4.3.

4.5. *The set of formulas is decidable.*

In fact, in our dialect, which has been simplified to include only $\neg, \rightarrow$, and $\forall$, we have

$$N^{-1}(k) \text{ is a formula}$$
$$\Leftrightarrow N^{-1}(k) \text{ is an atomic formula, or is of the form } \neg(P),$$
$$(P) \Rightarrow (Q), \text{ or} \forall x(P), \text{ where } P \text{ and } Q \text{ are formulas and x is a variable.}$$

Using the procedure in 4.3, we define the recursive functions

$$f_4(k) = \begin{cases} (k)_3 * \cdots * (k)_{l(k)-1}, & \text{if } l(k) \geqslant 4; \\ 1, & \text{otherwise;} \end{cases}$$

$$f_5(k) = \begin{cases} (k)_2 * \cdots * (k)_{\text{Par}(k,1)-1}, & \text{if } \text{Par}(k,1) \geqslant 3; \\ 1, & \text{otherwise;} \end{cases}$$

$$f_6(k) = \begin{cases} (k)_{\text{Par}(k,1)-3} * \cdots * (k)_{l(k)-1}, & \text{if } 3 \geqslant \text{Par}(k,1) \leqslant l(1) - 1; \\ 1, & \text{otherwise;} \end{cases}$$

$$f_7(k) = \begin{cases} (k)_4 * \cdots * (k)_{l(k)-1}, & \text{if } l(k) \geqslant 5; \\ 1, & \text{otherwise;} \end{cases}$$

$$\text{At}(k) = \begin{cases} 1, & \text{if } N^{-1}(k) \text{ is an atomic formula;} \\ 2, & \text{otherwise.} \end{cases}$$

The function

$$\text{Fm}(k) = \begin{cases} 1, & \text{if } N^{-1}(k) \text{ is a formula,} \\ 2, & \text{otherwise} \end{cases}$$

is computed using the following recursive relation (where $s(1) = 1$ and $s(k) = 2$ for $k \geqslant 2$):

$$\begin{aligned}\mathrm{Fm}(k) = s \circ \min\{&\mathrm{At}(k); \|(k)_1 = N(\text{“}\neg\text{”})\| \cdot \|(k)_2 = N(\text{“(”})\| \cdot \|l(k) \geqslant 4\| \\ &\cdot \mathrm{Fm}(f_4(k)); \\ &\|(k)_1 = N(\text{“(”})\| \cdot \|\mathrm{Par}(k,1) \geqslant 3\| \cdot \mathrm{Fm}(f_5(k)) \\ &\cdot \|(k)_{\mathrm{Par}(k,1)+1} = N(\text{“}\rightarrow\text{”})\| \cdot \|(k)_{\mathrm{Par}(k,1)+2} = N(\text{“(”})\| \\ &\cdot \|\mathrm{Par}(k, \mathrm{Par}(k,1)+2) = l(k)\| \cdot \mathrm{Fm}(f_6(k)); \\ &\|(k)_1 = N(\text{“}\forall\text{”})\| \cdot \|(k)_2 = N(\text{a variable})\| \cdot \|(k)_3 = N(\text{“(”})\| \\ &\cdot \|\mathrm{Par}(k,3) = l(k) \geqslant 5\| \cdot \mathrm{Fm}(f_7(k))\}.\end{aligned}$$

$\mathrm{Fm}(k)$ is now shown to be recursive using the device described in 4.3.

Corollary. *The sets of formulas of the form $\neg(P), (P) \rightarrow (Q)$, and $\forall x(P)$ are decidable.*

4.6. *The following function from $S(A) \times \mathbf{Z}^+ \times S(A)$ to $S(A)$ is computable: $\langle P, i, Q\rangle \mapsto$ the result of substituting P for the ith symbol in Q.*

We set

$$\mathrm{Sub}(k,i,m) = \begin{cases} (m)_1 * \cdots * (m)_{i-1} * k * (m)_i * \cdots * (m)_{l(m)}, & \text{if } i \leqslant l(m); \\ 1, & \text{otherwise.}\end{cases}$$

This function is clearly recursive, and coincides with the required map on the set of $\langle k, i, m\rangle$ with $k, m \in N^{-1}(S(A))$. □

4.7. *The following relation in $\mathbf{Z}^+ \times S(A) \times S(A)$ is decidable: "the one-letter expression x is a free variable in the ith place in the formula P."*

If fact, we set

$$\mathrm{Fr}(i,k,l) = \begin{cases} 1, & \text{if the condition in 4.7 holds for } p = N^{-1}(k) \\ & \text{and} \langle x\rangle = N^{-1}(l); \\ 2, & \text{otherwise.}\end{cases}$$

Then we have

$$\begin{aligned}N^{-1}(k) \text{ is not a formula,} \quad \text{or} \quad & N^{-1}(l) \text{ is not a variable,} \\ \text{or} \quad & i > l(k) \Leftrightarrow \mathrm{Fr}(i,k,l) = 2.\end{aligned}$$

Now suppose that $N^{-1}(k)$ is a formula, $N^{-1}(l)$ is a variable, and $i \leqslant l(k)$. Then the following alternatives remain:

$$\begin{aligned}&l \neq (k)_i \Rightarrow \mathrm{Fr}(i,k,l) = 2; \\ &l = (k)_i,\ \mathrm{At}(k) = 1 \Rightarrow \mathrm{Fr}(i,k,l) = 1; \\ &l = (k)_i,\ N^{-1}(k) \text{ has the form } \neg(P) \\ &\qquad\qquad \Rightarrow \mathrm{Fr}(i,k,l) = \mathrm{Fr}(i, f_5(k), l);\end{aligned}$$

$l = (k)_i$, $N^{-1}(k)$ has the form $(P) \to (Q)$, $i < \mathrm{Par}(k,1)$
$\Rightarrow \mathrm{Fr}(i,k,l) = \mathrm{Fr}(i, f_5(k), l)$
$l = (k)_i$, $N^{-1}(k)$ has the form $(P) \to (Q)$, $i > \mathrm{Par}(k,1)+2$
$\Rightarrow \mathrm{Fr}(i,k,l) = \mathrm{Fr}(i, f_6(k), l)$;
$l = (k)_i$, $N^{-1}(k)$ has the form $\forall x(P), (k)_2 = l \Rightarrow \mathrm{Fr}(i,k,l) = 2$;
$l = (k)_i$, $N^{-1}(k)$ has the form $\forall x(P), (k)_2 \neq l$
$\Rightarrow \mathrm{Fr}(i,k,l) = \mathrm{Fr}(i, f_7(k), l)$.

Here the functions f_5, f_6, and f_7 were defined in 4.5. The rest of the proof that Fr is recursive follows the same procedure as in 4.4 and 4.5. □

4.8. *The set* $\{\langle x, P, t\rangle | x$ *is a variable, P is a formula, t is a term, and x does not bind t in P*$\}$ *is decidable.*

In terms of the numbers $\langle i, k, m\rangle$, this condition means that

$$\forall j \leqslant l(k)\{\text{either } (k)_j \neq i, \text{or else } (k)_j = i \wedge \mathrm{Fr}(j,k,i) = 2,$$
$$\text{or else } (k)_j = i \wedge \mathrm{Fr}(j,k,i) = 1$$
$$\wedge\, \forall n \in [1, l(m)](\mathrm{Fr}(j+n \dot{-} 1, \mathrm{Sub}(m,j,k), \ \mathrm{Sub}(m,j,k)_{j+n\dot{-}1})$$
$$= \|(m)_n \text{ is a variable}\|)\}.$$

That is, if t is substituted in place of any free occurrence of x in P, all the variables in t remain free. □

4.9. *The following partial function is computable:* $\langle x, P, t\rangle \mapsto$ *the result of substituting t in place of all free occurrences of x in P.*

Let $\langle i, k, m\rangle$ be the numbers of x, P, and t. We set

$$f(j,k,i,m) = \begin{cases} (k)_j, & \text{if } \mathrm{Fr}(j,k,i) = 2; \\ m, & \text{if } \mathrm{Fr}(j,k,i) = 1. \end{cases}$$

This is a recursive function. We further set

$$\mathrm{Sub}\, t(i,k,m) = \mathop{*}_{j=1}^{l(k)} f(j,k,i,m).$$

This is the number of the expression obtained by substituting t in place of all free occurrences of x in P. □

5 Enumerability of Deducible Formulas

5.1. *General setup.* Let L be any language with a numbered countable alphabet A. We suppose that the following data is fixed:

(a) An enumerable set of "axioms" $\mathrm{Ax} \subset S(A)$.

(b) A partial recursive function $\mathrm{Inf}: \mathbf{Z}^+ \times S(S(A)) \to S(A)$, i.e., an enumerable family of "rules of deduction."

We shall say that an expression $P \in S(A)$ *is a direct consequence of the expressions* $P_1, \ldots, P_r$ *by the* ith *rule of deduction if* $\langle i, \langle P_1, \ldots, P_r \rangle \rangle \in D(\mathrm{Inf})$ and $\mathrm{Inf}(i, \langle P_1, \ldots, P_r \rangle) = P$. We shall call an expression P *deducible* (from the "axioms") if there exists a finite sequence of expressions $P_1, \ldots, P_n = P$ such that for each $j \leqslant n$ either $P_j \in \mathrm{Ax}$ or there exist $i \in \mathbf{Z}^+$ and $\{P_{k_1}, \ldots, P_{k_r}\} \subset \{P_1, \ldots, P_{j-1}\}$ such that P_j is a direct consequence of $P_{k_1}, \ldots, P_{k_r}$ by the ith rule of deduction. We let D denote the set of all deducible expressions.

5.2. **Proposition.** *D is enumerable.*

PROOF. Let $a : \mathbf{Z}^+ \to S(A)$ be a recursive function whose image coincides with Ax, and let $\inf : \mathbf{Z}^+ \to S(A)$ be the partial recursive function given by $\inf(n) = \mathrm{Inf}(t_1^{(2)}(n), N_1^{-1})(t_1^{(2)}(n)))$, where $N_1 : S(S(A)) \to \mathbf{Z}^+$ is any numbering of the texts that is compatible with the given numbering of the expressions.

We construct a recursive function $d : \mathbf{Z}^+ \to S(A)$ as follows:

$$d(2n-1) = a(n),$$
$$d(2n) = \inf(n), \qquad n \geqslant 1.$$

We claim that its image is D. In fact, it suffices to verify that (a) $\mathrm{Ax} \subset$ image of d; and (b) if $P_1, \ldots, P_r \in$ image of d and P is a direct consequence of $P_1, \ldots, P_r$ by the ith rule of deduction, then $P \in$ image of d.

But (a) is obvious, since all the axioms are written out in the odd-numbered places. To verify (b), we choose n such that

$$t_1^{(2)}(n) = i, \quad t_2^{(2)}(n) = N_1\left(\langle P_1, \ldots, P_r \rangle\right).$$

Then $d(2n) = P$. The proposition is proved. □

We now verify that the general setup in 5.1 can always be realized in languages of $\mathfrak{L}_1$.

5.3. *The rules of deduction* Gen *and* MP. We define the map $\mathrm{Inf} : \mathbf{Z}^+ \times S(S(A)) \to S(A)$ as follows:

$$D(\mathrm{Inf}) = \{\langle 1, \langle P, (P) \to (Q) \rangle \rangle | P \text{ and } Q \text{ are formulas}\}$$
$$\cup \{\langle i, \langle P \rangle \rangle | P \text{ is a formula}, i \geqslant 2\},$$

$$\mathrm{Inf}\langle 1, \langle P, (P) \to (Q) \rangle \rangle = Q,$$
$$\mathrm{Inf}\langle i, \langle P \rangle \rangle = \forall x_{i-1}(Q),$$

where x_j is the jth variable in L in any fixed numbering of the variables that has image $\mathbf{Z}^+$ and is compatible with the numbering of A. It is clear that Inf is recursive and exhausts the rules of deduction Gen and MP.

5.4. *The axioms.* We verify that the following sets are enumerable in any language in $\mathfrak{L}_1$:

(a) The tautologies.

(b) The logical quantifier axioms.
(c) The axioms of equality.

Two other sets we show to be enumerable are:

(d) The special axioms of L_1Ar.
(e) The special axioms of L_1Set.

Actually, using the methods of §4 it is not hard to prove that all of these sets are even decidable. But the proof of enumerability is somewhat shorter, and will suffice for our purposes.

5.5. *The tautologies.* In §5 of Chapter II we constructed a finite list of basis tautologies and showed that all the other tautologies can be deduced from them using MP. Thus, by Proposition 5.2, it is sufficient to verify that the basis tautologies are enumerable.

Each of the basis tautologies determines a set of formulas of the form

$$Q_1 P_{i_1} Q_2 P_{i_2} \cdots P_{i_r} Q_{r+1},$$

where the Q_i are fixed expressions that are nonempty (with the possible exception of Q_1 and Q_{r+1}), $i_1, \ldots, i_r \in \{1, \ldots, m\}$, and $\langle P_1, \ldots, P_m \rangle$ varies over all ordered m-tuples of formulas in L. Since the set of such m-tuples is decidable by 4.5 above, and since the operation of juxtaposition is recursive, it is clear that we obtain an enumerable set of formulas.

5.6. *The logical quantifier axioms.* In case our dialect of $\mathfrak{L}_1$ does not have $\exists$, these axioms can be expressed as the following two axiom schemes:

(a) $(\forall x(P(x))) \rightarrow (P(t))$, if x does not bind the term t in the formula P.
(b) $(\forall x((P) \rightarrow (Q))) \rightarrow ((P) \rightarrow (\forall x(Q)))$, if x does not occur freely in P.

By 4.8, the set of triples $\{\langle x, P, t\rangle | x \text{ does not bind } t \text{ in } P\}$ is decidable, and by 4.9, the map $\langle x, P, t \rangle \mapsto P(t)$ is recursive. Since juxtaposition is also recursive, the set of axioms (a) is the image of a decidable set under a recursive function, and so is enumerable.

We may similarly conclude that (b) is enumerable if we verify that the condition "x does not occur freely in P" is decidable. But this is equivalent to the following condition: "the formula obtained from P by substituting either of the variables x_1 and x_2 in place of all free occurrences of x in P coincides with P," where $\langle x_1, x_2 \rangle$ is any fixed pair of distinct variables. This condition is decidable by 4.9.

5.7. *The axioms of equality.* By the definition in 4.6 of Chapter II, it suffices to show that the set of formulas of the form

$$(x = y) \Rightarrow (P(x, x) \Rightarrow P(x, y))$$

is enumerable, where P runs through the atomic formulas in the language, x and y are variables, and $P(x, y)$ is obtained from P by replacing x by y in any subset of the occurrences of x in P. This set of formulas can be obtained, for example, as the image of the following function, which is partial recursive by the results in 4.4 and 4.6:

$S(A) \times A^1 \times A^1 \times S(\mathbf{Z}^+) \to S(A);$

$\langle P, \langle x \rangle, \langle y \rangle, \langle i_1, \ldots, i_r \rangle \rangle \mapsto$ the expression obtained by substituting y in the $i_1, \ldots, i_r$ places in the atomic formula P if x occurs in those places.

5.8. *The special axioms of* $\mathrm{L_1Ar}$ *and* $\mathrm{L_1Set}$. Most of these axioms contain only variables of the language, and not "metalanguage" variables for formulas. This is true of all the axioms of arithmetic except for induction and all the axioms of set theory except for replacement. Each set of axioms not containing variable formulas is decidable because it can be described by a condition such as "the set of formulas of length 40 in which "(" is in the first place, "∀" is in the second place, a variable is in the third place, "(" is in the fourth place, ..., ")" is in the 39th place, and ")" is in the 40th place; in which the variables in the 3rd, 8th, and 16th places are the same, in the 9th and 36th places are the same, and in the 17th and 37th places are the same; and in which these three variables are distinct." (This is the axiom of regularity in $\mathrm{L_1Set}$ in normalized notation.) Here we could also write down just one copy of each such axiom and generate the rest using Gen, the axiom of specialization, and MP.

The axioms of induction and replacement are shown to be enumerable using the same procedure as in the case of the basis tautologies and the quantifier axioms. We leave the details to the reader.

6 The Arithmetical Hierarchy

6.1. Using recursion on n, we define the classes Σ_n and Π_n of subsets of $(\mathbf{Z}^+)^m, m = 0, 1, 2, \ldots$, as follows:

(a) $\Sigma_0 = \Pi_0 = \{\text{decidable sets}\}$.
(b) $\Sigma_{n+1} = \{$projections of elements of Π_n having codimension $\geqslant 1.\}$
(c) $\Pi_{n+1} = \{$complements of elements of Σ_{n+1} in their ambient spaces $(\mathbf{Z}^+)^m\}$.

Obviously, Σ_1 consists of all enumerable sets (see Theorem 1.2 of Chapter VI), and Π_1 consists of their complements. The following result justifies calling $\{\Sigma_n, \Pi_n\}$ "the arithmetical hierarchy."

6.2. **Proposition.**

(a) $\forall n \geqslant 0, \Sigma_n \cup \Pi_n \subset \Sigma_{n+1} \cap \Pi_{n+1}$.
(b) $\bigcup_{n=0}^{\infty} \Sigma_n = \bigcup_{n=0}^{\infty} \Pi_n = \{$*arithmetical sets*$\}$, *i.e., all sets definable by formulas in* $\mathrm{L_1Ar}$.
(c) *For* $n \geqslant 1$ *the sets in* Σ_n *are precisely those that can be defined by formulas of the following* $\mathfrak{L}_1$ *type (where the quantifiers are taken over variables in*

$\mathbf{Z}^+$, *and E is a decidable set*):

$$\exists x_1 \, \forall x_2 \, \exists x_3 \cdots \forall x_n \neg(\langle x_1, \ldots, x_n, x_{n+1}, \ldots, x_m \rangle \in E), \quad n \text{ even};$$
$$\exists x_1 \, \forall x_2 \, \exists x_3 \cdots \exists x_n (\langle x_1, \ldots, x_n, x_{n+1}, \ldots, x_m \rangle \in E), \quad n \text{ odd}.$$

Similarly, for Π_n:

$$\forall x_1 \, \exists x_2 \, \forall x_3 \cdots \exists x_n (\langle x_1, \ldots, x_n, x_{n+1}, \ldots, x_m \rangle \in E), \quad n \text{ even};$$
$$\forall x_1 \, \exists x_2 \, \forall x_3 \cdots \forall x_n \neg(\langle x_1, \ldots, x_n, x_{n+1}, \ldots, x_m \rangle \in E), \quad n \text{ odd}.$$

(d) *The sets in Σ_n and Π_n are definable by the analogous formulas in* $\mathrm{L_1Ar}$ *with the following changes: instead of $\langle x_1, \ldots, x_m \rangle \in E$ we have any atomic formula, and the number of quantifiers is $\geqslant n$, with exactly $n-1$ alternations from $\exists$ to $\forall$ or $\forall$ to $\exists$.*

PROOF.

(a) We use induction on n. For $n = 0$ we have $\Sigma_0 \cup \Pi_0 = \Sigma_1 \cap \Pi_1$, by the definition of decidable sets. If $\Sigma_{n-1} \subset \Sigma_n$, then $\Sigma_n \subset \Sigma_{n+1}$ (since Σ_{n+1} consists of projections of the complements of elements of Σ_n, and Σ_n consists of projections of the complements of elements of Σ_{n-1}), and also $\Pi_n \subset \Pi_{n+1}$, by the definition of Π. Finally, we have $\Pi_n \subset \Sigma_{n+1}$, from which it trivially follows that $\Sigma_n \subset \Pi_{n+1}$. In fact, if $E \in \Pi_n$, then $E \times \mathbf{Z}^+ \in \Pi_n$ (since taking the product with $\mathbf{Z}^+$ commutes with complements and projections, and takes $\Sigma_0 = \Pi_0$ to itself), and hence $E =$ a projection of $E \times \mathbf{Z}^+ \in \Sigma_{n+1}$.

(b) It follows from (a) that $\bigcup_{n=0}^{\infty} \Sigma_n = \bigcup_{n=0}^{\infty} \Pi_n$. This class of sets is contained in the arithmetical sets, since all enumerable sets are arithmetical, and arithmeticality is preserved on taking projections and complements, which correspond to inserting $\exists$ and $\neg$, respectively.

In order to prove the converse {arithmetical sets} $\subset \bigcup_{n=0}^{\infty} \Sigma_n = \Sigma_\infty$, we first note that all sets definable by atomic formulas are decidable, and the rest of the arithmetical sets are obtained from them by taking projections, complements, unions, and intersections (see §2 of Chapter II). Thus, it suffices to show that Σ_∞ is closed with respect to (finite) unions and intersections. We claim that this is actually true for each Σ_n separately.

We prove this by induction on n. The result has already been proved for Σ_0. If Σ_n is closed with respect to $\cap$, then Π_n is closed with respect to $\cup$. Suppose $E_1, E_2 \in \Sigma_{n+1}, E_i =$ a projection of F_i, and $F_i \in \Pi_n$. We can then introduce dummy variables so as to identify the ambient spaces of the F_i, and the projection of these spaces onto an ambient space for both the E_i. Then $E_1 \cup E_2 =$ a projection of $F_1 \cup F_2$, so that $E_1 \cup E_2 \in \Sigma_{n+1}$. Thus, Σ_{n+1} is closed with respect to $\cup$.

Similarly, if Σ_n is closed with respect to $\cup$, it follows that Π_n is closed with respect to $\cap$, and an analogous argument shows that Σ_{n+1} is closed with respect to $\cap$. However, here we must embed the products $F_1 \times (\mathbf{Z}^+)^{m_2}$ and $(\mathbf{Z}^+)^{m_1} \times F_2$ for certain m_1 and m_2 in a single space in such a way that when we identify the two projections, we have $\mathrm{pr}(F_1 \times (\mathbf{Z}^+)^{m_2} \cap (\mathbf{Z}^+)^{m_1} \times F_2) = \mathrm{pr}\, F_1 \cap \mathrm{pr}\, F_2$. In terms of formulas this means that the variables bound by the $\exists$ quantifiers in the formulas corresponding to F_1 and F_2 must be renamed so that they form two disjoint sets.

(c) This assertion is proved by induction on n and a simple examination of the definitions. Here, whenever we take the complement, we must move the corresponding $\neg$ to the right of all the quantifiers by means of the usual commutation rule $\neg\forall = \exists\neg, \neg\exists = \forall\neg$. If we have a projection of codimension $m \geqslant 2$, which is defined by a series of quantifiers $\exists x_{i_1} \cdots \exists x_{i_m}$, we must reduce it to a projection of codimension 1 by replacing the set of variables $\langle x_{i_1}, \ldots, x_{i_m}\rangle$ by $\langle t_1^{(m)}(y), \ldots, t_m^{(m)}(y)\rangle$ in E and replacing the series of quantifiers by $\exists y$.

(d) The proof is analogous to that of (c). Here we use the fact that the sets in Σ_0 are Diophantine, and we observe that in general, $\exists \cdots \exists$ cannot be replaced by $\exists$ in this case.

The proposition is proved. □

6.3. **Theorem.** *For all* $n \geqslant 1$,

$$\Sigma_n \backslash \Pi_n \neq \varnothing, \quad \Pi_n \backslash \Sigma_n \neq \varnothing.$$

PROOF. The assertion that $\Sigma_1 \backslash \Pi_1 \neq \varnothing$ is precisely Theorem 5.8 of Chapter V on the existence of undecidable enumerable sets. We prove the general case by an analogous diagonal process applied to a versal family.

Let $\{E_k\}$ be a versal family of enumerable $(n+1)$-sets over $\mathbf{Z}^+$, and let E be its total space:

$$\langle k, x_0, \ldots, x_n\rangle \in E \Leftrightarrow \langle x_0, \ldots, x_n\rangle \in E_k.$$

To fix ideas, suppose n is even. We set

$$F = \{k | \exists x_1 \, \forall x_2 \cdots \forall x_n \neg(\langle k, k, x_1, \ldots, x_n\rangle \in E)\} \subset \mathbf{Z}^+.$$

By 6.2(c), we have $F \in \Sigma_n$. Since $\{E_k\}$ is versal, it follows by 6.2(c) that any subset of $\mathbf{Z}^+$ in Π_n can be represented in the form

$$F_{k_0} = \{x_0 \,|\neg \exists x_1 \, \forall x_2 \cdots \forall x_n \neg(\langle k_0, x_0, x_1, \ldots, x_n\rangle \in E))\}$$

for some $k_0 \in \mathbf{Z}^+$. It is clear that k_0 lies either in $F\backslash F_{k_0}$ or in $F_{k_0}\backslash F$. Hence $F \neq F_{k_0}$, and $F \in \Sigma_n \backslash \Pi_n$.

The other cases are handled analogously. □

6.4. *Remarks*

(a) From the point of view of the theorems of Tarski and Gödel, the results in 6.2 and 6.3 show us the tremendous distance from provability to truth: $D \in \Sigma_1$, while T falls not only outside Σ_1, but even outside Σ_∞. In the next section we indicate some mileposts along the way from D to T.

(b) Although not really formally justified by the above considerations, nevertheless it makes sense to classify arithmetic problems, i.e., questions "Is it true that $P \in T$?" according to the number of alternations between $\exists$ and $\forall$ when the closed formula P is written as in 6.2(c).

As we showed in §1 of Chapter I, the Fermat conjecture is expressed by a Π_1-formula, and the Riemann hypothesis is expressed by a Π_3-formula, although there is an assertion of type Π_1 that is equivalent to the RH.

H. Rogers writes that

> Almost all statements which (i) have been extensively studied by mathematicians and (ii) are known to be arithmetically expressible can be seen, from a relatively superficial examination, to have quite low level in the Σ_n classification. As has been occasionally remarked, the human mind seems limited in its ability to understand and visualize beyond four or five alternations of quantifier. Indeed, it can be argued that the inventions, subtheories, and central lemmas of various parts of mathematics are devices for assisting the mind in dealing with one or two additional alternations of quantifier.

7 Productivity of Arithmetical Truth

7.1. In this section we discuss a final feature of Gödel's theorem: the possibility, starting from any enumerable set of truths of arithmetic that we already know, effectively to enlarge this set by adding new truths. To see this more clearly, we examine the original version of the proof, in which the diagonal method is explicit, rather than hidden in the construction of an undecidable enumerable set. It is convenient to describe this version by comparing it with the proof of Tarski's theorem.

7.2. Suppose we are given a language of arithmetic ($\mathrm{L_1Ar}$, SAr, or an extension of one of them). Further suppose that we have chosen a fixed numbering of its alphabet, which determines a fixed numbering N of the formulas. (It is essential to note that the construction that follows is not invariant if we replace our numbering by an equivalent one.)

Both the Tarski and the Gödel arguments are based on the following "self-reference lemma":

7.3. **Lemma.** *Given any formula $P(x)$ in the language that has one free variable, we can effectively construct a closed formula Q_P that says, "my number does not belong to the set defined by P." In other words, Q_P is true if and only if $P(\overline{N}(Q_P))$ is false, where $\overline{N}(Q_P)$ is the term-name for $N(Q_P)$.*

PROOF. This lemma was proved for SAr in §11 of Chapter II. In $\mathrm{L_1Ar}$ we construct the formula Q_P as follows.

If $R(x)$ is a formula with one free variable, we call the formula $R(\overline{N}(R(x)))$ its *diagonalization*. Let diag : $\mathbf{Z}^+ \to \mathbf{Z}^+$ be the partial function

$$\text{the } N-\text{number of a formula with one free variable} \mapsto \text{the } N\text{-number of its diagonalization.}$$

It is easy to show, using the results and methods in §4, that diag is computable. Thus, its graph is definable by a formula in $\mathrm{L_1Ar}$ that can be explicitly constructed. We denote this formula by "$y =$ diag x," construct the formula $R(x) : \exists y($"$y =$ diag x"$\wedge P(y))$, and finally set

$$Q_p : \quad \neg R(\overline{N}(\neg R(x))) = \text{the diagonalization of } \neg R(x).$$

By the definitions, we then have

$$\begin{aligned} Q_p \text{ is true} &\Leftrightarrow \text{the number of } \neg R(x) \text{ does not satisfy } R(x) \\ &\Leftrightarrow \text{the number of diagonalization of } \neg R(x) \\ &\qquad \text{does not satisfy } P(x) \\ &\Leftrightarrow \text{the number of } Q_p \text{ does not satisfy } P(x). \end{aligned}$$

The lemma is proved. □

We note that it requires a large amount of technical work to verify that "$y =$diag x" is definable in $\mathrm{L_1Ar}$, which is why we used SAr instead in Chapter II.

7.4. The arguments of Tarski and Gödel now take the following parallel form:
Tarski:

(a) Suppose that truth is definable by a formula P.
(b) Then there is a formula Q_P that says "I am not true."
(c) The formula Q_P cannot be false (because of its semantics).
(d) The formula Q_P cannot be true (because of its semantics).
(e) Therefore, truth is not definable.

Gödel:

(a) Provability is definable by a formula P.
(b) There is a formula Q_P that says "I am not provable."
(c) The formula Q_P cannot be false (because of its semantics, since otherwise it would be provable, and hence true).
(d) Therefore, Q_P is true.
(e) Therefore, Q_P is not provable (because of its semantics).

We note that in the above paraphrasing of Gödel's argument, part (c) explicitly uses the stipulation that only true formulas are provable. When Gödel's paper appeared in 1931, specialists were very busy looking for finitistic proofs that the axioms of arithmetic are consistent, so that stipulating that $D \subset T$ would have run counter to the spirit of the times. Therefore, in Gödel's own original wording the argument looks somewhat different. This distinction is traditionally explained in great detail in all textbooks on logic. However, we shall be satisfied with remarking that if $D \not\subset T$, then $D \neq T$, and the incompleteness theorem is trivially true. But in that case we would be in such bad shape that we would no longer care about completeness or incompleteness.

The main point we are interested in is the following: given any fixed conception of provability that leads to an enumerable (or even to an arithmetical) set D of provable true formulas, we can effectively construct a new formula that is true but not provable. We now define more precisely what we mean by "effectively."

7.5. **Definition.** A set $F \subset \mathbf{Z}^+$ is said to be *productive* relative to a versal family $\{E_k\}$ of 1-sets if there exists a partial recursive function f such that for all $k \in \mathbf{Z}^+$ with $E_k \subset F$, we have $k \in D(f)$ and $f(k) \in F \backslash E_k$.

7.6. **Proposition.** *Under the conditions in 7.2, the set of numbers of true formulas is productive relative to the versal family $\{E_k\}$ constructed in §8 of Chapter VI.*

PROOF. To fix ideas, we shall work with the language $\mathrm{L_1Ar}$. We first construct an enumerable family $\{P_k(x_1)\}$ of formulas with one free variable x_1 such that P_k defines E_k. To do this, we define a sequence of terms $\bar{f}[k]$ in $\mathrm{L_1Ar}$ as in 8.1(a) of §8, Chapter VI, by setting

$$\begin{aligned}
&\bar{f}[4k] = \bar{k} = +(\cdots + (\bar{1}, \bar{1}) \cdots), \qquad k \text{ times;}\\
&\bar{f}[4k+1] = x_{k+1} = \text{the } (k+1)\text{st variable in } \mathrm{L_1Ar};\\
&\bar{f}[4k+2] = +(\bar{f}[t_1(k)], \bar{f}[t_2(k)]);\\
&\bar{f}[4k+3] = \cdot(\bar{f}[t_1(k)], \bar{f}[t_2(k)]).
\end{aligned}$$

We then write

$$P_k = \exists x_2 (\exists x_3 \cdots (\exists x_k (\bar{f}[t_1(k)] = \bar{f}[t_2(k)])) \cdots).$$

It is easy to see, using the methods in §4, that the function $k \mapsto N(P_k)$ is recursive. We next fix a translation of "$y = \text{diag } x$" and set

$$\begin{aligned}
R_k &= \exists x_{k+1} (\text{"}x_{k+1} = \text{diag } x_1\text{"}) \wedge (P_k(x_1))),\\
Q_{P_k} &= \neg(R_k(\overline{N}(\neg(R_k)))),
\end{aligned}$$

and finally

$$f(k) = N(Q_{P_k}).$$

This function is computable because $N(P_k)$ is computable. By Lemma 7.3, it satisfies the condition 7.5 with T in place of F. □

7.7. The concept of productivity gives us the following approach to the problem of exhausting T: we begin with the set D_0 of formulas that are provable in the Peano axiom system Ax_0, we define D_0 by a formula P_0; we set $\mathrm{Ax}_1 = \mathrm{Ax}_0 \cup \{Q_{P_0}\}$; and we similarly construct D_1, P_1, and $\mathrm{Ax}_2 = \mathrm{Ax}_1 \cup \{Q_{P_1}\}$, and so on. It follows from Gödel's theorem that as long as we do all this "uniformly effectively," we cannot obtain all of T even after transfinitely many steps. However, S. Feferman has shown that if we are willing to dispense with effectiveness, we can obtain all of TAr in this way. We conclude this section by formulating Feferman's result, which gives unexpected and philosophically interesting information about TAr. We omit the proof and the technical details (see Feferman's original article "Transfinite recursive progressions of axiomatic theories," *J. Symb. Logic* 27, no. 3 (1962), 259–316).

7.8. *Principles of extension.* In the first place, in order to exhaust TAr it is not enough to add Gödel's formula to Ax_i at every step. There are many other ways of constructing intuitively true formulas that in various ways formalize "having faith in the axioms Ax_i."

Feferman, in particular, uses the following construction. Suppose that we have already constructed the axiom system Ax_α (where α is an ordinal), and that the set of numbers of formulas deducible from Ax_α is defined by the formula D_α. For any formula $P(x)$ with one free variable, we construct a formula B_α^P that has the intuitive meaning "if $P(\bar{n})$ is provable (from Ax_α) for all term-names $\bar{n}$ of natural numbers, then $\forall x P(x)$ is true." These formulas B_α^P must lie in T, and we can set

$$\mathrm{Ax}_{\alpha+1} = \mathrm{Ax}_\alpha \cup \{B_\alpha^P | \text{all } P\};$$

$$\mathrm{Ax}_\beta = \bigcup_{\alpha<\beta} \mathrm{Ax}_\alpha, \quad \text{if } \beta \text{ is a limit ordinal.}$$

Here is a method for giving B_α^P explicitly. The function $n \mapsto N(P(\bar{n}))$ is computable as a function of n and $N(P)$. We define its graph by a formula $M(x, y, z)$, so that for $l, m, n \in \mathbf{Z}^+$,

$$M(\bar{l}, \bar{m}, \bar{n}) \text{ is true} \Leftrightarrow \begin{cases} l \text{ is the number of a formula } P \text{ with one free} \\ \text{variable } x, \text{and } m \text{ is the number of } P(\bar{n}). \end{cases}$$

We then set

$$B_\alpha^P = \forall y\, \forall z (M(\overline{N}(P), y, z) \Rightarrow D_\alpha(y)) \Rightarrow \forall x\ P(x).$$

7.9. *The problem of choosing D_α.* This is the subtlest part of the proof. Here it is crucial to show that D_β exists when β is a limit ordinal.

Feferman shows how the D_α a can be constructed for a suitable countable sequence of ordinals with limit γ not exceeding ${\omega_0}^{{\omega_0}^{\omega_0}}$ so that the following result will be true.

7.10. **Theorem.** *All true formulas in* $\mathrm{L_1 Ar}$ *are deducible from* $\cup_{\alpha<\gamma} \mathrm{Ax}_\alpha$.

Thus, suppose we have accepted the Peano axioms. Then, in order to attain the total truth in arithmetic, we must perform a transfinite sequence of acts of faith in our not having been led astray by the previous acts of faith.

8 On the Length of Proofs

8.1. The title of this section is taken from a short paper written by Gödel in 1936. His article consists of a precise formulation and proof of the following qualitative assertions.

Suppose we are given a formal language L together with some conception of deducibility of a formula P from a (variable) set of formulas $\mathfrak{a}$. Suppose, in addition, that we are actually given a function that estimates the "complexity of deduction" of a formula P from the set $\mathfrak{a}$. (In languages of $\mathfrak{L}_1$, this "complexity" could be the *minimal size of a deduction of P from* $\mathfrak{a}$, i.e., the number of signs of a fixed finite protoalphabet needed for such a deduction; note that the use of the word "complexity" here has nothing to do with the Kolmogorov complexity in §9 of Chapter VI.) We further assume that L contains a certain fragment of the logic of $\mathfrak{L}_1$, that L and $\mathfrak{a}$ are rich enough for the incompleteness principles to take effect, and that the "complexity of deduction" satisfies certain natural axioms. We then have the following facts:

(a) *There exist formulas deducible from* $\mathfrak{a}$ *whose deduction is arbitrarily more complex than the formula itself.*

Observation shows that this somewhat vaguely defined class includes, if not the most important, at least the most "prized" mathematical facts.

(b) *If we add any independent formula A to the axioms* $\mathfrak{a}$, *then we can find formulas deducible from* $\mathfrak{a}$ *whose deduction from* $\mathfrak{a} \cup \{A\}$ *is arbitrarily less complex than from* $\mathfrak{a}$ (*the principle of cutting down proofs*).

Compare with the great strength of "analytic" methods in comparison with "elementary"" methods in number theory.

The following more precise presentation of these ideas is based on a short article by Ehrenfeucht and Mycielski in *Bull. Amer. Math. Soc.* 17, No. 3 (1971), 366–367.

8.2. We consider the following set of data.

(a) A *countable alphabet* A with a fixed numbering $N : A \to \mathbf{Z}^+$.
(b) A subset $F \subset S(A)$ whose elements are called *formulas.*
(c) A partial function $\mathcal{D} : \mathcal{P}(F) \to \mathcal{P}(F)$ that to certain subsets $\mathfrak{a} \subset F$ associates sets $\mathcal{D}(\mathfrak{a})$ of formulas "*deducible from* $\mathfrak{a}$." We shall often write $\mathfrak{a} \vdash P$ instead of $P \in \mathcal{D}(\mathfrak{a})$.
(d) The *complexity of deduction*: this is a function $\mathrm{Cd}_{\mathfrak{a}}(P)$ that is defined for pairs $\mathcal{D} \subset F, P \in \mathcal{D}(\mathfrak{a})$, and takes values in $\mathbf{Z}^+$. It is convenient to take $\mathrm{Cd}_{\mathfrak{a}}(P) = \infty$ if $P \notin \mathcal{D}(\mathfrak{a})$.

We impose the following conditions on this data:

8.3. (a) *A contains* $\neg$, $\to$, (, *and*).
(b) *If P and $Q \in F$, then $\neg(P)$ and $(P) \to (Q) \in F$.* As usual, we shall write $P \to Q$ instead of $(P) \to (Q)$, and so on.
(c_1) $\mathfrak{a} \subset \mathcal{D}(\mathfrak{a})$; *if* $\mathfrak{a} \subset \mathfrak{a}'$ *and $\mathcal{D}$ is defined at* $\mathfrak{a}$, *then $\mathcal{D}$ is defined at* $\mathfrak{a}'$ *and* $\mathcal{D}(\mathfrak{a}) \subset \mathcal{D}(\mathfrak{a}')$.
(c_2) *If* $\mathfrak{a} \cup \{P\} \vdash Q$, *then* $\mathfrak{a} \vdash P \to Q$.
(c_3) $\mathfrak{a} \vdash P \to (\neg P \to Q)$ for any $P, Q \in F$.
(d_0) *If* $\mathfrak{a} \subset \mathfrak{a}'$, *then* $\mathrm{Cd}_{\mathfrak{a}'}(P) \leqslant \mathrm{Cd}_{\mathfrak{a}}(P)$.
(d_1) *The set* $\{\langle P, n\rangle | \mathrm{Cd}_{\mathfrak{a}}(P) \leqslant n\} \subset S(A) \times \mathbf{Z}^+$ *is decidable.*

Condition (d_1) does not actually have to hold for all $\mathfrak{a} \subset F$, but we shall consider only those $\mathfrak{a}$ for which it is true. In the case that $\mathrm{Cd}_{\mathfrak{a}}(P)$ is the size of the shortest $\mathfrak{L}_1$-deduction of P from $\mathfrak{a}$ in a finite protoalphabet, and $\mathfrak{a}$ is a decidable set of axioms, (d_1) holds for the following reason. We can write down all the texts in A having size $\leqslant n$—there are a finite number of them—and then verify for each one in turn whether it is a deduction of P from $\mathfrak{a}$.

(d_2) *There exists a general recursive junction $f(x,y,z)$ that is nondecreasing in x such that*

$$\mathrm{Cd}_{\mathfrak{a}\cup\{P\}}(Q) \leqslant f(\mathrm{Cd}_{\mathfrak{a}}(P \to Q),\ N(P), N(Q))$$

for all $Q \in \mathcal{D}(\mathfrak{a})$.

Both sides of this inequality are finite because of the previous conditions: since $\mathfrak{a} \vdash Q$, it follows by (c_1) that $\mathfrak{a}\cup\{P\} \vdash Q$, and then by ($c_2$) that $\mathfrak{a} \vdash P \to Q$. We have an estimate of the type in (d_2) in languages of $\mathfrak{L}_1$, because, starting with any deduction of $P \to Q$ from $\mathfrak{a}$, we can obtain a deduction of Q from $\mathfrak{a} \cup \{P\}$ by simply adding P and Q (by modus ponens). This increases the size of the deduction of $P \to Q$ by the sizes of P and Q.

(d_3) *There exists a general recursive function $g(x,y)$ such that*

$$\mathrm{Cd}_{\mathfrak{a}}(P \to (\neg P \to Q)) \leqslant g(N(P), N(Q)).$$

In languages of $\mathfrak{L}_1$, the formula $P \to (\neg P \to Q)$ is a logical axiom, and if $\mathfrak{a}$ contains this axiom, then the deduction has length 1 and size equal to the size of the formula itself. Of course, the size of this formula can be represented in the form $g(N(P), N(Q))$.

We now formulate Gödel's theorem on "cutting down proofs." We suppose that the conditions and conventions in 8.2–8.3 are fulfilled.

8.4. Theorem.

(a) *Suppose that $\mathfrak{a} \subset F$ and $\mathcal{D}(\mathfrak{a})$ is undecidable. Then for any general recursive function I there exist infinitely many formulas $P \in \mathcal{D}(\mathfrak{a}))$ such that*

$$\mathrm{Cd}_{\mathfrak{a}}(P) > l(N(P)).$$

(b) *Suppose that $\mathfrak{a}' = \mathfrak{a} \cup \{A\}$ and the formula A has the property that $\mathcal{D}(\mathfrak{a} \cup \{\neg A\})$ is undecidable. Then for any general recursive function r there exist infinitely many formulas $P \in \mathcal{D}(\mathfrak{a})$ such that*

$$\mathrm{Cd}_{\mathfrak{a}}(P) > r(\mathrm{Cd}_{\mathfrak{a}'}(P)).$$

PROOF.

(a) If the first assertion were false, then for a suitable l and for all $P \in \mathcal{D}(\mathfrak{a})$ we would have $\mathrm{Cd}_{\mathfrak{a}}(P) \leqslant l(N(P))$. But then the set

$$\mathcal{D}(\mathfrak{a}) = \{P | \mathrm{Cd}_R(P) \leqslant l(N(P))\} \subset S(A)$$

would be decidable by (d_1), since it is obtained by applying a bounded universal quantifier (in n) to the decidable set in (d_1). This contradicts the assumption.

(b) Let $P \in \mathcal{D}(\mathfrak{a} \cup \{\neg A\})$. By ($d_2$) we have

$$\mathrm{Cd}_{\mathfrak{a}\cup\{\neg A\}}(P) \leqslant f(\mathrm{Cd}_{\mathfrak{a}}(\neg A \to P),\ N(\neg A), N(P)).$$

If we now suppose that the second assertion of the theorem were false, then for a suitable nondecreasing general recursive function r we would obtain:

$$(\mathrm{Cd}_{\mathfrak{a}}(\neg A \to P) \leqslant r(\mathrm{Cd}_{\mathfrak{a}'}(\neg A \to P)),$$

or, by (d_2) and (d_3),

$$\begin{aligned}(\mathrm{Cd}_{\mathfrak{a}}(\neg A \to P) &\leqslant r \circ f(\mathrm{Cd}_{\mathfrak{a}}(A \to (\neg A \to P)), N(A), N(P))\\ &\leqslant r \circ f(g(N(A), N(P)), N(A), N(P)).\end{aligned}$$

Substituting this in the above inequality for $\mathrm{Cd}_{\mathfrak{a}\cup\{\neg A\}}(P)$, for fixed A we obtain an estimate of the form

$$\mathrm{Cd}_{\mathfrak{a}\cup\{\neg A\}}(P) \leqslant l(N(P)),$$

where l is general recursive and $P \in \mathcal{D}(\mathfrak{a} \cup \{\neg A\})$. But this contradicts the assumption that $\mathcal{D}(\mathfrak{a} \cup \{\neg A\})$ is undecidable by the first assertion of the theorem. □

VIII

Recursive Groups

1 Basic Result and Its Corollaries

1.1. We consider a countable "group alphabet"

$$A = \{a_1, a_2, \ldots; a_1^{-1}, a_2^{-1}, \ldots\}.$$

The expressions in the alphabet A, including the empty expression $\varnothing$, are traditionally called *words*. The word $a_i \cdots a_i$ ($m \geqslant 1$ times) will be written a_i^m; the word $a_i^{-1} \cdots a_i^{-1}$ ($m \geqslant 1$ times) will be written a_i^{-m}; and we agree to take $a_i^0 = \varnothing$. We call a word $a_{i_1}^{m_1} \cdots a_{i_r}^{m_r}$ *reduced* if either it is empty or there are no subwords of the form $a_i^{-1}a_i$ or $a_i a_i^{-1}$ when it is written in expanded form.

The operation of "joining and reducing" (by "reducing" we mean crossing out all subwords of the form $a_i a_i^{-1}$ or $a_i^{-1} a_i$) defines a group structure with unit $\varnothing$ (which we sometimes denote by 1) on the set of reduced words. This is a free group F with a countable set of generators $\{a_1, \ldots, a_n, \ldots\}$. We can also consider nonreduced words as elements in F: we identify such a word with the word obtained by reducing it.

We have a canonical numbering on $A : N(a_i) = 2i, N(a_i^{-1}) = 2i - 1$. All properties related to the computability of operations and the enumerability of subsets in A and $S(A)$ will be considered relative to any numbering of A equivalent to N and any numbering of $S(A)$ compatible with N (see the definitions in §1 of Chapter VII). We shall continually be making use of the following facts.

1.2. **Lemma.**

(a) *The set F of reduced words is decidable.*
(b) *The group operations in F are computable.*
(c) *A subgroup $G \subset F$ in enumerable in $S(A)$ if and only if it has an enumerable set of generators.*
(d) *A normal subgroup $H \subset G$ in an enumerable subgroup $G \subset F$ is enumerable if and only if it is generated as a normal subgroup by an enumerable set.*
(e) *A homomorphism $F \to F$ is recursive if and only if the induced map $\{a_1, \ldots, a_n, \ldots\} \to F$ is recursive.*

Yu. I. Manin, *A Course in Mathematical Logic for Mathematicians, Second Edition*, Graduate Texts in Mathematics 53, DOI 10.1007/978-1-4419-0615-1_8,

The proof is a good exercise in using the techniques of Chapter VII, and we leave it to the reader. It is convenient to begin by showing that the operation of reducing is computable; the rest goes through more or less automatically.

1.3. **Definition.** A group is called *recursive* if it is isomorphic to a quotient group of the form G/H, where $G \subset F$ is an enumerable subgroup and $H \subset G$ is an enumerable normal subgroup.

Here we could limit ourselves to subgroups $G \subset F$ that are generated by an enumerable subset of the standard generators $\{a_1, \ldots, a_n, \ldots\}$.

1.4. REMARKS AND EXAMPLES.

(a) Recursive groups have at most countably many elements.
(b) *Finitely presented* (f.p.) groups, i.e., those that have a finite number of generators and relations, are recursive. In particular, finite groups and *finitely generated* (f.g.) abelian groups are recursive.
(c) A subgroup H of an f.p. group G is not necessarily f.p. (or even f.g.). But *if it is finitely generated, then it is recursive.*

In fact, let $\{h_1, \ldots, h_m\}$ be generators of H. We add generators $\{h_{m+1}, \ldots, h_n\}$ of the group G that are connected by a finite number of relations, and we define a homomorphism $\phi : F \to G$ by setting $\phi(a_i) = h_i$ if $i \leqslant n$ and $\phi(a_i) = 1$ if $i > n$. The kernel E of ϕ is generated by a finite number of relations between $a_1, \ldots, a_n$ and by the set $\{a_{n+1}, a_{n+2}, \ldots\}$. Hence E is enumerable by Lemma 1.2(d). The subgroup $H \subset F$ generated by $a_1, \ldots, a_m$ is also enumerable, by Lemma 1.2(c). Therefore the set $\overline{H} \cap E$ is enumerable. But ϕ induces an isomorphism $\overline{H}/\overline{H} \cap E \Rightarrow H$. Consequently, H is recursive. □

The basic aim of this chapter is to prove the following remarkable theorem of Higman, which gives the converse of the simple assertion 1.4(c). (G. Higman, Subgroups of finitely presented groups, *Proc. Royal Soc.*, Ser. A, vol. 262 (1961), 455–475.)

1.5. **Theorem.**

(a) *Any recursive group G/H (in the notation of* 1.3) *can be embedded in a suitable* f.p. *group F/N.*
(b) *This embedding can be made effective, i.e., it can be induced by a suitable recursive map $G \to F$.*

Here are some applications of this theorem.

1.6. **Corollary** (Universal finitely presented groups). *There exists an* f.p. *group U such that any* f.p. *group G can be embedded in U (and hence, any recursive group can be embedded in U).*

In fact, any f.p. group is isomorphic to the quotient of F by a normal subgroup that is generated by a finite set of reduced words in F and by all a_i with $i \geqslant n$ for some n. We let $I \subset S(S(A)) \times \mathbf{Z}^+$ be the decidable set of pairs $\langle$a finite sequence of reduced words, $n\rangle$, and we let N_i (for $i \in I$) denote the corresponding normal subgroup. We construct the "doubly infinite" group alphabet

$\{a_{jk}, a_{jk}^{-1} | j, k \geqslant 1\}$, we identify I with $\mathbf{Z}^+$ by choosing a recursive numbering of I, and we define the group U_0 that has generators $\{a_{jk}\}$ and relations "N_j, written in the alphabet $\{a_{j1}, a_{j2}, \ldots\}$." It is clear that U_0 is recursive. It will also be clear from the results in the next section that U_0 is the free product of all the groups F/N_j, so that any f.p. group can be embedded in U_0. Thus, any f.p. group U in which we can embed U_0, using Higman's theorem, is universal. □

In M.K. Valiyev's article Examples of universal finitely presented groups, *Dok. AN SSSR*, 1973, vol. 211, no. 2, a universal group U is constructed that has 14 generators and 42 relations, and it is mentioned that such a group can be constructed with only 2 generators and 27 relations.

1.7. *F.p. groups with algorithmically undecidable word problem.*

Let G be the group with four generators a, b, c, d, and with the relations

$$b^{-m}ab^m = d^{-m}cd^m, \quad \text{for all } m \in E,$$

where $E \subset \mathbf{Z}^+$ is an undecidable enumerable set. It easily follows from the results in §2 that the equation

$$b^{-x}ab^x = d^{-x}cd^x$$

holds in G *only* if $x \in E$. (In fact, the elements $b^{-m}ab^m$ for $m \geqslant 1$ generate a free subgroup of G, so that G contains the free product of the subgroups generated by $\{b^{-x}ab^x | x \geqslant 1\}$ and by $\{d^{-x}cd^x | x \geqslant 1\}$ with amalgamation $\{b^{-x}ab^x = d^{-x}cd^x | x \in E\}$.) Hence, the question whether the equation $b^{-x}ab^x = d^{-x}cd^x$ holds is undecidable (as a mass problem indexed by x), and if we embed G effectively in an f.p. group, we may conclude that the word problem is unsolvable in this f.p. group.

The existence of such groups was first established by P.S. Novikov and W. Boone.

1.8. "*Natural*" *recursive groups.* In algebraic geometry over algebraic number fields, we find many examples of recursive groups that are not a priori finitely presented. We shall limit ourselves to one typical example.

Let $\mathcal{O}_n(\mathbf{Q})$ be the orthogonal group of automorphisms of an n-dimensional linear space L (over the rational numbers $\mathbf{Q}$) together with a quadratic form f. Let b be the corresponding bilinear form. The symmetry $\tau_x \in \mathcal{O}_n(\mathbf{Q})$ is defined for any vector $x \in L$ with $f(x) \neq 0$:

$$\tau_x(y) = y - \frac{b(x,y)}{f(x)}x$$

for all $y \in L$. The involutions $\tau_x \in \mathcal{O}_n(\mathbf{Q})$ give us an enumerable system of generators of $\mathcal{O}_n(\mathbf{Q})$, and all the relations are generated by the enumerable (indeed, decidable) system of relations

$$\tau_x^2 = 1, \qquad (\tau_x\tau_y\tau_z)^2 = 1, \quad \text{for all coplanar } \{x, y, z\}$$

(S. Becken).

The numbering of $L \cong \mathbf{Q}^n$ implicit here is taken to be compatible with any numbering of $\mathbf{Q}$ that is compatible with the standard numbering of $\mathbf{Z}^+$ and in which the field operations are computable.

1.9. Higman's theorem is related to the theorem that enumerable sets are Diophantine (Chapter VI), although it was first proved earlier than the latter result. Perhaps both facts are special cases of some general assertion about recursive algebraic structures.

In any case, the theorem on the Diophantine nature of enumerable sets can be used to simplify considerably the recursion-theoretic part of Higman's proof. This was shown by Valiyev, whose construction will be given in §§5–6 (cf. *Algebra i Logika,* vol. 7, No. 3 (1968)). §§2–4 will be devoted to the group-theoretic preliminaries; here we shall follow Higman.

2 Free Products and HNN-Extensions

2.1. Suppose we are given a family of groups $(G_i), i \in I$, and a family of group homomorphisms $\alpha_i : A \to G_i$. We consider the class of families (H, β_i) of homomorphisms $\beta_i : G_i \to H$ such that $\beta_i \circ \alpha_i : A \to H$ does not depend on $i \in I$. This class contains a *universal family* $\phi_i : G_i \to *_A G_k$ that is unique up to isomorphism: any other family (H, β_i) uniquely determines and is uniquely determined by the homomorphism $\gamma : *_A G_k \to H$ for which $\beta_i = \gamma \circ \phi_i$.

In what follows we shall need only the case in which all the α_i are embeddings. In this case $*_A G_k$ is called the *free product of the groups G_i with amalgamated subgroups $\alpha_i(A) \subset G_i$*. We shall generally denote the structure maps $G_i \to *_A G_k$ by ϕ_i, perhaps with additional indices. We let ϕ denote the structure homomorphism $\phi_i \circ \alpha_i : A \to *_A G_k$, which does not depend on i. If $A = \{1\}$, we write simply $*G_i$ instead of $*_A G_i$; if the set of indices is $\{1, \dots, n\}$, we write $G_1 * \cdots * G_n$, and so on. We shall continually be making use of the following structure lemma.

Let $\alpha_i : A \to G_i$ be embeddings, and let $S_i \subset G_i$, be subsets such that

$$G_i \backslash \alpha_i(A) = \bigcup_{s \in S_i} \alpha_i(A)s, \quad \text{and}$$

$$\alpha_i(A)s_1 \neq \alpha_i(A)s_2, \quad \text{for } s_1 \neq s_2 \in S_i.$$

2.2. **Proposition.** *Any element in the group $*_A G_i$ can be uniquely represented in the form*

$$\phi(a)\phi_{i_1}(s_1) \cdots \phi_{i_n}(s_n),$$

where $a \in A, s_k \in S_{i_k}, i_j \neq i_{j+1}$ for all j, and $n \geqslant 0$ depends on the element.

We shall call this the canonical expansion of an element.

For the proof of this fact and for further details, see, for example, Serre's lecture notes *Arbres, amalgames et SL_2.*

2.3. Corollaries

(a) *Under the conditions in 2.2, the structure homomorphisms ϕ and ϕ_i are embeddings.*

This allows us to identify A and G_i with subgroups of $*_A G_i$ using ϕ and ϕ_i. We shall do this in the statements that follow. However, in the several-step constructions in the later subsections, one and the same group will be embedded in another group in many different ways using various compositions of the structure maps, and it will be necessary to keep careful track of these embeddings.

(b) $G_i \cap G_j = A (in *_A G_i) \quad for\ i \neq j.$

In other words, $\phi_i(G_i) \cap \phi_j(G_j) = \phi(A)$. We can use Proposition 2.2 to prove $\subset$: otherwise we would have $\phi_i(s_i) = \phi_j(s_j)$, which would contradict the uniqueness.

(c) *Suppose we are given a family of embeddings $\beta_i : H_i \to G_i$ and a subgroup $B \subset A$ such that $\beta_i(H_i) \cap \alpha_i(A) = \alpha_i(B)$ for all i. Then the composition*

$$B \overset{\beta_i^{-1}\circ\alpha_i}{\Longrightarrow} H_i \overset{\phi_i\circ\beta_i}{\Longrightarrow} *_A G_i$$

does not depend on i, and therefore gives a canonical map $*_B H_i \to *_A G_i$. This map is an embedding. In particular, the subgroup of $*_A G_i$ generated by $\phi_i \circ \beta_i(H_i)$ is isomorphic to $*_B H_i$.

In fact, the canonical expansion in 2.2 of an element in $*_B H_i$ goes to the canonical expansion of the image of this element in $*_A G_i$.

(d) *With the same notation, we have*

$$(*_B H_i) \cap A = B \ in\ *_A G_i;$$
$$(*_B H_i) \cap G_j = H_j \ in\ *_A G_i.$$

2.4. *Generators and relations.* Let M be a set, and let R be a subset of the free group F_M that is freely generated by M. We let $|M : R|$ denote the quotient group $F_M/\overline{R}$, where $\overline{R}$ is the smallest normal subgroup of F_M containing R. This is what we mean by defining a group by generators (M) and relations (R).

We shall take the following liberties with notation:

(a) If M has a nonempty intersection with a group that has already been defined, then all relations coming from the relations in the earlier group are assumed to be included in R, even if they are not explicitly written out. We might completely omit any reference to R if there are no other relations besides those coming from the earlier group. For example, if E and $F \subset G$ are two subgroups, then $|E \cup F|$ is the subgroup they generate in G, and so on.

(b) Instead of writing, say, $a_1 a_2^{-1}$ is in R, we may write $a_1 = a_2$.

EXAMPLE. If the $\alpha_i : A \to G_i$ are embeddings, then $*_A G_i$ is defined by the following generators and relations:

$$\left| \bigcup_{i \in I} G_i : \alpha_i(a) = \alpha_j(a) \text{ for all } a \in A, i, j \in I \right|.$$

We now introduce a construction that will be fundamental for everything that follows (G. Higman, B. Neumann, H. Neumann).

Suppose we are given two embeddings of groups $\alpha, \beta : A \to G$.

2.5. **Definition.** The HNN-extension of the group G (relative to A, α, β) is the group

$$K = |G \cup \{t\} : t^{-1}\alpha(a)t = \beta(a) \text{ for all } a \in A|.$$

2.6. **Proposition.** *The following homomorphisms are embeddings:*

(a) $G \to K : g \mapsto$ *the class of* g *modulo the relations in* K.
(b) $G *_A t^{-1}Gt \to K$, *where the free product is taken relative to the embeddings* $a \mapsto \beta(a)$ *and* $a \mapsto t^{-1}\alpha(a)t$.

PROOF. In the group $G * \{u^n\}$, the subgroup U generated by G and $u^{-1}\alpha(A)u$ is isomorphic to $G * u^{-1}\alpha(A)u$. In fact, the canonical expansion of an element in $G * u^{-1}\alpha(A)u$ has the form $g_1 u^{-1}\alpha(a_1)ug_2 \cdots g_n u^{-1}\alpha(a_n)u$, where $g_1 \in G$, $g_2, \ldots, g_n \in G\backslash\{1\}, a_1, \ldots, a_{n-1} \in A\backslash\{1\}, a_n \in A$, and so this expansion also has the canonical form in $G * \{u^n\}$.

We construct the subgroup $V = G * v\beta(A)v^{-1} \subset G * \{v^n\}$ similarly.

We identify the group $W = G * w^{-1}Aw$ with U and V by means of the isomorphisms that are the identity on G and take $w^{-1}aw$ to $u^{-1}\alpha(a)u$ and $v\beta(a)v^{-1}$, respectively.

We now consider the group $(G * \{u^n\}) *_W (G * \{v^n\})$. The group $G \subset W$ is canonically embedded in it, and for all $a \in A$ the element $t = uv$ satisfies the relation

$$t^{-1}\alpha(a)t = \beta(a),$$

because we have made the identification $u^{-1}\alpha(a)u = v\beta(a)v^{-1}$. In addition, it is clear from Proposition 2.2 that in $(G * \{u^n\}) *_w (G * \{v^n\})$ the groups $u^{-1}\ Gu$ and vGv^{-1} generate a free product with amalgamation A embedded by means of the maps $a \mapsto u^{-1}\alpha(a)u$ and $a \mapsto v\beta(a)v^{-1}$, respectively. Hence, if we conjugate by v, we see that G and $t^{-1}Gt$ also generate a free product, as described in the statement of 2.6.

Therefore, the subgroup

$$K' = |G \cup \{t = u^v\}| \subset (G * \{u^n\}) *_W (G * \{v^n\})$$

is a homomorphic image of K, and assertions (a) and (b) hold for K'. Moreover, the canonical map $K \to K'$ is an isomorphism. To see this it suffices to note that there exists an isomorphism

$$K * \{v^n\} \stackrel{\sim}{\Rightarrow} (G * \{u^n\}) *_W (G * \{v^n\})$$

that takes $t \in K$ to uv. In particular, t has infinite order in K. The proposition is proved. □

We shall need to refine and generalize this result in two directions. In the first place, we want to consider iterated HNN-extensions; in the second place, we are interested in the connection between HNN-extensions of a group and a subgroup. We now bring together all the facts we need into a single statement.

Suppose that we are given an entire family of pairs of embeddings $\alpha_i, \beta_i : A_i \to G$ $(i \in I)$ and a subgroup $H \subset G$ with the property that $\alpha_i^{-1}(\alpha_i(A_i) \cap H) = \beta_i^{-1}(\beta_i(A_i) \cap H) = B_i \subset A_i$ are subgroups. Under these conditions we have the following result.

2.7 **Proposition.** *Let*

$$K_G = \left| G \cup \{t_i | i \in I\} : t_i^{-1}\alpha_i(a)t_i = \beta_i(a) \textit{ for all } i \in I, a \in A_i \right|;$$
$$K_H = \left| H \cup \{t_i' | i \in I\} : t_i'^{-1}\alpha_i(b)t_i' = \beta_i(b) \textit{ for all } i \in I, b \in B_i \right|.$$

Then

(a) *the $\{t_i\}$ freely generate a free subgroup in K_G;*
(b) *the natural maps $G \to K_G$ and $K_H \to K_G$ (the latter given by $t_i' \mapsto t_i$) are embeddings. In addition, $K_H \cap G = H$ in K_G.*

PROOF.

(a) If the relations in K_G implied a nontrivial relation between the t_i, this relation would be preserved in the quotient of K_G by the smallest normal divisor containing G. But in this quotient the relations $t_i^{-1}\alpha_i(a)t_i = \beta_i(a)$ become trivial $(1 = 1)$, and no restrictions are imposed on the images of the t_i. This proves (a).

(b) We first consider the case that I consists of one element. In the notation used in the proof of Proposition 2.6, we consider K_G as a subgroup of $(G * \{u^n\}) *_W (G * \{v^n\})$. By Proposition 2.2, in $G * \{u^n\}$ we have

$$H * \{u^n\} \cap G * u^{-1}\alpha(A)u = H * u^{-1}\alpha(B)u,$$

and similarly, in $G * \{v^n\}$ we have

$$H * \{v^n\} \cap G * v\beta(A)v^{-1} = H * v\beta(B)v^{-1}.$$

The above identifications of U and V with W identify these intersections with the subgroup

$$W_0 = H * w^{-1}Bw \subset G * w^{-1}Aw = W.$$

By Corollary 2.3(c), we have a canonical embedding

$$(H * \{u^n\}) *_{W_0} (H * \{v^n\}) \to (G * \{u^n\}) *_W (G * \{v^n\}).$$

But as at the end of the proof of 2.6, the group on the left is $K_H * \{v^n\}$ and the group on the right is $K_G * \{v^n\}$, so we obtain an embedding $K_H \to K_G$.

Furthermore (the intersection is taken in $(G * \{u^n\}) *_W (G * \{v^n\})$):

$$(H * \{u^n\}) *_{W_0} (H * \{v^n\}) \cap G * u^{-1}\alpha(A)u = H * u^{-1}\alpha(B)u,$$

so that if we now intersect with G, we obtain H. It follows a fortiori that $K_H \cap G = H$.

We prove (b) for finite I by an easy induction on n, and then for infinite I by passing to the inductive limit (which here is a union). We leave the details to the reader. □

3 Embeddings in Groups with Two Generators

In this section we prove a result that will be used later and that shows vividly in a simple situation how the number of generators can be decreased using embeddings.

3.1. Proposition.

(a) *Any countable or finite group G can be embedded in a group with two generators.*
(b) *If G is recursive, then there is such an embedding that is recursive.*

PROOF.
(a) The group $\mathbf{Z} * \mathbf{Z} = \{b^n\} * \{v^n\}$ has a free subgroup of countable rank, for example,

$$S = \big|\{b^{-1}vb^i | i \geqslant 0\}\big|.$$

It immediately follows from Proposition 2.2 that there are no relations between the generators $b^{-i}vb^i$.

Thus, if G is a free countable group, it embeds in $\mathbf{Z} * \mathbf{Z}$. If G is not, we could try to represent G in the form F/N, where F is countable and free, then embed F in $\mathbf{Z} * \mathbf{Z}$ and consider the induced homomorphism $F/N \to \mathbf{Z} * \mathbf{Z}/N'$, where N' is the normal subgroup in $\mathbf{Z} * \mathbf{Z}$ generated by N. Unfortunately, $N' \cap F$ may be strictly larger than N, so that this homomorphism does not have to be an embedding. The following construction shows how to deal with this problem.

Let $\{g_1, g_2, g_3, \ldots\}$ be a countable system of generators of G, where $g_i \neq 1$. We successively construct the following extensions of G:

(1) $G * \{u^n\}$;
(2) the HNN-extension of $G * \{u^n\}$,

$$\big|G * \{u^n\} \cup \{t_i | t_i^{-1}ut_i = ug_i, i = 1, 2, \ldots\}\big|$$

(note that u and the ug_i generate infinite cyclic subgroups in $G * \{u^n\}$);
(3) the free product P of this HNN-extension and the group $\{b^n\} * \{v^n\}$ with subgroups $|\{t_1, t_2, \ldots\}|$ and $|\{b^{-i}vb^i | i \geqslant 1\}|$ amalgamated by means of the isomorphism

$$t_i = b^{-i}vb^i, \qquad i \geqslant 1.$$

(4) P has the two rank-2 free subgroups $|\{b, v\}|$ and $|\{u, b\}|$. There are no relations between u and b because there can be no relations in the quotient by the smallest normal subgroup containing G, t_i, and v.

Finally, we construct the following HNN-extension of P:

$$Q = |P \cup \{a\} : a^{-1}ba = u, a^{-1}va = b|.$$

To complete the proof, it remains to verify that Q is *generated by the elements* a *and* b.

In fact, Q has the obvious system of generators $\{g_i, t_i (i \geqslant 1); u, v, a, b\}$. The relations $g_i = u^{-1}t_i^{-1}ut_i$ allow us to eliminate the g_i; the relations $t_i = b^{-i}vb^i$ allow us to eliminate the t_i; and the relations $u = a^{-1}ba$ and $v = aba^{-1}$ allow us to eliminate u and v. This proves the first part of the proposition. The following analysis of the construction establishes part (b).

If we express g_i in terms of a and b in Q using the above relations, we find that $g_i = e_i$, modulo the relations in Q, where

$$e_i = a^{-1}b^{-1}ab^{-i}ab^{-1}a^{-1}b^i a^{-1}bab^{-i}aba^{-1}b^i.$$

Hence, the subgroup $E = |\{e_i | i \geqslant 1\}|$ in the group $\{a^n\} * \{b^n\}$ has the following remarkable property: any normal subgroup $N \subset E$ generates a normal subgroup N' in $\{a^n\} * \{b^n\}$ such that $E \cap N' = N$ (compare with the remark at the beginning of the proof).

In particular, if $\{g_i\}$ is an enumerable system of generators of G that is connected by an enumerable set of relations, it follows that the map $g_i \mapsto e_i$ (mod the relations) induces a recursive embedding of G in the recursive group E/N', since N' is enumerable whenever N is. □

4 Benign Subgroups

4.1. **Definition-Lemma.** *Let G be a finitely presented group, and let $H \subset G$ be a subgroup. H is called* benign *if the following equivalent conditions are fulfilled:*

(a) *There exist a finitely presented group K, a finitely generated subgroup $L \subset K$, and an embedding $G \subset K$ such that $G \cap L = H$.*

(b) *The* HNN-*extension*

$$K_G = |G \cup \{t\} : t^{-1}ht = h, \ for\ all\ h \in H|$$

can be embedded in a finitely presented group.

(c) *$G *_H G$ can be embedded in a finitely presented group.*

PROOF OF THE EQUIVALENCE

(a) ⇒ (b). Suppose that $G \subset K$ and L satisfy (a). Then it follows by 2.6 that K_G is embedded in the HNN-extension

$$|K \cup \{t\} : t^{-1}lt = l, \text{ for all } l \in L|.$$

This group is finitely presented: we add t to the generators of K, and add the relations $t^{-1}l_i t = l_i$, for a finite system of generators $\{l_i\}$ of L, to the relations between the generators of K.

(b) $\Rightarrow$ (c). The group $G *_H G$ is embedded in K_G by 2.6(b), and K_G can be embedded in an f.p. group because we have assumed condition (b).

(c)$\Rightarrow$ (a). Suppose that $G *_H G \subset M$, where M is finitely presented. We set $K = M$, we set $L =$ the image of G under the composite embedding $\phi_2 : G \to G *_H G \to M$, and we embed G in K by means of $\phi_1 : G \to G *_H G \to M$. Since $\phi_1(G) \cap \phi_2(G) = H$, we have $G \cap L = H$ in K, as required. □

The basic goal of this section is to reduce Higman's theorem 1.5 to proving that all enumerable subgroups in $\mathbf{Z} * \mathbf{Z}$ are benign. For this purpose and for later uses we shall need the following lemma.

4.2. **Lemma.** *Let R be a benign subgroup of an f.g. free group F, and let $\overline{R}$ be the normal subgroup it generates. Then $F/\overline{R}$ can be embedded in an* f.p. *group.*

PROOF. Let i be an embedding of $F *_R F$ in an f.p. group K (see 4.1(c)), and let $\phi_1, \phi_2 : F \to F *_R F$ be the structure maps. We consider two embeddings of F in $K \times F/\overline{R}$:

$$\alpha : f \mapsto \langle i \circ \phi_1(f), f\overline{R} \rangle;$$
$$\beta : f \mapsto \langle i \circ \phi_2(f), 1 \rangle.$$

They obviously coincide on the subgroup $R \subset F$. Hence they are induced by a homomorphism

$$\gamma : F *_R F \to K \times F/\overline{R},$$

which has a trivial kernel, since the composition of γ with the projection onto K coincides with i.

We construct an HNN-extension that takes $i \times \{1\} : F *_R F \to K \times F/\overline{R}$ to γ:

$$L = \Big| K \times F/\overline{R} \cup \{t\} : t^{-1}\langle i \circ \phi_1(f), 1 \rangle t = \langle i \circ \phi_1(f), f\overline{R} \rangle,$$
$$t^{-1}\langle i \circ \phi_2(f), 1 \rangle t = \langle i \circ \phi_2(f), 1 \rangle \text{ for all } f \in F \Big|.$$

L obviously contains $F/\overline{R}$. We show that L is finitely presented.

Generators of L : $\{t\} \cup$ finite system of generators of $K \cup$ finite system of generators of F. This system is finite.

Relations in L :

(a) {the relations between the generators of K}.
(b) {the commutation relations between the generators of K and the generators of F}.

After imposing these relations, we may consider that we are working in $K \times F$.

(c) $t^{-1}\langle i\circ\phi_1(f),1\rangle t=\langle i\circ\phi_1(f),f\rangle$,
$t^{-1}\langle i\circ\phi_2(f),1\rangle t=\langle i\circ\phi_2(f),1\rangle$,
where f runs through the system of generators of F.

(d) The relations in R between the generators of F.

We can take the system of relations $R_0=$ (a) $\cup$ (b) $\cup$ (c) to be finite. We need only verify that the relations in (d) follow from R_0.

Let $R^{'}\subset F$ be the normal subgroup generated by R_0, i.e., the kernel of the natural homomorphism $F\to|K\cup F\cup\{t\}:R_0|$. We want to show that $R^{'}=\overline{R}$. The inclusion $R^{'}\subset\overline{R}$ is obvious. We verify the converse.

If $f\in F$, we set $f^{'}=f \bmod R^{'}$ and $f_{1,2}=i\circ\phi_{1,2}(f)\in K$. It then follows from the relations (b) and (c) that in $K\times F/R^{'}$ we have

$$t^{-1}\langle f_1,1\rangle t=\langle f_1,1\rangle\langle 1,f^{'}\rangle \quad\text{and}\quad t^{-1}\langle f_2,1\rangle t=\langle f_2,1\rangle.$$

On the other hand, if $f\in R$, then, since $F*_R F$ is embedded in K, it follows from the relations (a) that $f_1=f_2$. Hence $f^{'}=1$, so that $R\subset R^{'}$. □

This lemma gives us the following reduction step.

4.3. Proposition. *If all enumerable subgroups in* $\mathbf{Z}*\mathbf{Z}$ *are benign, then Higman's theorem is true.*

PROOF. Let G be the free group generated by an enumerable set of free generators $\{g_i\}, i=1,2,3,\ldots$, and let $N\subset G$ be an enumerable normal subgroup. We shall show how to embed G/N into an f.p. group.

We first consider the embedding $G\to\{a^n\}*\{b^n\}$ given by $g_i\mapsto e_i$, where the e_i are as defined at the end of §3. Let the image of N under this embedding generate the normal subgroup $N^{'}\subset\{a^n\}*\{b^n\}$. By the remark at the end of §3, G/N embeds in $\{a^n\}*\{b^n\}/N^{'}$. But $N^{'}$ is enumerable by Lemma 1.2(d), since it is generated by the image of an enumerable set under a recursive map. Therefore, $N^{'}$ is a benign normal subgroup. Lemma 4.2 then shows that $\{a^n\}*\{b^n\}/N^{'}$ can be embedded in an f.p. group. □

We conclude this section by establishing several basic properties of benign subgroups.

4.4. Lemma. *Let* $E,F\subset G$ *be benign subgroups of* G. *Then:*

(a) $E\cap F$ *is a benign subgroup;*
(b) $|E\cup F|$ ("*the sum of* E *and* F *in* G") *is a benign subgroup.*

PROOF. Let $\phi_1,\phi_2:G\to G*_E G$ and $\phi_1^{'},\phi_2^{'}:G\to G*_F G$ be the structure homomorphisms. Let M_1 and M_2 be f.p. groups such that $G*_E G\subset M_1$, and $G*_F G\subset M_2$. We identify $\phi_1(G)\subset M_1$ and $\phi_1^{'}(G)\subset M_2$ with G, and construct the group $M_1*_G M_2$. This group is finitely presented (since it suffices to add to the relations in M_1, and M_2 the relations $\phi_1(g_i)=\phi_1^{'}(g_i)$ for a finite system of generators of G). Let $\phi_1^{''},\phi_2^{''}:M_1,M_2\to M_1*_G M_2$ be the structure embeddings.

We set $K = M_1 *_G M_2$ and $L = \phi'' \circ \phi_2(G)$, and we embed G in K by means of $\phi_2'' \circ \phi_2'$.

We claim that $G \cap L = E \cap F$ (as a subgroup of G in K). In fact, $\phi_1''(M_1) \cap \phi_2''(M_2) = G$ with its canonical embedding in $M_1 *_G M_2$. If we take only $\phi_2(G)$ in M_1 and $\phi_2'(G)$ in M_2, then intersecting with the amalgamation G gives E and F, respectively, and intersecting $\phi_2(G)$ with $\phi_2'(G)$ gives $E \cap F$.

(b) The subgroups $\phi_1(|E \cup F|)$ and $\phi_2(G)$ have the same intersection with the amalgamation in $G *_E G$, since they actually contain it. Hence, by 2.3(d), we have $|\phi_1(|E \cup F|) \cup \phi_2(G)| \cap \phi_1(G) = |E \cup F|$ in $G *_E G$, i.e., since E is the amalgamation,

$$|\phi_1(F) \cup \phi_2(G)| \cap \phi_1(G) = |E \cup F|.$$

Similarly, we have

$$|\phi_1'(E) \cup \phi_2'(G)| \cap \phi_1'(G) = |E \cup F|$$

in $G *_F G$. The notation is compatible with the fact that these two intersections are identified in the amalgamation of the product $M_1 *_G M_2$, which is constructed as in part (a).

Applying 2.3(d) to this product, we find that

$$\left|\phi_1''\left(|\phi_1(F) \cup \phi_2(G)|\right) \cup \phi_2''\left(|\phi_1'(E) \cup \phi_2'(G)|\right)\right| \cap G = |E \cup F|.$$

But the group $|\phi_1'' \circ \phi_2(G) \cup \phi_2'' \circ \phi_2'(G)| \cap G$ obviously contains the right-hand side and is contained in the left-hand side of this equality, so that it also coincides with $|E \cup F|$.

Finally, $|\phi_1'' \circ \phi_2(G) \cup \phi_2'' \circ \phi_2'(G)|$ is a finitely generated subgroup of the finitely presented group $M_1 *_G M_2$. The proof is complete. □

4.5. **Lemma.** *Let G and H be f.g. subgroups of f.p. groups. Then any homomorphism from G to H takes benign subgroups of G to benign subgroups of H.*

Proof.

(a) If $A \subset G$ is benign, then $A \times \{1\} \subset G \times H$ is also benign, since, given an embedding of (G, A) in (K, L) as in 4.1(a), we can construct the obvious embedding of $(G \times H)$ in $(K \times M, L \times \{1\})$, where M is the f.p. group containing H, which also satisfies the conditions in 4.1(a). Conversely, if $A \times \{1\} \subset G \times H$ is benign, then from an embedding of $(G \times H, A \times \{1\})$ in (K, L) as in 4.1(a) we construct the corresponding embedding of (G, A) in $(K, L \cap G \times \{1\})$.

(b) Now let $\phi : G \to H$ be any homomorphism, let F be its graph, and let $A \subset G$ be a benign subgroup. Then in $G \times H$ we have

$$\{1\} \times \phi(A) = \left|\left(|A \times \{1\} \cup \{1\} \times H| \cap F\right) \cup G \times \{1\}\right| \cap \{1\} \times H.$$

It is clear from the assumptions regarding G and H that F is a benign subgroup in $G \times H$. By part (a), the other subgroups on the right in the formula are also benign. By Lemma 4.4, $\{1\} \times \phi(A)$ is a benign subgroup. Hence, $\phi(A)$ is also benign. □

5 Bounded Systems of Generators

5.1. Let $G' = |\{a_1, \ldots, a_n\}|, n \geqslant 1$, be the group freely generated by the a_i. We call a subset $R' \subset G'$ *bounded* if there exists an $r > 1$ such that any element in R' can be represented in the form $a_{i_1}^{x_1} \cdots a_{i_r}^{x_r}$, $x_i \in \mathbf{Z}$. In this section we prove the following special case of the hypothesis of Proposition 4.3:

5.2. **Proposition.** *If the subgroup $H' \subset G'$ is generated by a bounded enumerable subset $R' \subset G'$, then it is benign.*

Corollary. *The same is true if G' is an f.g. subgroup of an f.p. group* (using Lemma 4.5).

In the next section we show how the general case follows from this special case.

The proof of 5.2 consists of a series of reduction steps.

5.3. *First reduction.* In the free group $G = |\{a_1, b_1, c_1; \ldots; a_{rn}, b_{rn}, c_{rn}\}|$ we shall consider a set of "layered" words of the form

$$R = \{a_1^{x_1} b_1 c_1^{x_1} \cdots a_{rn}^{x_{rn}} b_{rn} c_{rn}^{x_{rn}}\}$$

and the subgroup $H \subset G$ it generates. We shall later show that *if R is enumerable, then H is benign.* This is a special case of 5.2 to which the general case reduces using the following technique.

Suppose we are given G' and R' as in 5.1. For each element $g' = a_{i_1}^{x_1} \cdots a_{i_r}^{x_r} \in R'$ we construct an element $g \in G$ as follows. We represent g' in the form

$$\prod_{i=1}^{n} a_i^{x_{1,i}} \prod_{i=1}^{n} a_i^{x_{2,i}} \cdots \prod_{i=1}^{n} a_i^{x_{r,i}},$$

where

$$x_{k,i} = \begin{cases} x_k, & \text{for } i = i_k, \\ 0, & \text{for } i \neq i_k. \end{cases}$$

We then set

$$g = \left(\prod_{i=1}^{n} a_i^{x_{1,i}} b_i c_i^{x_{1,i}}\right) \left(\prod_{i=1}^{n} a_{n+i}^{x_{2,i}} b_{n+i} c_{n+i}^{x_{2,i}}\right) \cdots \left(\prod_{i=1}^{n} a_{(r-1)n+i}^{x_{r,i}} b_{(r-1)n+i} c_{(r-1)n+i}^{x_{r,i}}\right).$$

If R' is enumerable, then the set R of all elements g obtained from all the $g' \in R'$ is enumerable.

We consider the surjective homomorphism $\phi : G \to G'$ given by $\phi(a_{nj+i}) = a_i \, (1 \leqslant i \leqslant n, 0 \leqslant j \leqslant r-1), \phi(b_i) = \phi(c_i) = 1$ for all $i = 1, \ldots, rn$. Clearly $\phi(R) = R'$, and hence $\phi(H) = H'$. It then follows from Lemma 4.5 that if R is benign in G, then R' is benign in G'. □

5.4. *Using the theorem that all enumerable sets are Diophantine.*

From this point on, we fix a pair (G, enumerable R), as in 5.3. We shall write $l \geqslant 1$ in place of rn. We define the set $E \subset \mathbf{Z}^{l+1}$ by the condition

$$R = \{a_0^{x_0} b_0 c_0^{x_0} \cdots a_l^{x_l} b_l c_l^{x_l} | \langle x_0, \ldots, x_l \rangle \in E\}.$$

It is not hard to see that R is enumerable if and only if E is enumerable.

We now show that E *can be represented as the projection onto the first* $l+1$ *coordinates of a set*

$$\bigcap_{s=1}^{N} E_s \subset \mathbf{Z}^{l+1} \times \mathbf{Z}^{m-1}, \qquad m \geqslant l+2,$$

where each of the E_s *is defined by an equation of one of the following forms*:

$$\begin{aligned} x_i &= c, & c &\in \mathbf{Z}; \\ x_i &= x_j, & 0 &\leqslant i, j \leqslant m; \\ x_k &= x_j + x_i, & l+1 &\leqslant k < j < i \leqslant m; \\ x_k &= x_j \cdot x_i, & l+1 &\leqslant k < j < i \leqslant m. \end{aligned}$$

In fact, let $\varepsilon_0, \ldots, \varepsilon_l \in \{1, -1\}$, and let $\bar{\varepsilon} = \langle \varepsilon_0, \ldots, \varepsilon_l \rangle$. We consider the enumerable sets

$$E^{\bar{\varepsilon}} = \{\langle x_0, \ldots, x_l \rangle \in (\mathbf{Z}^+ \cup \{0\})^{l+1} | \langle \varepsilon_0 x_0, \ldots, \varepsilon_l x_l \rangle \in E\}.$$

By the fundamental theorem in Chapter VI, there exist polynomials $P^{\bar{\varepsilon}}$ with integral coefficients such that

$$\begin{aligned} E^{\bar{\varepsilon}} = {} & \text{the projection of the 0-level of } P^{\bar{\varepsilon}} \text{ in } (\mathbf{Z}^+ \cup \{0\})^{l+1} \\ & \times (\mathbf{Z}^+)^{n-l} \text{ onto the first } l+1 \text{ coordinates } \langle x_0, \ldots, x_1 \rangle. \end{aligned}$$

Here we can take n large enough that the sets of variables that actually occur in $P^{\bar{\varepsilon}'}$ and in $P^{\bar{\varepsilon}''}$ and that "drop out" in the projection do not intersect if $\bar{\varepsilon}' \neq \bar{\varepsilon}''$. If we add the $(n+1)2^{l+3}$ new variables $y_{ij\bar{\varepsilon}}$ $(0 \leqslant i \leqslant n, j = 1, 2, 3, 4)$ to the variables that drop out in the projection, we find that E can be represented as the projection onto the first $l+1$ coordinates of the 0-level of the following polynomial, where the 0-level is now in $\mathbf{Z}^{l+1} \times \mathbf{Z}^{n+(n+1)2^{l+3}-l}$:

$$Q = \prod_{\bar{\varepsilon}} \Bigg[\left(P^{\bar{\varepsilon}}(\varepsilon_0 x_0, \ldots, \varepsilon_l x_l, x_{l+1}, \ldots, x_n)\right)^2 + \sum_{i=0}^{l} \left(\varepsilon_i x_i - \sum_{j=1}^{4} y_{ij\bar{\varepsilon}}^2\right)^2 + \sum_{i=l+1}^{n} \left(x_i - 1 - \sum_{j=1}^{4} y_{ij\bar{\varepsilon}}^2\right)^2 \Bigg].$$

Finally, in order to represent the set $Q = 0$ as a projection of an intersection $\bigcap_{s=1}^{N} E_s$ of the required type, we introduce additional variables as follows. Let $x_0, \ldots, x_t$ be the variables that occur in Q. Instead of $Q = 0$ we write $Q_1 = Q_2$, where Q_1 is the sum of the monomials in Q with positive coefficients, and Q_2 is the sum of the monomials with negative coefficients. Then

$$\text{0-level of } Q = \text{a projection of } (x_{t+1} = Q_1) \cap (x_{t+2} = Q_2) \cap (x_{t+1} = x_{t+2}).$$

If Q_1 and Q_2 are constants or variables, this gives us the desired representation. Otherwise, we write, say, Q_1 in the form $Q_1^{'} + Q_1^{''}$ or $Q_1^{'} \cdot Q_1^{''}$, and after introducing two more variables, we have, for example,

$$(x_{t+1} = Q_1^{'} + Q_1^{''}) = \text{a projection of } (x_{t+3} = Q_1^{'}) \cap (x_{t+4} = Q_1^{''}) \cap (x_{t+1} = x_{t+3} + x_{t+4}).$$

We complete the proof by induction on the sum of the absolute values of the coefficients and on the degree of Q. □

5.5. *Second reduction.* We now assume that along with the pair (G, R) described in 5.3, we have fixed a representation of E in the form $\bigcap_{s=1}^N E_s$ as in 5.4. In this subsection we show that *the subgroup $H \subset G$ generated by R is benign if all of the following subgroups $\overline{H}_s \subset \overline{G}, s = 1, \ldots, N$, are benign:*

$$\overline{G} = \left|\{a_0, b_0, c_0; \ldots; a_m, b_m, c_m; \bar{a}_1, \bar{b}_1, \bar{c}_1, \ldots, \bar{a}_l, \bar{b}_l, \bar{c}_l, \}\right|;$$

$$\overline{H}_s = \left|\left\{\left(\prod_{i=l+1}^m a_i^{x_i} b_i c_i^{x_i}\right)^{-1} \left(\prod_{i=1}^l \bar{a}_i^{x_i} \bar{b}_i \bar{c}_i^{x_i}\right)^{-1} \prod_{i=0}^m a_i^{x_i} b_i c_i^{x_i}; \langle x_0, \ldots, x_m\rangle \in E_s\right\}\right|.$$

To show this, we first set

$$a(x_0, \ldots, x_m) = \left(\prod_{i=l+1}^m a_i^{x_i} b_i c_i^{x_i}\right)^{-1} \left(\prod_{i=1}^l \bar{a}_i^{x_i} \bar{b}_i \bar{c}_i^{x_i}\right)^{-1} \prod_{i=0}^m a_i^{x_i} b_i c_i^{x_i}. \quad (1)$$

The set of words $\{a(x_0, \ldots, x_m); \langle x_0, \ldots, x_m\rangle \in \mathbf{Z}^{m+1}\}$ is free, since when we join two such words (or when we join such a word with the inverse of another such word), any cancellation cannot involve the "middle part" of each word, which consists of the symbols $\bar{a}_i, \bar{b}_i, \bar{c}_i$.

It hence follows that

$$\bigcap_{s=1}^N \overline{H}_s = \left|\left\{a(x_0, \ldots, x_m), \langle x_0, \ldots, x_m\rangle \in \bigcap_{i=1}^N E_s\right\}\right|,$$

and the subgroup $\overline{H} = \bigcap_{s=1}^N \overline{H}_s \subset \overline{G}$ is benign if all of the $\overline{H}_s$ are benign. Finally, we have

$$\left|\overline{H} \cup \{a_{l+1}, b_{l+1}, c_{l+1}, \ldots, a_m, b_m, c_m; \bar{a}_1, \bar{b}_1, \bar{c}_1, \ldots, \bar{a}_l, \bar{b}_l, \bar{c}_l\}\right|$$
$$= \left|\left\{\prod_{i=0}^l a_i^{x_i} b_i c_i^{x_i}, \langle x_0, \ldots, x_l\rangle \in E = \text{projection of } \bigcap_{s=1}^N E_s\right\} \cup \{a_{l+1}, b_{l+1}, c_{l+1}, \ldots, \bar{a}_l, \bar{b}_l, \bar{c}_l\}\right|,$$

so that

$$H = \left|\overline{H} \cup \{a_{l+1}, b_{l+1}, \ldots, \bar{b}_l, \bar{c}_l\}\right| \cap |\{a_0, \ldots, b_l, c_l\}|.$$

Therefore, H is benign whenever $\overline{H}$ is benign. □

5.6. *Construction of the group K.* We use the criterion 4.1(a) to verify that the $\overline{H}_s \subset \overline{G}$ are benign subgroups. That is, we explicitly construct a finitely presented group $K \supset \overline{G}$ and finitely generated subgroups $L_s \subset K$ such that $L_s \cap \overline{G}$ for all $s = 1, \ldots, N$. We construct K as a multiple HNN-extension of $\overline{G}$.

(a) *The first* HNN-*extension.* We set

$$K_0 = \left|\overline{G} \cup \{t_0, \ldots, t_m\} : R_0\right|,$$

where R_0 is the set of relations

$$\{t_i^{-1} b_i t_i = a_i b_i c_i \quad \text{and} \quad t_i^{-1} \bar{b}_i t_i = \bar{a}_i \bar{b}_i \bar{c}_i, \text{ for } i = 0, \ldots, m;$$
$$\text{the } t_i \text{ commute with all the other generators of } \overline{G}\,\}. \tag{2}$$

(b) *The second* HNN-*extension.* We set

$$K = \left|K_0 \cup \{t_{ijk};\, l+1 \leqslant k < i, k < j, i \neq j; i, j, k \leqslant m\} : R\right|,$$

where R is the set of relations

$$\{t_{ijk}^{-1} b_i t_{ijk} = a_i b_i c_i, t_{ijk}^{-1} c_j t_{ijk} = t_k c_j;$$
$$\text{the } t_{ijk} \text{ commute with the } t_k \text{ and with the other generators of } \overline{G}\}. \tag{3}$$

Unlike what we saw in 5.6(a), here it is not completely obvious that K is an HNN-extension of K_0. To check this it suffices to show that the map ϕ_{ijk} (i, j, k fixed $i \neq k, j \neq k$) from the set $\{\text{generators of } \overline{G}\} \cup \{t_k\}$ to itself that takes

$$b_i \mapsto a_i b_i c_i, \qquad c_j \mapsto t_k c_j, \qquad t_k \mapsto t_k,$$

and leaves the other generators of $\overline{G}$ fixed, extends to an automorphism of the subgroup $|\overline{G} \cup \{t_k\}| \subset K_0$. We have

$$\left|\overline{G} \cup \{t_k\}\right| = \left|\overline{G} * \{t_k^n\} : t_k^{-1} b_i t_k = b_i, t_k^{-1} c_j t_k = c_j, \ldots\right|,$$

where the $\cdots$ stands for relations that do not involve b_i and c_j, and so are taken to themselves under ϕ_{ijk}. On the other hand, the two relations that are written out are taken to relations that follow from the defining relations in K_0: the first goes to

$$t_k^{-1} a_i b_i c_i t_k = a_i b_i c_i,$$

and the second goes to

$$t_k^{-1} t_k c_j t_k = t_k c_j.$$

It remains to use the stipulation that $i \neq k$ and $j \neq k$.

It is clear from the definition of K that K is finitely presented. It follows from the properties of HNN-extensions that $\overline{G} \subset K$.

5.7. *Construction of the subgroups* $L_s \subset K$. The form of L_s will depend on the equation defining the set E_s (see 5.4). We define a large number of groups, which will include all the L_s:

$$L_i^c = \left|\{a(\underbrace{0\cdots 0}_{i}\, c0\cdots 0), t_r(r \neq i)\}\right|,$$

$$L_{ij}^{=} = \left|\{a(0\cdots 0), t_i t_j, t_r(r \neq i, j)\}\right|,$$
$$L_{ijk}^{+} = \left|\{a(0\cdots 0), t_i t_k, t_j t_k, t_r(r \neq i, j, k)\}\right|,$$
$$L_{ijk}^{\times} = \left|\{a(0\cdots 0), t_{ijk}, t_{jik}, t_r(r \neq i, j, k)\}\right|,$$

and analogously, in the notation of 5.5,

$$\overline{H}_i^c = \left|\{a(x_0, \ldots, x_m), x_i = c\}\right|,$$
$$\overline{H}_{ij}^{=} = \left|\{a(x_0, \ldots, x_m), x_i = x_j\}\right|,$$
$$\overline{H}_{ijk}^{+} = \left|\{a(x_0, \ldots, x_m), x_k = x_j + x_i\}\right|,$$
$$\overline{H}_{ijk}^{\times} = \left|\{a(x_0, \ldots, x_m), x_k = x_j \cdot x_i\}\right|.$$

The L_s are clearly finitely generated. It remains to perform one final series of verifications:

5.8. $\overline{H}_i^c = \overline{G} \cap L_i^c, \overline{H}_{ij}^{=} = \overline{G} \cap L_{ij}^{=}$, *and so on.*

First of all, it follows from (1), (2), and (3) that

$$t_i^{-1} a(x_0, \ldots, x_m) t_i = a(x_0, \ldots, x_{i-1}, x_i + 1, x_{i+1}, \ldots, x_m), \tag{4}$$
$$t_{ijk}^{-1} a(x_0, \ldots, x_m) t_{ijk} = a(y_0, \ldots, y_m), \tag{5}$$

where $y_i = x_i + 1, y_k = x_k + x_j$, and $y_s = x_s$ for $s \neq i, k$. (To verify (5) recall that since $k \geqslant l + 1$, it follows that t_k commutes with the middle part of the word $a(x_0, \ldots, x_m)$, which consists of $\bar{a}_i, \bar{b}_i, \bar{c}_i, i \leqslant l$.)

It hence follows that

$$L_i^c = \left|\overline{H}_i^c \cup \{t_r | r \neq i\}\right|,$$
$$L_{ij}^{=} = \left|\overline{H}_{ij}^{=} \cup \{t_i t_j, t_r | r \neq i, j\}\right|,$$
$$L_{ijk}^{+} = \left|\overline{H}_{ijk}^{+} \cup \{t_i t_k, t_j t_k, t_r | r \neq i, j, k\}\right|,$$
$$L_{ijk}^{\times} = \left|\overline{H}_{ijk}^{\times} \cup \{t_{ijk}, t_{jik}, t_r | r \neq i, j, k\}\right|.$$

In fact, the inclusions $\subset$ are obvious. Next, if we begin with $a(x_0, \ldots, x_m)$ and conjugate by t_r, it follows by (4) that we can vary the rth coordinate arbitrarily. This immediately gives the inclusion $L_i^c \supset H_i^c$, and hence the first required equality. The second equality is obtained analogously.

The third equality: conjugating by $t_i t_k$ increases the ith and kth coordinates by 1, and conjugating by $t_j t_k$ increases the jth and kth coordinates by 1, so that we can obtain any vector with $x_k = x_j + x_i$ starting from a vector with zeros in these places.

The fourth equality: conjugating by t_{ijk} increases x_i by 1 and increases x_k by x_j, and conjugating by t_{jik} increases x_j by 1 and x_k by x_i. Hence, we can obtain any vector with $x_k = x_j \cdot x_i$ starting from the zero vector.

This new characterization of the groups L_s shows that $L_s \cap \overline{G} \supset H_s$ for all s. It remains to prove the converse.

To do this, we note that using (4) and (5), we can represent any element in L_s in the form Th, where $T \in |\{t_i, t_{ijk}\}|$ (here the set of admissible indices i and ijk depends on s) and $h \in H_s$. This follows by the same argument as above. But by Proposition 2.7(a), all the $\{t_i, t_{ijk}\}$ generate a free subgroup that has a trivial intersection with $\overline{G}$ (see the proof of 2.7(a)). Consequently, if $Th \in \overline{G}$, it follows that $T = 1$ and $h \in H_s$, which completes the proof. □

6 End of the Proof

6.1. In this section we finish the verification of Proposition 4.3, and hence the proof of Higman's theorem.

Let $G = |\{a, b\}|$, and let $H \subset G$ be an enumerable subgroup. We shall show that H is benign. The first step is to reduce the problem to proving that a certain special subgroup

$$H' \subset G' \cong \mathop{*}_{1}^{7} \mathbf{Z},$$

which does not depend on H, is benign. To define H', we first introduce the following recursive enumeration $\gamma : \mathbf{Z}^+ \to G$ (which covers each $g \in G$ infinitely many times):

$$\gamma(2^{m_0} 3^{m_1} \cdots p_r^{m_r} \cdots) = \prod_{i=0}^{\infty} a^{m_{4i} - m_{4i+1}} b^{m_{4i+2} - m_{4i+3}}.$$

We then set

$$G' = |\{a, b, t, v, c, d, e\}|;$$

$$\tau : S(\{a, b, a^{-1}, b^{-1}\}) \to G' : \prod_{i \geqslant 0} a^{m_{2i}} b^{m_{2i+1}} \mapsto t \prod_{i \geqslant 0} (v^{-i} a v^i)^{m_{2i}} (v^{-1} b v^i)^{m_{2i+1}};$$

$$H' = |\{\tau(g) c^n d e^n | g \in S(\{a, b, a^{-1}, b^{-1}\}), n \in \mathbf{Z}^+, g = \gamma(n)\}| \subset G'.$$

The formula for τ defines τ on words that are not necessarily reduced, and reducing a word can change its image under τ. Also note that a generator $\tau(g) c^n d e^n$ of H' is uniquely determined by n.

6.2. Lemma. *If $H' \subset G'$ is a benign subgroup, then any enumerable subgroup $H \subset G$ is benign.*

PROOF.

(a) We set

$$H'' = |\{\tau(h)c^n de^n | \text{image of } h \in H, n \in \mathbf{Z}^+, h = \gamma(n)\}| \subset H'.$$

Then

$$H'' = H' \cap |\{a, b, t, v, c^n de^n | n \in \gamma^{-1}(H)\}|.$$

In fact, the inclusion $\subset$ is obvious. The converse follows because the set of images of the elements $c^n de^n, n \geqslant 1$, in the quotient of G' by the kernel generated by a, b, t, and v, is free. Hence, in any reduced word in the generators $\tau(g)c^n de^n$, the sequence of n's can be uniquely recovered from the word, and if all the n's lie in $\gamma^{-1}(H)$, it follows that the word lies in H''.

Thus, H'' is the intersection of H' with the subgroup generated by a bounded enumerable set of generators (since $\gamma^{-1}(H)$ is enumerable whenever H is). Consequently, H'' is benign if H' is benign.

(b) We set

$$\overline{H} = |\{\tau(h) | h \in H\}| \subset G'.$$

It is easy to see that

$$|\overline{H} \cup \{c, d, e\}| = |H'' \cup \{c, d, e\}|.$$

Hence,

$$\overline{H} = |H'' \cup |\{c, d, e\}|| \cap |\{a, b, v, t\}|.$$

By Lemma 4.4, $\overline{H}$ is benign if H'' is benign.

(c) Finally, we consider the homomorphism $\phi : G' \to G$ that takes a to a, b to b, and t, v, c, d, and e to 1. Obviously, $\phi(\overline{H}) = H$. By Lemma 4.5, H is benign if $\overline{H}$ is benign. □

6.3. We now prove that the subgroup $H' \subset G'$ is benign. To do this, we construct a commutative diagram of group embeddings

$$\begin{array}{ccccc} G' & \to & K' & \to & K \\ \uparrow & & \uparrow & & \uparrow \\ H' & \to & L' & \to & L \end{array}$$

with the following properties:

(a) K is defined by a finite set of generators and a bounded enumerable set of relations; L is generated by a bounded enumerable set of words in the generators of K.
(b) $L' \stackrel{\sim}{\to} L$ is an isomorphism.
(c) $H' = G' \cap L'$ in K'.

It will then follow that H' is benign. In fact, let $K = F/\overline{R}$, where F is the free group generated by a finite system of generators of K, R_0 is a bounded enumerable set of relations between these generators, and $\overline{R}$ is the normal subgroup generated by these relations. It follows from Proposition 5.2 that R_0 generates a benign subgroup R in F, and then Lemma 4.2 implies that $K = F/\overline{R}$ can be embedded in an f.p. group M. When we embed K in M, the bounded enumerable set of generators of L remains a bounded enumerable set in M (relative to the generators of M), and hence $L \subset M$ is benign by the corollary to Proposition 5.2. Therefore, by (b) and (c) we have that the subgroup $H' = G' \cap L$ is benign as a subgroup of M whenever G and L are benign. Hence, there is an embedding of (M, H') in $(\overline{M}, \overline{H})$ such that $\overline{M}$ is finitely presented, $\overline{H}$ is finitely generated, and $H' = \overline{H} \cap M$. This embedding induces an embedding of the pair (G', H') in $(\overline{M}, \overline{H})$ with the same properties. Consequently, H' is also benign in G'.

It remains to construct the diagram of embeddings with properties (a), (b), and (c).

6.4. *The group* K'. This will be a multiple HNN-extension of G', which, as in Proposition 2.7, we define using four countable sequences of nontrivial isomorphisms of the subgroup $|\{t, c, d, e, v^{-i}av^i, v^{-i}bv^i | i \geqslant 0\}| \subset G'$ with G'. Since the elements listed here freely generate this subgroup, it is sufficient to indicate where our isomorphisms take these elements. These isomorphisms will be induced in K' by conjugation by four sequences of generators $x_i, \bar{x}_i, y_i$, and $\bar{y}_i, i \geqslant 0$ (instead of the t_i, $i \in I$, in §2). The following table gives the action of these generators. We use the notation $a_i = v^{-i}av^i, b_i = v^{-i}bv^i, p_j =$ the jth prime number. The element in the table, say, in the c-row and the $\bar{x}_i$-column, is $\bar{x}_i^{-1}cx_i$.

	x_i	$\bar{x}_i$	y_i	$\bar{y}_i$
t	ta_i	ta_i^{-1}	tb_i	tb_i^{-1}
c	$c^{p_{4i}}$	$c^{p_{4i+1}}$	$c^{p_{4i+2}}$	$c^{p_{4i+3}}$
d	d	d	d	d
e	$e^{p_{4i}}$	$e^{p_{4i+1}}$	$e^{p_{4i+2}}$	$e^{p_{4i+3}}$
a_j	$\begin{cases} a_i^{-1}a_ja_i, j \leqslant i \\ a_j, \quad j \geqslant i \end{cases}$	$\begin{cases} a_ia_ja_i^{-1}, j < i \\ a_j, \quad j \geqslant i \end{cases}$	$\begin{cases} b_i^{-1}a_jb_i, j < i \\ a_j, \quad j \geqslant i \end{cases}$	$\begin{cases} b_ia_jb_i^{-1}, j < i \\ a_j, \quad j \geqslant i \end{cases}$
b_j	$\begin{cases} a_i^{-1}b_ja_i, j < i \\ b_j, \quad j \geqslant i \end{cases}$	$\begin{cases} a_ib_ja_i^{-1}, j < i \\ b_j, \quad j \geqslant i \end{cases}$	$\begin{cases} b_i^{-1}b_jb_i, j < i \\ b_j, \quad j \geqslant i \end{cases}$	$\begin{cases} b_ib_jb_i^{-1}, j < i \\ b_j, \quad j \geqslant i \end{cases}$

We finally set

$$K' = |G' \cup \{x_i, \bar{x}_i, y_i, \bar{y}_i | i \geqslant 0\} : \text{ the relations in the table}|,$$

and we take $G' \to K'$ to be the natural embedding.

6.5. *The group L'.* We set

$$L' = \left|\{tcde, x_i, \bar{x}_i, y_i, \bar{y}_i | i \geqslant 0\}\right| \subset K',$$

and we take $L' \to K'$ to be the natural embedding. In Section 6.7 we shall verify that H' is embedded in L' (as a subgroup of K', in view of the commutativity of the diagram).

6.6. *The groups K and L.* We set

$$K = \left|G' \cup \{u_1, u_2, u_3, u_4, v_1, v_2, v_3, v_4\} : R\right|,$$

where the relations R and the embedding $K' \to K$ are both defined by the conditions

R = the image of the relations in the table after making the substitutions

$$x_i \mapsto u_1^{-i} v_1 u_1^i, \qquad \bar{x}_i \mapsto u_2^{-i} v_2 u_2^i, \qquad y_i \mapsto u_3^{-i} v_3 u_3^i, \quad \bar{y}_i \mapsto u_4^{-i} v_4 u_4^i;$$

$K' \to K$ is the homomorphism that is the identity on G' and acts by these substitutions on the other generators.

The homomorphism $K' \to K$ is an embedding. In fact, the elements $u_j^{-i} v_j u_j^i$ are free in $|\{u_j, v_j\}|$, so that K can be considered as the free product of K' and $|\{u_j, v_j | 1 \leqslant j \leqslant 4\}|$ with the amalgamation given by the above substitutions (here we take into account Proposition 2.7(a)).

Finally, we set

$$L = \text{ the image of } L' \text{ under the embedding } K' \to K.$$

6.7. The diagram has now been constructed. It follows immediately from the definitions that it satisfies 6.4(a) and (b). It remains to show that $H' = G' \cap L'$ in K'.

(a) We set $[n] = \tau(g)c^n de^n$ for $n \in \mathbf{Z}^+$ and $g = \gamma(n)$ in the notation of 6.1. We recall that H' is generated by all the $[n]$ in G', and hence in K' as well.

The table of relations in K' was composed in such a way that the following relations would be fulfilled:

$$x_i^{-1}[n]x_i = [p_{4i}n], \qquad \bar{x}_i^{-1}[n]\bar{x}_i = [p_{4i+1}n],$$
$$y_i^{-1}[n]y_i = [p_{4i+2}^n], \qquad \bar{y}_i^{-1}[n]\bar{y}_i = [p_{4i+3}n].$$

For example, we verify the first relation. Let $n = \Pi_{p_j}{}^{m_j}$. Then, according to the definitions,

$$\gamma(n) = \prod_j a^{m_{4j} - m_{4j+1}} b^{m_{4j+2} - m_{4j+3}},$$

$$[n] = t \prod_j a_j^{m_{4j} - m_{4j+1}} b_j^{m_{4j+2} - m_{4j+3}} c^n de^n,$$

so that by the first column of the table in 6.4,

$$x_i^{-1}[n]x_i = ta_i a_i^{-1} \prod_{j<i}(\cdots); a_i \prod_{j\geqslant i}(\cdots); c^{p_{4i}n} d e^{p_{4i}n} = [p_{4i}n].$$

If we further take into account that $[1] = tcde \in L^{'}$, we may conclude from these conjugation formulas that $[n] \in L^{'}$ for all n, and that $H^{'} \subset L^{'}$, as promised in 6.5. Moreover, $|H^{'} \cup \{x_i, \bar{x}_i, y_i, \bar{y}_i | i \geqslant 0\}| = L$, since the inclusion $\subset$ has been verified, and the inclusion $\supset$ is obvious.

(b) We now show that in $K^{'}$ we have

$$|H^{'} \cup \{x_i, \bar{x}_i, y_i, \bar{y}_i | i \geqslant 0\}| \cap G^{'} = H^{'}.$$

Since $K^{'}$ is an HNN-extension of $G^{'}$, it suffices to show that we are in the situation of Proposition 2.7 (as described in the paragraph preceding the proposition, at the end of 2.6), and then to apply 2.7(b).

We verify these conditions, for example, for the first series of isomorphisms of the subgroup of $G^{'}$, as described at the beginning of 6.4. This series corresponds to conjugating by x_i in $K^{'}$. The conditions take the following form in our case:

$$\begin{aligned} x_i^{-1}\Big[H^{'} \cap \big|\{t, c, d, e; a_j, b_j | j \geqslant 0\}\big|\Big] x_i \\ = H^{'} \cap x_i^{-1}\big|\{t, c, d, e; a_j, b_j | j \geqslant 0\}\big| x_i; \end{aligned}$$

i.e., if we use the definition of $H^{'}$ and the table,

$$x_i^{-1} H^{'} x_i = H^{'} \cap \big|\{t, c^{p_{4i}}, d, e^{p_{4i}}; a_j, b_j | j \geqslant 0\}\big|.$$

Since $x_i^{-1}[n]x_i = [p_{4i}n]$, the inclusion $\subset$ is obvious. Conversely, suppose we are given an element in $H^{'}$ that is written as a reduced word in the $[n]$: $\Pi_{j\geqslant 0}[n_j]^{\varepsilon_j}, \varepsilon_j = \pm 1$. We consider the corresponding reduced word g in $G^{'}$. We show that if all the powers of c and d that occur in g are divisible by p_{4i}, then all the n_j with nonzero ε_i in the above product are divisible by p_{4i}, i.e., $[n_j] \in x_i^{-1} H^{'} x_i$.

In fact, let $\bar{g}$ = the image of g in $\{c, d, e\}|$ under the homomorphism that takes t, a_j, and b_j to 1. Since $[\bar{n}] = c^n d e^n$, it follows that all the $[\bar{n}]$ are free, and that $\bar{g}$ uniquely determines the sequence $\{\varepsilon_j n_j\}$. It is not hard to see that the formulas that express $\varepsilon_j n_j$ in terms of the powers of c and e that occur in the reduced word $\bar{g}$ are linear with integer coefficients (more precisely, they are a disjunction of linear formulas accompanied by inequality conditions). Therefore, if all these powers are divisible by p_{4i}, then so is n_j.

This completes the proof. □

IX

Constructive Universe and Computation

1 Introduction: A Categorical View of Computation

1.1. **Words and integers: two constructive worlds.** (a) In Chapters I and II we have studied *alphabets, words* (finite sequences of letters of an alphabet), *expressions* (certain syntactically well formed words such as *terms* and *formulas* defined in I.2.3), *deductions* (finite sequences of formulas defined in II.5.1).

Let us fix an alphabet of a first-order language and denote by $\mathcal{W} \supset \mathcal{F}$ the sets of words and formulas respectively.

Studying deducibility, we have implicitly introduced the set $\mathcal{D} \subset \mathcal{F}$ of all formulas deducible from, say, a fixed finite set of formulas (axioms). This whole set $\mathcal{D}$ can be systematically generated and well ordered following a finitely describable procedure that, say, first totally orders the alphabet, then totally orders elementary steps of deductions etc., prescribing in what order to apply them iteratively to the axioms and already deduced formulas.

In this way we get a bijection $\mathbf{Z}^+ \to \mathcal{D}$ that is intuitively "computable," together with the inverse bijection. Of course, it is a simple particular case of *numbering* defined in VII.1.2 and studied later on in VII.1. See also II.11 for a useful numbering of all formulas in the Smullyan language.

Having achieved in this way the encoding of certain linguistic constructions by arithmetic ones, we have been able in Part III to reduce many problems of syntax (and partly semantics) of formal languages to number theory.

(b) We could have considered $\mathbf{Z}^+$ as a set of certain words in a finite alphabet as well, for example, as the set of binary strings whose first bit is 1. Then the whole theory of computability in Chapter V could have been based on the notion of *Turing machine(s)*, in place of elementary arithmetic. This viewpoint, leading to the "same" notion of computability and the same supply of computable (partially recursive) functions, nevertheless enriches our intuition in two essential respects.

(i) Whereas before Alan Turing, the most common mental image of mathematical reasoning was related to some form of (written) *language,* Turing represented computation as the dynamical evolution of an idealized *physical system.*

Yu. I. Manin, *A Course in Mathematical Logic for Mathematicians, Second Edition*,
Graduate Texts in Mathematics 53, DOI 10.1007/978-1-4419-0615-1_9,

This dethroning of the linguistic metaphor and its replacement by a metaphor grounded in science was a great breakthrough, and a premonition of the age of computers.

Among other developments, Turing's metaphor broke the ground for (at first mental) replacement of the classical computing machine by a quantum one. The burgeoning theory of quantum computers owes Turing this debt of gratitude.

(ii) Turing's insight allowed him to undertake a *microscopic* analysis of the intuitive idea of algorithmic computation. In a sense, he found its genetic code. The atom of information is one bit, the atomic operators can be chosen to act upon one/two bits, and to produce changes in the output of the same restricted size. Finally, the sequence of operations at each step is strictly determined by the local environment of bounded size, again several bits. Needless to say, mathematically "the same" idea can be described in purely linguistic terms. In fact, Markov's normal algorithms do just that. But as we argued above, this would constitute a philosophical regression.

One goal of this chapter is to go in the reverse direction, and to present *a "macrocosm"* of the classical theory of computation.

The sets $\mathbf{Z}^+$, $\mathcal{W}$, $\mathcal{F}$, $\mathcal{D}$ are examples of what we will call below *constructive worlds.* Elements of these sets—integers, words, formulas, deducible formulas—are *constructive structures* of the respective kind. Other examples include worlds of finite graphs, finite groups, finite rings (up to isomorphism, or "all" in a fixed countable universe of sets).

Each of these worlds is countably infinite, but it is natural to allow also *finite* constructive worlds, such as all binary strings of restricted length.

In Sections 2 and 3 below we will unite different constructive worlds into a *constructive universe.* It will be a *category,* with constructive worlds as objects, and semicomputable functions as morphisms. Church's thesis will get a very natural reformulation:

Categorical Church's Thesis: *Any two constructive universes are equivalent.*

For more detailed explanations, see Section 2 below, especially Comments 2.3.

1.2. **Languages as categories.** In Sections 4 and 5 of this chapter, we explain that there exist natural constructive worlds that are themselves *categories,* and at the same time *languages*, that are more convenient for describing morphisms between constructive worlds than conventional languages, discussed in Chapters 1 and 2 of this book.

Roughly speaking, we can base the theory of recursive functions on a constructive world of *descriptions* of these functions, whereas the set of functions themselves *does not form a constructive world.*

This raises a challenge: to find a well-structured world of descriptions faithfully reflecting properties of recursive functions as *morphisms.*

Our suggestion elaborated in Section 3 is motivated, on the one hand, by progress in general algebra, the theory of (generalized) operads, and on the other hand, by the recent paper by N. Yanofsky (math.LO/0602053), who has constructed a specific operad acting on primitive recursive functions.

We may and will treat operads as functors on appropriate categories of decorated graphs. Such graphs themselves form constructive worlds, with effectively computable finite sets of morphisms. If we admit these categories as new types of languages, then a *functor* defined on such a category becomes the categorical version of a *model* of this language.

The decorated graphs are idealized versions of flowcharts, which are quite popular in the description of various computational processes. Already in the 1960s, Dana Scott, among others, used an appropriately formalized version of them. He united them into a lattice which can be treated as a category satisfying strong additional restrictions: see his survey paper "The lattice of flow diagrams" in Springer Lecture Notes in Math, vol. 188 (1971).

This, and the return to the Turing philosophy, complemented by the progress of quantum physics, motivates the last subject matter of this chapter: *Introduction to the theory of quantum computation.*

1.3. **Why quantum computation?** Information processing (computation) is the dynamical evolution of a highly organized physical system produced by technology (computer) or nature (brain). The initial state of this system is (determined by) its *input*; its final state is the *output.*

Physics describes nature in two complementary modes: classical and quantum. Up to the 1990s, the basic mathematical models of computing mimiced classical automata, although the first suggestions for studying quantum models date back at least to 1980.

Roughly speaking, the motivation to study quantum computing comes from several sources: physics and technology, cognitive science, and mathematics. We will briefly discuss them in turn.

(i) Physically, the quantum mode of description is more fundamental than the classical one. In the 1970s and 1980s it was remarked that because of the superposition principle, or quantum entanglement, it is computationally infeasible to simulate quantum processes on classical computers. Roughly speaking, in quantizing a classical system with N states we obtain a quantum system whose state space is an $(N-1)$-dimensional complex projective space whose volume grows *exponentially* with N. One can argue that the main preoccupation of quantum chemistry is the struggle with the resulting difficulties. Reversing this argument, one might expect that quantum computers, if they can be built at all, will be considerably more powerful than classical ones.

Serious preoccupation with quantum computing has also been stimulated by rapid progress in the microfabrication techniques of modern computers. It has already led us to the level where quantum noise becomes an essential hindrance to the error-free functioning of microchips. It is only logical to start *exploiting* the essential quantum-mechanical behavior of small objects in devising computers, instead of *neutralizing* it.

(ii) As another motivation, one can invoke highly speculative, but intriguing, conjectures that the "wetware" of brains in fact somehow relies upon quantum computations.

Even without subscribing to this idea wholeheartedly until more experimental data are generated, we must be aware of the great quantitative discrepancy between the information processing capacity of the brain and our understanding of how it might do what it does.

For example, the IBM Deep Blue chess computer, which in 1996–1997 played at the level of the world champion Kasparov, could evaluate about 10^8 positions per second and search the game tree to a depth of about 10 moves/countermoves, and up to 40 in exceptional cases.

Since the characteristic time of neuronal processing is about 10^{-3} sec, it is very difficult to explain how the classical brain could possibly do the job and play chess as successfully. Existing models of neural networks cannot pass this test by very wide margin.

A less spectacular, but no less a resource-consuming task, is speech generation and perception, which is routinely done by billions of human brains, but still presents a formidable challenge for modern computers using classical algorithms.

Computational complexity of cognitive tasks has several sources: basic variables can be fields; a restricted number of small blocks can combine into exponentially growing trees of alternatives; databases of incompressible information have to be stored and searched.

Two paradigms have been developed to cope with these difficulties: logic-like languages and combinatorial algorithms, on the one hand, and statistical matching of observed data to an unobserved model, on the other.

In many cases, the second strategy efficiently supports acceptable performance, but usually cannot achieve the excellence of the Deep Blue level. Both paradigms require huge computational resources, and it is not clear how they can be organized, unless hardware allows fast and massive parallel computing.

The idea of "quantum parallelism" (see Section 7 below) is an appealing theoretical alternative. However, it is not at all clear that it can be made compatible with the available experimental evidence, which depicts the central nervous system as a distinctly classical device.

The following way out might be worth exploring. The implementation of efficient quantum algorithms that have been studied so far can be provided by one, or several, quantum chips (registers) controlled by a classical computer. A very considerable part of the overall computing job, besides controlling quantum chips, is also assigned to the classical computer. Analyzing a physical device of such architecture, we would have direct access to its classical component (electrical or neuronal network), whereas locating its quantum components might constitute a considerable challenge. For example, quantum chips in the brain might be represented by macromolecules of the type that were considered in some theoretical models for high-temperature superconductivity.

The difficulties are seemingly increased by the fact that quantum measurements produce nondeterministic outcomes. Actually, one could try to use this to one's advantage, because there exist situations in which we can distinguish the quantum randomness from the classical case by analyzing the probability distributions and using Bell-type inequalities. With hindsight, one recognizes in Bell's setup the first example of the game-like situation in which quantum players can behave demonstrably more efficiently than classical ones.

(ii) Finally, we turn to mathematics. One can argue that nowadays one does not even need additional motivation to study quantum automata, given the predominant mood prescribing the quantization of "everything that moves." Quantum groups, quantum cohomology, quantum invariants of knots, etc., come to mind. This actually seemed to be the primary motivation before 1994 when P. Shor devised the first significant quantum algorithm showing that prime factorization can be done on quantum computers in polynomial-time, that is, considerably faster than by any known classical algorithm.

Shor's paper gave a new boost to the subject. Another beautiful result, due to L. Grover, is that a quantum search among N objects can be done in $c\sqrt{N}$ steps. We briefly present these ideas in Sections 8 and 9.

Last, but not least, large-scale quantum computers do not exist as yet. The quantum algorithms invented and studied up to now will stimulate the search for a technological implementation that—if successful—will certainly correct our present understanding of quantum computing and quantum complexity.

2 Expanding Constructive Universe: Generalities

In this chapter, given a category $\mathcal{C}$ and two of its objects X, Y, we will denote by $\mathcal{C}(X, Y)$ the set of morphisms $X \to Y$ in $\mathcal{C}$.

All our objects will be sets endowed with an additional structure, and sets will lie in the initial layers of the Gödel universe $\mathcal{L}$ of constructible sets (cf. IV.1).

Morphisms will be partial maps.

We choose once and for all some concrete sets, representatives of natural numbers and $\mathbf{Z}^+$ in $\mathcal{L}$, such as $0 = \emptyset, 1 = \{\emptyset\}, 2 = \{\emptyset, 1\}, \dots$ and $\mathbf{Z}^+ = \{1, 2, 3, \dots\}$.

We will first discuss some peculiarities of categories whose morphisms are partial maps of sets.

2.1. **Category of sets and partial maps: two approaches.** (a) In the first approach, partial maps from a set X to a set Y are pairs $(f, D(f))$ where $D(f)$ is a subset of X (possibly, empty), and $f : D(f) \to Y$ is an actual map. Denote by $Par\,(X, Y)$ the set of partial maps. The composition is defined exactly as was done for a particular case in V.2.3:

$$(g, D(g)) \circ (f, D(f)) := (g \circ f, f^{-1}(D(g))).$$

One easily sees that in this way we get a category, say $ParSets$.

Notice that each set of morphisms $Par(X,Y)$ is *pointed*, in the sense that it has a canonical element "empty map," say, $\varnothing_{X,Y}$. Its composition with any other morphism is again the respective empty map.

(b) This last remark motivates the consideration of another category: that of *pointed sets PSets*. An object of $PSets$ is a pair $(X, *_X)$, where $*_X \in X$ (so that X cannot be empty). A morphism $(X, *_X) \to (Y, *_Y)$ is an everywhere defined map $\varphi : X \to Y$ such that $\varphi(*_X) = *_Y$. The composition is evident.

Deleting marked points, we get a functor $PSets \to ParSets$:

$$X \mapsto X^\circ := X \setminus \{*_X\}, \quad \varphi \mapsto \varphi^\circ := (f, D(f)),$$

where for $\varphi : X \to Y$, $D(f)$ is defined as $\varphi^{-1}(Y^\circ)$, and f as the restriction of φ to $D(f)$.

This functor turns out to be *an equivalence of categories.*

In fact, a quasi-inverse functor can be constructed by formally adding an extra marked point $*_X$ to each object X in $ParSets$, and extending each partial map $(f, D(f))$ from X to Y by sending $X \setminus D(f)$ to $*_Y$.

This formal completion of sets and partial maps by adding "improper," "infinite" elements was reinvented many times, in particular, in topology (one-point compactification) and in theoretical computer science. I am grateful to A. Beilinson, who drew my attention to the good categorical properties of this operation.

The basic category of sets is endowed by the symmetric monoidal structure: Cartesian product. It is naturally extended to $ParSets$ and to $PSets$. In $PSets$ one can put

$$(X, *_X) \times (Y, *_Y) := (X^\circ \times Y^\circ) \cup \{(*_X, *_Y)\},$$

so that the equivalence above becomes monoidal equivalence.

An equivalent (functorially isomorphic) definition uses "reduced product." Namely, $(X, *_X) \times (Y, *_Y)$ can be defined as $X \times Y$ with the "coordinate cross" $X \times \{*_Y\} \cup \{*_X\} \times Y$ contracted to the base point.

There is another symmetric monoidal structure on $Sets$: *disjoint union* $\coprod$.

It is not canonical and requires choices: what is the disjoint union of a set with itself? For a construction, see, e.g., F. Borceux, *Handbook of Categorical Algebra 2* (Cambridge UP, 1994), Example 6.1.9.

This structure, as soon as it is chosen, can be directly extended to $ParSets$ and $PSets$.

Below, we will use both points of view on partial maps interchangeably, as equivalent ones.

2.2. Definition. A subcategory C of $ParSets$ as above is called *a constructive universe* if it contains the constructive world $\mathbf{Z}^+$ of all integers ≥ 1, and also finite sets $\emptyset$, $\{1\}, \ldots, \{1, \ldots, n\}, \ldots$ and satisfies the following conditions (a)–(d):

(a) $C(\mathbf{Z}^+, \mathbf{Z}^+)$ is defined as the set of all partially recursive functions.
(b) Any infinite object of C is isomorphic in C to $\mathbf{Z}^+$.

(c) If U is finite, $\mathcal{C}(U,V)$ consists of all partial maps $U \to V$. If V is finite, $\mathcal{C}(U,V)$ consists of f such that $D(f)$ and inverse images of all elements of V are enumerable.

(d) $\mathcal{C}$ inherits from *ParSets* two compatible symmetric monoidal structures: Cartesian product $\times$ and disjoint sum $\coprod$.

2.3. Comments. (i) The statement (b) is a version of the Church thesis. In V.2.4 we stated Church's thesis in the context of functions from $(\mathbf{Z}^+)^m$ to $(\mathbf{Z}^+)^n$.

Here we make it simultaneously broader and vaguer. Imagine that we want to speak about algorithmic processing of variable finite objects of a given type U into similar objects of possibly different type V. U and V might be words, graphs, groups, finite and finitely describable Bourbaki structures, ... We postulate that one always can translate such a processing into the calculation of values of a recursive function. The main step in the reduction is the choice of two "computable numberings," those of U and V.

Formally, such an numbering is an isomorphism $\mathbf{Z}^+ \to U$ in $\mathcal{C}$. Two such different numberings of the same constructive world can differ only by a recursive permutation of numbers, that is, by an automorphism of $\mathbf{Z}^+$ in $\mathcal{C}$. We will call such numberings *equivalent ones.*

In practice, a numbering of a set-theoretically defined constructive world U, embedding it into $\mathcal{C}$, is chosen in such a way that some "natural" constructions on constructive objects of the type U given a priori become obviously computable.

For example, we can renumber U in an eminently theoretically important and sophisticated way, ordering U by the growing Kolmogorov complexity of its constructive objects. But then the simplest operations would become noncomputable. Generally, such a Kolmogorov numbering *will not* be an isomorphism in $\mathcal{C}$: cf. further discussion in Section 10.

Returning to (b), we see that each infinite constructive world, that is, an object of $\mathcal{C}$, is endowed with a well-defined class of enumerable subsets. This fact is used in the statement (c). The axiom (c) is justified by the fact that partial recursive functions on $\mathbf{Z}^+$ taking only a finite number of values are characterized by the stated properties.

Similarly, decidable subsets are well defined.

(ii) Notice that because of (c), two finite constructive worlds are isomorphic iff they have the same cardinality, and the automorphism group of any finite U consists of all permutations of U. Therefore, the whole category $\mathcal{C}$ is equivalent to its full subcategory, whose objects are $\mathbf{Z}^+$ and finite sets, one of each cardinality.

However, this subcategory is too small to accommodate even our standard definition of partial recursive functions in V.2: we have to extend it by Cartesian products. For many constructions, it is also convenient to have disjoint sums. This is the reason we completed the definition by the requirement (d). It implies that canonical projections of Cartesian products and structure embeddings into disjoint sums are computable.

(iii) In view of the previous remark, any two constructive universes are equivalent (even as monoidal categories). Nevertheless, as a matter of principle, *we always consider $\mathcal{C}$ as an open category, and at any moment allow ourselves to add to it new constructive worlds.* If some infinite V is added to $\mathcal{C}$, it must come together with a class of equivalent numberings.

In this way, we may declare the world of a decidable subset of any object of $\mathcal{C}$ to be an object of $\mathcal{C}$.

Here is another example. The world U^* of finite sequences of elements of a constructive world U ("words in the alphabet U") is endowed with a canonical class of numberings. Hence we may assume that $\mathcal{C}$ is closed with respect to the construction $U \mapsto U^*$. All natural functions, such as length of the word $U^* \to \mathbf{Z}^+$, or the ith letter of the word $U^* \to U$, are computable. Moreover, if $f : U \to V$ is a morphism in $\mathcal{C}$, then the partial function f^* sending $(u_1, \dots, u_n)$ to $(f(u_1), \dots, f(u_n))$, whenever all $f(u_i)$ are defined, is a morphism $U^* \to V^*$, and $(g \circ f)^* = g^* \circ f^*$. Hence $U \mapsto U^*$ extends to a covariant endofunctor $\mathcal{C} \to \mathcal{C}$.

(iv) Some (or even "all"?) infinite constructive worlds U come together with a natural class of bijective numberings $u : \mathbf{Z}^+ \to U$ such that any two numberings u, v in this class have one of the following properties:

$u^{-1} \circ v$ is a primitive recursive permutation;

or even

$u^{-1} \circ v$ is a *polynomial-time* computable permutation (cf. 6.5 below).

If a version of $\mathcal{C}$ includes only objects satisfying the first (resp. the second) condition, one can define a subcategory $\mathcal{C}_{prim}$ (resp. $\mathcal{C}_{pol}$) having the same objects, but only primitive recursive (resp. polynomial-time computable) morphisms.

The assumption that "all" constructive worlds do in fact satisfy one of the two requirements could be called the "primitive recursive," resp. "polynomial-time" Church's thesis.

2.4. A natural numbers object. We could have replaced $\mathbf{Z}^+$ in the above discussions by an abstract natural numbers object in an unspecified category $\mathcal{B}$. Its definition conforms to a general spirit of categorical reasoning: *sets of morphisms rather than objects* should be bearers of additional structures.

More precisely, assume that $\mathcal{B}$ admits a terminal object $\mathbf{1}$. *A triple* $(\mathcal{N}, z, s)$ *in* $\mathcal{B}$, consisting of an object $\mathcal{N}$ and two morphisms

$$z : \mathbf{1} \to \mathcal{N},\ s : \mathcal{N} \to \mathcal{N},$$

is called *a natural numbers object* if for any other pair of morphisms in $\mathcal{B}$ of the form

$$f : \mathbf{1} \to X,\ g : X \to X$$

there exists a unique morphism $h : \mathcal{N} \to X$ such that

$$h \circ z = f,\ h \circ s = g \circ h,$$

that is, the diagram is commutative. Of course, the leftmost arrow can only be $id_{\mathbf{1}}$.

$$\begin{array}{ccccc} \mathbf{1} & \xrightarrow{z} & \mathcal{N} & \xrightarrow{s} & \mathcal{N} \\ \downarrow & & \downarrow h & & \downarrow h \\ \mathbf{1} & \xrightarrow{f} & X & \xrightarrow{g} & X \end{array}$$

This is the simplest form of categorical recursion: values of the morphism h on the categorical points $s^{\circ n} \circ z \in \mathcal{B}(\mathbf{1}, \mathcal{N})$ are given by $g^{\circ n} \circ f \in \mathcal{B}(\mathbf{1}, X)$. Thus, f is the initial condition (value at $n = 0$), and g corresponds to one iterative step applied to the previous value.

Clearly, $\mathbf{Z}^+$ together with

$$z : \mathbf{1} \mapsto 1 \in \mathbf{Z}^+, \ s : n \mapsto n+1$$

is a natural numbers object in the category of sets.

We will return to this philosophy, discussing *normal models of computation* in 6.1 below.

In Sections 3–5, we however, we stick to the more down-to-earth approach, sketched at the beginning of this section.

3 Expanding Constructive Universe: Morphisms

3.1. **Programming methods.** We now turn to the computability properties of the sets of morphisms $\mathcal{C}(U, V)$. Again, it is a matter of principle that $\mathcal{C}(U, V)$ itself, and even $\mathcal{C}_{prim}$, *is not a constructive world if* U *is infinite.*

Indeed, otherwise we would have an intuitively computable bijective numbering of all partial recursive (resp. primitive recursive) functions $\mathbf{Z}^+ \to \mathbf{Z}^+$. Using numbers of such functions as their descriptions, we could algorithmically distinguish them. But the latter problem is not algorithmically solvable.

In order to compensate this by a sample of positive statements, let us consider the following situation.

Any diagram in $\mathcal{C}$

$$\mathrm{ev_P} : P \times U \to V$$

(evaluation morphism) defines a partial map $P \to \mathcal{C}(U, V)$, $p \mapsto \overline{p}$, where $\overline{p}(u) := \mathrm{ev}_P\,(p, u)$.

3.2. **Definition.**

(a) We will say that a constructive world $P = P(U, V)$ together with the evaluation map $\mathrm{ev_P}$ as above is *a programming method.* Elements of P are called *programs.*

(b) A programming method $(Q = Q(U, V), \mathrm{ev}_Q)$ is called *versal* (resp. *primitive versal*) if two conditions are satisfied.

First, the map $Q \to \mathcal{C}(U,V) : q \mapsto \overline{q}$ is surjective (resp. its image consists of all primitive recursive morphisms).

Second, for any programming method $(P = P(U,V), \mathrm{ev}_P)$ with the same source U and target V (resp. for any (P, ev_P) producing only primitive recursive morphisms) there is at least one *compilation morphism* in $\mathcal{C}$

$$\mathrm{comp} : P(U,V) \to Q(U,V),$$

that is, an everywhere defined, computable map $P \to Q$ such that if $\mathrm{comp}(p) = q$, then $\overline{p} = \overline{q}$.

3.3. **Claim.** *Versal programming methods exist.*

PROOF. For brevity, we will consider only the case of infinite U, V. Then P is infinite as well. Since any infinite object is isomorphic to $\mathbf{Z}^+$, we will identify U, V with $\mathbf{Z}^+$, but for convenience we will keep the notation P for the world of programs. Thus we may restrict ourselves to considering only evaluation morphisms $\mathrm{ev} : P \times \mathbf{Z}^+ \to \mathbf{Z}^+$.

Such a morphism computes all recursive functions $\mathbf{Z}^+ \to \mathbf{Z}^+$ iff it is a versal family in the sense of V.5.7.

Now consider another versal family, that of recursive functions of two variables $P \times \mathbf{Z}^+ \to \mathbf{Z}^+$. Let $P^{'}$ be its base:

$$\mathrm{Ev} : P^{'} \times P \times \mathbf{Z}^+ \to \mathbf{Z}^+.$$

We now affirm that *the programming method* $(Q := P^{'} \times P, \mathrm{Ev})$ *is versal.*

In fact, versality of Ev implies that for any $\mathrm{ev} : P \times U \to V$, there exists $p^{'} \in P^{'}$ such that $\mathrm{Ev}\,(p^{'}, p, u) = \mathrm{ev}\,(p, u)$ for all $(p, u) \in P \times \mathbf{Z}^+$. Therefore, the map

$$\mathrm{comp} : P \to Q : p \mapsto (p^{'}, p)$$

is a compilation morphism for (P, ev).

Remark. We can now make precise the statement made at the beginning of 3.1. Namely, it means that for any programming method $P(U,V)$, the canonical map $P(U,V) \to \mathcal{C}(U,V)$ cannot be bijective if U is infinite. In fact, if it is surjective, then it is essentially the same as a versal family; but the equivalence relation on the base of a versal family induced by $p \mapsto \overline{p}$ is not decidable (or even recursively enumerable).

3.4. **Composition of morphisms at the level of programming methods.** Let U_1, U_2, U_3 be three objects of $\mathcal{C}$, and $(Q_{ij}, \mathrm{ev}_{ij})$ three versal programming methods, for $\mathcal{C}(U_i, U_j)$, $ij = 12, 13, 23$ respectively.

Then $(Q_{23} \times Q_{12}, \mathrm{ev}_{23} \circ (id_{Q_{23}} \times \mathrm{ev}_{12}))$ is a programming method for $\mathcal{C}(U_1, U_3)$. It calculates the composition of morphisms $U_1 \to U_2 \to U_3$.

Since Q_{13} is versal for morphisms $U_1 \to U_3$, there exists a compilation morphism

$$\mathrm{comp} : Q_{23} \times Q_{12} \to Q_{13}$$

that reproduces composition of morphisms on the level of programs.

Notice that even if we restrict ourselves to the full subcategory with one object $U_1 = U_2 = U_3 = \mathbf{Z}^+$ and fix a choice of Q and comp, *the composition of morphisms on the level of programs generally will not be associative.* Moreover, a program calculating identical morphisms generally will not be the identity for program composition.

This motivates the following definition.

3.5. **Definition.** *A category of algorithms* over a constructive universe $\mathcal{C}$ is a pair consisting of a category $\mathcal{A}$ and a functor $J : \mathcal{A} \to \mathcal{C}$ with the following properties:

(a) $\mathcal{A}$ is enriched over $\mathcal{C}$.

This means in particular that morphism sets in $\mathcal{A}$ are objects of $\mathcal{C}$, and the composition maps $\mathcal{A}(U,V) \times \mathcal{A}(V,W) \to \mathcal{A}(U,W)$, as well as identities, are morphisms in $\mathcal{C}$ fitting into standard commutative diagrams.

(b) J identifies $Ob\,\mathcal{A}$ with a subset of $Ob\,\mathcal{C}$. We will make no distinction between U and $J(U)$.

(c) For any objects U, V of $\mathcal{A}$, $\mathcal{A}(U,V)$ is a programming method. In particular, it comes together with the evalution morphism in $\mathcal{C}$

$$\mathrm{ev}_{U,V} : \mathcal{A}(U,V) \times U \to V.$$

This morphism must satisfy the following condition: for all $f \in \mathcal{A}(U,V)$ and $u \in U$,

$$J(f)(u) = ev_{U,V}(f,u).$$

3.6. **Comments.** (i) The notion of a category of algorithms formalized in the previous definition was introduced (in a somewhat less explicit form) by N. Yanofsky in math.LO/0602053. The same paper contains a construction of such a category in which J defines surjections $J : \mathcal{A}(U,V) \to \mathcal{C}_{prim}(U,V)$.

(ii) Since $\mathcal{A}$ is enriched over $\mathcal{C}$, we actually work here in a 2-categorical context: morphisms in $\mathcal{A}$, being objects of $\mathcal{C}$, are connected by 2-morphisms. In particular, the associativity of composition is not a literal family of identities $h \circ (f \circ g) = (h \circ f) \circ g$ but rather a family of canonical isomorphisms

$$a_{h,f,g} : h \circ (f \circ g) \to (h \circ f) \circ g$$

interconnected by the standard coherence conditions.
A similar remark applies to left and right identities.

(iii) Given a category $\mathcal{A}$ as above, we will call programs $p \in \mathcal{A}(U,V)$ *algorithms.* In fact, N. Yanofsky reserves this name for a category satisfying stronger coherence properties, which is in a certain sense canonical. A part of his constructions will be described in Section 5.

4 Operads and PROPs

In this section, we will consider a somewhat reduced version $\mathcal{C}_0$ of the constructive universe with two monoidal structures $(\mathcal{C}, \times, \coprod)$ defined in 3.2. First, we will exclude all finite objects of cardinality ≥ 2.

4.1. **Definition.** $(\mathcal{C}_0, \times)$ is a full monoidal subcategory of $(\mathcal{C}, \times)$ such that each object of $\mathcal{C}_0$ is either infinite or has cardinality 1.

4.2. **Reduction.** From Definition 3.2 it follows that $\mathcal{C}_0$ is equivalent to its full subcategory consisting of Cartesian powers $(\mathbf{Z}^+)^m$, $m \geq 0$, and partial recursive functions. Moreover, $(\mathbf{Z}^+)^m \times (\mathbf{Z}^+)^n$ can be canonically identified with $(\mathbf{Z}^+)^{m+n}$, so that the category will become strict. The zeroth Cartesian power is a one-point set $\{*\}$, the unit for the monoidal structure.

The family of morphisms $\mathcal{C}((\mathbf{Z}^+)^m, (\mathbf{Z}^+)^m)$, and in fact similar families of morphisms in any symmetric or enriched symmetric monoidal category, are naturally endowed with structures, known under the names *collections* and PROPs.

4.3. **Definition.** (a) *A collection* $\mathcal{P}$ in a category $\mathcal{B}$ is a family of objects $\mathcal{P}(m,n)$, $m, n \geq 0$ in $\mathcal{B}$, together with group homomorphisms

$$\mathbf{S}_m \times \mathbf{S}_n^{op} \to Aut_{\mathcal{B}}\, \mathcal{P}(m,n).$$

We interpret such a homomorphism as a pair consisting of a left action of the symmetric group $\mathbf{S}_m$ and a right action of $\mathbf{S}_n$ on $\mathcal{P}(m,n)$ that commutes with it.

(b) *A morphism* of collections $f : \mathcal{P} \to \mathcal{Q}$ is a family of morphisms $f_{m,n} : \mathcal{P}(m,n) \to \mathcal{Q}(m,n)$ commuting with the action of symmetric groups.

4.4. **Endomorphism collections.** Let $(\mathcal{E}, \times)$ be a symmetric monoidal category with unit object e. For $U \in Ob\, \mathcal{E}$, put

$$Coll\, End\, (U)(m,n) := \mathcal{E}(U^n, U^m).$$

The action of $\mathbf{S}_m$ (resp. $\mathbf{S}_n^{op}$) is induced by permutations of factors in the Cartesian powers U^m (resp. U^n). The zeroth power is interpreted as e.

Whenever $\mathcal{E}$ is an enriched category, one must first make sense of permutation groups acting on objects in the category of morphisms. This does not present any additional difficulties.

A PROP is a collection, endowed with additional composition laws mutually compatible with the actions of the symmetric groups.

4.5. **Vertical and horizontal products in endomorphism collections.** Endomorphism collections are naturally endowed with two additional structures:

(a) *Vertical products*

$$\mathcal{E}(U^m, U^n) \times \mathcal{E}(U^n, U^l) \to \mathcal{E}(U^m, U^l) : \quad (f,g) \mapsto g \circ f.$$

(b) *Horizontal products*

$$\mathcal{E}(U^{m_1}, U^{n_1}) \times \cdots \times \mathcal{E}(U^{m_s}, U^{n_s}) \to \mathcal{E}(U^{m_1+\cdots+m_s}, U^{n_1+\cdots+n_s}).$$

The latter are induced by the monoidal structure in $\mathcal{E}$:

$$(f_1, \ldots, f_s) \mapsto f_1 \times \cdots \times f_s .$$

If $\mathcal{E}$ is enriched, the category of morphisms must be strict monoidal, and its monoidal structure must be compatible with that of $\mathcal{E}$ in the standard way, so that the horizontal products still make sense.

In a constructive universe, a vertical product is the composition/substitution of partial maps.

These structures in endomorphism collections satisfy a number of cumbersome but straightforward universal conditions, which we only list here:

(i) Associativity of vertical products; units for them in $\mathcal{E}(m, m)$.
(ii) Compatibility of vertical products with actions of symmetry groups.
(iii) Associativity of horizontal products.
(iv) Compatibility of horizontal and vertical products.
(v) Compatibility of horizontal products with actions of symmetric groups.

Assuming that these conditions have been written formally, we can now give a general definition:

4.6. (Tentative) definition.

(a) A PROP in a category $\mathcal{B}$ is a collection in $\mathcal{B}$, endowed with horizontal and vertical compositions as in 5.3, enjoying the universal properties 4.5 (i)–(v).
(b) *An operad* in a category $\mathcal{B}$ is a collection whose only nontrivial terms are $\mathcal{P}(1, n)$, endowed with a right action of $\mathbf{S}_n$ and vertical products that satisfy 4.5 (i),(ii).

The collection *Coll End* (U) as above is denoted by *Prop End* (U) when it is endowed with its natural structures

Any PROP produces a collection if compositions are forgotten; this functor under quite general conditions can be proved to have a left adjoint functor: *free PROP generated by a collection.* This gives a rise to the notion of *subcollection of generators* of a PROP similar to, say, generators of a monoid.

We are most interested in *Prop End*$_{\mathcal{C}}(\mathbf{Z}^+)$ as an algebraic approximation to the constructive universe $\mathcal{C}$. We might also try to restrict ourselves to its primitive recursive version. However, it turns out that the preceding framework, even we if take the trouble to formalize it by supplying all commutative diagrams implicit in Definition 4.6, is too narrow for our goals.

4.7. Example: the collection of basic recursive functions. Working now in $\mathcal{C}$, we can define the collection of basic recursive functions $\mathcal{R} \subset$ *Prop End*$_{\mathcal{C}}$ $(\mathbf{Z}^+)$, using the notation of V.2.2 . The respective terms of the collection are

$$\mathcal{R}(1,0) := \{1^{(0)}\},$$
$$\mathcal{R}(1,1) := \{\text{suc}, 1^{(1)}, \text{id}^{(1)}\},$$

$$\mathcal{R}(1,n) := \{1^{(n)}, \mathrm{pr}_i^n\} \quad \text{for} \quad n \geq 2.$$

The remaining components of $\mathcal{R}$ will be empty.

The action of the symmetric groups is induced by that in $Prop\,End_{\mathcal{C}}\,(\mathbf{Z}^+)$. In fact, it is not identical only on $\mathcal{R}(1,n)$: the pr_i^n are permuted as the i's are, $i \in \{1,\ldots,n\}$.

We would like to have an algebraic structure reflecting our knowledge that basic functions "generate" all primitive/partial recursive functions. But to do this, we lack some necessary operators iteratively acting on basic functions. In fact, composition V.2.3 (a) is accommodated in the general definition of PROP, and juxtaposition can be dealt with if we add the diagonal $\Delta : \mathbf{Z}^+ \to \mathbf{Z}^+ \times \mathbf{Z}^+$, but the recursion and μ-operator are very specific for $\mathcal{C}$, and we lack general means to deal with them.

In the next section, we will introduce the constructive world of graphs, and its extensions, worlds of *decorated graphs*. We will turn these worlds into categories, and will explain how they provide very convenient *linguistic tools* for speaking about PROPs and similar structures, in particular, about the PROP of recursive functions.

Later we will see that similar constructions naturally arise in the computation theory as well.

The relevant graphs will be (geometric versions of) *Boolean circuits*, finite automata for processing binary input data.

5 The World of Graphs as a Topological Language

5.1. **Introduction.** Generally, each constructive world comes with its own supply of "natural operations." Although any two constructive worlds of the same cardinality are connected by a computable isomorphism, this does not mean that, say, a natural numbering of formulas in a language of arithmetic provides convenient tools for their syntactic analysis or for thinking about their interpretations in a model.

In particular, when we replace nonconstructive sets of morphisms, say $\mathcal{C}(U^m, U^n)$, by a constructive world of respective programming methods, we have to deal with *two different sets of natural operations* in this constructive world:

(a) *Evaluations* (see 3.1), where a programming method being fixed, the main operation consists in calculating values of, say, a partial recursive function.
(b) Operations, *producing new programming methods from old ones*, such as composition, compilation, recursion.

In principle, the latter are not qualitatively different from evaluations, since we can think about programming methods whose inputs and outputs are programming methods as well.

What is needed for efficient constructivization of programming methods is a good encoding scheme, simultaneously intuitive and accommodating natural operations.

We already mentioned two mental worlds in which various encoding schemes can crystallize:

(i) World of expressions in a language (to which we appealed in previous chapters).
(ii) World based on scientific/engineering imagery, such as Turing's machines, or Boolean circuits (cf. below).

In this section, we will describe the third, topological one:

(iii) World of (decorated) *graphs*: geometric images of information flows and hubs where the flows merge, get processed, and diverge again to flow further.

Moreover, we will formalize and endow this world by the structure of a constructive category.

Looking at graphs as a replacement of formulas in a language, we define *models/interpretations* as functors on various categories of decorated graphs.

5.2. **Graphs.** One usually imagines a graph as a picture, or better, a topological space, consisting of several points (*vertices*) pairwise connected by several (curvi)linear segments (*edges*).

We will consider each edge as consisting of two "halves" (*flags*), issuing from their respective vertices and joined at the edge's midpoint. Moreover, we will allow certain flags not to be paired into edges; they will be called *tails*.

A combinatorial graph is a collection of two abstract sets and two incidence relations. Here is a formal definition.

5.3. **Definition.** A *combinatorial graph*, or simply *graph*, τ is a quadruple $(F_\tau, V_\tau, \partial_\tau, j_\tau)$, where F_τ, V_τ are finite sets (elements of a constructive world), and (∂_τ, j_τ) are maps. Elements of F_τ are called *flags* of τ, elements of V_τ are called *vertices* of τ; vertices and flags are disjoint. The map $\partial_\tau : F_\tau \to V_\tau$ associates to each flag a vertex, its *boundary.* The map $j_\tau : F_\tau \to F_\tau$ is an involution: $j_\tau^2 = id$.

(a) *Marginal cases.* If V_τ is empty, F_τ must be empty as well. This defines an *empty graph.* In contrast, F_τ might be empty whereas V_τ is not.
(b) *Corollas, tails, edges.* One-vertex graphs with identical j_τ are called *corollas.* Let v be a vertex of τ, $F_\tau(v) := \partial_\tau^{-1}(v)$. Then $\tau_v := (F_\tau(v), \{v\},$ evident ∂, identical j) is a corolla, which is called by the corolla of v in τ.

Flags fixed by j_τ form the set of *tails* of τ denoted by T_τ.

Two-element orbits of j_τ form the set E_τ of *edges* of τ. Elements of such an orbit are called *halves* of the respective edge.

5.4. **Geometric realization of a graph.** First, let τ be a corolla. If its set of flags is empty, its *geometric realization* $|\tau|$ is, by definition, a point. Otherwise construct a disjoint union of segments $[0, 1/2]$ bijectively indexed by flags, and identify in it all points 0. This is $|\tau|$. The image of all 0's thus becomes the geometric realization of the unique vertex of τ.

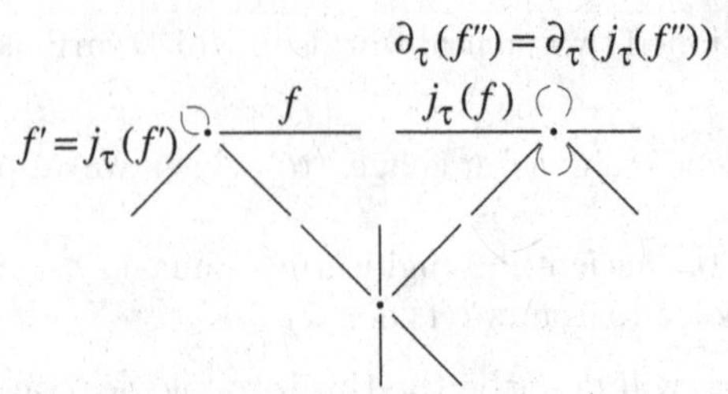

Generally, to construct $|\tau|$ take a disjoint union of geometric realizations of corollas of all vertices and identify points $1/2$ of any two flags forming an orbit of j_τ, that is, an edge.

A graph τ is called *connected* (resp. *simply connected*, resp. *tree* etc) iff its geometric realization is such. In the same vein, we can speak about connected components of a graph, etc. Vertices v with empty $F_\tau(v)$ are considered connected components.

5.5. **Decorations.** We will not try to aximatize a general notion of decoration, and only list some classes of them most useful for describing flowcharts.

(a) *Orientations.* Any map $F_\sigma \to \{in, out\}$ such that halves of any edge are oriented by different labels is called *an orientation of* σ. On the geometric realization, a flag marked by *in* (resp. *out*) is oriented toward (resp. away from) its vertex.

Tails of σ oriented *in* (resp. *out*) are called (*global*) *inputs* (resp. (*global*) *outputs*) of σ. Similarly, $F_\sigma(v)$ is partitioned into inputs and outputs of the vertex v.

Consider an orientation of σ. Its edge is called an *oriented loop* if both its halves belong to the same vertex. Otherwise, an oriented edge starts at a source vertex and ends at a different target vertex.

More generally, a sequence of distinct edges $e_1, \dots, e_n$, is called a *simple path* of length n if e_i and e_{i+1} have a common vertex and the $n-1$ vertices obtained in this way are distinct. If, moreover, e_1 and e_n also have a common vertex distinct from the mentioned ones, this path is a *wheel* of length n. A loop is a wheel of length one. Edges in a wheel are endowed only with a cyclic order up to inversion.

Clearly, all edges in a path (resp. a wheel) can be oriented so that the source of e_{i+1} is the target of e_i.

If the graph is already oriented, the induced orientation on any path (resp. wheel) either has this property or does not. Respectively, the path is called oriented or not.

(b) *Directed graphs.* An oriented graph σ is called *directed* if it satisfies the following condition:

On each connected component of the geometric realization, one can define a continuous real-valued function ("height") in such a way that moving in the direction of orientation along each flag decreases the value of this function.

In particular, a directed graph has no oriented wheels.

Notice that, somewhat counterintuitively, a directed graph is not necessarily oriented "from its inputs to its outputs" as is usually shown on illustrating pictures. In effect, take a corolla with only *in* flags and another corolla with only *out* flags, and graft one input to one output. The resulting graph is directed (check this) although its only edge is oriented from global outputs to global inputs.

This is one reason why it is sometimes sensible to consider only those directed graphs that have at least one input and atleast one output at each vertex.

(c) *Labeling of vertices.* A labeling of vertices by a set S is a map $V_\tau \to S$. As above, S may consist, e.g., of names of basic functions.

(d) *Coloring of flags.* A coloring of flags by a set I is a map $F_\tau \to I$. In the context of flowcharts, we can imagine, for example, that we start with a family of objects $\{U_i \,|\, i \in I\}$, and want to describe morphisms between products of such objects. Then the color i of an input/output will specify that this input/output must be taken from U_i. In this case halves of an edge must have the same color.

Even if we have only one object in this family, we may want to totally order the sets of inputs/outputs of each vertex. This is what is needed to present the vertex as encoding a map $U^m \to U^n$ rather than a map $U^{\{inputs\}} \to U^{\{outputs\}}$, and make a direct connection with the world of descriptions, using traditional notation for functions, such as $(f_1(u_1,\dots,u_m),\dots,f_n(u_1,\dots,u_m))$. Such a total ordering of, say, inputs is equivalent to their coloring by $\{1,\dots,m\}$. This is the case when an ordering is not intrinsically needed, but used only in the comparison of flowcharts with descriptions.

We will now explain that after introducing morphisms of graphs, we will be able to efficiently use them to encode operations and identities between operations.

5.6. **Isomorphisms of graphs.** The notion of isomorphism is (almost) straightforward: *an isomorphism $h: \tau \to \sigma$ consists of two bijections*

$$h_V : V_\tau \to V_\sigma, \quad h^F : F_\sigma \to F_\tau$$

commuting with boundary and involution maps. Composition is composition of maps.

Notice, however, one peculiarity: h_V is covariant, whereas h^F is *contravariant.* This choice can be explained using the intuition behind flowcharts: a change of arguments produces the lift of functions *in the reverse direction.*

5.7. **Groupoid of corollas** Cor. Consider first *the category (groupoid) of oriented corollas with isomorphisms preserving orientation.*

It is equivalent to the groupoid whose objects are ordered pairs of sets $\{\{1,\dots,m\},\ \{1,\dots,n\}\}$ and morphisms are permutations acting on two sets separately.

5.8. **Claim.** A collection $\mathcal{P}$ in a category $\mathcal{B}$ (cf. Definition 4.3) is "the same as" a $\mathcal{B}$-valued functor $\mathcal{P}$ on the groupoid of oriented corollas.

In fact, $\mathcal{P}(n,m)$ can be identified with the value of $\mathcal{P}$ on a corolla with inputs $\{1,\dots,m\}$ and outputs $\{1,\dots,n\}$. The action of $\mathbf{S}_m \times \mathbf{S}_m^{op}$ is determined by values of $\mathcal{P}$ on the automorphisms of this corolla.

5.9. **Disjoint sums of corollas and mergers.** A graph $\tau = (F_\tau, V_\tau, \partial_\tau, j_\tau)$ is called a *disjoint sum of corollas* if its set of edges is empty. Equivalently, all flags are tails.

Let τ, σ be disjoint sums of corollas. Define *a merger morphism* $\tau \to \sigma$ as a pair of maps, compatible with boundaries,

$$h_V : V_\tau \to V_\sigma, \quad h^F : F_\sigma \to F_\tau$$

such that h_V is a surjection and h^F is a bijection. Composition of mergers is obviously a merger. If σ is a corolla, h is called a *total merger.*

We will assume that a monoidal structure disjoint union $\coprod$ on $\mathcal{C}$ is chosen and fixed; it can be naturally extended to graphs and then restricted to the category of disjoint sums of corollas.

Denote by $DCor$ the category of disjoint sums of corollas with compositions of mergers and automorphisms as morphisms.

5.10. **Claim.** A collection $\mathcal{P}$ in a symmetric monoidal category $(\mathcal{B}, \times)$, endowed with horizontal products 5.3.(b) satisfying the associativity conditions 5.3(iii) and compatibility with action of symmetric groups 5.3(v), is "the same as" a *symmetric monoidal functor*

$$\mathcal{P} : (DCor, \coprod) \to (\mathcal{B}, \times).$$

In fact, horizontal products as given in 4.5 are simply values of $\mathcal{P}$ on obvious total mergers.

A stylistic remark: the quotation marks around the expression "the same as" are supposed to alert the reader to the fact that Claim 5.10 must in fact be understood as the first *definition* of a collection with horizontal compositions. Having avoided a precise statement of the compatibility conditions 4.5 (iii) and 4.5 (v), we now simply hide them in the standard definition of a (symmetric monoidal) functor and implicit combinatorics of mergers and isomorphisms.

We still do not have enough morphisms to give a definition of *PROPs* as functors. We will now supply them, by introducing contraction morphisms.

5.11. **Definition.** (a) *A contraction morphism* $h : \tau \to \sigma$ is a pair of maps

$$h_V : V_\tau \to V_\sigma, \quad h^F : F_\sigma \to F_\tau$$

such that h^F is an injection bijective on tails, h_V is a surjection, and any two vertices in a fiber $h_V^{-1}(v)$ can be connected by a path consisting of edges whose halves lie in $F_\sigma \setminus h^F(F_\tau)$.

(b) If σ, τ are oriented, h^F must be compatible with orientation.

5.12. **Application to** *PROPs.* In geometric realizations, a contraction morphism induces a map that boils down to the geometric contraction of a subgraph of τ consisting of edges in $F_\sigma \setminus h^F(F_\tau)$.

Let us show how combined grafting and contraction of flowcharts allows us to interpret functorially the composition of morphisms in $Prop\,End\,(U)$, that is, vertical products in 4.5 (a).

Namely, first we interpret $\mathcal{E}(U^m, U^n)$ as the value of a functor $\mathcal{P} : DCor \to Sets$ on sums of oriented corollas endowed with automorphisms and mergers. Now extend the category *DCor* to include morphisms that can be obtained as graftings followed by contractions (and, of course, products of such morphisms). Our functor $\mathcal{P}$ has a natural extension to this larger category. In particular, if we take the union of two oriented corollas, graft bijectively outputs of the first one to inputs of the second one, and then contract all edges obtained in this way, we will get a morphism in the extended *DCor*, and the value of $\mathcal{P}$ on it will be the composition map 4.5 (a).

We will now present another category of decorated graphs that can be used to generate descriptions of (primitive) recursive functions. This is a modified version of a part of Yanofsky's preprint math. CT/0609748.

5.13. **The constructive world of decorated graphs** *Prim.* Elements of *Prim* are disjoint unions of trees τ in which each vertex is the boundary of at least two flags. Moreover, τ must be endowed with an *admissible decoration.* The latter consists of the following data. They can be chosen independently on each connected component so that in the following discussion we speak about trees if we have not explicitly mentioned the general case.

(a) *A marked tail,* which is called the *root,* or *the (global) output* of τ. Its vertex is called the *root vertex.* The remaining tails are called *(global) inputs* of τ. Global inputs form a set $F_\tau^{in} \subset F_\tau$, and we consider the global output as an one-element subset $F_\tau^{out} \subset F_\tau$.

A choice of root determines (and is equivalent to) the choice of a specific *orientation*: a map $F_\tau \to \{in, out\}$. Namely, in each shortest path (sequence of flags) from a global input to the root, assign *out* to the flag that leaves its vertex, and *in* to the flag that enters it. This defines the partition of all flags into two subsets: (local) inputs and outputs.

We will say that τ with such a decoration is an *oriented tree.* We repeat that by definition, each oriented tree must have exactly one global output and at least one global input.

(b) All corollas of an oriented tree are also oriented trees. The next part of a decoration is a choice of *total order on the set of inputs of each corolla of* τ, and, if τ is not connected, a choice of total order on the set of its connected components.

(c) A map *arity/coarity*: $F_\tau \to \mathbf{N} : f \mapsto (a(f), c(f))$. If two flags are halves of an edge, they must be assigned the same arity/coarity.

(d) A map $op: V_\tau \to \{\mathbf{c}, \mathbf{b}, \mathbf{r}\}$. The value $op\,(v)$ assigned to a vertex is called the respective *operator*: $\mathbf{c}, \mathbf{b}, \mathbf{r}$ stand respectively for *composition, bracketing, recursion.*

(e) A map $in: F_\tau^{in} \to$ {*basic recursive functions*} such that for each $i \in F_\tau^{in}$, $in(i)$ is a basic function of arity $a(i)$ and coarity $c(i)$.

All these data must be *compatible.* A part of the compatibility conditions was already included in the description. We will now formally introduce the remaining set, and simultaneously explain an interpretation of graphs in *Prim* (without decoration 5.11 (e)) as operations acting on families of input functions.

5.14. **Objects of *Prim* as flowcharts.** Given an oriented tree τ with a decoration as above, we interpret the whole of τ as a symbol of an *operation* $Op(\tau)$ that can be performed over families of functions, indexed by global inputs of τ.

More precisely, let $f = \{f_i \,|\, i \in F_\tau^{in}\}$ be a family of functions (or even partial functions) such that $f_i: (\mathbf{Z}^+)^{a(i)} \to (\mathbf{Z}^+)^{c(i)}$. Then

$$Op\,(\tau)(f) = g: \ (\mathbf{Z}^+)^a \to (\mathbf{Z}^+)^c,$$

where (a, c) is the arity/coarity of the root.

The prescription for getting g, given f, runs as follows.

One-vertex case. Let τ be a corolla whose vertex is decorated by $\mathbf{c}, \mathbf{b}$, or $\mathbf{r}$. Then g is obtained by applying to the family $\{f_i\}$, $i \in F_\tau^{in}$, the respective elementary operation: composition, bracketing, or recursion. This requires the following compatibilities, which vary depending on the label of the vertex.

(a) *Composition.* Let $(a_1, c_1), \dots, (a_r, c_r)$ be the family of arities/coarities of inputs ordered as the respective flags. They must then be constrained by the condition $c_1 = a_2, \dots, c_{r-1} = a_r$, and the arity/coarity of the output must be (a_1, c_r).

For a general τ, these compatibility conditions must be satisfied for all corollas τ_v of all vertices decorated by $\mathbf{c}$.

In the flowchart interpretation, such a corolla transforms an input family $(f_1, \dots, f_r)$, $f_i: (\mathbf{Z}^+)^{a_i} \to (\mathbf{Z}^+)^{c_i}$, into the composition $f_r \circ f_{r-1} \circ \cdots f_1$.

Notice an essential difference in treating compositions in the context of *PROPs*, resp. *Prim*: for *PROPs*, we graft and contract, whereas for *Prim*, we endow a vertex with the task of composing.

This is because the corollas for PROPs are flowcharts accepting arguments from, say, $(\mathbf{Z}^+)^m$ and producing a vector in $(\mathbf{Z}^+)^n$, whereas decorated trees in *Prim* accept and produce arguments that are themselves vectors of functions, and we want to compose these functions rather than programs producing them.

(b) *Bracket.* With the same notation as in (a), the compatibility condition reads $a_\bullet := a_1 = \cdots = a_r$, and the arity/coarity of the output must be $(a_\bullet, c_1 + \cdots + c_r)$.

For a general τ, these compatibility conditions must be satisfied for all corollas τ_v of all vertices decorated by $\mathbf{b}$ and respective orderings.

In the flowchart interpretation, such a corolla transforms an input family $(f_1,\dots,f_r)$, $f_i:(\mathbf{Z}^+)^{a_\bullet}\to(\mathbf{Z}^+)^{c_i}$, into the map

$$\langle f_1,\dots,f_r\rangle:(\mathbf{Z}^+)^{a_\bullet}\to(\mathbf{Z}^+)^{c_1+\dots+c_r}.$$

It was called juxtaposition in V.2.3 (b).

(c) *Recursion.* If a vertex is decorated by $\mathbf{r}$, it must have exactly *two local inputs.* If the arity/coarity of the first one (in their structure order) is (a,c), for the second one it must be $(a+c,c)$, and for the local output it must be $(a+1,c)$. This is our compatibility condition.

In the flowchart interpretation, such a vertex takes as input two arbitrary maps $f_1:(\mathbf{Z}^+)^a\to(\mathbf{Z}^+)^c$, $f_2:(\mathbf{Z}^+)^{a+c}\to(\mathbf{Z}^+)^c$ and produces the output

$$g:(\mathbf{Z}^+)^{a+1}\to(\mathbf{Z}^+)^c$$

defined recursively as

$$g(x,1):=f_1(x),$$
$$g(x,k+1):=f_2(x,f_1(x,k))$$

for each $x\in(\mathbf{Z}^+)^a$, $k\in\mathbf{Z}^+$.

This form of recursion is more restrictive than the one that is often used: it does not allow f_2 to depend explicitly on the recursion parameter k. However, R. M. Robinson proved in 1947 that it suffices to use it in order to get all primitive recursive functions if an extension of the list of basic functions is allowed. Afterward, M. D. Gladstone showed that such an extension is unnecessary (*Jour. Symb. Logic*, 32:4 (1967), 505–508). I am grateful to N. Yanofsky for these references.

General case. First consider a connected graph τ. Assume that it has ≥ 2 vertices. We define the operation $Op\,(\tau)$ by induction on the number of vertices.

Namely, for a vertex v that is the boundary of a global input, consider the subfamily $f_v:=\{f_i\,|\,\partial_\tau(i)=v\}$. Denoting by τ_v the corolla of v (an *in-corolla*), calculate $g_v:=Op\,(\tau_v)(f_v)$ as specified above.

One can check that this prescription produces the result independent of arbitrary choices.

Now consider the maximal decorated subtree τ^0 of τ whose flags and vertices do not belong to this *in*-corolla. Its global inputs consist of all global inputs of τ not adjacent to v, and $j_\tau(r)$, where r is the root of our corolla. Decoration of τ^0 is the restriction of that of τ; global inputs of τ retain also their input functions f_i. Decorate the input $j_\tau(r)$ by g_v and put

$$Op\,(\tau)(\{f_i\}):=Op\,(\tau^0)(\{f_i,g_v\,|\,\partial(i)\neq v\}).$$

The right-hand side is defined due to the inductive assumption.

Finally, if τ is the disjoint union of connected components $\coprod_{a\in A}\tau_a$, we put

$$Op\,(\coprod_{a\in A}\tau_a):=\times_{a\in A}Op\,(\tau_a)$$

in the sense that $Op(\tau)$ acts on the family, naturally indexed by A, of (families of) global inputs of connected components, and produces the family of outputs, as well indexed naturally by A.

As we implied in the previous discussion, we can apply $Op(\tau)$ to families consisting not necessarily of basic, or even recursive, functions.

But if we want to define programming methods based upon *Prim*, then we must decorate global inputs by some basic functions, and interpret the resulting decorated tree as as a program producing one concrete recursive function.

Here the choice becomes ambiguous: we may change the list of basic functions, and we may allow the application of $\mathbf{c}, \mathbf{b}, \mathbf{r}$ to some restricted class of subfamilies, getting the more general cases from trees larger than corollas.

For $\mathbf{c}$ and $\mathbf{b}$, we allowed arbitrary natural families, implicitly using associativity of intended interpretations. Yanofsky allows only two inputs. For $\mathbf{r}$, we essentially adhered in 5.12 (c) to the choice made by Yanofsky.

5.15. ***Prim* as a world of programming methods.** We now define $Prim(m,n)$ as the subset of *Prim* consisting of graphs whose outputs (roots of connected components) have the total arity/coarity (m,n).

The evaluation morphism in $\mathcal{C}$

$$\mathrm{ev}_{P(m,n)}: P(m,n)\times(\mathbf{Z}^+)^m \to (\mathbf{Z}^+)^n$$

we have already essentially described. Namely,

$$\mathrm{ev}_{P(m,n)}(\tau,(x_1,\dots,x_m)) := f_\tau(x_1,\dots,x_m),$$

where f_τ is the total output of the flowchart τ, which we formerly denoted by $Op(\tau)$, applied to the input decorations of τ.

A computable multiple composition morphism (cf. 3.4 above)

$$\mathrm{comp}:\ P(m_{r-1},m_r)\times\dots\times P(m_2,m_3)\times P(m_1,m_2)\to P(m_1,m_r)$$

can be constructed as follows. For simplicity, we will describe only the composite $\mathrm{comp}(\tau_r,\tau_{r-1},\dots,\tau_1)$ for an r-tuple of decorated trees $\tau_1,\tau_2,\dots,\tau_r$.

Consider a corolla with vertex decorated by $\mathbf{c}$, r inputs decorated by the arities $(m_1,m_2),\dots,(m_{r-1},m_r)$, and an output decorated by (m_1,m_r). Graft inputs of this corolla to the roots of $\tau_1,\dots,\tau_r$ respectively. The resulting tree represents the composition.

Of course, on the combinatorial level, we will have to make a stupid choice of some "concrete" vertex and flags of this corolla, but the result will be unique up to unique isomorphism identical on the component trees τ_i.

However, if we iterate partial compositions that on the level of maps correspond, say, to $h\circ g\circ f$, $(h\circ g)\circ f$, and $h\circ(g\circ f)$ respectively, we will get three different decorated trees, say $\sigma_{123},\sigma_{12,3},\sigma_{1,23}$.

On the combinatorial/geometric level these trees are interconnected by two contraction morphisms (cf. 5.11) $\sigma_{12,3}\to\sigma_{123}$ and $\sigma_{1,23}\to\sigma_{123}$ that contract the edges entering the root vertices, whose ends are marked by $\mathbf{c}$. One can simply declare that such contractions generate an equivalence relation on the elements of *Prim*, and that algorithms encoded by *Prim* are actually such (or even bigger) equivalence classes rather than isomorphism classes of the decorated trees.

However, since we work in a categorical context, and strive to produce a category of algorithms in the sense of Definition 3.5, a better way to act is to organize *Prim* into a constructive category, and then to *localize* it with respect to those morphisms $\tau \to \sigma$ that produce a natural identification $Op(\tau)$ and $Op(\sigma)$.

Recall that the *localization of a category $\mathcal{B}$ with respect to a set of its morphisms S* is a functor $L : \mathcal{B} \to \mathcal{B}[S^{-1}]$ that makes all morphisms in S invertible and that is the initial object among all functors with this property.

Here is a simple version of this construction.

5.16. **Definition–Claim.** Consider the category *Pr* whose set of objects is the set *Prim*, and morphisms are compositions of the following maps of decorated graphs:

(i) Isomorphisms.
(ii) Contractions of subtrees of the following type: all vertices of such a subtree are decorated by **c**. After the contraction, the resulting vertex must be marked by **c**. The remaining decorations do not change.
(iii) Contractions of subtrees, all of whose all vertices are decorated by **b**. After the contraction, the resulting vertex must be marked by **b**. The remaining decorations do not change.

Denote by P the localization of *Pr* with respect to all morphisms. It has the natural structure of a category of programming methods for which composition and bracket operations become associative.

One can similarly accommodate more sophisticated equivalence relations between decorated trees, studied by Yanofsky.

To this end one can extend the category *Pr* by some extra morphisms, and then localize with respect to them as well.

6 Models of Computation and Complexity

In this section we are gradually zooming, passing from the macroscopic view of the constructive universe to "human scale" to microscopic (Boolean and Turing's) level.

6.1. **Normal models.** Let U be an infinite set. In this subsection we will be considering partial functions $U \to U$ that can be constructed by iteration. In other contexts, they might be called dynamical systems with discrete time, or cascades.

A normal model of computation M is the structure (P, U, I, F, s) consisting of four sets and a map

$$I, F \subset P \times U, \quad s : P \times U \to P \times U.$$

Here s is an everywhere defined function such that $s(p,u) = (p, s_p(u))$ for any $(p,u) \in P \times U$. Intuitively, p is a program, u is a configuration of the deterministic discrete-time computing device, and $s_p(u)$ is the new configuration obtained from u after one unit of time (clock tick). The subset I is that of initial data, or inputs. The subset $F \subset P \times U$ (final configurations, outputs) must be a part of the set of fixed points of s: if $(p,u) \in F$, then $s(p,u) = (p,u)$.

In this setting, we denote by f_p the partial function $f_p : U \to U$ such that we have $u \in D(f_p)$, $f_p(u) = v$, if and only if

$$(p,u) \in I, \text{ and for some } n \geq 0,\ (p, s_p^n(u)) \in F \text{ and } s_p^n(u) = v.$$

The minimal such n will be called the *time* (number of clock ticks) needed to calculate $f_p(u)$ using the program p.

Any finite sequence

$$(p, u, s_p(u), \ldots, s_p^m(u)),\ u \in I,$$

will be called a *protocol of computation of length* m for the model M.

We now add the constructivity conditions.

We require P, U to be constructive worlds, s computable. In addition, we require I, F to be decidable subsets of $P \times U$. Then f_p are computable, and protocols of given length (resp. of arbitrary length, resp. or those stopping at F) form constructive worlds. If we denote by Q_M the world of protocols stopping at F and by $\mathrm{ev} : Q_M \times U \to U$ the map $(p,u) \mapsto s_p^{\max}(u)$, we get a programming method.

Such a model M is called *versal* if the respective programming method Q_M is versal.

The notion of normal model of computation includes both normal algorithms and Turing machines.

Consider, for example, the standard description of the constructive world $\mathcal{T}$ of Turing machines T slightly adapted to our conventions. It includes the following data:

(a) The constructive world $U = \{0,1\}^*$ of, say, binary words that can be written on the tape of any T from our world.

(b) For each T, a finite set of *internal states* J_T, containing *initial state, accepting state rejecting state,* and remaining *intermediate states* J_T^0. All J_T must be elements of a constructive world of states J, and the map $T \mapsto J_T$ must be computable.

(c) The computable partial map $\tau : J \times \mathbf{N} \times U \to J \times \mathbf{N} \times U$, where $\mathbf{N}$ are natural numbers (including 0). For each T, it must send the subset $J_T \times \mathbf{N} \times U$ into itself.

A triple $(i,n,u) \in J_T \times \mathbf{N} \times U$ is the configuration of T in which T is in state i, and the head is scanning the nth square of the tape (the initial bit of u is counted as the first square, the square to the left of it is the zeroth square). The domain of definition of τ_T consists only of those triples for which $n \leq |u| + 1$, where $|u|$ is the length of u: the head must scan either one of the bits of u, or one of the next-door neighbors. The triple $\tau_T(i,n,u) = (i_1, n_1, u_1)$ depicts the next internal state of the machine, position of the head, and the new word on the tape. The usual restrictions on the τ_T are $n_1 = n \pm 1$, and u_1 may differ from u only at the nth bit.

The fixed points of τ are triples for which i = accepting or rejecting state.

We can reduce such a description to our normal form by putting $U = \{0,1\}^*$,

$$P := J \times \mathbf{N},\ I := \{\text{initial states}\} \times \{1\} \times U.$$

States F are those triples (accepting state,n,u) that can be reached from some point of I after a finite iteration of τ. Finally, to get an everywhere defined s coinciding with τ on its definition domain, we can extend τ to a computable map in some trivial way. For example, starting with some triple (i, n, u) not in I, we can prescribe s to move the head to the left until it reaches the first nonempty tape square, to continue moving until it reaches the next empty square, and then move one square to the right.

Turing machines have one feature that we did not keep in our definition of normal models. It is sometimes called *locality* of the iteration map, which depends only on the restricted number of bits in of the current position and changes only a restricted number of bits in moving to the next position. Discussing complexity later, we will suggest a useful and sufficiently general weakening of this requirement.

6.2. **Boolean circuits.** Boolean circuits are classical models of computation well suited for studying maps between the finite sets whose elements are encoded by binary words. Discussing them, we will identify the alphabet $\{0,1\}$ with the 2-element field $\mathbf{F}_2$.

Consider the commutative polynomial algebra generated over $\mathbf{F}_2$ by a countable sequence of independent variables, say $x_1, x_2, x_3, \ldots$. Define the Boolean algebra $\mathbf{B}$ as the quotient algebra of $\mathbf{F}_2[x_1, x_2, \ldots]$ modulo the ideal generated by polynomials $x_i^2 - x_i$. Each Boolean polynomial, element of $\mathbf{B}$, determines a function on $\oplus_{i=1}^{\infty} \mathbf{F}_2$ with values in $\mathbf{F}_2 = \{0,1\}$.

We start with the following simple fact.

6.3. **Claim.** *Any map $f : \mathbf{F}_2^m \to \mathbf{F}_2^n$ can be represented by a unique vector of Boolean polynomials.*

PROOF. It suffices to consider the case $n = 1$. Then this map is surjective, because f is represented by

$$F(x_1, \ldots, x_m) := \sum_{y=(y_i)\in \mathbf{F}_2^m} f(y) \prod_i (x_i + y_i + 1).$$

In fact, the product at $f(y)$ is the Kronecker delta $\delta_{x,y}$.

Moreover, the vector spaces of such maps and of Boolean polynomials over $\mathbf{F}_2$ have the common dimension 2^m. In fact, Boolean polynomials are represented by linear combinations of monomials $x_{i_1} \cdots x_{i_k}$, one for each subset $\{i_1, \ldots, i_k\} \subset \{1, \ldots, m\}$. This completes the proof.

Now we can calculate any vector of Boolean polynomials by iterating operations from a small finite list, which is chosen and fixed, e.g., $\mathcal{B} := \{x, 1, x+y, xy, (x,x)\}$. Such operators are called *classical gates*. A sequence of such operators, together with an indication of their arguments from the previously computed bits, is called a *Boolean circuit*. The number of steps in such a circuit is considered (a measure of) the time of computation.

As the word *circuit* suggests, one may consider even better representations by flowcharts, which are oriented graphs, with vertices decorated by the names of gates.

When the relevant finite sets are not $\mathbf{F}_2^m$, and perhaps have a wrong cardinality (not a power of 2), we encode their elements by finite sequences of bits and consider the restriction of Boolean polynomials to the relevant subset.

As above, a protocol of computation in this model can be represented as the finite table consisting of rows (generally of variable length) that accommodate sequences of 0's and 1's. The initial line of the table is the input. Each subsequent line must be obtainable from the previous one by the application of one the basic functions in $\mathcal{B}$ to the sequence of neighboring bits (the remaining bits are copied unchanged). The last line is the output. The exact location of the bits that are changed in each row and the nature of change must be a part of the protocol.

Physically, one can implement the rows as the different registers of the memory, or else as the consecutive states of the same register (then we have to make a prescription for how to cope with the variable length, e.g., using blank symbols).

6.4. **Turing machines vs. Boolean circuits.** Any protocol of the Turing computation of a function can be treated as such a protocol of an appropriate Boolean circuit, and in this case we have only one register (the initial part of the tape) whose states are consecutively changed by the head/processor. We will still use the term "gate" in this context.

A computable function f with infinite domain is the limit of a sequence of functions f_i between finite sets whose graphs extend each other. A Turing program for f furnishes a computable sequence of Boolean circuits, which compute all f_i in turn. Such a sequence is sometimes called *uniform*.

6.5. **Size, complexity, and polynomial-time computability.** The quantitative theory of computational models deals simultaneously with the space and time dimensions of protocols. The preceding subsection focused on time; here we introduce space. For Boolean (and Turing machine) protocols this is easy: the length of each row of the protocol plus specifications for the next step is the space required at that moment. The maximum of these lengths, up to a multiplicative constant, bounds the total space required from above and from below.

The case of normal models and infinite constructive worlds U is more interesting.

Generally we will say that a *a size function* $U \to \mathbf{N}$ is any function such that for every $H \in \mathbf{N}$, there are only finitely many objects of size $\leq H$. Thus the number of bits $|n| = [\log_2 n] + 1$ and the identical function $\|n\| = n$ are both size functions on $\mathbf{Z}^+$. Using a numbering, we can transfer them to any constructive world. In these two examples, the number of constructive objects of size $\le H$ grows as $\exp cH$, resp. cH. Such a count in more general cases allows one to make a distinction between the *bit size,* measuring the length of a description of the object, and the *volume* of the object.

In most cases we require computability of size functions. However, there are exceptions: for example, Kolmogorov complexity is a noncomputable size function with very important properties: see VI.9.

Given a size function (on all relevant worlds) and a versal normal model of computations M, we can consider the following complexity problems:

(A) *For a given morphism (computable map) $f : U \to V$, estimate the smallest bit size $K_M(f)$ of the program p such that $f = f_p$.*

According to V.9, there exists an *optimal* universal model of computation $\mathcal{U}$ such that with $P = \mathbf{N}$ and the bit size function, for any other model $\mathcal{S}$ there exists a constant c such that for any f,

$$K_{\mathcal{U}}(f) \le K_M(f) + c.$$

When $\mathcal{U}$ is chosen, $K_{\mathcal{U}}(f)$ is called the Kolmogorov complexity of f. With a different choice of $\mathcal{U}$ we will get the same complexity function up to $O(1)$-summand.

This complexity measure is highly nontrivial (and especially interesting) in the case of one-point U. It measures, then, the size of the most compressed description of a variable constructive object in V. This complexity is quite "objective," being almost independent of arbitrary choices. Being uncomputable, it cannot be directly used in computer science. However, it furnishes some basic restrictions on computability, strikingly similar to those provided by conservation laws in physics.

Recall that on $\mathbf{N}$ we have $K_{\mathcal{U}}(n) \le |n| + O(1) = \log_2 \|n\| + O(1)$. The first inequality "generically" can be replaced by equality, but infinitely often $K_{\mathcal{U}}(n)$ becomes much smaller that $|n|$.

(B) *For a given morphism (recursive map) $f : U \to V$, estimate the time needed to calculate $f(u), u \in D(f)$, using the program p and compare the results for different p and different models of computation.*

(C) *The same for the function "maximal size of intermediate configurations in the protocol of the computation of $f(u)$ using the program p" (space, or memory).*

In the last two problems, we have to compare functions rather than numbers: time and space depend on the size of input. Here a cruder polynomial scale appears naturally. Let us show how this happens.

Fix a computational model $\mathcal{S}$ with the transition function s computing functions $U \to U$, and choose a bit size function $u \mapsto |u|$ on U satisfying the following crucial assumption, a weakening of the locality requirement valid for Turing machines:

(i) $|u| - c \le |s_p(u)| \le |u| + c$, *where the constant* c *may depend on* p *but not on* u.

In this case we have $|s_p^m(u)| \le |u| + c_p m$: the required space grows no more than linearly with time.

Let now $(\mathcal{S}', s')$ be another model such that $s_p = s'_q$ for some q. For example, such q always exists if $\mathcal{S}'$ is versal. Assume that s' satisfies (i) as well, and moreover,

(ii) s *can be computed in the model* $\mathcal{S}'$ *in time bounded by a polynomial* F *in the bit size of input.*

This requirement is certainly satisfied for Turing and Markov models, and is generally reasonable, because an elementary step of an algorithm deserves its name only if it is computationally tractable.

Then we can replace one application of s_p to $s_p^m(u)$ by $\le F(|u| + cm)$ applications of s'_q. And if we needed $T(u)$ steps in order to calculate $f_p(u)$ using $\mathcal{S}$, we will need no more than $\le \sum_{m=1}^{T(u)} F(|u| + cm)$ steps to calculate the same function using $\mathcal{S}'$ and q. In a detailed model, there might be a small additional cost of merging two protocols. This is an example of the compilation morphism lifted to the worlds of protocols.

Thus, from the assumptions (i) and (ii) it follows that functions computable in polynomial-time by $\mathcal{S}$ have the same property for all reasonable models. Notice also that for such functions, $|f(u)| \le G(|u|)$ for some polynomial G and that the domain $D(f)$ of such a function is decidable: if after $T(|u|)$ iterations of s_p we are not in a final state, then $u \notin D(f)$.

Thus we can define the class PF of functions, say $\mathbf{N}^k \to \mathbf{N}$, computable in polynomial-time using a fixed universal Turing machine and arguing as above that this definition is model-independent.

If we want to extend it to a constructive universe $\mathcal{C}$, however, we will have to postulate additionally that any constructive world U comes together with a natural class of numberings that together with their inverses are *computable in polynomial-time.* The bit size will be defined in terms of one of these numberings.

This postulate, accepted for "all constructive worlds," seems to be a part of the content of the *"polynomial Church thesis"* invoked by M. Freedman in his talk at the Berlin ICM, 1998.

If we take this strengthening of Church is thesis for granted, and take two bit-size functions determined by two polynomial numberings, then the quotient of two such size functions is bounded from above and away from zero.

Below we will be considering only the universes $\mathcal{C}$ and worlds U with these properties, and $|u|$ will always denote a computable bit size. Gödel's numbering for $\mathbf{N} \times \mathbf{N}$ shows that that such $\mathcal{C}$ is still closed with respect to finite products. (Notice, however, that the beautiful numbering of $\mathbf{N}^*$ using primes is not polynomial-time computable; it may be replaced by another one that is in PF).

6.6. P/NP problem. Let U be a constructive world. By definition, a subset $E \subset U$ *belongs to the class* P if its characteristic function χ_E (equal to 1 on E and 0 outside) belongs to the class PF.

Furthermore, $E \subset U$ *belongs to the class* NP if there exists a subset $E' \subset U \times V$ belonging to P and a polynomial G such that

$$u \in E \iff \exists\, (u, v) \in E' \text{ with } |v| \le G(|u|).$$

Here V is another constructive world (which may coincide with U). We will say that E is obtained from E' by a *polynomially truncated projection.*

Such a v can be called *a witness* of the inclusion $u \in E$. The polynomial-time calculation establishing that $\chi_{E'}(u, v) = 1$ is *a short proof* that $u \in E$.

The discussion above establishes in what sense this definition is model-independent.

Clearly, P$\subset$ NP.

The question whether these two classes coincide is the celebrated *P/NP problem.*

A naive algorithm calculating χ_E from $\chi_{E'}$ by searching for v with $|v| \le G(|u|)$ and $\chi_{E'}(u, v) = 1$ will generally take exponential time v (because $|u|$ is a bit-size function). Of course, if one can treat all such v simultaneously, using massive parallellism, the required time will be polynomial: time will be traded for space. Or else, if an oracle tells you that $u \in E$ and supplies an appropriate v, you can convince yourself that this is indeed so in polynomial-time, by computing $\chi_{E'}(u, v) = 1$.

Notice that enumerable sets can be alternatively described as projections of decidable ones, and that in this context projection does create undecidable sets. Nobody as yet has been able to translate the diagonalization argument used to establish this to the P/NP domain.

It has long been known that the P/NP problem can be reduced to checking whether some very particular sets—*NP-complete ones*—belong to P.

6.7. Definition. The set $E \subset U$ is called *NP-complete* if, for any other set $D \subset V, D \in$ NP, there exists a function $f : V \to U, f \in$ PF, such that $D = f^{-1}(E)$, that is, $\chi_D(v) = \chi_E(f(v))$.

We will sketch the classical argument (due to S. Cook, L. Levin, R. Karp) showing the existence of NP-complete sets. In fact, the reasoning is constructive: it furnishes a polynomially computable map producing f from the descriptions of $\chi_{E'}$ and the truncating polynomial G.

In order to describe one NP-complete problem, we will define an infinite family of Boolean polynomials b_u indexed by the following data, constituting objects u of the constructive world U. One u is a collection

$$m \in \mathbf{N};\quad (S_1, T_1), \ldots, (S_N, T_N),$$

where S_i, $T_i \subset \{1, \ldots, m\}$, and b_u is defined as

$$b_u(x_1, \ldots, x_m) = \prod_{i=1}^{N} \left(1 + \prod_{k \in S_i} (1 + x_k) \prod_{j \in T_i} x_j \right).$$

We choose the bit size of u as $|u| = mN$.

Put

$$E = \{u \in U \,|\, \exists v \in \mathbf{F}_2^m,\, b_u(v) = 1\}.$$

Using the language of Boolean truth values, one says that v *satisfies* b_u if $b_u(v) = 1$, and E is called the *satisfiability problem*, or SAT.

6.8. **Proposition.** SAT $\in$ NP.

PROOF. In fact, let

$$E^{'} = \{(u, v) \,|\, b_u(v) = 1\} \subset U \times (\oplus_{i=1}^{\infty} \mathbf{F}_2).$$

Clearly, E is the full projection of $E^{'}$. A bit of contemplation will convince the reader that $E^{'} \in$ P. In fact, we can calculate $b_u(v)$ performing $O(Nm)$ Boolean multiplications and additions. The projection to E can be replaced by a polynomially truncated projection, because we have to check only v of bit size $|v| \le m$.

6.9. **Theorem.** SAT is NP-*complete*.

PROOF (*sketch*). In fact, let $D \in$ NP, $D \subset A$, where A is some constructive world. Take a representation of D as a polynomially truncated projection of some set $D^{'} \subset A \times B, D^{'} \in$ P. Choose a normal, say Turing, model of computation and consider the Turing protocols of computation of $\chi_{D'}(a, b)$ with fixed a and variable polynomially bounded b. As we have explained above, for a given a, any such protocol can be imagined as a table of a fixed polynomially bounded size whose rows are the consecutive states of the computation. In the "microscopic" description, the positions in this table can be filled only by 0 or 1. In addition, each row is supplied by the specification of the position and the inner state of the head/processor. Some of the arrangements are valid protocols, others are not, but the local nature of the Turing computation allows one to produce a Boolean polynomial b_u in appropriate variables such that the valid protocols are recognized by the fact that this polynomial takes value 1. This defines the function f reducing D to E. The construction is so direct that the polynomial-time computability of f is straightforward.

Many natural problems are known to be NP-complete, in particular 3-SAT. It is defined as the subset of SAT consisting of those u for which $\mathrm{card}\,(S_i \cup T_i) = 3$ for all i.

6.10. **Remark.** Most Boolean functions are not computable in polynomial-time. Several versions of this statement can be proved by simple counting.

First of all, fix a finite basis $\mathcal{B}$ of Boolean operations as in 6.3, each acting on $\le a$ bits. Then sequences of these operations of length t generate $O((bn^a)^t)$ Boolean functions $\mathbf{F}_2^n \to \mathbf{F}_2^n$, where $b = \mathrm{card}\,\mathcal{B}$. On the other hand, the number of all functions 2^{n2^n} grows as a double exponential of n and for large n cannot be obtained in time t polynomially bounded in n.

The same conclusion holds if we consider not all functions but only permutations: Stirling's formula for $\mathrm{card}\, S_{2^n} = 2^n!$ involves a double exponential.

Here is one more variation of this problem: define the time complexity of a conjugacy class in S_{2^n} as the minimal number of steps needed to calculate some permutation in this class. This notion arises if we are interested in calculating automorphisms of a finite world of cardinality 2^n that is not supplied with a specific encoding by binary words. Then it can happen that a judicious choice of encoding will drastically simplify the calculation of a given function. However, for most functions we still will not be able to achieve polynomial-time computability, because the asymptotic formula for the number of conjugacy classes (partitions)

$$p(2^n) \sim \frac{\exp\left(\pi\sqrt{\frac{2}{3}(2^n - \frac{1}{24}}\right)}{4\sqrt{3}(2^n - \frac{1}{24})}$$

again displays double exponential growth.

7 Basics of Quantum Computation I: Quantum Entanglement

In this section we will discuss the basics: how to use the superposition principle in order to accelerate (certain) classical computations.

For a minimal physics background, the reader may wish to reread II. 12.1–12.9.

7.1. **Description of the problem.** *Let N be a large number, $F:\{0,\dots,N-1\}\to\{0,\dots,N-1\}$ a function such that the computation of each particular value $F(x)$ is tractable, that is, can be done in time polynomial in $\log x$. We want to compute (to recognize) some property of the graph $(x, F(x))$, for example:*

(i) Find the least period r of F, i.e., the least residue $r \bmod N$ such that $F(x+r \bmod N) = F(x)$ for all x (the key step in the factorization problem.)
(ii) Find some x such that $F(x) = 1$ or establish that such x does not exist (search problem.)

As we already mentioned, a direct attack on such a problem consists in compiling the complete list of pairs $(x, F(x))$ and then applying to it an algorithm recognizing the property in question. Such a strategy requires at least exponential time (as a function of the bit size of N), since already the length of the list is N. Barring a theoretical breakthrough in understanding such problems (for example a proof that P=NP), a practical response might be in exploiting the possibility of parallel computing, i.e., calculating simultaneously many—or even all—values of $F(x)$. This takes less time but uses (dis)proportionally more hardware.

A remarkable suggestion due to D. Deutsch consists in using a quantum superposition of the classical states $|x\rangle$ as the replacement of the union of N classical registers, each in one of the initial states $|x\rangle$. To be more precise, here is a mathematical model formulated as a definition.

7.2. **Quantum parallel processing: version I.** Keeping the notation above, assume moreover that $N = 2^n$.

(i) *The quantum space of inputs/outputs* is the 2^n-dimensional complex Hilbert space H_n with the orthonormal basis $|x\rangle$, $0 \leq x \leq N-1$. Vectors $|x\rangle$ are called classical states.

(ii) *The quantum version of* F is the unique unitary operator $U_F : H_n \to H_n$ such that $U_F|x\rangle = |F(x)\rangle$.

Quantum parallel computing of F is (a physical realization of) a quantum system with the state space H_n and the evolution operator U_F.

Naively speaking, if we apply U_F to the initial state which is a superposition of all classical states, with, say, equal amplitudes, we will get simultaneously all classical values of F (i.e., their superposition):

$$U_F\left(\frac{1}{\sqrt{N}}\sum|x\rangle\right) = \frac{1}{\sqrt{N}}\sum|F(x)\rangle.$$

Now, this does not look very promising. In fact, U_F exists *only if F is a permutation,* and in this case the left hand side is simply identical to the right-hand side!

To get a more workable version, we will have to take superpositions with different weights. We will also have to devise tricks for replacing, say, search functions (1 on desirable elements, 0 elsewhere) by permutations. For this, see Section 7.3 below.

For the time being, we will start discussing various issues related to our preliminary picture, before passing to its more realistic modification.

(A) We put $N = 2^n$ above because we are imagining the respective classical system as an n-bit register: cf. the discussion of Boolean circuits. Every number $0 \leq x \leq N-1$ is written in the binary notation $x = \sum_i \epsilon_i 2^i$ and is identified with the pure (classical) state $|\epsilon_{n-1}, \dots, \epsilon_0\rangle$, where $\epsilon_i = 0$ or 1 is the state of the ith register. The quantum system H_1 is called a *qubit.* We have $H_n = H_1^{\otimes n}$, $|\epsilon_{n-1}, \dots, \epsilon_0\rangle = |\epsilon_{n-1}\rangle \otimes \cdots \otimes |\epsilon_0\rangle$.

This conforms to the general principles of quantum mechanics. The Hilbert space of the union of systems can be identified with the tensor product of the Hilbert spaces of the subsystems. Accordingly, *decomposable vectors* correspond to the states of the compound for which one can say that *the individual subsystems are in definite states.*

In a general state of the register, the individual bits do not store any definite values: this is the essence of *quantum entanglement.*

(B) Pure quantum states, strictly speaking, are points of the *projective space* $P(H_n)$, that is, complex lines in H_n. Traditionally, one considers instead vectors of norm one. This leaves undetermined an overall phase factor $\exp i\varphi$. If we have two state vectors, individual phase factors have no objective meaning, but *the difference of their phases does have one.* This difference can be measured by observing effects of *quantum interference.*

Quantum interference is highly important and is used for implementing efficient quantum algorithms.

(C) If a quantum system S is *isolated from its environment*, its dynamical evolution with time t is described by the unitary operator acting on its Hilbert space, $U(t) = \exp iHt$, where H is the Hamiltonian, t is time. Therefore one option for implementing U_F physically is to design a device for which U_F would be a fixed time evolution operator. However, this seemingly contradicts many deeply rooted notions of the algorithm theory. For example, calculating $F(x)$ for different inputs x takes different times, and it would be highly artificial to try to equalize them already in the design.

Instead, one can try to implement U_F as the result of a sequence of brief interactions, *carefully controlled by a classical computer,* of S with the environment (say, laser pulses). Mathematically speaking, U_F is represented as a product of some standard unitary operators $U_m, \dots, U_1$ each of which acts only on a small subset (two, three) of classical bits. These operators are called *quantum gates.*

The complexity of the respective quantum computation is determined by its length (the number m of the gates) and by the complexity of each of them.

The latter point is a subtle one: continuous parameters, e.g., phase shifts, on which U_i may depend, makes the information content of each U_i potentially infinite and leads to a suspicion that a quantum computer will in fact perform an analog computation, only implemented in a fancy way.

This point has been discussed and refuted on several occasions by displaying those features of quantum computation that distinguish it from both analog and digital classical information processing. Philosophically, all arguments are variations on the theme of von Neumann's theorem on the impossibility of hidden parameters (cf. II.12).

One more problem related to the necessity to renounce the image of an isolated quantum register is that of stability, or *fault tolerance.* Even very weak, but uncontrolled, interactions with the environment will quickly lead to the spreading of quantum noise, destroying the useful information. This is called *quantum decoherence.*

One defense strategy is the technique of fault-tolerant computation using quantum codes for producing continuous variables highly protected from external noise.

7.3. **Reducing general functions to permutations.** As we have already remarked, the requirement that F must be a permutation is highly restrictive: for instance, in the search problem F takes only two values.

There is nothing justifying this restriction in the schemes of classical computation, but in our quantum model, only permutations F extend to unitary operators (*"quantum reversibility"*).

The standard way out consists in introducing *two* n-bit registers instead of one, for keeping the value of the argument as well as that of the function. This also conforms with our initial idea that we want to learn something about *the graph of* F.

More precisely, if $F(|x\rangle)$ is an arbitrary function of classical bits, we can replace it by the permutation $\widetilde{F}(|x, y\rangle) := |x, F(x) \oplus y\rangle$, where $\oplus$ is the Boolean (bitwise) sum. This involves no more than a polynomial increase of the classical complexity, and the restriction of $\widetilde{F}$ to $y \doteq 0$ produces the graph of F, which we need anyway for the type of problems we are interested in.

In the quantum Boolean circuit version this trick must be applied to all gates.

More precisely, in order to process a classical algorithm (sequence of Boolean gates) for computing F into the quantum one, we replace each classical gate by the respective reversible quantum gate, i.e., by the unitary operator corresponding to it tensored with the identical operator. Besides two registers for keeping $|x\rangle$ and $F(|x\rangle)$ we will have to introduce as well extra qubits in which we are not particularly interested. The corresponding Hilbert space and its content is sometimes referred to as "scratchpad," "garbage," etc. Besides ensuring reversibility, additional space and garbage can be introduced as well for considering functions $F : \{0, \dots, N-1\} \to \{0, \dots, M-1\}$, where N, M are not powers of two (then we extend them to the nearest power of two). For more details, see the next section.

Notice that the choice of gate array (Boolean circuit) as the classical model of computation is essential in the following sense: *a quantum routine cannot use conditional instructions.* Indeed, to implement such an instruction we must observe the memory in the midst of calculation, but the observation generally will *change its current quantum state.*

In the same vein, we must *avoid copying instructions,* because the classical copying operator $|x\rangle \to |x\rangle \otimes |x\rangle$ is not linear. In particular, each output qubit from a quantum gate can be used only in one gate at the next step (if several gates are used in parallel): *cloning is not allowed.*

These examples show that the basics of quantum code writing will have a very distinct flavor.

We now pass to the problems posed by the input/output routines.

Input, or initialization, in principle can be implemented in the same way as a computation: we produce an input state starting, e.g., from the classical state $|0\rangle$ and applying a sequence of basic unitary operators: see the next section. Output, however, involves an additional quantum-mechanical notion: that of *observation.*

7.4. **Quantum observation.** The simplest model of observation of a quantum system with the Hilbert space H is that of interaction with another system, and their subsequent disentanglement.

Possible results of such an interaction will form an orthonormal basis $|\chi_i\rangle$ of H (depending on the physical details of observation). If our system was in some entangled state $|\psi\rangle$ at the moment of observation, it will be observed in some state $|\chi_i\rangle$ *with probability* $|\langle\chi_i|\,\psi\rangle|^2$.

This means first of all that every quantum computation is inherently probabilistic. Observing (a part of) the quantum memory is not exactly the same as "printing the output." We must plan a series of runs of the same quantum program and the subsequent classical processing of the observed results, and we can hope only to get the desired answer with probability close to one.

Furthermore, this means that by implementing quantum parallelism simple-mindedly as at the beginning of this section, and then observing the memory as if it were the classical n-bit register, we will simply get some value $F(x)$ with probability $1/N$. This does not use the potential of the quantum parallelism. Therefore we formulate a corrected version of this notion, allowing more flexibility and stressing the additional tasks of the designer, each of which eventually contributes to the complexity estimate.

7.5. **Quantum parallel processing: version II.** *To solve efficiently a problem involving properties of the graph of a function F, we must design:*

(i) An auxiliary unitary operator U carrying the relevant information about the graph of F.
(ii) A computationally feasible realization of U with the help of standard quantum gates.
(iii) A computationally feasible realization of the input subroutine.
(iv) A computationally feasible classical algorithm processing the results of many runs of quantum computation.

All of this must be supplemented by quantum error-correcting encoding, which we will not address here. In the next section we will discuss some standard quantum subroutines.

8 Selected Quantum Subroutines

8.1. **Initialization.** Using the same conventions as in Section 7 and the subsequent comments, in particular the identification $H_n = H_1^{\otimes n}$, we have

$$\frac{1}{\sqrt{N}}\sum_{x=0}^{N-1}|x\rangle = \frac{1}{\sqrt{N}}\sum_{\epsilon_i=0,1}|\epsilon_{n-1}\cdots\epsilon_0\rangle = \left(\frac{1}{\sqrt{2}}(|0\rangle+|1\rangle)\right)^{\otimes n}.$$

In other words,

$$\frac{1}{\sqrt{N}}\sum_{x=0}^{N-1}|x\rangle = U_1^{(n-1)}\cdots U_1^{(0)}|0\cdots 0\rangle,$$

where $U_1 : H_1 \to H_1$ is the unitary operator

$$|0\rangle \mapsto \frac{1}{\sqrt{2}}(|0\rangle+|1\rangle),\ |1\rangle \mapsto \frac{1}{\sqrt{2}}(|0\rangle-|1\rangle),$$

and $U_1^{(i)} = \mathrm{id}\otimes\cdots\otimes U_1\otimes\cdots\otimes\mathrm{id}$ acts only on the ith qubit.

Thus making the quantum gate U_1 act on each memory bit, one can in n steps initialize our register in the state that is the superposition of all 2^n classical states with equal weights.

8.2. **Quantum computations of classical functions.** Let $\mathcal{B}$ be a finite basis of classical gates containing the one-bit identity and generating all Boolean circuits, and $F: \mathbf{F}_2^m \to \mathbf{F}_2^n$ a function. We will describe how to turn a Boolean circuit of length L calculating F into another Boolean circuit of comparable length consisting only of reversible gates, and calculating a modified function, which, however, contains all information about the graph of F. Reversibility means that each step is a bijection (actually, an involution) and hence can be extended to a unitary operator, that is, a quantum gate. For a gate f, define $\widetilde{f}(|x,y\rangle) = |x, f(x)+y\rangle$ as in 7.3 above.

8.3 **Claim.** *A Boolean circuit $\mathcal{S}$ of length L in the basis $\mathcal{B}$ can be processed into the reversible Boolean circuit $\widetilde{\mathcal{S}}$ of length $O((L+m+n)^2)$ calculating a permutation $H: \mathbf{F}_2^{m+n+L} \to \mathbf{F}_2^{m+n+L}$ with the following property:*

$$H(x,y,0) = (x, F(x)+y, 0) = (\widetilde{F}(x,y), 0).$$

Here x, y, z have sizes m, n, L respectively.

PROOF. We will understand L here as the sum of sizes of the outputs of all gates involved in the description of $\mathcal{S}$. We first replace in $\mathcal{S}$ each gate f by its reversible counterpart $\widetilde{f}$. This involves inserting extra bits, which we put side by side into a new register of total length L. The resulting subcircuit will calculate a permutation $K: \mathbf{F}_2^{m+L} \to \mathbf{F}_2^{m+L}$ such that $K(x,0) = (F(x), G(x))$ for some function G (garbage).

Now add to the memory one more register of size n keeping the variable y. Extend K to the permutation $\overline{K}: \mathbf{F}_2^{m+L+n} \to \mathbf{F}_2^{m+L+n}$ keeping y intact: $\overline{K}: (x,0,y) \mapsto (F(x), G(x), y)$. Clearly, $\overline{K}$ is calculated by the same Boolean circuit as K, but with extended register.

Extend this circuit by the one adding the contents of the first and the third registers: $(F(x), G(x), y) \mapsto (F(x), G(x), F(x)+y)$. Finally, build the last extension that calculates $\overline{K}^{-1}$ and consists of reversed gates calculating $\overline{K}$ in reverse order. This clears the middle register (scratchpad) and produces $(x, 0, F(x)+y)$. The whole circuit requires $O(L+m+n)$ gates if we allow the application of them to not necessarily neighboring bits. Otherwise we must insert gates for local permutations, which will replace this estimate by $O((L+m+n)^2)$.

8.4. **Fast Fourier transform.** Finding the least period of a function of one real variable can be done by calculating its Fourier transform and looking at its maxima. The same strategy is applied by Shor in his solution of the factorization problem. We will show now that the discrete Fourier transform Φ_n is computationally easy (quantum polynomial-time). We define $\Phi_n : H_n \to H_n$ by

$$\Phi_n(|x\rangle) = \frac{1}{\sqrt{N}} \sum_{c=0}^{N-1} |c\rangle \exp(2\pi i c x/N).$$

In fact, it is slightly easier to implement directly the operator

$$\Phi_n^t(|x\rangle) = \frac{1}{\sqrt{N}} \sum_{c=0}^{N-1} |c^t\rangle \exp(2\pi i c x/N),$$

where c^t is c read from right to left. The effects of the bit reversal can then be compensated at a later stage without difficulty.

Let $U_2^{(kj)} : H_n \to H_n$, $k < j$, be the quantum gate that acts on the pair of the kth and jth qubits in the following way: it multiplies $|11\rangle$ by $\exp(i\pi/2^{j-k})$ and leaves the remaining classical states $|00\rangle, |01\rangle, |10\rangle$ intact.

8.5. **Lemma.** *We have*

$$\Phi_n^t = \prod_{k=0}^{n-1} \left(U_1^{(k)} \prod_{j=k+1}^{n-1} U_2^{(kj)} \right).$$

By our rules of the game, this presentation has polynomial length in the sense that it involves only $O(n^2)$ gates. However, implementation of $U_2^{(kj)}$ requires controlling variable phase factors that tend to 1 as $k - j$ grows. Moreover, arbitrary pairs of qubits must allow quantum-mechanical coupling, so that for large n, the interaction between qubits must be nonlocal. The contribution of these complications to the notion of complexity cannot be estimated without going into the details of the physical arrangement. Therefore we will add a few words on this subject.

One possible implementation of a quantum register consists of a collection of ions (charged atoms) in a linear harmonic trap (optical cavity). Two of the electronic states of each ion are denoted by $|0\rangle$ and $|1\rangle$ and represent a qubit. Laser pulses transmitted to the cavity through the optical fibers and controlled by the classical computer are used to implement gates and readout. The Coulomb repulsion keeps ions apart (spatial selectivity), which allows the preparation of each ion separately in any superposition of $|0\rangle$ and $|1\rangle$ by timing the laser pulse properly and preparing its phase carefully. The same Coulomb repulsion allows for collective excitations of the whole cluster, whose quanta are called phonons. Such excitations are produced by laser pulses as well under appropriate resonance conditions. The resulting resonance selectivity combined with the spatial selectivity implements a controlled entanglement of the ions that can be used in order to simulate two- and three-bit gates.

Another recent suggestion is to use a single molecule as a quantum register, representing qubits by nuclear spins of individual atoms, and using interactions through chemical bonds in order to perform multiple-bit logic. The classical technique of nuclear magnetic resonance developed since the 1940s, which allows one to work with many molecules simultaneously, provides the startup technology for this project.

8.6. **Quantum search.** All the subroutines described up to now have boiled down to some identities in the unitary groups involving products of not too many operators acting on subspaces of small dimension. They did not involve output subroutines and therefore did not "compute" anything in the traditional sense of the word. We will now describe the beautiful quantum search algorithm due to L. Grover, which produces a new identity of this type, but also demonstrates the effect of observation and the way one can use quantum entanglement in order to exploit the potential of quantum parallelism.

We will treat only the simplest version. Let $F : \mathbf{F}_2^n \to \{0,1\}$ be a function taking the value 1 at exactly one point x_0. We want to compute x_0. We assume that F is computable in polynomial-time, or else that its values are given by an oracle. Classical search for x_0 requires on the average about $N/2$ evaluations of F where $N = 2^n$.

In the quantum version, we will assume that we have a quantum Boolean circuit (or quantum oracle) calculating the unitary operator $H_n \to H_n$,

$$I_F : |x\rangle \mapsto e^{\pi i F(x)}|x\rangle.$$

In other words, I_F is the reflection inverting the sign of $|x_0\rangle$ and leaving the remaining classical states intact.

Moreover, we put $J = -I_\delta$, where $\delta : \mathbf{F}_2^n \to \{0,1\}$ takes the value 1 only at 0, and $V = U_1^{(n-1)} \cdots U_1^{(0)}$, as in 8.1.

8.6. **Claim.** (i) *The real plane in H_n spanned by the uniform superposition ξ of all classical states and by $|x_0\rangle$ is invariant with respect to $T := VJVI_F$.*

(ii) *T restricted to this plane is the rotation (from ξ to $|x_0\rangle$) by the angle φ_N, where*

$$\cos\varphi_N = 1 - \frac{2}{N}, \ \sin\varphi_N = 2\,\frac{\sqrt{N-1}}{N}.$$

The check is straightforward.

Now, φ_N is close to $2/\sqrt{N}$, and for the initial angle φ between ξ and $|x_0\rangle$ we have

$$\cos\varphi = -\frac{1}{\sqrt{N}}.$$

Hence in $[\varphi/\varphi_N] \approx \pi\sqrt{N}/4$ applications of T to ξ we will get the state very close to $|x_0\rangle$. Stopping the iteration of T after as many steps and measuring the outcome in the basis of classical states, we will obtain $|x_0\rangle$ with probability very close to one.

One application of T replaces in the quantum search one evaluation of F. Thus, thanks to quantum parallelism, we achieve a polynomial speedup in comparison with the classical search. The case in which F takes the value 1 at several points and we want to find only one of them can be treated by an extension of this method. If there are n such points, the algorithm requires about $\sqrt{N/n}$ steps, and n need not be known a priori.

Still, this does not help solving NP-complete problems, because the square root of an exponential is still an exponential.

9 Shor's Factoring Algorithm

Efficient factorization of large integers became in the last decades an important applied problem, because standard public key cryptosystems rely on the perceived difficulty of this problem. At least in 2000, it was practically impossible to factorize a product of two 150-decimal-digit primes: estimated running times of the best existing factorization algorithms were in the billions years.

Producing such public key cryptosystems on an industrial scale requires mass production of large primes. This last problem recently was shown to be in the class P (M. Agrawal, N. Kayal, N. Saxena). Existing practical algorithms can prove primality of a 10000-bit number in several weeks.

For this reason, when P. Shor demonstrated that a quantum algorithm can efficiently solve the factorization problem, and thus provide means for systematically breaking the public key cryptosystems, his discovery attracted much public attention. We will sketch his algorithm in this section.

9.1. **Notation.** Let M be a natural number to be factored. We will assume that it is odd and is not a power of a prime number.

Denote by N the volume of the basic memory register we will be using (not counting scratchpad). Its bit size n will be about twice that of M. More precisely, choose $M^2 < N = 2^n < 2M^2$. Finally, let $1 < t < M$ be a random parameter with $\gcd(t, M) = 1$. This condition can be checked classically in time polynomial in n.

Below we will describe one run of Shor's algorithm, in which t (and of course, M, N) is fixed. Generally, polynomially many runs will be required, in which the value of t can remain the same or be chosen anew. This is needed in order to gather statistics. Shor's algorithm is a probabilistic one, with two sources of randomness that must be clearly distinguished. One is built into the classical probabilistic reduction of factoring to the finding of the period of a function. Another stems from the necessity of observing quantum memory, which, too, produces random results.

More precise estimates than those given here show that a quantum computer that can store about $3n$ qubits can find a factor of M in time of order n^3 with probability close to 1. On the other hand, it is widely believed that no recursive function of the type $M \mapsto$ *a proper factor of* M belongs to PF.

9.2. **A classical algorithm.** Put

$$r := \min\{\rho \,|\, t^\rho \equiv 1 \bmod M\},$$

which is the least period of $F : a \mapsto t^a \bmod M$.

Claim. *If one can efficiently calculate r as a function of t, one can find a proper divisor of M in time polynomial in $\log_2 M$ with probability $\geq 1 - M^{-m}$ for any fixed m.*

In fact, choose such t for which the period r satisfies

$$r \equiv 0 \bmod 2, \ t^{r/2} \not\equiv -1 \bmod M.$$

Then $\gcd(t^{r/2}+1, M)$ is a proper divisor of M. Notice that gcd is computable in polynomial-time.

The probability that this condition holds is $\geq 1 - 1/2^{k-1}$, where k is the number of different odd prime divisors of M, hence $\geq 1/2$ in our case. Therefore we will find a good t with probability $\geq 1-M^{-m}$ in $O(\log M)$ tries. The longest calculation in one try is that of $t^{r/2}$. The usual squaring method performs this in polynomial-time as well.

9.3. **Quantum algorithm calculating r.** Here we describe one run of the quantum algorithm that purports to compute r, given M, N, t. We will use the working register that can keep a pair consisting of a variable $0 \leq a \leq N-1$ and the respective value of the function $t^a \bmod M$. One more register will serve as the scratchpad needed to compute $|a, t^a \bmod M\rangle$ reversibly. When this calculation is completed, the content of the scratchpad will be reversibly erased: cf. 8.3 above. In the remaining part of the computation the scratchpad will no longer be used, so we may decouple it and forget about it.

The quantum computation consists of four steps, three of which were described in Section 8:

(i) Partial initialization produces from $|0,0\rangle$ the superposition

$$\frac{1}{\sqrt{N}} \sum_{a=0}^{N-1} |a, 0\rangle.$$

(ii) Reversible calculation of F processes this state into

$$\frac{1}{\sqrt{N}} \sum_{a=0}^{N-1} |a, t^a \bmod M\rangle.$$

(iii) Partial Fourier transform then furnishes

$$\frac{1}{N} \sum_{a=0}^{N-1} \sum_{c=0}^{N-1} \exp\left(2\pi i a c/N\right) |c, t^a \bmod M\rangle.$$

(iv) The last step is the observation of this state with respect to the system of classical states $|c, m \bmod M\rangle$. This step produces some concrete output

$$|c, t^k \bmod M\rangle$$

with probability

$$\left| \frac{1}{N} \sum_{a:\, t^a \equiv t^k \bmod M} \exp\left(2\pi i a c/N\right) \right|^2 .$$

The remaining part of the run is assigned to the classical computer and consists of the following steps.

(A) *Find the best approximation (in lowest terms) to* $\frac{c}{N}$ *with denominator* $r' < M < \sqrt{N}$*:*

$$\left| \frac{c}{N} - \frac{d'}{r'} \right| < \frac{1}{2N}.$$

As we will see below, we may hope that r' will coincide with r in at least one run among at most polynomially many. For this reason, we will try r' in the role of r right away:

(B) *If* $r' \equiv 0 \bmod 2$, *calculate* $\gcd(t^{r'/2} \pm 1, M)$.

If r' is odd, or if r' is even, but we did not get a proper divisor of M, repeat the run $O(\log\log M)$ times with the same t. In case of failure, change t and start a new run.

9.4. **Justification.** We will now show that given t, from the observed values of $|c, t^k \bmod M\rangle$ we can find in $O(\log\log M)$ runs the correct value of r with probability close to 1.

Let us call the observed value of c *good* if

$$\exists l \in \left[-\frac{r}{2}, \frac{r}{2}\right], \quad rc \equiv l \bmod N.$$

In this case there exists d such that

$$-\frac{r}{2} \le rc - dN = l \le \frac{r}{2},$$

so that

$$\left|\frac{c}{N} - \frac{d}{r}\right| < \frac{1}{2N}.$$

Hence if c is good, then r' found in 9.3 (A) in fact divides r.

Now call c *very good* if $r' = r$.

Estimating the exponential sum in 9.3 (iv), we can easily check that the probability of observing a good c is $\ge 1/3r^2$. On the other hand, there are $r\varphi(r)$ states $|c, t^k \bmod M\rangle$ with very good c. Thus to find a very good c with high probability, $O(r^2 \log r)$ runs will suffice.

10 Kolmogorov Complexity and Growth of Recursive Functions

Consider general functions $f : \mathbf{Z}^+ \to \mathbf{Z}^+$. Computability theory uses several growth scales for such functions, of which two are most useful: f may be majorized by some recursive function (e.g., when it is itself recursive), or by a polynomial (e.g., when it is computable in polynomial-time). Linear growth does not seem particularly relevant in this context. However, this impression is quite misleading, at least if one allows one most important *uncomputable* reordering of $\mathbf{Z}^+$. In fact, we make the following claim:

10.1. **Claim.** *There exists a permutation* $\mathbf{K} : \mathbf{Z}^+ \to \mathbf{Z}^+$ *such that for any partially recursive function* $f : \mathbf{N} \to \mathbf{N}$ *there exists a constant* c *with the property*

$$\mathbf{K} \circ f \circ \mathbf{K}^{-1}(n) \le cn \text{ for all } n \in \mathbf{K}(D(f)).$$

Moreover, $\mathbf{K}$ *is bounded by a linear function, but* $\mathbf{K}^{-1}$ *is not bounded by any recursive function.*

PROOF. We will use the Kolmogorov complexity measure of integers, as was explained in VI.9. We first recall its definition.

For a recursive function $u : \mathbf{Z}^+ \to \mathbf{Z}^+$, $x \in \mathbf{Z}^+$, put $C_u(x) := \min\{k \,|\, f(k) = x\}$, or ∞ if such k does not exist. Call such a function u *optimal* if for any other recursive function v, there exists a constant $c_{u,v}$ such that $C_u(x) \le c_{u,v} C_v(x)$ for all x. Optimal functions do exist (see Theorem VI.9.2); in particular, they take all positive integer values (however, they certainly are not everywhere defined). Fix one such u and call $C_u(x)$ the (exponential) complexity of x. By definition, $\mathbf{K} = \mathbf{K}_u$ rearranges $\mathbf{Z}^+$ in order of increasing complexity. In other words,

$$\mathbf{K}(x) := 1 + \mathrm{card}\,\{y \,|\, C_u(y) < C_u(x)\}.$$

We first show that

$$\mathbf{K}(x) = \exp\left(O(1)\right) C_u(x).$$

Since C_u takes each value at most once, we have $\mathbf{K}(n) \le C_u(n)$. In order to show that $C_u(x) \le c\,\mathbf{K}(x)$ for some c it suffices to check that

$$\mathrm{card}\,\{k \le N \,|\, \exists\, x,\, C_u(x) = k\} \ge b\,N$$

with some $b > 0$. In fact, at least half of the numbers $x \le N$ have complexity that is no less than $x/2$.

Now, VI.9.7(b) implies that for any recursive function f and all $x \in D(f)$, we have $C_u(f(x)) \le \mathrm{const}\, C_u(x)$. Since $C_u(x)$ and $\mathbf{K}(x)$ have the same order of growth up to a bounded factor, our claim follows.

10.2. **Corollary.** *Denote by* S_∞^{rec} *be the group of recursive permutations of* $\mathbf{Z}^+$*. Then* $\mathbf{K}\, S_\infty^{\mathrm{rec}}\, \mathbf{K}^{-1}$ *is a subgroup of permutations of no more than linear growth.*

Actually, appealing to Proposition VI.9.6, one can considerably strengthen this result. For example, let σ be a recursive permutation, $\sigma^{\mathbf{K}} = \mathbf{K}\sigma\mathbf{K}^{-1}$. Then $\sigma^{\mathbf{K}}(x) \le cx$, so that $(\sigma^{\mathbf{K}})^n(x) \le c^n x$ for $n > 0$. But actually the last inequality can be replaced by

$$(\sigma^{\mathbf{K}})^n(x) \le c' n$$

for a fixed x and variable n. With both x and n variable one gets the estimate $O(xn \log{(xn)})$.

Recall that finite permutations appear in the quantum versions of Boolean circuits, because we must treat any function with the help of an appropriate unitary operator: cf. the discussion in 7.3 above.

For the same reason, infinite (computable) permutations might naturally appear in models of quantum Turing machines and normal computation models. In fact, if one assumes that the transition function s is a permutation, and then extends it to the unitary operator U_s in the infinite-dimensional Hilbert space, one might be interested in studying the spectral properties of such operators. But the latter depend only on the conjugacy class. Perhaps the universal conjugation $U_{\mathbf{K}}$ will be a useful theoretical tool in this context.

10.3. **Final comments.** Finally, we would like to comment on the hidden role of Kolmogorov complexity in the real life of classical computing.

The point is that in a sense (which is difficult to formalize), we are interested only *in the calculation of sufficiently nice functions,* because a random Boolean function will have (super)exponential complexity anyway.

A nice function, at the very least, has a short description and therefore a small Kolmogorov complexity. Thus, dealing with practical problems, we actually work *not* with small numbers, graphs, circuits, ..., but rather with an initial segment of the respective constructive world *reordered with the help of* **K**. We systematically replace a large object by its short description.

But then the "natural operations" that can be performed on our objects lose computability when we have replaced the objects by their short descriptions.

This inherent tension, incompatibility of shortest descriptions with most-economic algorithmic processing, is the central issue of any computation theory.

The place-value notation of numbers that played such a great role in the development of human civilizations is the ultimate system of short descriptions that bridges the abyss. Kolmogorov complexity goes far beyond this point.

Part IV
MODEL THEORY

X

Model Theory

Model theory the part of logic that studies structures (in Bourbaki's sense) in relation to their descriptions in formal languages, usually first-order ones. The study of structures and classes of structures is essentially a subject of algebra or universal algebra, but model theory is different in its approach in that it places a special emphasis on the question of language and definability in the structures. This approach has paid off with applications in various parts of concrete mathematics.

1 Languages and Structures

Given a language L, an *L-structure* (or just *structure*) is essentially the same thing as an interpretation of L as explained in Section II.2. But the stress now is rather on algebra than on logic, so instead of the notation ϕ, which realized the interpretation of the symbols of L in a set A, we will refer to the structure $\mathbf{A} = (A, L)$, which provides an interpretation for the symbols of L. We write, e.g., $\mathbf{A} = (A, +, \cdot, 0, 1)$ when $L = \{+, \cdot, 0, 1\}$. We call A *the domain* of the structure $\mathbf{A}$.

Unless stated otherwise, we deal in this chapter with first-order languages. For an L-formula P one writes $\mathbf{A} \models P$ to say that the value of P under the interpretation is "true." Usually, in the above notation we will assume that P is a *sentence*, that, is a formula with no free variables.

According to this notation, $T_\phi L$ of II.6.1, for an interpretation ϕ of L, becomes

$$\mathrm{Th}(\mathbf{A}) := \{P : \mathbf{A} \models P\},$$

the theory of structure $\mathbf{A}$, where $\mathbf{A}$ is the structure given by ϕ.

Often, for a formula $P(x_1, \ldots, x_n)$ with free variables $x_1, \ldots, x_n$ and elements $a_1, \ldots, a_n \in A$ we say $\mathbf{A} \models P(a_1, \ldots, a_n)$, meaning that we have extended the interpretation given by $\mathbf{A}$ to the interpretation of variables $x_i \mapsto a_i$.

We also assume, as is standard in model theory, that every language contains the symbol $=$ and its interpretation is always equality, that is, *structures are normal models*.

Yu. I. Manin, *A Course in Mathematical Logic for Mathematicians, Second Edition*, Graduate Texts in Mathematics 53, DOI 10.1007/978-1-4419-0615-1_10,

1.1 **Embeddings.** If $\mathbf{A}$ and $\mathbf{B}$ are L-structures with domains A and B correspondingly, an *embedding* $h : \mathbf{A} \to \mathbf{B}$ is a map $A \to B$ that *preserves* the symbols of relations, operations, and constants of L, that is,

(i) for any n-ary relation symbol $p \in L$ and $a_1, \dots, a_n \in A$, $\mathbf{A} \vDash p(a_1, \dots, a_n)$ iff $\mathbf{B} \vDash p(h(a_1), \dots, h(a_n))$;
(ii) for any n-ary operation symbol $f \in L$ and $a_1, \dots, a_n, a \in A$, $\mathbf{A} \vDash f(a_1, \dots, a_n) = a$ iff $\mathbf{B} \vDash f(h(a_1), \dots, h(a_n)) = h(a)$;
(iii) for any constant symbol $c \in L$ and $a \in A$, $\mathbf{A} \vDash c^{\mathbf{A}} = a$ iff $\mathbf{B} \vDash c^{\mathbf{B}} = h(a)$, where $c^{\mathbf{A}}$ stands for the interpretation of c in the structure $\mathbf{A}$.

1.2 **Exercise.** *Any embedding is injective.*

A surjective embedding is called an *isomorphism.*

1.3 **Definable sets.** Recall that for an L-structure $\mathbf{A}$ and an L-formula $P(x_1, \dots, x_n)$ one defines (definition II.2.8) the set

$$P(\mathbf{A}) = \{\bar{a} \in A^n : \mathbf{A} \vDash P(\bar{a})\}.$$

Sets of this form are called *definable.*

Since any subset of A^n can be viewed as an n-ary relation, $P(\bar{v})$ determines also an *L-definable relation.* If a $P(\mathbf{A})$ coincides with a graph of an operation $f : A^{n-1} \to A$, we say then that f is an *L-definable operation.*

1.4 **Exercise.**

(i) *An embedding $h : \mathbf{A} \to \mathbf{B}$ of L-structures preserves atomic L-formulas, i.e., for any atomic $P(x_1, \dots, x_n)$ for any $\bar{a} \in A^n$,*

$$\mathbf{A} \vDash P(\bar{a}) \text{ iff } \mathbf{B} \vDash P(h(\bar{a})).$$

(ii) *Given an $\forall$-formula $P(\bar{a})$, that is, one of the form $\forall x_1 \cdots \forall x_m\, Q(x_1, \dots, x_m, \bar{a})$ with Q quantifier-free, and an embedding $h : \mathbf{A} \to \mathbf{B}$, $\bar{a}$ in $\mathbf{A}$,*

$$\mathbf{B} \vDash P(\bar{a}) \text{ implies } \mathbf{A} \vDash P(h(\bar{a})).$$

(iii) *An isomorphism $h : \mathbf{A} \to \mathbf{B}$ between L-structures preserves any L-formula $P(x_1, \dots, x_n)$, i.e., for any $\bar{a} \in A^n$,*

$$\mathbf{A} \vDash P(\bar{a}) \text{ iff } \mathbf{B} \vDash P(h(\bar{a})).$$

1.5 **Corollary.** *For definable subsets (relations)*

$$h(P(\mathbf{A})) = P(\mathbf{B});$$

in particular, definable subsets in a given structure $\mathbf{A}$ are invariant under the action of $\mathrm{Aut}(\mathbf{A})$.

The invariance under Aut is often useful in checking nondefinability of some subsets or relations.

1.6 Exercise. *Multiplication is not definable in* $\mathbf{R}^+_{\text{group}} := \langle \mathbf{R}, + \rangle$, *the additive group of reals.*

The test of invariance works for $\mathbf{R}^+_{\text{group}}$ because the group $\text{Aut}(\mathbf{R}^+_{\text{group}})$ is large; in fact, the structure is *homogeneous* in the sense that two n-tuples satisfy the same formulas (have the same *type*) if and only if there is an automorphism taking one to another, and also every possible type is realized in the given model of the theory. This is not the case in general. For example, for the structure $\mathbf{R}_{\text{field}} := \langle \mathbf{R}, +, \cdot, 0, 1 \rangle$ the automorphism group is trivial. We get much better understanding of definability in this structure by looking into a *nonstandard saturated* model of the corresponding theory (see 2.13, 3.11, and 4.6).

1.7 Definability of a structure. The notion of a definable set can be extended to that of a definable structure.

Let L_0 and L_1 be languages and for the sake of brevity assume that L_1 is a relational language. One says that the language L_1 is interpreted in an L_0-structure $\mathbf{A}$ if for some n,

there is given an L_0-formula $Q(\bar{x})$ with n free variables,
there is given an L_0-formula $E(\bar{x}, \bar{y})$ with $2n$ free variables,
for every m-ary predicate symbol p_i in L_0 there is given an L_0-formula $P_i(\bar{x}_1, \ldots, \bar{x}_m)$ with mn free variables,
such that $E(\mathbf{A})$ is an equivalence relation on the set $Q(\mathbf{A})$ and the $P_i(\mathbf{A})$ are relations on $Q(\mathbf{A})$ preserved by the equivalence $E(\mathbf{A})$.

Under these assumptions one considers the domain $Q(\mathbf{A})/E(\mathbf{A})$ and the interpretation of the symbols p_i on the domain given by $P_i(\mathbf{A})$.

One says that *an* L_1*-structure* $\mathbf{M}$ *is definable* (*interpretable*) *in an* L_0*-structure* $\mathbf{A}$ if the above L_1-structure on the domain $Q(\mathbf{A})/E(\mathbf{A})$ is isomorphic to $\mathbf{M}$.

It is clear from the definition that assuming that $\mathbf{M}$ is defined in $\mathbf{A}$, every definable set in $\mathbf{M}$ can be rewritten as a definable quotient set in $\mathbf{A}$ and every L_1-sentence holding in $\mathbf{M}$ can be rewritten into an appropriate L_0-sentence holding in $\mathbf{A}$.

1.8 Example. Let $\text{F} = (\text{F}, +, \cdot, 0, 1)$ be a field and $\text{GL}_n(\text{F})$ a group on the domain $\text{GL}_n(\text{F})$ of $n \times n$ nondegenerate matrices in the language $(*, e)$ of groups.

The natural interpretation of $\text{GL}_n(\text{F})$ is on the domain

$$D := \{X = (x_{ij}) \in F^{n^2} \colon i, j = 1, \ldots, n, \ \det X \neq 0\},$$

with the interpretation of e as the element of D with $x_{ii} = 1$, $x_{ij} = 0$ for all $i, j \leq n$, $i \neq j$, and the operation $X * Y = Z$ interpreted on D by the known polynomial equations.

1.9 Definition. Given two L-structures $\mathbf{A}$ and $\mathbf{B}$ and an embedding $h : \mathbf{A} \to \mathbf{B}$, we say that the *the embedding is elementary* if for any L-formula $P(x_1, \ldots, x_n)$ and any $a_1, \ldots, a_n \in A$,

$$(*) \quad \mathbf{A} \vDash P(a_1, \ldots, a_n) \text{ iff } \mathbf{B} \vDash P(h(a_1), \ldots, h(a_n)).$$

In this situation $\mathbf{A}$ is also said to be *an elementary substructure* of $\mathbf{B}$ and $\mathbf{B}$ *an elementary extension* of $\mathbf{A}$, written $\mathbf{A} \preccurlyeq \mathbf{B}$ or $\mathbf{A} \preccurlyeq_h \mathbf{B}$.

We say that $\mathbf{A}$ is *elementarily equivalent to* $\mathbf{B}$, written $\mathbf{A} \equiv \mathbf{B}$, if for any L-sentence P,

$$\mathbf{A} \vDash P \text{ iff } \mathbf{B} \vDash P.$$

It is also useful to consider partial $h : \mathbf{A} \to \mathbf{B}$ with $\operatorname{dom} h = X \subset A$ and $\operatorname{range} h = Y \subset B$. Provided $(*)$ holds for any $a_1, \ldots, a_n \in X$ and any L-formula P, such an h is said to be an *elementary monomorphism* $\mathbf{A} \to \mathbf{B}$.

Before proceeding further we want to make a note on the notion of deducibility used in model theory. It is semantic, in distinction to the syntactic one elaborated in Chapter II. In the first-order context these notions are equivalent due to the Gödel completeness theorem, but in general the semantic approach is more flexible and can be used when no formal system of rules of deduction is available.

Let $\mathcal{E}$ be a set of L-sentences. We write $\mathbf{A} \vDash \mathcal{E}$ if for any $S \in \mathcal{E}$, $\mathbf{A} \vDash P$.

1.10 **Definition.** An L-sentence S is said to be *a logical consequence of* a finite $\mathcal{E}$, written $\mathcal{E} \vDash S$, if $\mathbf{A} \vDash \mathcal{E}$ implies $\mathbf{A} \vDash S$ for every L-structure $\mathbf{A}$. For $\mathcal{E}$ infinite, $\mathcal{E} \vDash S$ means that there is a finite $\mathcal{E}^0 \subset \mathcal{E}$ such that $\mathcal{E}^0 \vDash S$.

S is called *logically valid*, written $\vDash S$, if $\mathbf{A} \vDash S$ for every L-structure $\mathbf{A}$.

1.11 **Definition.** $\mathcal{E}$ is said to be *finitely satisfiable* (*f.s.*) if any finite subset of $\mathcal{E}$ is satisfiable, that is, has a model.

$\mathcal{E}$ is said to be *deductively closed* if for any L-sentence $\mathcal{E}$, $\mathcal{E} \vDash S$ implies $S \in \mathcal{E}$.

Clearly, a complete satisfiable $\mathcal{E}$ is deductively closed. In model-theoretic constructions one often moves between variations of a given language.

1.12 **Definition.** Let $\mathbf{A} = (A, L)$ be an L structure and L' a language whose nonlogical symbols of that are in L, that is, $L' \subseteq L$. The structure $\mathbf{A}' = (A, L')$ on the domain A with the symbols of L' interpreted as in $\mathbf{A}$ is called *the L'-reduct of* $\mathbf{A}$. Conversely, $\mathbf{A}$ is an *expansion* of $\mathbf{A}'$ to the language L.

Obviously, under the notation above for an L'-formula $P(v_1, \ldots, v_n)$ and $a_1, \ldots, a_n \in A$,

$$\mathbf{A}' \vDash P(a_1, \ldots, a_n) \text{ iff } \mathbf{A} \vDash P(a_1, \ldots, a_n).$$

1.13 A special and broadly used form of expansion of a structure $\mathbf{A} = (A, L)$ is the expansion by constant symbols naming elements in $\mathbf{A}$. For $C \subseteq A$ let $L_A = L \cup \{c_a : a \in C\}$ be the extension of the language by the constant symbols and $\mathbf{A}_C$ the natural expansion of $\mathbf{A}$ to L_C assigning to c_a the element a. L_C-formulas are then called *formulas with parameters in* C.

2 The Compactness Theorem

This section discusses the compactness theorem and its various immediate applications. This theorem was implicit in Gödel's completeness theorem and was proved independently by A. Mal'tsev in 1936. Its later proofs based on Henkin's method produce more specialized models with more refined applications.

2.1 **Compactness theorem.** *Let $\mathcal{E}$ be a finitely satisfiable set of L-sentences. Then $\mathcal{E}$ is satisfiable; moreover, $\mathcal{E}$ has a model of cardinality less than or equal to $|L| + \aleph_0$.*

Below we discuss three proofs of the theorem. Note that each of them uses the axiom of choice; that is, the construction of the model is in general ineffective.

The first proof is an application of Gödel's completeness theorem II.6.2 and uses the deduction system of II.2.2– II.2.5.

2.2 **Lemma.** *$\mathcal{E}$ is consistent.*

PROOF. Suppose $\mathcal{E} \vdash P$. Then $\mathcal{E}^0 \vdash P$ for some finite $\mathcal{E}^0 \subseteq \mathcal{E}$, since only finitely many formulas are involved in the proof. In particular, if $\mathcal{E}$ were inconsistent, already its finite subset $\mathcal{E}^0$ would be. But then $\mathcal{E}^0$ could not be satisfiable, contradicting the assumption.

2.3 **Lemma (Lindenbaum's theorem).** *$\mathcal{E}$ can be completed; that is, there is a complete f.s. set of L-sentences $\mathcal{E}^\#$ such that $\mathcal{E} \subseteq \mathcal{E}^\#$.*
PROOF. (Uses the axiom of choice). Let

$$\mathcal{S} = \{\mathcal{E}' : \mathcal{E} \subseteq \mathcal{E}' \text{ an f.s. set of } L\text{-sentences}\}.$$

Clearly $\mathcal{S}$ satisfies the hypothesis of Zorn's lemma, so it contains a maximal element, $\mathcal{E}^\#$ say. This is complete, for otherwise, say $S \notin \mathcal{E}^\#$ and $\neg S \notin \mathcal{E}^\#$. By maximality neither $\{S\} \cup \mathcal{E}^\#$ nor $\{\neg S\} \cup \mathcal{E}^\#$ is f.s. Hence there exist finite $\mathcal{E}_1 \subseteq \mathcal{E}^\#$ and $\mathcal{E}_2 \subseteq \mathcal{E}^\#$ such that neither $\{S\} \cup \mathcal{E}_1$ nor $\{\neg S\} \cup \mathcal{E}_2$ is satisfiable. However, $\mathcal{E}_1 \cup \mathcal{E}_2 \subseteq \mathcal{E}^\#$, finite, so has a model, $\mathbf{A}$ say. But either $\mathbf{A} \models S$, so $\mathbf{A} \models \{S\} \cup \mathcal{E}_1$, or $\mathbf{A} \models \neg S$, so $\mathbf{A} \models \{\neg S\} \cup \mathcal{E}_2$, a contradiction.

Clearly, $\mathcal{E}^\#$ of the lemma is Gödelian so has a model by II.6.2. This model is also a model of $\mathcal{E}$. This finishes the first proof

2.4 **Exercise.** *Let $\alpha, \alpha_1, \ldots, \alpha_n, \beta, \beta_1, \ldots, \beta_n, \gamma$ be closed L-terms, p, f L-symbols for n-ary predicate and n-ary operation, correspondingly, and $P(v_0, v_1, \ldots, v_n)$ an L-formula with free variables $v_0, v_1, \ldots, v_n$. Prove that*

(a) $\alpha = \beta \models \beta = \alpha$;
(b) $\alpha = \beta, \beta = \gamma \models \alpha = \gamma$;
(c) $\models \alpha = \alpha$;
(d) $\alpha_1 = \beta_1, \ldots, \alpha_n = \beta_n, P(\alpha_1, \ldots, \alpha_n) \models P(\beta_1, \ldots, \beta_n)$;
(e) $\alpha = \beta, \alpha_1 = \beta_1, \ldots, \alpha_n = \beta_n, f(\alpha_1, \ldots, \alpha_n) = \alpha \models f(\beta_1, \ldots, \beta_n) = \beta$;
(f) $P(\beta, \alpha_1, \ldots, \alpha_n) \models \exists v_0 P(v_0, \alpha_1, \ldots, \alpha_n)$.

A set $\mathcal{E}$ of L-sentences is said to be *with witnesses* if for any sentence in $\mathcal{E}$ of the form $\exists v P(v)$ there is a closed L-term λ such that $P(\lambda) \in \mathcal{E}$.

2.5 **Exercise.** *There exists a closed L-term if there exists a set of L-sentences that is complete, with witnesses, and f.s.* (Consider the L-sentence $\exists v\, v = v$.)

2.6 **Lemma.** *For some extension $\tilde{L} \supseteq L$ of the language, with $|\tilde{L}| = |L| + \aleph_0$, there is an extension $\tilde{\mathcal{E}} \supseteq \mathcal{E}$ of $\mathcal{E}$ to a complete f.s. set of $\tilde{L}$-sentences with witnesses.*

PROOF. We are going to obtain $\tilde{L}$ just by adding $|L|+\aleph_0$ new constant symbols. We introduce new languages L_i and a complete set of L_i-sentences $\mathcal{E}_i$ ($i = 0, 1, \dots$). Let $L_0 = L$. By Lindenbaum's theorem there exists $\mathcal{E}_0 \supseteq \mathcal{E}$, a complete set of L_0-sentences.

Given an f.s. $\mathcal{E}_i$ in language L_i, introduce the new language

$$L_{i+1} = L_i \cup \{c_Q : \ Q \ \text{ a one-variable } L_i\text{-formula}\}$$

and the new set of L_{i+1} sentences

$$\mathcal{E}_i^* = \mathcal{E}_i \cup \{(\exists v Q(v) \to Q(c_Q)) : \ Q \text{ a one-variable } L_i\text{-formula}\}.$$

Claim. $\mathcal{E}_i^*$ is finitely satisfiable. Indeed, given a finite $\mathcal{E}' \subseteq \mathcal{E}_i^*$, let $\mathcal{E}'' = \mathcal{E}' \cap \mathcal{E}_i$ and take a model $\mathbf{A}$ of $\mathcal{E}''$ with a domain A, which we assume well-ordered. Assign constants to symbols c_Q as follows:

$$c_Q = \begin{cases} \text{the first element in } Q(\mathbf{A}), & \text{if } Q(\mathbf{A}) \neq \varnothing, \\ \text{the first element in } A, & \text{otherwise.} \end{cases}$$

Denote the expanded structure by $\mathbf{A}^*$. By definition, for all $Q(v)$, $\mathbf{A}^* \models (\exists v Q(v) \to Q(c_Q))$. So $\mathbf{A}^* \models \mathcal{E}'$. This proves the claim.

Let $\mathcal{E}_{i+1}$ be a complete f.s. set of L_{i+1}-sentences containing $\mathcal{E}_i^*$.

Take $\mathcal{E}^* = \bigcup_{i \in \mathbf{N}} \mathcal{E}_i$. This is finitely satisfiable. By construction one sees immediately that $\mathcal{E}^*$ is with witnesses and is complete in the language $\bigcup_{i \in \mathbf{N}} L_i = L + \{\text{new constants}\}$.

An L-structure $\mathbf{A}$ is called *named* if for every $a \in A$ there is a closed L-term λ such that $\lambda^{\mathbf{A}} = a$.

2.7 **Proposition.** *For any complete f.s. set $\mathcal{E}$ of L-sentences with witnesses there is a named model.*

PROOF. Let Λ be the set of closed terms of L. This is nonempty by 2.5. For $\alpha, \beta \in \Lambda$ define $\alpha \backsim \beta$ iff $\alpha = \beta \in \mathcal{E}$.

This is an equivalence relation by 2.4.1–2.4.3.

For $\alpha \in \Lambda$, let $\tilde{\alpha}$ denote the $\backsim$-equivalence class containing α. Let

$$A = \{\tilde{\alpha} : \ \alpha \in \Lambda\}.$$

This will be the domain of our model $\mathbf{A}$. We want to define relations, operations, and constants of L on A.

Let p be an n-ary relation symbol of L and $\alpha_1, \dots, \alpha_n \in \Lambda$. Define

$$\mathbf{A} \models p(\tilde{\alpha}_1, \dots, \tilde{\alpha}_n) \text{ if } p(\alpha_1, \dots, \alpha_n) \in \mathcal{E}.$$

By 2.4.4 the definition does not depend on the choice of representatives in the $\backsim$-classes.

For a unary operation symbol f of L of arity m and $\alpha_1, \ldots, \alpha_m \in \Lambda$, set

$$\mathbf{A} \models f(\tilde{\alpha}_1, \ldots, \tilde{\alpha}_m) = \tilde{\tau}, \text{ where } \tau = f(\alpha_1, \ldots, \alpha_m).$$

By 2.4.5 the operation f in $\mathbf{A}$ is well-defined.

Finally, for a constant symbol c, $c^{\mathbf{A}}$ is just $\tilde{c}$.

We now prove by induction on the complexity of an L-formula $Q(v_1, \ldots, v_n)$ that

$$(*) \quad \mathbf{A} \models Q(\tilde{\alpha}_1, \ldots, \tilde{\alpha}_n) \text{ iff } Q(\alpha_1, \ldots, \alpha_n) \in \mathcal{E}.$$

For atomic formulas we have this by definition.

If $Q = (Q_1 \wedge Q_2)$ then $\mathbf{A} \models (Q_1(\tilde{\alpha}_1, \ldots, \tilde{\alpha}_n) \wedge Q_2(\tilde{\alpha}_1, \ldots, \tilde{\alpha}_n))$ iff $\mathbf{A} \models Q_1(\tilde{\alpha}_1, \ldots, \tilde{\alpha}_n)$ and $\mathbf{A} \models Q_2(\tilde{\alpha}_1, \ldots, \tilde{\alpha}_n)$ iff (by induction hypothesis) $Q_1(\alpha_1, \ldots, \alpha_n), Q_2(\alpha_1, \ldots, \alpha_n) \in \mathcal{E}$ iff (by deductive closedness) $(Q_1(\alpha_1, \ldots, \alpha_n) \wedge Q_2(\alpha_1, \ldots, \alpha_n)) \in \mathcal{E}$, which proves $(*)$ in this case.

The case $Q = \neg P$ is proved similarly.

In case $Q = \exists v P$, $\mathbf{A} \models \exists v P(v, \tilde{\alpha}_1, \ldots, \tilde{\alpha}_n)$ iff there is $\beta \in \Lambda$ such that $\mathbf{A} \models P(\tilde{\beta}, \tilde{\alpha}_1, \ldots, \tilde{\alpha}_n)$ iff there is $\beta \in \Lambda$ such that $P(\beta, \alpha_1, \ldots, \alpha_n) \in \mathcal{E}$. The latter implies, by 2.4.6 and deductive closedness, that $\exists v P(v, \alpha_1, \ldots, \alpha_n) \in \mathcal{E}$, and the converse holds because $\mathcal{E}$ is with witnesses. This proves $(*)$ for the formula and finishes the proof of $(*)$ for all formulas.

2.8 **The second proof** of the compactness theorem.

By 2.6, $\mathcal{E} \subseteq \tilde{\mathcal{E}}$, for some complete f.s. set $\tilde{\mathcal{E}}$ of $\tilde{L}$-sentences with witnesses, $|\tilde{L}| = |L| + \aleph_0$. By 2.7 this has a named model, say $\mathbf{A}$. By definition, $|A| = |L| + \aleph_0$, and clearly the reduct of $\mathbf{A}$ to the language L is a model of $\mathcal{E}$.

2.9 **The third proof** of the compactness theorem uses ultraproducts of models.

Let B be a Boolean algebra. A *filter in* B is a subset $U \subseteq B$ such that

(i) $\varnothing \notin U$;
(ii) $X \in U,\ X \subseteq Y \in B \Rightarrow Y \in U$;
(iii) $X, Y \in U \Rightarrow X \cap Y \in U$.

A filter U is called an *ultrafilter* if also

(iv) for all $Y \in B$, either $Y \in U$ or $I \setminus Y \in U$.

A filter U on B is said to be *principal* if there is $X_0 \in B$ such that $X_0 \subseteq X$ for all $X \in U$. Otherwise, we say that U is *nonprincipal*.

In this section we deal with the case that B is the Boolean algebra of all subsets of a given set I. Then U is said to be a *filter on* I.

Now let $\mathbf{A}_i = (A_i, L)$, $i \in I$, be a set of L-structures and U a filter on I. We are going to construct a new structure, denoted by $\prod_{i \in I} \mathbf{A}_i / U$, using the data.

Let $\prod_{i \in I} A_i$ stand for the Cartesian product of the sets, that is, the set of all functions $\varphi : I \to \bigcup_{i \in I} A_i$ with $\varphi(i) \in A_i$. Define an equivalence relation (check it) on the set $\prod_{i \in I} A_i$:

$$\varphi \approx_U \psi \quad \text{iff} \quad \{i \in I : \varphi(i) = \psi(i)\} \in U$$

(we say "φ is equal to ψ almost everywhere modulo U").

Now we denote by $\prod_{i\in I} A_i/U$ the quotient of $\prod_{i\in I} A_i$ by the equivalence $\approx_U$. This is going to be the domain of the structure under construction; an element of it represented by $\varphi \in \prod_{i\in I} A_i$ will be denoted by $\tilde{\varphi}$.

We interpret a symbol p of an n-ary relation on $\prod_{i\in I} A_i/U$ by assuming $p(\tilde{\varphi}_1,\dots,\tilde{\varphi}_n)$ true if $\{i : \mathbf{A}_i \vDash p(\varphi_1(i),\dots,\varphi_n(i))\} \in U$, that is, $\mathbf{A}_i \vDash p(\varphi_1(i),\dots,\varphi_n(i))$ for almost all i. It is easy to check that this is well defined.

The same principle is used to interpret the meaning of $f(\tilde{\varphi}_1,\dots,\tilde{\varphi}_n) = \tilde{\varphi}_{n+1}$ for a symbol f of an n-ary operation, and similarly interpretation of $c = \tilde{\varphi}$ for a symbol of constant c. This defines the L-structure $\prod_{i\in I} \mathbf{A}_i/U$, *a filtered product of L-structures along U*. When U is an ultrafilter, $\prod_{i\in I} \mathbf{A}_i/U$ is called an *ultraproduct*. In case $\mathbf{A}_i = \mathbf{A}$ for all $i \in I$, the ultraproduct is called an *ultrapower*, written $\mathbf{A}^I/U$.

2.10 **Los's theorem.** *Let $\mathbf{A}_i = (A_i, L)$, $i \in I$, be a set of L-structures, U an ultrafilter on I, and $\prod_{i\in I} \mathbf{A}_i/U$ the ultraproduct along U.*

For every L-formula $P(x_1,\dots,x_n)$ with free variables $x_1,\dots,x_n$ and every $\tilde{\varphi}_1,\dots,\tilde{\varphi}_n \in \prod_{i\in I} A_i/U$,

$$\prod_{i\in I} \mathbf{A}_i/U \vDash P(\tilde{\varphi}_1,\dots,\tilde{\varphi}_n) \textit{ iff } \{i:\ \mathbf{A}_i \vDash P(\varphi_1(i),\dots,\varphi_n(i))\} \in U.$$

PROOF. Induction on the complexity of P. For $P(x_1,\dots,x_n)$ of the form $p(x_1,\dots,x_n)$, $f(x_1,\dots,x_{n-1}) = x_n$, $c = x_n$, for symbols of predicate, operation, or constant, the statement holds by definition.

Assuming the statement of the theorem for formulas P_1 and P_2 of a given complexity, one gets it for the formula $P_1 \,\&\, P_2$ by the property (iii) of a filter.

For a formula of the form $\exists x_n\, P_1$, if $\prod_{i\in I} \mathbf{A}_i/U \vDash \exists x_{n+1} P(\tilde{\varphi}_1,\dots,\tilde{\varphi}_n, x_{n+1})$, then by definition, there exists $\tilde{\varphi}_{n+1}$ in the structure such that $\prod_{i\in I} \mathbf{A}_i/U \vDash P_1(\tilde{\varphi}_1,\dots,\tilde{\varphi}_n,\tilde{\varphi}_{n+1})$. By induction, $\mathbf{A}_i \vDash P_1(\varphi_1(i),\dots,\varphi_n(i),\varphi_{n+1}(i))$ for almost all $i \in I$ modulo U. This implies $\mathbf{A}_i \vDash \exists x_{n+1} P_1(\varphi_1(i),\dots,\varphi_n(i), x_{n+1})$ for almost all $i \in I$. In the reverse direction, the latter implies the existence of a function φ_{n+1} such that $\mathbf{A}_i \vDash P_1(\varphi_1(i),\dots,\varphi_n(i),\varphi_{n+1}(i))$ for the same values of $i \in I$. This proves the inductive step in the case in question.

Since every formula up to logical equivalence can be written in terms of &, $\exists$, and $\neg$, to complete the proof of the theorem it suffices to check the statement for a formula of the form $\neg P_1$. This case is immediate by property (iv) defining an ultrafilter.

End of the proof of the compactness theorem. Third version. Without loss of generality we assume that $\mathcal{E}$ is deductively closed, in particular, if $S_1,\dots,S_n \in \mathcal{E}$ then $(S_1 \,\&\cdots\&\, S_n) \in \mathcal{E}$.

By the assumptions, for every sentence $S \in \mathcal{E}$ there exists a model $\mathbf{A}_S$. Now we introduce an ultrafilter on $\mathcal{E}$. For every $S \in \mathcal{E}$ set $X_S = \{Q \in \mathcal{E} :\ Q \vDash S\}$. Clearly $X_{S_1 \& S_2} = X_{S_1} \cap X_{S_2}$. It follows that the set

$$U_0 = \{Y \subseteq \mathcal{E} :\ X_S \subseteq Y, \text{ for some } S \in \mathcal{E}\}$$

is a filter. By Zorn's lemma, U_0 is contained in a maximal filter U, equivalently, an ultrafilter.

Now by Los's theorem the ultraproduct $\prod_{S\in\mathcal{E}} \mathbf{A}_S/U$ is a model of any $S \in \mathcal{E}$, so a model for $\mathcal{E}$.

2.11 **Topological interpretation.** Consider the set S of all L-structures of bounded cardinality; card $L+\aleph_0$ will do. Consider the quotient $\mathcal{S} = S/\equiv$, where $\equiv$ stands for elementary equivalence between L-structures. Every L-sentence P singles out a subset

$$[P] = \{\mathbf{A} \in S : \mathbf{A} \vDash P\}$$

of S and a corresponding subset of $\mathcal{S}$. Consider the topology on $\mathcal{S}$ with an open basis given by sets of the form $[P]$. The statement of the compactness theorem can be reformulated as follows:

The topological space $\mathcal{S}$ of L-structures is compact.

Now let $I \subseteq \mathcal{S}$ be a set of points in the space and U an ultrafilter on I. In a compact Hausdorff space there exists a unique limit point along the given ultrafilter, $\lim_U I$. This point is provided by the ultrapoduct construction and Los's theorem. Namely, $\lim_U I$ is given by the equivalence class represented by $\prod_{i\in I} \mathbf{A}_i/U$, with $\mathbf{A}_i \in S$ representing corresponding points $i \in I$.

2.12 **Ultrapowers.** Once ultraproducts were discovered it was noticed that ultrapowers $\mathbf{A}^I/U$ by a nonprincipal ultrafilter provide a special kind of model of a given complete theory. Note that by Los's theorem,

$$\mathbf{A} \prec \mathbf{A}^I/U.$$

And this elementary extension of $\mathbf{A}$ has a remarkable property: every sequence $\{a_i : i \in I\}$ of elements of $\mathbf{A}$ has a "limit" $\mathbf{a}$ in $\mathbf{A}^I/U$ along the ultrafilter. Just take $\mathbf{a}$ to be $\tilde{\varphi}$ for $\varphi : i \mapsto a_i$.

The limit in question can be defined properly in a topology on A similar to that of 2.11. Consider the topology τ_{Def} on A whose basic closed subsets that are the definable subsets of $\mathbf{A}$ (in later sections we will add to these the subsets *definable with parameters*). Our $\mathbf{a}$ is a limit point of the sequence in this topology.

Much more can be said about an ultrapower, but in general, its properties depend essentially on the choice of the ultrafilter and on set-theoretic assumptions. The simplest case is one of a nonprincipal ultrafilter on a countable set I assuming also CH. We also assume the language L and the structure $\mathbf{A}$ to be countable. Under these assumptions every countable sequence in $\mathbf{A}^I/U$ has a τ_{Def}-limit point in $\mathbf{A}^I/U$. This important property is called *saturation* and will be discussed in detail later. Here we only quote one of the remarkable corollaries of saturation of ultrapowers.

The Kiesler–Shelah theorem. *For L-structures $\mathbf{A}$ and $\mathbf{B}$,*

$$\mathbf{A} \equiv \mathbf{B} \quad \textit{iff for some } I \textit{ and an ultrafilter } U \textit{ on } I,\ A^I/U \cong \mathbf{B}^I/U.$$

H. Keisler proved this theorem in 1961 assuming CH. In fact, under CH, for countable L, $\mathbf{A}$ and $\mathbf{B}$, one can restrict I to be a countable set and U any nonprincipal ultrafilter on I.

Later, in the 1970s, S. Shelah produced a clever combinatorial proof avoiding CH.

Finally, we remark here that ultrapowers and, more generally, ultraproducts found many applications (e.g., a construction of Gromov's asymptotic cones by van den Dries and Wilkie), but nowadays the preference in most cases is given to an equivalent treatment, via saturated models.

2.13 **Nonstandard models of classical theories.** A very simple application of the compactness theorem establishes the existence of *nonstandard models* of such theories as arithmetic, real analysis, and others.

Let $\mathbf{N} = (\mathrm{N}, +, \cdot, 0, 1)$ be the usual structure on nonnegative integers in the language of arithmetic, $\mathrm{L_1Ar}$ (which is also used as a language for fields). The theory $\mathrm{Th}(\mathbf{N})$ is called *complete arithmetic* (to be distinguished from *Peano arithmetic*, given by a system of axioms that is incomplete).

Any model of $\mathrm{Th}(\mathbf{N})$ distinct from (not isomorphic to) $\mathbf{N}$ is called a *nonstandard* arithmetic. The existence of one such is immediate by the compactness theorem once one considers the set of $\mathrm{L_1Ar}(c)$-sentences

$$\mathcal{E} = \mathrm{Th}(\mathbf{N}) \cup \{\neg c = n : \ n = 0, 1, \ldots\},$$

where $L(c)$ stands for the extension of the language L by a constant symbol (or a set of constant symbols).

Clearly, $\mathcal{E}$ is finitely satisfiable and any of its models, reduced to the language $\mathrm{L_1Ar}$, is nonstandard. One easily sees (prove it) that necessarily $c \geq \bar{n}$ for every n (for the given theory $x_1 \leq x_2$ replaces $\exists y\, x_1 + y = x_2$); that is, nonstandard elements of arithmetic are "infinite integers."

One can be more creative in constructing nonstandard integers in nonstandard models by choosing a more interesting $\mathcal{E}$ and ending up with, say, a nonstandard integer that is divisible by any standard n.

It is useful to see how a nonstandard model can be obtained using ultraproducts. Let U be a nonprincipal ultrafilter on an infinite set I, and

$$^*\mathbf{N} = \mathbf{N}^I/U,$$

the ultrapower of $\mathbf{N}$, that is, by Los's theorem, a model of the complete arithmetic. Let $\varphi : I \mapsto \mathbf{N}$ be a function that is not constant on any $X \in U$. Clearly, $\tilde{\varphi}$ is a nonstandard integer. In particular, for $I = \mathbf{N}$ and $\varphi : n \mapsto n!$, the nonstandard integer $\tilde{\varphi}$ is divisible by any standard one.

Let us introduce now a first-order formalism for real analysis, which is weaker than $\mathrm{L_2Real}$ of III.2 but powerful enough to express many interesting problems. The language $\mathrm{L_1Real}$ consists of symbols of operations, one for each n-ary function $f : \mathbf{R}^n \to \mathbf{R}$. Observe that this is enough to express the relation $\langle x_1, \ldots, x_n \rangle \in S$ for any given subset $S \subseteq \mathbf{R}^n$; just use the characteristic function of S. In particular, any real number is named by a symbol of operation. We reserve the standard notation for symbols of operations $+, \cdot, -, /$ as well as for standard relations on $\mathbf{R}$.

Let $\mathbf{R}_{\mathrm{analysis}}$ be the obvious $\mathrm{L_1Real}$-structure on $\mathbf{R}$. This we assume to be the *standard model of real analysis*. Correspondingly, any model of $\mathrm{Th}(\mathbf{R}_{\mathrm{analysis}})$

other than the standard one is nonstandard, we say in a short form a *nonstandard model of the reals.*

We claim that any nonstandard model $^*\mathbf{R}$ of the reals contains an element α such that

$$0 < \alpha < \frac{1}{n} \tag{1}$$

for every positive integer n. Indeed, there must be a new, unnamed element, say γ, in $^*\mathbf{R}$. Let

$$[\gamma]^- = \{q \in \mathbf{Q} : q \leq \gamma\}, \quad [\gamma]^+ = \{q \in \mathbf{Q} : q > \gamma\}$$

be the corresponding Dedekind cut, where we allow one of the sets to be empty. If, say $[\gamma]^- = \varnothing$, set $\alpha := -\gamma^{-1}$, which satisfies (1). Similarly in the other case. So, we may assume that both parts of the cut are nonempty. Let r be the unique (standard) real number defined by the cut.

We have either $r < \gamma$ or $r > \gamma$. Assuming the first, set $\alpha := \gamma - r$. This satisfies (1). In the second case set $\alpha := r - \gamma$. This proves the claim.

Call α satisfying (1) *a positive infinitesimal.* An *infinitesimal* is a nonstandard real that is equal to α or $-\alpha$ for a positive infinitesimal α.

Call a nonstandard γ *infinite* if $[\gamma]^+$ or $[\gamma]^-$ is empty. Otherwise γ is said to be *bounded.*

It can now easily be checked that the subset $B \subseteq {}^*\mathbf{R}$ of bounded elements forms a ring, and its subset $\mu \subseteq B$ of infinitesimals is its maximal ideal. The rule $\mathrm{st} : r + \alpha \mapsto r$, for $r \in \mathbf{R}$, $\alpha \in \mu$, determines a well-defined surjective homomorphism of rings $B \to \mathbf{R}$, called *the standard part map.* Obviously, when identified with the (partial) map $^*\mathbf{R} \to \mathbf{R}$, this is exactly the residue map corresponding to the unique valuation on $^*\mathbf{R}$ with the valuation ring B.

Now let $f : \mathbf{R} \to \mathbf{R}$ be a function. By assumption, our language contains a symbol of operation $\bar{f}$ interpreted as f. Let $^*f : {}^*\mathbf{R} \to {}^*\mathbf{R}$ be the function in the nonstandard model corresponding to $\bar{f}$. Similarly for notations of subsets.

The following is easy to check:

f is continuous in the interval (r_1, r_2) iff $^*f(x+\alpha) - {}^*f(x)$ is infinitesimal, for any $x \in (r_1, r_2)$ and any infinitesimal α.

g is a derivative of f on (r_1, r_2) iff $g(x) = \mathrm{st}(^*f(x+\alpha) - {}^*f(x)/\alpha)$ for any standard real $x \in (r_1, r_2)$ and an infinitesimal α.

and so on.

One can also extend the definitions of nonstandard analysis to analysis in Hilbert and Banach spaces, to measure theory, and indeed to any part of mathematics that deals with limits.

Nonstandard analysis provides a solid foundation to Leibniz's idea of infinitesimal calculus. It allows a convenient graphical formalism for operating with limits and infinities and as such leads to a number of beautiful proofs, sometimes new. Yet in its general form the method has obvious limits; after all, it is just a reformulation of analysis in metamathematical terms based on the compactness theorem. A much deeper mathematics based on understanding definability has been developed in concrete cases for *tame* theories, such as the theory of the field of reals $(\mathbf{R}, +, \cdot, 0, 1)$ or $(\mathbf{R}, +, \cdot, 0, 1, \exp)$, the field of reals with exponentiation. The way forward is in classifying definable relations in a given structure and, eventually, understanding the structure of saturated models of the corresponding theory. This method is called *elimination of quantifiers*; see Section 3.18.

3 Basic Methods and Constructions

3.1 **Definition.** We will call a set T of L-sentences an *L-theory,* or simply *theory,* if T is satisfiable and deductively closed.

A subset $\mathcal{E}$ of T such that T is the set of all logical consequences of $\mathcal{E}$ is said to be *a set of axioms of* T.

3.2 **Method of diagrams.** For an L-structure $\mathbf{A}$ let $L_A = L \cup \{c_a : a \in A\}$ be the expansion of the language, $\mathbf{A}_A$ the natural expansion of $\mathbf{A}$ to L_A assigning to c_a the element a. Define the *diagram of* $\mathbf{A}$ to be $\mathrm{Diag}(\mathbf{A}) = \{S\colon$ atomic or negation of atomic L_A-sentence, s.t. $\mathbf{A}_A \models S\}$ and the *complete diagram of* $\mathbf{A}$ to be

$$\mathrm{CDiag}(\mathbf{A}) = \{S : L_A\text{-sentence such that } \mathbf{A}_A \models S\}.$$

Theorem (Method of Diagrams). *For an L-structure* $\mathbf{B}$,

(*i*) *there is an expansion* $\mathbf{B}_A$ *to the language* L_A *such that* $\mathbf{B}_A \models \mathrm{Diag}(\mathbf{A})$ *iff* $\mathbf{A} \subseteq \mathbf{B}$.

(*ii*) *there is an expansion* $\mathbf{B}_A$ *to the language* L_A *such that* $\mathbf{B}_A \models \mathrm{CDiag}(\mathbf{A})$ *iff* $\mathbf{A} \preccurlyeq \mathbf{B}$.

Proof. By definition, $a \to c_a^{\mathbf{B}_A}$ is an embedding iff $\mathbf{B}_A \models \mathrm{Diag}(\mathbf{A})$.

The same holds for an elementary embedding and $\mathrm{CDiag}(\mathbf{A})$. □

Corollary. *Given an L-structure* $\mathbf{A}$ *and an L-theory* T,

(*i*) *the set* $T \cup \mathrm{Diag}(\mathbf{A})$ *is finitely satisfiable iff there is a model* $\mathbf{B}$ *of* T *such that* $\mathbf{A} \subseteq \mathbf{B}$.

(*ii*) *the set* $T \cup \mathrm{CDiag}(\mathbf{A})$ *is finitely satisfiable iff there is a model* $\mathbf{B}$ *of* T *such that* $\mathbf{A} \preccurlyeq \mathbf{B}$.

3.3 **Application. Local theorems of Mal'tsev.** In the 1940s A. Mal'tsev proved a number of theorems dealing with embeddings of some algebraic structures into others using the compactness theorem, or, more specifically, the method of diagrams. He called this type of theorem *local* in the sense that it used the fact that if a certain property holds for finitely generated subalgebras (holds locally) then it holds for the algebra itself. We present an example of such a theorem.

Recall that a group G is said to be *linear of rank n* if it is isomorphic to a subgroup of $\mathrm{GL}_n(\mathrm{F})$ for some field F.

Theorem (A. Mal'tsev). *G is linear of rank n if every finitely generated subgroup of G is.*

PROOF. We use the notion of definability of structures explained in 1.7 and the example 1.8. Observe that the interpretation of $\mathrm{GL}_n(\mathrm{F})$ in a field F is independent of F.

Let $G = (G, *, e)$ be a locally linear group of rank n, that is, with the property that every finitely generated subgroup of it is embeddable into $\mathrm{GL}_n(\mathrm{F})$ for some field F.

Consider the theory T_F stating the axioms of fields in the language $(+, \cdot, 0, 1)$. Consider the diagram $\mathrm{Diag}(G)$ of the group.

Let $D((x_{ij}))$ be the formula in n^2 variables in the language of fields defining the set

$$\{(x_{ij}) \in F^{n^2} : i, j = 1, \ldots, n,\ \det(x_{ij}) \neq 0\}.$$

Now we want to rewrite the diagram of G by a diagram $\mathrm{Diag}^F(G)$ in the language of fields extended by contant symbols. For each constant symbol c^g naming an element g of G we introduce n^2 contant symbols c^g_{ij}, $i, j \in \{1, \ldots, n\}$, and include in $\mathrm{Diag}^F(G)$ the formula $D((c^g_{ij}))$ for every $g \in G$. For each subformula in the diagram of the form $c^g * c^h = c^{gh}$, include in $\mathrm{Diag}^F(G)$ the formula

$$\sum_k c^g_{ik} c^h_{kj} = c^{gh}_{ij}.$$

Consider the set of sentences

$$T = T_F \cup \mathrm{Diag}^F(G).$$

The assumption that every finitely generated subgroup of G is isomorphic to a subgroup of a $\mathrm{GL}_n(\mathrm{F})$ guarantees that T is finitely satisfiable. By 3.2 the theorem follows.

3.4 **Löwenheim–Skolem theorem.** *Suppose T is an L-theory having an infinite model* **A**. *Then for every $\kappa \geq \operatorname{card} L + \aleph_0$ there is a model* **B** *of T of cardinality equal to κ.*

PROOF. In case $\operatorname{card} A \leq \kappa$ we will construct **B** such that $\mathbf{A} \preccurlyeq \mathbf{B}$. This is called *the upward Löwenheim–Skolem theorem.*

Consider the extension of the language L_A by the new constant symbols c_α, $\alpha < \kappa$, and consider the set of sentences

$$\mathrm{CDiag}(\mathbf{A}) \cup \{\neg c_\alpha = c_\beta :\ \alpha < \beta < \kappa\}.$$

This is finitely satisfiable because **A** is infinite. So it has a model $\mathbf{B} \succcurlyeq \mathbf{A}$ of cardinality not bigger than that of the language, that is, $\leq \kappa$. But each c_α is interpreted by a different element of **B**, so $\operatorname{card} B = \kappa$.

In case $\operatorname{card} A \geq \kappa$ one proves *the downward Löwenheim–Skolem theorem,* which provides a **B** of cardinality κ as an elementary substructure of **A**.

Start with a nonempty subset $B_0 \subseteq A$ of cardinality κ. Fix some $a_0 \in B_0$. For each L-formula $P(v_1, \ldots, v_n)$ define a function $g_P : A^{n-1} \to A$ by

$$g_P(a_1, \ldots, a_{n-1}) = \begin{cases} \text{an element } a \in A : \mathbf{A} \models P(a_1, \ldots, a_{n-1}, a) \\ \qquad\qquad\qquad\qquad\qquad\qquad \text{if such exists,} \\ a_0 \quad \text{if not} \end{cases}$$

(g_P are called *Skolem functions*).

Let B be the closure of B_0 under all the g_P. This is closed under all the L-operations f, since any such $(n-1)$-ary f coinsides with the Skolem function $g_{f(v_1 \ldots v_{n-1}) = v_n}$. Let $\mathbf{B}$ be the structure on B induced from $\mathbf{A}$. It is easy to prove, by induction on the complexity of formulas, that for any L-formula $Q(v_1, \ldots, v_n)$ and any $b_1, \ldots, b_n \in B$,

$$\mathbf{B} \models Q(b_1, \ldots, b_n) \Leftrightarrow \mathbf{A} \models Q(b_1, \ldots, b_n),$$

that is, $\mathbf{B} \preccurlyeq \mathbf{A}$, of cardinality κ, as required.

3.5 **Elementary chains of models.** Let, for an ordinal κ,

$$\mathbf{A}_0 \subseteq \mathbf{A}_1 \subseteq \cdots \subseteq \mathbf{A}_\alpha \subseteq \cdots \qquad (\alpha < \kappa) \tag{2}$$

be a κ-sequence of L-structures forming a chain with respect to embeddings, with $\mathbf{A}_\delta$ for limit ordinals $\delta \leq \kappa$ defined as follows:

the domain $A_\delta = \bigcup_{\alpha<\delta} A_\alpha$,
predicate $p^{\mathbf{A}_\delta} = \bigcup_{\alpha<\delta} p^{\mathbf{A}_\alpha}$, for each predicate symbol p of L,
operation $f^{\mathbf{A}_\delta} : A_\delta^m \to A_\delta$ maps $\bar{a}$ to b iff $\bar{a}$ is in A_α for some α and $f^{\mathbf{A}_\alpha}(\bar{a}) = b$, for each operation symbol f of L,
and $c^{\mathbf{A}_\delta} = c^{\mathbf{A}_0}$, for each constant symbol from L.

The chain (2) is said to be *elementary* if for each α,

$$\mathbf{A}_\alpha \preccurlyeq \mathbf{A}_{\alpha+1}.$$

3.6 **Lemma.** *For an elementary chain (2), $\mathbf{A}_\alpha \preccurlyeq \mathbf{A}_\delta$ for any $\alpha < \delta \leq \kappa$.*

PROOF. Clearly, it suffices to prove the statement for all limit ordinals $\delta \leq \kappa$. By induction we may assume that $\mathbf{A}_\alpha \preccurlyeq \mathbf{A}_\beta$, for all $\alpha < \beta < \delta$.

Now, in order to prove $\mathbf{A}_\alpha \preccurlyeq \mathbf{A}_\delta$, we prove

$$A_\alpha \models Q(\bar{a}) \Leftrightarrow \mathbf{A}_\delta \models Q(\bar{a}) \tag{**}$$

for all L-formulas $Q(\bar{x})$ and $\bar{a}$ in A_α by induction on the complexity of Q.

We may assume that Q is constructed from atomic formulas using &, $\neg$, and $\exists$ only.

For Q atomic, (**) follows from the fact that $\mathbf{A}_\alpha \subseteq \mathbf{A}_\delta$, an embedding. The cases of $Q = Q_1 \wedge Q_2$ and $Q = \neg Q_1$ are easy. In the case $Q(\bar{x}) = \exists y\, P(\bar{x}, y)$ the $\Rightarrow$ side of (**) follows immediately from the induction hypothesis and the meaning of $\exists$.

Proof of $\Leftarrow$: $\mathbf{A}_\delta \models \exists y\, P(\bar{a}, y)$ implies $\mathbf{A}_\delta \models P(\bar{a}, b)$, for some $b \in \mathbf{A}_\delta$, so $b \in \mathbf{A}_\beta$ for some $\alpha < \beta < \delta$. By the induction hypothesis $\mathbf{A}_\beta \models P(\bar{a}, b)$. The latter implies $\mathbf{A}_\beta \models \exists y\, P(\bar{a}, y)$ and so $\mathbf{A}_\alpha \models \exists y\, P(\bar{a}, y)$, since $\mathbf{A}_\alpha \preccurlyeq \mathbf{A}_\beta$. □

3.7 **Types.** We fix a complete L-theory T. A set τ of L-formulas $P(\bar{x})$ with n free variables $\bar{x} = (x_1, \ldots, x_n)$ is called an *n-type* (in T) if for any $P_1(\bar{x}), \ldots, P_k(\bar{x}) \in \tau$,

$$T \models \exists \bar{x} \bigwedge_{i \leq k} P_i(\bar{x}).$$

Type τ is called *complete* if also for any $P(\bar{x})$ either $P(\bar{x}) \in \tau$ or $\neg P(\bar{x}) \in \tau$.

A type τ is called *principal* if there is $P(\bar{x})$ such that $T \models \exists \bar{x}\, P(\bar{x})$ and for any $Q(\bar{x}) \in \tau$, $T \models \forall \bar{x}(P(\bar{x}) \to Q(\bar{x}))$.

P is called then a *principal formula for type* τ.

A type that is not principal is called *nonprincipal.*.

Example. The set of formulas $\{0 < x < \frac{1}{n} :\ 0 < n \in \mathbf{N}\}$ is a 1-type in the theory of reals $\mathrm{Th}(\mathbf{R}_{\text{field}})$. (Here, $0 < x < \frac{1}{n}$ stands for $0 < x \ \&\ \bar{n} \cdot x < 1$, where $x < y$ is written for $\exists z\,(z \neq 0 \ \&\ y = x + z^2)$.)

Suppose $\bar{a} \in A^n$. Then we define *the L-type of* $\bar{a}$ *in* $\mathbf{A}$,

$$\mathrm{tp}_{\mathbf{A}}(\bar{a}) = \{P(\bar{x}) :\ \mathbf{A} \models P(\bar{a})\}.$$

Clearly, $\mathrm{tp}_{\mathbf{A}}(\bar{a})$ is a complete n-type.

Remarks.

(i) When $\mathbf{A} \subseteq \mathbf{B}$ then $\mathrm{tp}_{\mathbf{A}}(a)$ and $\mathrm{tp}_{\mathbf{B}}(a)$ may be different. But it follows immediately from the definitions that

$$\mathbf{A} \preccurlyeq \mathbf{B} \text{ implies } \mathrm{tp}_{\mathbf{A}}(a) = \mathrm{tp}_{\mathbf{B}}(a).$$

(ii) If $\pi : \mathbf{A} \to \mathbf{B}$ is an isomorphism, $\bar{a} \in A^n$, $\bar{b} \in B^n$, and $\pi : \bar{a} \to \bar{b}$, then $\mathrm{tp}_{\mathbf{A}}(\bar{a}) = \mathrm{tp}_{\mathbf{B}}(\bar{b})$.

We say that an n-type p is *realized* in $\mathbf{A}$ if there is $\bar{a} \in A^n$ such that $p \subseteq \mathrm{tp}_{\mathbf{A}}(\bar{a})$.

If there is no such $\bar{a}$ in $\mathbf{A}$ we say that p is *omitted* in $\mathbf{A}$.

3.8 **Exercise.** A principal type p is realized in any model $\mathbf{A}$ of T.

3.9 **Lemma.** *Given a set $\mathcal{T} = \{\tau^\alpha : \alpha < \kappa\}$ of n-types, an L-structure $\mathbf{A}$, and a cardinal $\kappa \geq \max\{|\mathbf{A}|, |L|\}$, there is a $\mathbf{B} \succcurlyeq \mathbf{A}$ of cardinality κ such that all types from $\mathcal{T}$ are realized in $\mathbf{B}$.*

Proof. In view of 3.6 it suffices to prove the statement for $\mathcal{T}$ consisting of just one n-type τ. Consider the expansion L_{Ac} of L_A by new constants $c_1, \ldots, c_n$

and the theory

$$T_{Ac} = \mathrm{CDiag}(\mathbf{A}) \cup \{P(c_1, \ldots, c_n) : \; P(x_1, \ldots, x_n) \in \tau\}.$$

It is immediate from the definition of type that T_{Ac} is finitely satisfiable.

By the compactness theorem there is a model $\mathbf{B}_{Ac} \models T_{Ac}$ of cardinality at most $\operatorname{card} A + \aleph_0$. Since $\mathbf{B}_{Ac} \models \mathrm{CDiag}(\mathbf{A})$, the L-reduct $\mathbf{B}$ of $\mathbf{B}_{Ac}$ is an elementary extension of $\mathbf{A}$.

3.10 **Example.** Any proper elementary extension of the standard model $\mathbf{R}$ of the reals (in the language containing $+$ and $\cdot$) realizes the infinitesimal type, by the argument in 2.13. This remarkable property is equivalent to the statement that $\mathbf{R}$ is complete in the standard metric.

3.11 **Saturation.** Given an infinite cardinal κ, a structure $\mathbf{A}$ is called *κ-saturated* if for any cardinal $\lambda < \kappa$ and for any expansion $\mathbf{A}_C$ of $\mathbf{A}$ by constant symbols $C = \{c_i : i \leq \lambda\}$, every 1-type in $\mathrm{Th}(\mathbf{A}_C)$ is realized in $\mathbf{A}_C$.

We say just *saturated* instead of κ-saturated when $\kappa = \operatorname{card} A$.

Remark. A finite structure $\mathbf{A}$ is κ-saturated for every κ.

Theorem. *Let T be a complete theory.*

(*i*) *For every $\kappa \geq \operatorname{card} T$ there exists a κ-saturated model of T of cardinality $\leq \kappa^+$.*

(*ii*) *Any two saturated models of T of the same cardinality are isomorphic.*

PROOF. (i) We use here a standard construction.

We assume that T has infinite models. Let $\mathbf{A}$ be a model of T of cardinality κ. By 3.9 there is an elementary extension $\mathbf{A}' \succcurlyeq \mathbf{A}$ such that any 1-type in $\mathrm{Th}(\mathbf{A})$ over any $C \subseteq A$ with $\operatorname{card} C < \kappa$ is realized in $\mathbf{A}'$.

Denote $\mathbf{A}$ by $\mathbf{A}^{(0)}$ and then construct, using 3.9 repeatedly, an elementary chain of models

$$\mathbf{A}^{(0)} \preccurlyeq \mathbf{A}^{(1)} \preccurlyeq \cdots \preccurlyeq \mathbf{A}^{(\alpha)} \cdots$$

of length μ, for $\mu \geq \kappa$ a regular cardinal ($\mu = \kappa^+$ will always do) such that $\mathbf{A}^{(\alpha+1)}$ realizes all 1-types over subsets of $\mathbf{A}^{(\alpha)}$ of cardinality less than κ. Then the union $\mathbf{A}^* = \bigcup_{\alpha<\kappa^+} \mathbf{A}^{(\alpha)}$ of the elementary chain, by Lemma 3.6, is an elementary extension of $\mathbf{A}$, and indeed of each $\mathbf{A}^{(\alpha)}$. By construction, for any subset C of the domain A^* of cardinality $< \kappa$ one can find $\lambda < \mu$ such that $C \subseteq \bigcup_{\alpha<\lambda} A^{(\alpha)} \subseteq A^{(\lambda)}$. It follows that $\mathbf{A}^*$ is a κ-saturated model of T. This proves (i).

(ii) We use the above method in combination with *the back-and-forth method.* Let

$$A = \{a_i : 0 \leq i \leq \kappa\}, \quad B = \{b_i : 0 \leq i \leq \kappa\}$$

be the domains of saturated models $\mathbf{A}$ and $\mathbf{B}$ of cardinality κ, with ordinal orderings. We construct by induction on $\alpha < \kappa$ the subsets $A_\alpha \subset A$ and $B_\alpha \subset B$ with orderings

$$A_\alpha = \{a^j : j < \alpha\}, \quad B_\alpha = \{b^j : j < \alpha\}$$

satisfying the conditions

$$\mathrm{tp}(a^{j_1}, \cdots, a^{j_m}) = \mathrm{tp}(b^{j_1}, \ldots, b^{j_m}) \tag{3}$$

for any finite sequences $0 \le j_1 < \cdots < j_m < \alpha$;

$$\text{if } \delta + 2n < \alpha,\ \delta \text{ limit}, n \in \omega, \text{then } a_{\delta+n} \in A_\alpha \tag{4}$$

$$\text{if } \delta + 2n + 1 < \alpha,\ \delta \text{ limit}, n \in \omega, \text{then } b_{\delta+n} \in B_\alpha \tag{5}$$

Clearly, (3) implies that $a^j \mapsto b^j$ is an elementary monomorphism $A_\alpha \rightarrowtail B_\alpha$. When we reach $\alpha = \kappa$, this together with (4) and (5) will give us an isomorphism $\mathbf{A} \cong \mathbf{B}$.

For $\alpha = 1$, take $a^0 := a_0$ and choose b^0 to be the first element among the b_i satisfying the type $\mathrm{tp}(a^0)$.

Now assume that A_α and B_α have been constructed. We introduce constant symbols c^j naming the a^j in $\mathbf{A}$ and b^j in $\mathbf{B}$. Set $C_\alpha = \{c^j : j < \alpha\}$.

If α is of the form $\delta + 2n$ and $a_{\delta+n} \notin A_\alpha$, we choose $a^\alpha := a_{\delta+n}$. If already $a_{\delta+n} \in A_\alpha$, we skip the step. Then we choose b^α to be the first element among the b_i satisfying the type $\mathrm{tp}(a^\alpha/C_\alpha)$. Such a b_i does exist since $\mathrm{card}\, C_\alpha < \kappa$ and $\mathbf{B}$ is κ-saturated.

If α is of the form $\delta+2n+1$ and $b_{\delta+n} \notin B_\alpha$, we choose $b^\alpha := b_{\delta+n}$. Then we choose a^α to be the first element among the a_i satisfying the type $\mathrm{tp}(b^\alpha/C_\alpha)$.

In each case, (3)–(5) are satisfied for $\alpha + 1$.

On limit steps λ of the construction we take

$$A_\lambda = \bigcup_{\alpha<\lambda} A_\alpha, \quad B_\lambda = \bigcup_{\alpha<\lambda} B_\alpha.$$

This has the desired properties.

3.12 In case κ is regular, e.g., $\kappa = 2^\lambda = \lambda^+$ for some cardinal λ, the construction in the proof of 3.11(i) produces a κ-saturated model of cardinality κ. In particular, assuming GCH, saturated models exist, and assuming CH, there exist saturated models of countable theories of cardinality the continuum or less.

3.13 The back-and-forth method used in the proof of (ii) above is a universal tool in model theory, apparently first used by G. Cantor in his construction of the isomorphism between countable dense orders. In fact, Cantor's theorem is a special case of 3.11(ii), since a dense linear order is $\aleph_0$-saturated.

It follows from 3.11 that if T_1 and T_2 are complete theories in the same language having saturated models $\mathbf{A}_1$ and $\mathbf{A}_2$, respectively, of the same cardinality, then $T_1 = T_2$ iff $\mathbf{A}_1 \cong \mathbf{A}_2$. This is a powerful criterion of elementary equivalence in case the existence of saturated models can be established (see also 2.12). In general a saturated model may not exist without assuming some form of generalized continuum hypothesis, but there are ways, using set-theoretic analysis, around this problem.

In fact, there is a way, less algebraic but more universal, to apply a back-and-forth procedure to establish completeness of theories.

3.14 *A back-and-forth system* between L-structures $\mathbf{A}$ and $\mathbf{B}$ is a nonempty set I of isomorphisms of substructures of $\mathbf{A}$ and substructures of $\mathbf{B}$ such that

$a \in \operatorname{Dom} f_0$ and $a' \in \operatorname{Range} f_0$, for some $f_0 \in I$, and

(forth) for every $f \in I$ and $a \in A$ there is a $g \in I$ such that $f \subseteq g$ and $a \in \operatorname{Dom} g$;

(back) For every $f \in I$ and $b \in B$ there is a $g \in I$ such that $f \subseteq g$ and $b \in \operatorname{Range} g$.

It is easy to adjust the proof above to prove the following.

3.15 Theorem (Ehrenfeucht–Fraisse criterion for saturated models). *Given $\aleph_0$-saturated L-structures* $\mathbf{A}$ *and* $\mathbf{B}$, $\mathbf{A} \equiv \mathbf{B}$ *if and only if there exists a back-and-forth system between the two structures.*

In view of this theorem and similar facts, in model theory one often operates under the principle that *there is no harm in assuming GCH.*

Saturated structures play an important role in model theory. The reader familiar with algebraic geometry could compare it with the role played by a *universal domain* in the sense of A. Weil, that is, a field of infinite transcendence degree. In fact, it is convenient in a concrete context of a given complete theory T to fix a κ-saturated model $\mathbf{M}$ with a κ "large enough" (to all intents and purposes). Such a model is often called *the universal domain* for T. In model-theoretic slang one more often refers to $\mathbf{M}$ as *the monster model.*

3.16 **Homogeneity.** One says that a *structure* $\mathbf{A}$ *is homogeneous* if for any subset X of A of cardinality strictly less than $\operatorname{card} A$ an elementary monomorphism $h : \mathbf{A} \to \mathbf{A}$ with domain X can be extended to an automorphism of $\mathbf{A}$.

A standard application of the back-and-forth method furnishes the following fact: *A saturated structure is homogeneous.*

3.17 **Omitting types.** Despite the importance of saturated models, the ability to construct a model in which certain types are omitted is key in the analysis of the variety of models and technically much more difficult (the model theorists' folklore of 1960s put it: *any fool can realize a type but it takes a model-theorist to omit one*). For example, there is a model of the theory $\operatorname{Th}(\mathbf{R})$ of the field of reals that omits types of all transcendental reals. This follows from results in 4.6 below. Using Henkin's construction of models, R. Vaught proved that *if T is a theory in a countable language then any countable collection of nonprincipal types can be omitted in some countable model of T.*

We would also like to mention the following important result.

Theorem (Ehrenfeucht–Mostowski). *Let T be a complete theory of a countable language and assume that T has infinite models. Given an infinite cardinal λ, there is a model* $\mathbf{A}$ *of cardinality λ that realizes at most $\aleph_0$ complete n-types for every $n \in$* $\mathbf{N}$. *Moreover, every two n-tuples satisfying the same complete type are conjugated by an automorphism of* $\mathbf{A}$.

To prove the theorem one uses the known Ramsay theorem of infinite combinatorics in combination with more traditional methods. We skip the proof, which can be found elsewhere.

3.18 **Quantifier elimination.** The criterion of elementary equivalence above can be adopted for classifying elementary equivalent n-tuples in a given structure and, moreover, classifying definable subsets of a given structure.

Proposition. *Given a saturated L-structure* $\mathbf{A}$ *and two n-tuples* $\bar{a}$ *and* $\bar{b}$ *in* $\mathbf{A}$,

$$\mathrm{tp}(\bar{a}) = \mathrm{tp}(\bar{b}) \quad \textit{iff there is } \pi \in Aut(\mathbf{A}) \textit{ s.t. } \pi : \bar{a} \mapsto \bar{b}.$$

PROOF. We need to prove only the left-to-right implication. Extend the language by n constant symbols to name $\bar{a}$, in the first case, and to name $\bar{b}$, in the second one. We obtain two expansions $\mathbf{A}_a$ and $\mathbf{A}_b$ of $\mathbf{A}$ to the extended language, both still saturated. The proposition follows by 3.11.

Define the *quantifier-free type of a tuple* $\bar{a}$ *in* $\mathbf{A}$,

$$\mathrm{qftp}_A(\bar{a}) := \{Q(\bar{x}) : \text{quantifier-free}, \mathbf{A} \models Q(\bar{a})\}.$$

Theorem. *Given a saturated model* $\mathbf{A}$ *of a complete theory* T*, the following two conditions are equivalent:*

(*i*) *for any two n-tuples* $\bar{a}$ *and* $\bar{b}$ *in* $\mathbf{A}$,

$$\mathrm{qftp}(\bar{a}) = \mathrm{qftp}(\bar{b}) \quad \textit{iff there is } \pi \in \mathrm{Aut}(\mathbf{A}) \textit{ s.t. } \pi : \bar{a} \mapsto \bar{b};$$

(*ii*) *any L-formula with n free variables is equivalent to a quantifier-free L-formula.*

PROOF. Assuming (ii), any n-type is equivalent to a quantifier-free one. So, (ii) $\Rightarrow$ (i) by the proposition.

We prove the converse. Let $Q(\bar{x})$ be an L-formula with free variables $\bar{x}$,

$$\tau_Q = \{P(\bar{x}) : \text{ quantifier-free}, \mathbf{A} \models \forall \bar{x}(Q(\bar{x}) \to P(\bar{x}))\}.$$

Claim. $\tau_Q \cup \{\neg Q\}$ is inconsistent.

Indeed, otherwise in the saturated $\mathbf{A}$ there is a realization $\bar{b}$ of the type τ_Q together with Q. Then $\mathrm{qftp}(\bar{b})$ will be consistent with Q, for otherwise $\neg R(\bar{x})$ is in τ_Q for some $R \in \mathrm{qftp}(\bar{b})$. Then there exists $\bar{c}$ realizing $\mathrm{qftp}(\bar{b})\&Q$, a contradiction.

It follows from the claim, by the compactness theorem, that for some $S(\bar{x})$, a conjunction of finitely many formulas of τ_Q, $\mathbf{A} \models \forall \bar{x}(S(\bar{x}) \to Q(\bar{x}))$. But by definition, also $\mathbf{A} \models \forall \bar{x}(Q(\bar{x}) \to S(\bar{x}))$, so in $\mathbf{A}$ and in T, Q is equivalent to a quantifier-free formula S.

3.19 **Remark.** The quantifier elimination criterion above may look somewhat restricted by the assumption of the existence of a saturated model. In fact, using 3.15 one can drop the restriction at the cost of having a more complex condition in (i).

4 Completeness and Quantifier Elimination in Some Theories

4.1 **The theory of an algebraically closed field.** ACF_p, the theory of *an algebraically closed field of characteristic* $p > 0$, is given by the following axioms in the language $\mathrm{L_1Ar}$ with the binary operations $+$, $\cdot$ and constant symbols 0 and 1:

I. Axioms of fields.
II. The axioms of algebraic closedness: for each positive $n \in \mathbf{N}$,

$$\forall y_1, \ldots, y_n \exists x\, x^n + y_1 x^{n-1} + \cdots + y_n = 0.$$

III. The axiom of characteristic p:

$$\underbrace{1 + \cdots + 1}_{p} = 0.$$

The theory ACF_0 of algebraically closed fields of characteristic zero is given by axioms I, II and negations of axioms III for all prime p:

$$\neg \underbrace{1 + \cdots + 1}_{p} = 0.$$

Remark. It is immediate by the axioms that the ultraproduct $\prod_{p \in \mathrm{Primes}} K_p/U$ of models K_p of ACF_p along a nonprincipal ultrafilter is a model of ACF_0.

Moreover, if an $\mathrm{L_1Ar}$-sentence P holds in all but finitely many K_p, $p \in$ Primes, then P holds on an algebraically closed field of characteristic zero.

4.2 **Theorem (Tarski).** ACF_p *is complete and allows quantifier elimination.*

Proof. In essence the theorem follows from the well-known Steinitz theorem: Given two algebraically closed fields $\mathbf{A}$ and $\mathbf{B}$ of the same characteristic p and their common subfield k,

$$\mathbf{A} \cong_k \mathbf{B} \text{ if and only if } \mathrm{trd}\,(\mathbf{A}/k) = \mathrm{trd}\,(\mathbf{B}/k),$$

were trd is the transcendence degree of the field over the subfield, the cardinality of a maximal algebraically independent subset of the field over the subfield.

Consider two $\aleph_0$-saturated models $\mathbf{A}$ and $\mathbf{B}$ of ACF_p of the same uncountable cardinality κ and let $\bar{a}$ be an n-tuple in $\mathbf{A}$, $\bar{b}$ an n-tuple in $\mathbf{B}$ such that for every polynomial $p(x_1, \ldots, x_n)$ over the prime field k_0,

$$p(\bar{a}) = 0 \text{ iff } p(\bar{c}) = 0.$$

Note that under the assumptions, the fields $k_0(\bar{a})$ and $k_0(\bar{b})$ are isomorphic by the unique isomorphism π_0 sending $\bar{a}$ to $\bar{b}$. So, we may assume that $k_0(\bar{a}) = k_0(\bar{b}) = k$ is a common subfield of $\mathbf{A}$ and $\mathbf{B}$.

Clearly $\mathrm{trd}\,(\mathbf{A}/k) = \kappa = \mathrm{trd}\,(\mathbf{B}/k)$, so by Steinitz there is an isomorphism $\pi : \mathbf{A} \to \mathbf{B}$ such that $\pi(\bar{a}) = \bar{b}$. This proves, by 3.15, that ACF_p is complete. When we consider $\mathbf{A} = \mathbf{B}$, the existence of the automorphism π affirms elimination of quantifiers, by 3.18.

Corollary (Strong Lefshetz principle). *For an* $\mathrm{L_1Ar}$*-sentence* P *the following are equivalent:*

(*i*) $\mathbf{C} \vDash P$;
(*ii*) $\mathrm{F} \vDash \mathrm{P}$, *for any algebraically closed field* F *of characteristic* 0;
(*iii*) $\tilde{\mathbf{F}}_p \vDash P$, *for all but finitely many primes* p.

Here and below $\tilde{\mathbf{F}}_p$ is the algebraic closure of the p-element field $\mathbf{F}_p$.

The original Lefshetz principle is an informally established fact known to algebraic geometers: an algebrogeometric statement proven in the context of complex algebraic geometry holds also for any abstract algebraically closed field of characteristic zero.

4.3 **Constructible sets.** The quantifier elimination statement in the language $\mathrm{L_1Ar}$ can be translated into the following form, also known as Chevalley's theorem: given an algebraically closed field F, the family of $\mathrm{L_1Ar}$-definable (using parameters) subsets of F^n, for all n, coincides with the family of constructible subsets.

Here *constructible* means a set representable as a Boolean combination of zero-sets of polynomials.

Note also that the family of $\mathrm{L_1Ar}$-definable sets is the same as the family of sets obtained from zero-sets of polynomials by applying Boolean operations (union, intersection, complement) and projections $\mathrm{F}^{n+1} \to \mathrm{F}^n$.

4.4 **Definable functions.** An easy analysis of constructible sets and constructible functions (those with constructible graphs) yields the following:

Let F be an algebraically closed field of characteristic 0, $V \subseteq \mathrm{F}^n$ a constructible subset, and $f : V \to \mathrm{F}$ a constructible function defined everywhere on V. Then there is a constructible partition $V = V_1 \cup \cdots \cup V_k$ such that for each $i \in \{1, \ldots, k\}$,

$$f_{|V_i}(\bar{v}) = \frac{p_i(\bar{v})}{q_i(\bar{v})} \quad \text{for all } \bar{v} \in V_i,$$

p_i, q_i polynomials over F, q_i not vanishing on V_i.

A corollary of the above is this: Let $V \subseteq \mathrm{F}^n$, $W \subseteq \mathrm{F}^m$ be constructible subsets and $f : V \to W$ a constructible map, $\mathrm{Dom}\, f = V$, $\mathrm{Range}\, f = W$, in an algebraically closed field F of characteristic 0. Then there are constructible partitions

$$V = V_1 \cup \cdots \cup V_k, \quad W = W_1 \cup \cdots \cup W_k,$$

such that for each $i \in \{1, \ldots, k\}$, $f(V_i) = W_i$ and $f_{|V_i}$ coincides with a rational map (given by $\langle g_{i1}(\bar{v}), \ldots, g_{im}(\bar{v}) \rangle$, each $g_{ij}(\bar{v})$ a rational function with a denominator not vanishing on V_i).

4.5 **Application (J. Ax).** Let $V = V(\mathbf{C})$ be an abstract algebraic variety and $f : V \to V$ a regular injective map. Then f is surjective.

Proof. First we note that the abstract algebraic variety V is *definably equivalent* to a constructible subset $W \subseteq \mathbf{C}^n$ for some n. By this we mean that V, given as an atlas of charts $V = \bigcup_{i \leq k} V_i$, with each V_i in a bijective correspondence ϕ_i with an affine variety U_i, glued together by regular maps ϕ_{ij}, $i, j \in \{1, \dots, k\}$, can instead be put in a bijective correspondence $\psi : V \to W$ with the definable set in such a way that the induced maps $U_i \to W$ and the corresponding gluing maps are definable using parameters. (Of course, the Zariski topology in this representation is ignored.)

As a result, we reformulate the data:

$f : W \to W$ is an injective definable map on a definable subset $W \subseteq \mathbf{C}^n$, both using parameters $a_1, \dots, a_m \in \mathbf{C}$. So, we write $f(w)$ as $F(\bar{a}, w)$ and W as W_a, which are now written in terms of $+, \cdot, 0$, and 1. Importantly, $F(\bar{a}, w)$ is a piecewise rational map, by 4.4.

The condition on $\bar{a}$ expressing the fact that $F(\bar{a}, w)$ is an injective map $W_a \to W_a$ defined on the whole of W_a can be written as an $\forall$-formula (check it). Call this formula $\mathrm{Inj}^F(\bar{a})$.

Now suppose toward a contradiction that f is not surjective. Then the sentence

$$P\colon\ \exists \bar{z} \exists u\, (\mathrm{Inj}^F(\bar{z})\ \&\ u \in W_z\ \&\ \forall x \in W_z F(\bar{z}, x) \neq u)$$

holds in $\mathbf{C}$. Hence by the strong Lefshetz principle, for some prime p, $\tilde{\mathrm{F}}_p \models P$. So, for some $\bar{b}$ and c in $\tilde{\mathrm{F}}_p$,

$$\tilde{\mathrm{F}}_p \models \mathrm{Inj}^F(\bar{b})\ \&\ c \in W_b\ \&\ \forall x \in W_b\, F(\bar{b}, x) \neq c.$$

The formula in question is clearly equivalent to an $\forall$-formula. Hence, by 1.4(ii) for any subfield $k \subset \tilde{\mathrm{F}}_p$ containing $\bar{b}$ and c,

$$k \models \mathrm{Inj}^F(b)\ \&\ c \in W_b\ \&\ \forall x \in W_b\, F(\bar{b}, x) \neq c.$$

We can choose k to be a finite subfield and thus get a statement that $F(\bar{b}, x)$ defines an injective map $W_b(k)$ into itself that is not surjective. This contradicts the fact that $W_b(k)$ is finite.

James Ax observed this, by then unknown, fact in his paper "The elementary theory of finite fields", *Ann. of Math.* 88 (1968), 239–271. Later, G. Shimura gave a proof of Ax's theorem by means of reduction mod p. A. Borel published a third proof based on cohomology with compact supports, Injective endomorphisms of algebraic varieties, Archiv der Mathematatik, 1968.

4.6 **The theory of real closed fields.** A natural language for this theory is $\mathrm{L}_{RCF} = \{+, \cdot, \leq, 0, 1\}$. The axioms of the theory RCF, the theory of real closed fields are:

I. The axioms of ordered fields.
II. $\forall x\, (0 \leq x \to \exists y\, y^2 = x)$.
III. For every odd n the axiom

$$\forall y_1, \dots, y_n \exists x\, x^n + y_1 x^{n-1} + \cdots + y_n = 0.$$

We note that II and III together are equivalent to the *sign change statement:* for every polynomial $f(x)$ over the field, if for some $a < b$, $f(a) \cdot f(b) < 0$, then there is c, $a < c < b$ with $f(c) = 0$.

Among standard algebraic facts about real closed fields are the following.

Lemma. Let $\mathbf{A}, \mathbf{B}$ be real closed fields and $\mathbf{A}_0 \subseteq \mathbf{A}$, $\mathbf{B}_0 \subseteq \mathbf{B}$ subfields such that $\mathbf{A}_0 \cong_\varphi \mathbf{B}_0$. Then

(i) the isomorphism φ can be extended to an isomorphism $\psi : \tilde{\mathbf{A}}_0 \to \tilde{\mathbf{B}}_0$ between the relative algebraic closures of the respective subfields in $\mathbf{A}$ and $\mathbf{B}$.

(ii) assuming that $\mathbf{A}_0$ and $\mathbf{B}_0$ are respectively algebraically closed in $\mathbf{A}$ and $\mathbf{B}$, $a_0 \in A \setminus A_0$, $b_0 \in B \setminus B_0$ such that for any $a \in A$, $a_0 \le a$ if and only if $b_0 \le \varphi(a)$, then φ can be extended to an isomorphism of ordered fields $\psi : \mathbf{A}_0(a_0) \to \mathbf{B}_0(b_0)$.

(iii) assuming that $\mathbf{A}_0$ is algebraically closed in $\mathbf{A}$, a finite system of inequalities $f(x) \le 0$, for $f(x) \in \mathbf{A}_0[x]$, has a solution in $\mathbf{A}$ if and only if it has a solution in $\mathbf{A}_0$.

4.7 **Theorem (Tarski–Seidenberg).** The theory RCF is complete and allows quantifier elimination.

Proof. We use the same method as in 4.2. Let $\mathbf{A}$ and $\mathbf{B}$ be two real closed κ-saturated fields.

Claim. Suppose A_0, B_0 are respectively subfields of cardinality less than κ of $\mathbf{A}$ and $\mathbf{B}$, and suppose $A_0 \cong_\varphi B_0$ as ordered rings. Then for every $a \in A$ there are $b \in B$ and an extension ψ of the isomorphism φ such that $A_0(a) \cong_\psi B_0(b)$.

For a algebraic over A_0 the claim follows by (i) of the lemma in 4.6.

For a transcendental we first consider the quantifier-free $\mathrm{L}_{RCF}(A_0)$-type τ_A of a, that is, the set of formulas $f(x) > 0$ for polynomial $f(x)$ over A_0, holding for $x = a$. We obtain a quantifier-free $\mathrm{L}_{RCF}(B_0)$-type τ_B by replacing parameters from A_0 in every $f(x)$ by corresponding parameters in B_0. Note that τ_B is a type, that is, it is consistent in the theory of $\mathbf{B}$, by (iii) of the lemma in 4.6. Now we use the assumption of κ-saturation and find an element $b \in B$ realizing the type τ_B. Define $\psi : A_0(a) \to B_0(b)$ as the unique isomorphism of fields with $\psi(a) = b$. By construction ψ also preserves the order. Claim proved.

The completeness of RCF is now immediate from the claim by 3.15.

To establish quantifier elimination consider in a given κ-saturated $\mathbf{A}$ two n-tuples $\bar{a}$ and $\bar{b}$ satisfying the same quantifier-free L_{RCF}-formulas. It follows that the subfields $\mathbf{Q}(\bar{a})$ and $\mathbf{Q}(\bar{b})$ are isomorphic as ordered fields by the isomorphism sending $\bar{a}$ to $\bar{b}$. Now the above claim allows a construction of a back-and-forth system between $\mathbf{A}_a$, that is, $\mathbf{A}$ with $\bar{a}$ named by constants, and $\mathbf{A}_b$, with $\bar{b}$ named by the same constants. By 3.18 quantifier elimination follows.

4.8 **Semialgebraic sets and semialgebraic functions.** Semialgebraic sets are the solution sets of equations $p(x_1, \ldots, x_n) = 0$ and inequalities $q(x_1, \ldots, x_n) > 0$, for p, q polynomials over $\mathbf{R}$, and those obtained from such solution sets by means of finite intersections and unions. Clearly, by definition these are exactly the quantifier-free sets definable in $\mathbf{R}$. So the Tarski–Seidenberg theorem 4.7 in effect says that sets definable (using parameters) in $\mathbf{R}$ are precisely the semialgebraic sets. In particular, the projection of a semialgebraic set is semialgebraic.

Note that a solution set of a one-variable polynomial inequality $f(x) > 0$ is a finite union of open intervals (a, b), where $-\infty \leq a < b \leq +\infty$. It follows that any definable subset of $\mathbf{R}$ is a finite union of open intervals and points. This property of an ordered structure is called *o-minimality* (order-minimality) and will be discussed in later sections.

A *semialgebraic function* is a function with a semialgebraic graph. Again by the Tarski–Seidenberg theorem this is the same as definable in $\mathbf{R}$ using parameters.

Suppose $g(x)$ is a semialgebraic function. By the above, we may assume $\operatorname{Dom} g = (a, b)$. We may also assume that the graph $g(x) = y$ is defined just by a conjunction of polymomial equations and inequalities over $\mathbf{R}$. Clearly, at least one of these must be an equation $p(x, y) = 0$. Write $p(x, y)$ as $a_0(x)y^n + a_1(x)y^{n-1} + \cdots + a_n(x)$, for some $n > 0$ and $a_i(x) \in \mathbf{R}[x]$. It follows that y is one of the roots of the polynomial $a_0(x)y^n + a_1(x)y^{n-1} + \cdots + a_n(x)$. On a subinterval y can coincide with the greatest root of $p(x, y)$, second greatest one, and so on

This proves the following.

Fact. Given a definable (i.e., semialgebraic) function $g : \mathbf{R} \to \mathbf{R}$, its domain can be divided into finitely many open intervals and points such that on each interval or point, $g(x)$ is equal to the kth-greatest root of a polynomial $p(x, y)$, for some $k \leq \deg_y p$.

4.9 **Application (L. Hörmander).** Let $p(x_1, \ldots, x_n)$ be a polynomial over $\mathbf{R}$ and $f_p(r)$ the function of the nonnegative real variable r defined as follows:

$$f_p(r) := \min\{p(x_1, \ldots, x_n) : |x_1| + \cdots + |x_n| = r\}.$$

Assume that for any given positive real R there is r_R such that for $r > r_R$, $f_p(r) > R$.

Then there are a positive rational number a and a positive real c such that

$$\lim_{r \to \infty} r^{-a} f_p(r) = c.$$

Proof. First note that f_p is definable in the field of reals. So, by the fact in 4.8, f is defined piecewise, on finitely many intervals, by the formulas

$$f_p(r) = g_k(r) = \text{ the } k\text{th root of the polynomial } q_0(r)y^m + q_1(r)y^{m-1} + \cdots + q_m(r)$$

for $q_0(r), \ldots, q_m(r)$ polynomials in r.

We are interested in the interval (d, ∞), some large enough $d \in \mathbf{R}$, and may assume that no $q_i(r)$ vanishes in the interval.

Consider a nonstandard model $^*\mathbf{R}$ of the theory and the following preordering on $^*\mathbf{R}$:

$$\alpha \ll \beta \text{ if } \frac{\alpha}{\beta} \text{ is infinitesimal.}$$

Set $\alpha \approx \beta$ if neither $\alpha \ll \beta$ nor $\beta \ll \alpha$, equivalently, $(\alpha\beta^{-1} - c)$ is infinitesimal, for some $c \in \mathbf{R}$ (see 2.13).

Let $\gamma \in {}^*\mathbf{R}$ be positive infinite, that is, $\gamma > r$ for any standard real r. Denote $\delta = f_p(\gamma)$.

Then δ is a root of the polynomial $q_0(\gamma)y^m + q_1(\gamma)y^{m-1} + \cdots + q_m(\gamma)$, and clearly this can not happen unless

$$q_i(\gamma)\delta^{m-i} \approx q_j(\gamma)\delta^{m-j}$$

for some $0 \le i < j \le m$.

Hence

$$\delta^{j-i} \approx \frac{q_j}{q_i}(\gamma) \approx \gamma^N,$$

for $N = \deg q_j - \deg q_i$. It follows that $\delta \approx \gamma^a$, for $a = N/j - i$. By definition, this means

$$\gamma^{-a} f_p(\gamma) = c + \alpha \tag{6}$$

for some $c \in \mathbf{R}$ and an infinitesimal α. It remains to show that (6) holds for the same a and c for every nonstandard infinite γ.

Note that (6) implies that for $x = \delta$ the following L_1Real-definable property holds:

$$|x^{-a} f_p(x) - c| < 1. \tag{7}$$

Hence, again by o-minimality, for all $x \in (d, \infty)$, for some $d \in \mathbf{R}$, (7) holds. If for another choice of γ we had different a or c, then we would have (7) with the different parameters holding on (d', ∞), for some $d' \in \mathbf{R}$, clearly a contradiction.

It remains to see that $c > 0$ and $a > 0$. This is immediate from the assumption on f_p.

4.10 **Remark.** A comment on the cause of efficiency of the method of proof above and in other similar cases is in order. Quantifier elimination is in fact a powerful calculus designed to translate complex formal expressions (L-formulas) into something simple and, in many cases, geometrically meaningful. An example of such an expression is the definition of the function f_p in 4.9. Its conversion into a semialgebraic function, if carried out "by hand," is a painful process, difficult to see through.

Note also that modern methods of elimination of quantifiers demonstrated in 3.18, 4.1, and 4.7 are more efficient and more "mathematical" than those of the 1950s. The initial instinct was to analyze the syntax of an arbitrary L-formula and get rid of quantifiers in the formula one by one in an inductive process.

4.11 **Decidability**. The theories ACF_p, for each p prime or equal to 0, and the theory RCF are decidable.

PROOF. These are just special cases of the following general statement: *A complete theory T axiomatizable by a recursively enumerable set of axioms is decidable.* This is easy to see. Indeed, if there is an algorithm listing axioms of T then it is easy to compile an algorithm that lists all consequences of the axioms, that is, enumerates, T. Now, given a sentence P one can decide whether P is in T by the following algorithm: list by the above algorithm formulas Q_i of T and check at each step whether $P = Q_i$ or $\neg P = Q_i$. By completeness at some step one or the other must happen and this obviously decides whether P is in T.

The explicit axiomatizations of ACF_p and RCF are clearly recursive; hence decidability follows.

The easy argument above can be adapted to prove decidability for some incomplete theories, such as ACF, the theory of all algebraically closed fields. This is axiomatized by the recursive set of axioms 4.1, I and II. And we also know that if P is not deducible from ACF, then $\neg P$ is consistent with some ACF_p, p prime or 0. Note that there is an obvious enumeration of the family ACF_p, $p \in \mathrm{Primes} \cup \{0\}$: for each p and number n we can effectively produce an axiom $S_{pn} \in \mathrm{ACF}_p$, listing eventually all of the axioms. This can be extended to an algorithm that for each $n \in \mathbf{N}$ produces formulas $P_{p,k}$, $p = 0$ or prime, $k \in \mathbf{N}$, $p, k \leq n$, such that $\{P_{p,k} : k \in \mathbf{N}\} = \mathrm{ACF}_p$.

Now, given a sentence P, turn on an algorithm that for a given $n \in \mathbf{N}$, produces

(i) $Q_1, \ldots, Q_n$, the first n consequences of ACF;
(ii) $P_{0,1}, \ldots, P_{0,n}$, $P_{2,1}, \ldots, P_{2,n}, \ldots,$ $P_{p,1}, \ldots, P_{p,n}$, for $p = 0$ or prime, $p \leq n$.

We check at each step n whether P is in (i) or $\neg P$ is in (ii). One of these two must happen at some stage n, and this decides correspondingly whether P is deducible from ACF.

4.12 The theory of p-adic numbers.

The symbols of the language for valued fields L_{valf} has the symbols of field theory, namely $0, 1, +, \cdot$, and a unary relation symbol V.

The theory $\mathbf{TQ}_p$ in this language is axiomatized as follows.

I. A model F of the axioms carries a structure of a field of characteristic zero.

II. Axioms stating that V singles out a maximal subring of the field F (the *valuation ring*), so $V(\mathrm{F})$ is a local ring with a unique maximal ideal $M(\mathrm{F})$. We stipulate that $V(\mathrm{F})/\mathrm{M}(\mathrm{F}) \cong \mathbf{F}_{\mathrm{p}}$, the p-element field. The canonical homomorphism is denoted by res.

III. The *value group* $\mathrm{F}^{\times}/V^{\times}(\mathrm{F}) = \Gamma(\mathrm{F})$ is a $\mathbf{Z}$-group, i.e., written additively, has the same $(+, <)$ theory as the ordered group of integers $\mathbf{Z}$ ($<$ is the order relation definable using the valuation ring). The infinitely many axioms for a $\mathbf{Z}$-group say that there exists a minimal positive element and $n\Gamma$ is a subgroup of Γ of index precisely n.

We denote by $v(x)$ the image of $x \in \mathrm{F}^{\times}$ under the canonical homomorphism. We add the axiom stating that $v(p)$ is the minimal positive element of Γ.

IV. Axiom stating that *Hensel's lemma holds:* for any polynomial $f(x)$ over $V(\mathrm{F})$, if there is $a \in V(\mathrm{F})$ such that $\mathrm{res} f(a) = 0$ and $\mathrm{res} f'(a) \neq 0$, then are $a' \in V(\mathrm{F})$, $f(a') = 0$, and $v(a' - a) > v(f'(a))$.

Thus one gets an axiomatization of the *theory of p-adically closed fields*, namely, all the fields that are elementarily equivalent to $\mathbf{Q}_p$ in the language of valued fields.

For quantifier-elimination purposes one needs an extension of the language. A. Macintyre introduced the extension by countably many unary predicates p_n, $n \geq 2$; we call this extension $L_{\text{valf}}^{\text{Mac}}$. The axiomatic description of the new predicates is V:

$$\forall x[p_n(x) \leftrightarrow \exists y(y^n = x)].$$

That is, each p_n singles out the set of nth powers in F.

Obviously, the last set of axioms does not impose any extra conditions on the valued field.

4.13 Theorem (J. Ax–S. Kochen, Yu. Ershov, A. Macintyre). The theory $\mathrm{T}\mathbf{Q}_p$ is complete, decidable and allows elimination of quantifiers in the language $L_{\text{valf}}^{\text{Mac}}$.

We do not give a proof of the theorem here. The model-theoretic methods of known proofs are essentially the same as above but the algebra is much more involved. The first proofs of completeness, decidability, and elimination of quantifiers (in a different language) were given by Ax–Kochen in 1965. Independently, Yu. Ershov proved completeness and decidability.

Macintyre proved elimination of quantifiers in the present form in 1976. Note that in general, the choice of a language for quantifier elimination may be essential for applications. The introduction of the predicates p_n made the quantifier elimination statement much more useful and powerful. The first consequence of this quantifier elimination is the manifestation of similarities between the theory of the reals and the theory of the p-adics. Recall that in the reals the predicate $p_2(x)$, which of course means $x \geq 0$ in this context, is used for the quantifier elimination statement. It is also useful to remark that this predicate is basic for describing the topology and geometry over the reals.

4.14 p-adic integration. Let $\mathbf{Z}_p$ denote the ring of p-adic integers. Let $f_l(\bar{x}), \ldots, f_r(\bar{x})$ be polynomials in m variables $\bar{x} = \langle x_1, \ldots, x_m \rangle$ over $\mathbf{Z}_p$. For $n \in \mathbf{N}$, let $\tilde{N}_n$ be the number of elements in the set

$$\{\bar{x} \bmod p^n : \bar{x} \in \mathbf{Z}^m \text{ and } f_i(\bar{x}) = 0 \bmod p^n, \text{ for } i = 1, \ldots, r\},$$

and let N_n be the number of elements in the set

$$\{\bar{x} \bmod p^n : \bar{x} \in \mathbf{Z}^m \text{ and } f_i(\bar{x}) = 0, \text{ for } i = 1, \ldots, r\}.$$

To these data one can associate the following Poincaré series:

$$\tilde{P}(T) = \sum_{n=0}^{\infty} \tilde{N}_n T^n, \quad P(T) = \sum_{n=0}^{\infty} N_n T^n.$$

Borevich and Shafarevich conjectured that $\tilde{P}(T)$ is a rational function of T. This was proved by Igusa, in the case $r = 1$, and Meuser for arbitrary r, by adapting Igusa's method. Serre and Oesterlé asked whether also $P(T)$ is a rational function. Denef proved the rationality of both series using p-adic quantifier elimination. This method was extended later for further applications.

For $a \in \mathbf{Q}_p$, let $|a| = p^{-v(a)}$. Let $|d\bar{x}| = |dx_1| \cdot |dx_2| \cdots |dx_m|$ be the Haar measure on $\mathbf{Q}_p^m$ such that the measure of $\mathbf{Z}_p^m$ is 1.

Igusa's original proof starts by establishing a rational relation between the integral

$$J(s) = \int_{\mathbf{Z}_p} |f(x)|^s |dx|,$$

as a function of p^{-s}, for $s \in \mathbf{R}$, $s > 0$, and $\tilde{P}(p^{-1-s})$. The calculation of the integral is elementary in the case that $f(x)$ is a monomial, using the fact that the function $|f(x)|$ is then constant on $p^n\mathbf{Z}_p \setminus p^{n+1}\mathbf{Z}_p$. In general, though, $|f(x)|$ is still piecewise constant; it is quite hard to determine the absolute value on the pieces. Here Igusa uses the embedded resolution of singularities of Hironaka.

Meuser's proof extends these calculations to a similar integral over the domain $\mathbf{Z}_p^m$.

A similar idea in the case of $P(T)$ leads to a p-adic integral over a more complex domain. Denef considers the domain

$$\begin{aligned} D_f =\{\langle x_1,\dots,x_m,w\rangle \in \mathbf{Z}_p^{m+1} : \exists \bar{y} \in \mathbf{Z}_p^m\ \bar{x} \equiv \bar{y} \bmod w \ \text{ and} \\ f_i(\bar{y}) = 0 \text{ for } i = 1,\dots,r\} \end{aligned}$$

and the integral

$$I_f(s) = \int_{D_f} |w|^s |d\bar{x}||dw|.$$

Again, by elementary calculation

$$I_f(s) = \frac{p-1}{p} P(p^{-(m+1)}p^{-s}).$$

So to prove that $P(T)$ is rational we need to prove that $I(s)$ is a rational function of p^s. The main new difficulty here is the nonelementary shape of D_f, but this is overcome by the use of Macintyre's quantifier-elimination theorem. It is sufficient to prove that an integral

$$\int_S |g(\bar{x})|^s |d\bar{x}|,$$

for a polynomial $g(\bar{x})$ and a semialgebraic subset S of $\mathbf{Z}_p^m$, is a rational function of p^s. This can be done by essentially Igusa's method. As was mentioned above, this uses Hironaka's resolution of singularities. But later Denef noticed that a more thorough characterization of p-adic semialgebraic sets based on earlier work by P. Cohen (the cell decomposition, widely used in the analysis of real semialgebraic sets) allows one to prove the theorem without referring to Hironaka.

5 Classification Theory

The term classification theory usually refers to the body of work around Shelah's *Classification Theory*, the main idea of which being to place every complete theory into a node of a hierarchical tree of *stability theory.* In its broadest meaning the degree of stability is an indicator of a tameness, or in other words, the degree to which a structural classification of models of a given complete theory can be developed. The study of o-minimal theories (now also extended to the study of c-minimal, v-minimal, and others), another very important part of model theory, is usually treated separately. But we include o-minimality in this survey, seeing a justification of this both in its importance and in its interactions with stability theory seen in recent years.

5.1 **Categorical Theories.** Classification theory has at its center *theories categorical in uncountable powers.* Unless stated otherwise we assume throughout in this section that our languages are countable.

Recall that a theory T is (*absolutely*) *categorical* if T has a unique, up to isomorphism, model. By Löwenheim–Skolem this can be the case only when the unique model of T is finite, while one really is interested in infinite structures. So a more flexible notion of categoricity has been considered. We say that a theory T is *categorical in power* μ (μ-categorical) if T has a unique, up to isomorphism, model of cardinality μ. It is easy to see that if for an infinite cardinal μ, a theory T has no finite models and is μ-categorical, then T is complete. So, μ-categoricity is a stronger form of completeness.

It is interesting and appropriate to look at the phenomenon of μ-categoricity from the algebraic point of view. Suppose we are given an L-structure $\mathbf{A}$ of cardinality μ such that the L-theory $\mathrm{Th}(\mathbf{A})$ is μ-categorical. This can be translated into a more suggestive characterization: the sentences of $\mathrm{Th}(\mathbf{A})$ together with the cardinal μ comprise a complete set of invariants for $\mathbf{A}$. An especially interesting case is that in which L is small (countable) and μ is large. This, in effect, could be taken as a mathematical form of *algorithmic compressibility,* the property of nature that some philosophers of science believe makes the laws of the universe and science itself possible.

J. Los conjectured in the 1950s that if a theory T of a countable language is μ-categorical for some uncountable μ, then it is μ-categorical in all uncountable powers (uncountably categorical). A decade later, M. Morley published a seminal paper with a proof of Los's conjecture.

One of the main new tools in Morley's paper was the notion of a rank, a function with certain properties assigning an ordinal number to each definable set, which Morley proved exists for every uncounably categorical theory.

5.2 **Stability.** The point of departure in Morley's analysis of a κ-categorical T is the fact that the number of complete 1-types in T must be countable, and moreover, given a set C of new constant symbols naming some elements in a model of T and the complete theory T_C of this model in the extended language, the number of complete types in the theory T_C is at most $\operatorname{card} C + \aleph_0$. This follows from the Ehrenfeucht–Mostowski theorem (see 3.17), immediately in the case of types over T, and with a little more work in general. This property of a theory T is called $\aleph_0$-*stability.* The term "stability" should be taken here as the opposite to "diversity" of types of elements in models of T. The actual terminology used to express this "diversity" is *forking* (also *dividing, splitting,* and some others), and stability guarantees that forking does not go too far.

More generally, given an infinite cardinal number κ, a theory T is said to be κ-*stable* if the expanded theory T_C has at most κ complete 1-types for every C of cardinality κ.

A theory T is said to be *stable* if it is κ-stable for some infinite κ.

Shelah's theory distinguishes several cases of stability. $\aleph_0$-stability is the strongest one and implies κ-stability for all κ. Another possibility for a stable theory T is that it is κ-stable for all $\kappa \geq 2^{\operatorname{card} T}$. In this case T is said to be *superstable.* In remaining cases T is stable in all cardinals except for those of low cofinality.

Note that the definitions above remain equivalent if one replaces 1-types by n-types.

5.3 **Morley rank.** The following definition makes sense for $\aleph_0$-stable theories.

Let $\mathbf{M}$ be a universal domain for T and $\operatorname{Def}_n(\mathbf{M})$ the collection of all nonempty subsets of M^n definable with parameters in $\mathbf{M}$. *Morley rank* is the minimal function $\operatorname{rk} : \operatorname{Def}_n(\mathbf{M}) \to \operatorname{Ord}$ (ordinal numbers) satisfying the following:

$\operatorname{rk} S \geq \alpha+1$ if and only if there is a countable family $\{S_i : i \in \mathbf{N}\}$ of pairwise disjoint subsets of S with $\operatorname{rk} S_i \geq \alpha$;
for a limit δ, $\operatorname{rk} S \geq \delta$ if and only if $\operatorname{rk} S \geq \alpha$ for all $\alpha < \delta$.

In effect, since $\mathbf{M}$ is at least $\aleph_0$-saturated, the definition does not depend on $\mathbf{M}$. The fact that one can assign an ordinal $\operatorname{rk} S$ with the above property to every definable set S is due to the bound on the number of possible types, that is, $\aleph_0$-stability of T. A simple combinatorial argument proves that a priori $\operatorname{rk} S < \aleph_1$. A much more difficult theorem (J. Baldwin) established later says that for uncountably categorical T, $\operatorname{rk} S$ is always a finite number. Moreover, in this case the rank enjoys the following *addition formula:*

Let $\operatorname{pr} : M^{n+m} \to M^m$ be the projection $\langle x_1, \ldots, x_n, \ldots, x_{n+m}\rangle \mapsto \langle x_1, \ldots, x_n\rangle$ and $S \in \operatorname{Def}_{n+m}(\mathbf{M})$. Then

$$\operatorname{rk}\operatorname{pr}\mathrm{S} + \min_{\mathrm{a}\in\operatorname{pr}\mathrm{S}} \operatorname{rk}\mathrm{S}_{\mathrm{a}} \leq \operatorname{rk}\mathrm{S} \leq \operatorname{rk}\operatorname{pr}\mathrm{S} + \max_{\mathrm{a}\in\operatorname{pr}\mathrm{S}} \operatorname{rk}\mathrm{S}_{\mathrm{a}},$$

where S_a is a fiber over a.

5.4 **Example.** The theory ACF_p of algebraically closed fields of characterictic p is μ-categorical for uncountable μ, since the isomorphism type of a model F of ACF_p is, by Steinitz's theorem, determined by $\operatorname{trd}\mathrm{F}$, the transcendence degree of F, and $\operatorname{trd}\mathrm{F} = \operatorname{card}\mathrm{F}$, for uncountable F. Recall that definable sets in this structure are just constructible sets. So algebrogeometric dimension, $\dim S$, is well defined for any definable set S. One can easily check (by induction on $\dim S$) that

$$\operatorname{rk} S = \dim S$$

in this case.

Stability is inherited by a structure definable in a stable structure is itself stable. Following 3.3, the group $\mathrm{GL}_n(\mathrm{F})$ (in the language of groups) is definable in the field F. So, the theory of $\mathrm{GL}_n(\mathrm{F})$ is $\aleph_0$-stable and rk is well defined in this theory. In fact, as in the previous example, every definable set S in this theory is constructible and $\mathrm{rk}\, S = \dim S$.

There are many more structures definable in ACF_p. A natural class of examples is that of algebraic varieties as structures in *the natural language for algebraic varieties*: let $V = V(\mathrm{F})$ be the set of F-points of an algebraic variety defined over some $C \subseteq \mathrm{F}$, F a model of ACF_p. For each n and each C-definable subvariety $W \subseteq V^n$ introduce the symbol p_W of an n-ary predicate on V. The natural language for the algebraic variety V consists of all the p_W for all W as above. The structure V on the domain $V(\mathrm{F})$ with the obvious interpretation of the predicates of the natural language is definable in the field F. Its theory is $\aleph_0$-stable and, for C big enough (e.g., if C contains an algebraically closed subfield), Morley rank coincides with dimension. In general $\mathrm{rk}\, S \leq \dim S$ for constructible sets S definable in V.

5.5 On the other hand, RCF, the theory of the field of reals $\mathbf{R}$, is not stable. In fact, *every theory T with an order relation definable on an infinite subset in its model is not stable.* Indeed, first note that by the method of diagrams, we can embed into a model of T any ordered set $(C, <)$. It is known that for every infinite κ there is an ordered set $(C, <)$ of cardinality κ with more than κ Dedekind cuts in it. Distinct cuts give rise to distinct complete types in the theory T_C, which shows that T is not κ-stable.

In the above case one says that the theory T has *the strict order property.*

It is not difficult to see that the theory $\mathrm{T}\mathbf{Q}_p$ of the p-adically valued field has the strict order property (use the order on the value group). With more analysis of definability one can see that the theory of the field $\mathbf{Q}_p$ in the language of fields alone has also the strict order property, so is unstable.

5.6 Another pattern of nonstability can be seen in the example of a *pseudofinite field.* By definition this is any infinite field F that is elementarily equivalent to an ultraproduct $\prod_{i \in I} \mathrm{F_i}/\mathrm{U}$ of finite fields F_i. The study of such fields was started by Ax and Kochen and it is known that they do not have the strict order property but do satisfy another property that implies nonstability.

One says that a complete theory T has the *independence property* if there is a formula $P(\bar{x}, \bar{y})$ in the language of the theory such that for every n in some model $\mathbf{A}$ of the theory one can find n tuples $\bar{c}_j$, $j \in \{1, \ldots, n\}$ and 2^n tuples $\bar{b}_J$, $J \subseteq \{1, \ldots, n\}$ such that

$$\mathbf{A} \vDash P(\bar{b}_J, \bar{c}_j) \text{ iff } j \in J.$$

Clearly, the number of complete m-types (m = length $\bar{x}$) over parameters $\bar{c}_1, \ldots, \bar{c}_n$ is at least 2^n. Using compactness one finds in a universal domain for T a subset C of cardinality κ with at least 2^κ complete types over C.

5.7 **Indiscernibles and orthogonality.** A subset I in an L-structure $\mathbf{A}$ is said to be *indiscernible* over a set of parameters C (C-indiscernible) if for any

L_C-formula $P(x_1, \ldots, x_n)$ either $\mathbf{A} \models P(i_1, \ldots, i_n)$ for all distinct $i_1, \ldots, i_n \in I$ or $\mathbf{A} \models \neg P(i_1, \ldots, i_n)$ for all distinct $i_1, \ldots, i_n \in I$.

An instructive example of an indiscernible set I is an algebraically independent subset of an algebraically closed field.

A characteristic feature of a stable theory T is that in saturated models of T, indiscernibles (over a given small set of parameters) are ubiquitous.

In particular, choosing $C = A$ to be a set of parameters naming all elements of a small (say, countable) model $\mathbf{A}$ and $\mathbf{A} \preccurlyeq \mathbf{B}$, a saturated enough model, every 1-type p over A can be defined by an A-indiscernible set $I \subseteq B$:

(i) every $a \in I$ realizes p;
(ii) for every C, $A \subseteq C \subseteq B$, the *average type of I over C*,

$$\mathrm{Av}_C(I) = \{Q(x) : L_C\text{-formula s.t. } \mathbf{B} \models Q(i) \text{ for all but finitely many } i \in I\}$$

is a nonforking extension of p.

Consequently, in general the cardinality of a maximal A-indiscernible subset $I = I_p \subseteq B$ such that $p = \mathrm{Av}_A(I_p)$ is an important cardinal invariant of $(p, \mathbf{B})$ over $\mathbf{A}$. For instance, in the above example with fields, $\mathrm{card}\, I_p$ is exactly $\mathrm{trd}\,(\mathbf{B}/\mathbf{A})$, the transcendence degree of the field $\mathbf{B}$ over $\mathbf{A}$. The average type of an algebraically independent subset of a field is called the *generic type of the field.*

Two types p and q over A are said to be *nonorthogonal* if $\mathrm{card}\, I_p = \mathrm{card}\, I_q$ in every model $\mathbf{B}$, $\mathbf{A} \preccurlyeq \mathbf{B}$. Clearly, the nonorthogonality is an equivalence relation. A theory is called *unidimensional* if any two 1-types over a model are nonorthogonal. Otherwise, the types are said to be *orthogonal.*

Every uncountably categorical theory is unidimensional. On the other hand, it is easy to construct a stable theory with "many dimensions." For example, the theory of the direct product $\mathbf{A}_1 \times \mathbf{A}_2$ of two algebraically closed fields is ω-stable but "two-dimensional." It has models with any combination of $\mathrm{trd}\, \mathbf{A}_1$ and $\mathrm{trd}\, \mathbf{A}_2$.

Particularly interesting and essential is the analysis of the orthogonality relation in the theory of differentially closed fields.

5.8 **Differentially closed fields.** A structure $(K, +, \cdot, 0, 1, D)$ in the language of fields extended by an operation symbol $D : K \to K$ is called a *differential field* if K is a field of characteristic 0 and D satisfies the Leibniz rule: $Dxy = xDy + yDx$. A differential polynomial of order $\leq n$ in the variable y is an expression $f(D^n y, D^{n-1}y, \ldots, y)$, where $f(x_0, x_1, \ldots, x_n)$ is a polynomial over K. A differential field is said to be *differentially closed* if for every $n > 0$ and every differential polynomial $g(y)$ of order n and a nonzero differential polynomial $h(y)$ of order $< n$ there is an $s \in K$ such that $g(s) = 0$ and $h(s) \neq 0$. (Note that the field $(K, +, \cdot)$ is algebraically closed then.) This is easily axiomatized in the first-order language, and the corresponding theory is called DCF_0. This theory was studied by A. Robinson, who proved that it is complete and has elimination of quantifiers. Later, C. Wood observed that the theory is ω-stable of Morley rank ω. The fact that the rank is infinite agrees well with the intuition that K is an infinite-dimensional space over the one-dimensional field of constants $C_K = \{y \in K : Dy = 0\}$. Indeed, $\mathrm{rk}\, C_K = 1$. Moreover, in general the solution space $S_f = \{y \in K : f(D^n y, D^{n-1}y, \ldots, y) = 0\}$ of a differential equation of order n is of rank n.

Now let us compare the generic type of S_f (appropriately defined) with the generic type of the field of constants. The nonorthogonality of the two types can be translated into the statement that the solution space S_f is parametrized by the field of constants C_K. A typical example is given by a linear differential equation $f(D^n y, D^{n-1}y, \ldots, y) = 0$, where S_f is in a definable bijective correspondence with the linear space C_K^n.

On the other hand, for a generically chosen differential equation f, the definable set S_f is orthogonal to the constants.

5.9 **Example. Algebraically closed difference fields.** A difference field is a structure $(K, +, \cdot, 0, 1, \sigma)$, with $(K, +, \cdot)$ a field and σ an automorphism of the field. A difference field $(K, +, \cdot, \sigma)$ is said to be algebraically closed (also existentially closed) if any finite set of quantifier-free formulas over K that has a solution in some extension of K has a solution in K. Hrushovski proved that this definition is axiomatizable and that the theory of a given algebraically closed difference field of characteristic zero, although unstable, is simple.

It is useful to observe the many similarities between this theory (also called ACFA, algebraically closed field with an automorphism) and the theory DCF_0. The *fixed field* $F = \{x \in K : \sigma x = x\}$ is a direct analogue of the field of constants, and is known to be of rank 1 (so-called SU-rank, in the case of simple theories). Given a polynomial f over K, the solution set $S_f = \{y \in K : f(\sigma^n y, \sigma^{n-1}y, \ldots, y) = 0\}$ of a difference equation of order n is of rank n.

For some definable sets more can be said, e.g., the solution set T_m for the equation $\sigma y = y^m$, for $m > 1$, is of Morley rank 1 and the set is orthogonal to the fixed field. This set contains important Diophantine information: in an algebraically closed difference field K any root of unity of order n, prime to m, belongs to the set T_m. Indeed, the equations $y^n = 1$ and $\sigma y = y^m$ have a solution in a differentially closed field, since there is a Galois automorphism taking a root y of order n to y^m.

5.10 **Shelah's criterion of stability.** A complete theory T is stable if and only if it does not have the strict order property or the independence property.

We saw already that any of the properties imply nonstability. The converse is a nontrivial and powerful statement proved by Shelah using beautiful infinite combinatorics, characteristic of many proofs in this field.

Negation of any of the two properties for a theory T is seen as an indicator of tameness of T. A theory T is said to be *simple* if it does not have the strict order property.

The theory of a pseudofinite field is simple.

A theory T is said to be *dependent* (or NIP, nonindependence property) if it does not have the independence property.

The theory $\mathrm{T}\mathbf{Q}_p$ of the p-adics is dependent (Shelah–Hrushovski). A large class of dependent theories is the class of o-minimal ones.

5.11 **o-minimal theories.** A complete theory T is said to be *o-minimal* if any model $\mathbf{A}$ of T is linearly ordered by a definable relation $<$ and every subset of A definable with parameters is a union of finitely many open intervals and points (A.Pillay, C.Steinhorn, and L. van den Dries).

The property of o-minimality implies a rich structural theory. One of the consequences of the theory is the fact that *an o-minimal theory is dependent.*

We have mentioned above that the theory RCF is o-minimal. A seminal theorem of A. Wilkie establishes o-minimality of the theory $\mathbf{R}_{\exp} = (\mathbf{R}, +, \cdot, \exp, 0, 1)$, the field of reals with exponentiation. One of the corollaries of this theorem is the fact proved earlier by A. Khovanski that a zero-set of a system of exponential-polynomial equations has finitely many connected components.

Many more expansions of $\mathbf{R}$ by classical analytic functions have been proved to be o-minimal, and o-minimal analysis today has become a broadly used tool of real analytic geometry.

6 Geometric Stability Theory

6.1 **Strongly minimal sets and pregeometries.** In analyzing models of uncountably categorical theories (and more generally) and their definable substructures with regard to the nonorthogonality relation, one realizes the special role played by the minimal ones.

A structure $\mathbf{M}$ is said to be *minimal* if $\operatorname{rk} M = 1$ and for any partition $M = S_1 \dot{\cup} S_2$ into subsets definable using parameters, $\operatorname{rk} S_1 = 0$ or $\operatorname{rk} S_2 = 0$. $\mathbf{M}$ is said to *strongly minimal* if every $\mathbf{M}' \equiv \mathbf{M}$ is minimal. This is also applicable when M is a definable subset in an ambient structure $\mathbf{A}$. One treats the set M as the domain of a structure $\mathbf{M}$ with relations on M induced from $\mathbf{A}$. In this case one usually calls M a *strongly minimal set.* In algebraic geometry, or rather the theory ACF_p, the strongly minimal subsets of F^n are (irreducible) algebraic curves with a finite number of points added or removed.

It is not difficult to prove that the theory of a strongly minimal $\mathbf{M}$ is uncountably categorical.

In an arbitrary L-structure $\mathbf{A}$ one defines the notion of an (*abstract*) *algebraic closure* cl.

Given a subset $U \subseteq A$ and a point $v \in A$ we say that $v \in \operatorname{cl}(U)$ (v belongs to the algebraic closure of U) if there is an L_U-formula $P(x)$ such that the definable set $P(\mathbf{A})$ is finite and contains the point v.

Again, in ACF_p and in RCF the abstract algebraic closure is the usual field-theoretic algebraic closure.

It is easy to check that in any structure the following properties hold:

(i) $U \subseteq V$ implies $U \subseteq \operatorname{cl}(U) \subseteq \operatorname{cl}(V)$;
(ii) $\operatorname{cl}(\operatorname{cl}(U)) = \operatorname{cl}(U)$.

Less obvious is the following property, *the exchange principle,* which holds in any strongly minimal structure $\mathbf{M}$:

(iii) For any $U \subseteq M$ and elements $v, w \in M$,

$$w \in \operatorname{cl}(U, v) \setminus \operatorname{cl}(U) \to v \in \operatorname{cl}(U, w).$$

(Here and below, $\mathrm{cl}(U,v) := \mathrm{cl}(U \cup \{v\})$.)

Note also that the operator cl is finitary, in the sense that

$$\mathrm{cl}(U) = \bigcup\{\mathrm{cl}(U') : \text{ finite } U' \subseteq U\}.$$

We say that (M, cl) is a (*combinatorial*) *pregeometry* if cl is a finitary operator satisfying (i)–(iii).

A (*combinatorial*) *geometry* is a pregeometry (M, cl) such that $\mathrm{cl}(u) = \{u\}$ for any $u \in M$. This notion, known also under the names *matroid* and *dependence relation*, was used in combinatorics and in algebra by van der Waerden to develop a unified theory of dependence relations such as linear dependence and algebraic dependence in fields.

Given a pregeometry (M, cl) one associates the geometry $(\bar{M}, \bar{\mathrm{cl}})$ with it by setting $\bar{M}$ to be $M \setminus \mathrm{cl}(\varnothing)$ factored by the equivalence relation $u \sim v \Leftrightarrow \mathrm{cl}(u) = \mathrm{cl}(v)$.

On the other hand, one can modify a pregeometry (M, cl) by replacing the closure operator cl with cl_a, for a fixed element $a \in M$, defined as

$$\mathrm{cl}_a(X) := \mathrm{cl}(U, a) \text{ for any } U \subseteq X.$$

The new pregeometry (M, cl_a) is called then the *localization* of (M, cl) at a. The model-theoretic meaning of localization is just the extension of the language by a symbol for a.

6.2 **Dimension in a pregeometry.** A set $U \subseteq M$ is said to be *independent* if $\mathrm{cl}(U) \neq \mathrm{cl}(U')$, for any proper subset $U' \subset U$.

A maximal independent subset of M is said to be a *basis of* M.

It is easy to prove that any two bases of a pregeometry $\mathbf{M}$ are of the same cardinality, which is called the *dimension* of $\mathbf{M}$. More generally, we denote by $d(X)$, for $X \subseteq M$, the dimension of the subspace $\mathrm{cl}(X)$ of the pregeometry $\mathbf{M}$. When working with strongly minimal and more generally stable structures, it is important to distinguish this notion from other notions of dimension, such as the Morley rank. For these reasons we sometimes say *combinatorial dimension* for the dimension of a pregeometry. Note, however, that there is a deep relationship between the combinatorial dimension and ranks, in particular the Morley rank.

For a definable set $S \subseteq M^n$ *in a strongly minimal structure* $\mathbf{M}$,

$$\mathrm{rk}\, S = \max\{d(x_1, \ldots, x_n) : \langle x_1, \ldots, x_n \rangle \in S\}.$$

Pregeometries (M, cl) induced by strongly minimal structures $\mathbf{M}$ have the following crucial property, called *homogeneity*:

Every bijection between two bases of (M, cl) *can be extended to an automorphism of the pregeometry.*

6.3 **Examples**

(1) Let $\mathbf{M}$ be a trivial infinite structure, that is, an infinite set considered as a structure in the trivial language (the only predicate is equality). This is a strongly minimal structure with the pregeometry given by the the trivial closure operator, $\mathrm{cl}(U) = U$ for every set U.

(2) Let $\mathbf{A} = (A, +, 0)$ be an abelian divisible group satisfying the assumption that for each positive integer n the equation $nx = 0$ has finitely many solutions in $\mathbf{A}$. This structure is strongly minimal and its theory has elimination of quantifiers. The closure operator cl is the same as the linear closure, that is, $\{u_1, \ldots, u_k\}$ is dependent in the sense of cl if and only if $m_1u_1 + \cdots + m_ku_k = 0$ for some nonzero string of integers $m_1, \ldots, m_k$. This example can be generalized by considering K-modules for arbitrary division rings K instead of $\mathbf{Q}$.

Observe that the geometry associated with the pregeometry (A, cl) is the projective geometry over K (projective space $\mathbf{P}^\kappa(K)$), where κ is the cardinal number equal to the dimension of $\mathbf{A}$.

(3) An algebraically closed field F is a strongly minimal structure that is a pregeometry with respect to the (field-theoretic) algebraic closure.

The pregeometries (1) and (2) satisfy the property called *modularity*:

$$w \in \mathrm{cl}(U, v) \Leftrightarrow \exists u \in \mathrm{cl}(U) : w \in \mathrm{cl}(u, w).$$

Example (3) is not modular. One says that (M, cl) is *locally modular* if a localization (M, cl_a), for some $a \in M$, is modular.

An example of a locally modular but not modular pregeometry is an *affine geometry* over a field K, that is, a K-vector space V with a set $\{v_0, v_1, \ldots, v_n\} \subseteq V$ considered dependent if and only if $\{v_1 - v_0, \ldots, v_n - v_0\}$ are K-lineraly dependent.

All the pregeometries listed above are homogeneous. Note that, for example, the pregeometry of algebraic dependence on the reals $\mathbf{R}$ is not homogeneous. As a matter of fact, it is very hard to find a homogeneous pregeometry not reducible to (1)–(3) in an obvious way. The only examples known today come from a construction by E. Hrushovski, which will be discussed below.

6.4 **Weak trichotomy theorem.** *Let* $\mathbf{M}$ *be a strongly minimal structure and* (M, cl) *the pregeometry induced by it. Then one and only one of the following holds:*

(*i*) *the geometry associated with* (M, cl) *is trivial;*
(*ii*) *the geometry associated with* (M, cl_a), *a localization of* (M, cl), *is isomorphic to a projective geometry over a (countable) division ring;*
(*iii*) *there is a pseudoplane definable in* $\mathbf{M}$.

We need to explain (iii). A pseudoplane (first considered by A. Lachlan) is a structure on two infinite domains P and L with a binary relation I between the domains. Elements of P are called *points*, elements of L *lines*, and I is called an *incidence relation*. We may associate with any $\ell \in L$ the set of points incident to ℓ, and one of our assumptions is that distinct lines correspond to distinct such sets. Our definition here is more narrow than Lachlan's original one. The assumptions are:

- the structure (P, L, I) is ω-stable with $\mathrm{rk}\, P = 2 = \mathrm{rk}\, L$.
- the set of points incident to a given line is of rank 1;

- the set of lines incident to a given point is of rank 1;
- every two lines intersect in at most finitely many points;
- through any two points pass finitely many lines.

An example of a pseudoplane is an algebraic surface P (not necessarily closed) with a 2-dimensional family L of curves on it (F-points of these, for F algebraically closed). Removing, if necessary, exceptional 1-dimensional subsets from P and L one can always get the above conditions satisfied.

A special case of a pseudoplane is an abstract affine (or projective) plane, well known to combinatorial geometers. It is a classical theorem that any such plane, if it satisfies a combinatorial *Desargues theorem*, is definably equivalent to a division ring F. Under the assumptions of ω-stability such a division ring has to be an algebraically closed field. Thus, the weak trichotomy theorem suggests that the pregeometries of the three examples in 6.3 are the only ones possible. This was proposed as *the trichotomy conjecture* by the present author. Observe that when one assumes *local finiteness* of a strongly minimal structure **M**, that is, that $\mathrm{cl}(U)$ finite for finite $U \subset M$, then the type 6.3(3) pregeometry is excluded: algebraically closed fields are not locally finite. So the following supports the trichotomy conjecture.

6.5 **Theorem.** *An infinite locally finite homogeneous geometry is isomorphic to one of the following:*

(*i*) *trivial geometry;*
(*ii*) *projective geometry over a finite field;*
(*iii*) *affine geometry over a finite field.*

Note that (*ii*) *and* (*iii*) *are special cases of 6.4*(*ii*).

The proof of the theorem is based, as is the proof of 6.4, on a combinatorial-geometric analysis, using delicate calculations with model-theoretic ranks. The main target of the proof is to exclude the possibility of a pseudoplane. One develops an intersection theory on a pseudoplane (akin to Bézout's theorem) and arrives at a numerical contradiction.

A refinement of this method lead to a similar classification of all finite homogeneous geometries starting from dimension 7.

An alternative proof of the theorem has been derived from the classification of finite simple groups and ensuing classification of finite 2-transitive groups (Cherlin, Mills).

Nevertheless, the general trichotomy conjecture was refuted in a series of examples engineered by Hrushovski.

6.6 **The trichotomy principle.** The weak trichotomy theorem was just one, technical, motivation for the trichotomy conjecture. There are more serious, conceptual, reasons to hope for a form of the trichotomy conjecture to be true. The main one is the undying intuition that the reality around us can be reduced to basic simple forms. A large structure that has a categorical description in a countable language may well be considered as one of those "simple forms" (see also a short discussion in 5.1), and one would expect that all such are known.

So the artificial counterexamples constructed by Hrushovski in 1988 raised the question whether this intuition is fundamentally wrong, or there is a way to amend the conjecture or at least find a less alarming explanation of the facts. Fortunately, the developments of the last 20 years strongly support the latter notion. First, a very productive way to correct the initial conjecture has been found (Zariski geometries, see below), and second, the counterexamples have been to a great extent explained in terms of mainstream mathematical structures, much in the spirit of the trichotomy conjecture.

6.7 **Diophantine geometry.** The model-theoretic geometric concepts introduced above are crucial for many applications.

Consider a field K and a subgroup Γ of a commutative algebraic group $\mathbf{A}(K)$. We say that Γ has the *Lang property* if for every algebraic variety $V \subseteq \mathbf{A}$, the intersection $\Gamma \cap V(K)$ is a union of finitely many cosets of subgroups of the form $\Gamma \cap \mathbf{B}(K)$, for $\mathbf{B}$ an algebraic subgroup of $\mathbf{A}$.

Now suppose $\Gamma \subseteq \Delta^n$, where Δ is a strongly minimal group definable in some expansion of the field K (e.g., a differentially closed field or a difference field). For Δ either (ii) or (iii) of 6.4 must hold, and provided that Δ satisfies (ii) (is locally projective), it is easy to deduce that any definable subset of Δ^n is a finite union of cosets of definable subgroups (a more general version of this proved by Hrushovski and Pillay). It follows that Δ^n, and hence Γ, has the Lang property. In fact, the converse is also true: the Lang property of Γ is equivalent to a more general version of (ii), called *one-basedness.* In particular, Faltings' theorem stating that any finitely generated subgroup Γ of a semiabelian variety $\mathbf{A}(K)$, for K of characteristic zero, has the Lang property is equivalent to the statement that the theory of an algebraically closed field K expanded with the predicate for Γ is superstable, with the geometry of Γ one-based (A. Pillay).

6.8 **Zariski geometries.** The original aim was to reformulate and strengthen the idea of a "simple form" behind the trichotomy conjecture. This is done by adding a topological component to what originally was a concept of pure logic. We now want to distinguish positively definable sets (definable without using the logical negation) from arbitrary ones. In an L-structure $\mathbf{M}$ we call a subset $S \subseteq M^n$ *closed* if it is positively quantifier-free definable (using parameters). We denote by pr the projection $M^{n+1} \to M^n$, for any n, m, and write $\overline{S}$ for the closure of a subset $S \subseteq M^n$, the minimal closed set containing S, when such exists. We denote by $S(a)$ the fiber of S over $a \in \mathrm{pr}\, \mathrm{S}$ under the projection.

A one-dimensional *Zariski structure* (also often *Zariski geometry*) is a strongly minimal structure $\mathbf{M}$ satisfying the following:

(Z0) the closed sets form a Noetherian topology on M^n, for all $n \geq 1$.
(Z1) $\mathrm{pr}\, \mathrm{S} \supseteq \overline{\mathrm{pr}\, (\mathrm{S})} \setminus \mathrm{F}$, for some proper closed subset $F \subset \overline{\mathrm{pr}\, (\mathrm{S})}$.
(Z2) For S a closed subset of M^{n+1}, there is m such that for all $a \in M^n$, $S(a) = M$ or $|S(a)| \leq m$.
(Z3) Given a closed irreducible $S \subseteq M^n$, every irreducible component of the diagonal $S \cap \{x_i = x_j\}$ $(i < j \leq n)$ is of Morley rank at least $\mathrm{rk}\, S - 1$.

In fact, one can equivalently reformulate this definition without assuming that the dimension in $\mathbf{M}$ is the Morley rank, that is, without assuming a priori that $\mathbf{M}$ is strongly minimal.

Similarly, but with a bit more work, one introduces the notion of a general (multidimensional) Zariski structure as a topological structure with a nice dimension notion. A key basic theorem then states that the theory of a Zariski structure **M** allows elimination of quantifiers, is ω-stable, and the Morley rank of **M** is finite.

Obvious examples of Zariski structures are smooth algebraic varieties $M(\mathrm{F})$, for F an algebraically closed field, in the natural language for algebraic varieties (see 5.4).

A less obvious class of Zariski structures is the class of compact complex manifolds M in the language L_{an}, whose basic m-ary relations correspond to analytic subsets $S \subseteq M^m$. Note that this class is essentially nonalgebraic and very diverse. The fact that each of the structures in this class is ω-stable of finite Morley rank is quite surprising and is a good illustration of the power of the notion of a Zariski structure.

One more class of examples comes from the theory DCF_0, of differentially closed fields. A solution space for a differential equation $f(y) = 0$ in one variable of order n is a Zariski structure of dimension n (Hrushovski for $n = 1$, Pillay in general). A similar but more delicate statement is true for appropriate theories in positive characteristic (Hrushovski). The differentiation in this case is understood as the *Hasse differentiation*, a sequence of operators corresponding to orders of differentiation.

The theory ACFA (see 5.9) is a source of another class of Zariski geometries. The structure induced on any strongly minimal subset of an algebraically closed difference field is Zariski (Hrushovski–Sokolovich).

We say that a Zariski structure **M** is *nonlinear* if there is a strongly minimal subset in **M** of type 6.4(iii).

6.9 **Classification theorem for Zariski structures.** (Hrushovski, Zilber 1993) *For any nonlinear Zariski geometry* **M** *there are an algebraically closed field* F *and a nonconstant continuous function*

$$f : \mathbf{M} \to \mathrm{F}.$$

In particular, for a one-dimensional Zariski structure **M** *there are a smooth algebraic curve* C_M *and a continuous finite covering map*

$$\mathbf{p} : \mathbf{M} \to C_M(\mathrm{F});$$

the image of any relation on **M** *is just a Zariski closed (algebraic) relation on* C_M.

The proof is in fact a reconstruction of algebraic geometry in **M**. We start in a universe "without numbers," but with nicely interacting geometric objects such as curves, surfaces, and so on. It is possible then to develop in this universe a good intersection theory and an analysis of singularities, so that the notion of "a given branch of a curve a at the point p is tangent to a given branch of a curve b at p" is well defined.

Now we look at a family of curves passing through a given point on the surface $X \times X$, where X is a fixed curve, so that the curves from the family, or rather their branches, give rise to local functions $X \to X$ around a point. Composing the local functions and factorizing by the tangency relation, we get a one-dimensional group $(F, \cdot)$ with a Zariski structure on it. A similar construction with F in place of X gets us a one-dimensional Zariski field $(F, +, \cdot)$, which has to be algebraically closed by Liouville's argument.

We then continue with the intersection theory and prove a form of Bézout's theorem, which is used to prove the generalization of Chao's theorem: every closed subset S of the projective space $\mathbf{F}\mathbf{P}^{\mathrm{n}}$ is a zero-set of a system of homogeneous polynomial equations.

The latter translates into the final statement of the classification theorem: the only relations on F induced from $\mathbf{M}$ are the constructible ones.

6.10 **Applications.** A consequence of the classification theorem is that *the trichotomy principle holds for strongly minimal structures definable in:*

(*a*) *differentially closed fields of characteristic zero,*
(*b*) *Hasse-differentially closed fields of positive characteristic p,*
(*c*) *algebraically closed difference fields,*
(*d*) *compact complex manifolds.*

Hrushovski used (a) to give a new proof of the Mordell–Lang conjecture for function fields in characteristic 0, (b) to formulate and prove the analogue of the Mordell–Lang conjecture for function fields of positive characteristic, and (c) to produce a new proof, with better than previously known numerical estimates, of the Lang property for torsion points of semiabelian varieties (the Manin–Mumford conjecture).

Pillay and Ziegler used (d) to establish a useful connection between the classification theory of compact complex manifolds and the theory of differential fields.

6.11 **"New" stable structures.** As mentioned above, the trichotomy conjecture is false in general. Hrushovski in 1988 introduced a construction that produced a series of unexpected strongly minimal, and more general stable, structures for which the trichotomy principle fails.

Suppose we have a class of strongly minimal L-structures $\mathcal{H}$ with the (combinatorial) dimension $d(X)$ for finite subsets of the structures. We want to introduce a new function or relation on $\mathbf{M} \in \mathcal{H}$ so that the new structure gets a good notion of dimension.

Hrushovski observed that this can be done using the principle of *free fusion.* That is, the new function should be related to the old structure in as free a way as possible. A more precise form of this principle states that *the number of explicit dependencies in X in the new structure must not be greater than $d(X)$.*

The explicit L-dependencies on X can be counted as the L-codimension, $|X| - d(X)$. The explicit dependencies induced by a new relation are those given by simplest "equations," that is, basic formulas.

So, for example, if we want a *new unary function* f *on a field,* the condition should be

$$\operatorname{trd}(X \cup f(X)) - |X| \geq 0, \tag{8}$$

since in the set $Y = X \cup f(X)$ the number of explicit field dependencies is $|Y| - \operatorname{trd}(Y)$, and the number of explicit dependencies in terms of f (those of the form $f(x_1) = x_2$) is $|X|$. We call the counting function $\delta(X) = \operatorname{trd}(X \cup f(X)) - |X|$ a *predimension in* $(\mathbf{M}, f)$.

In general, we think of a *fusion* between two structures, (M, L_1) and another one that lives on the same domain, say (M, L_2). Both structures carry a combinatorial pregeometry, with notions of dimension $d_1(X)$ and $d_2(X)$ respectively. Then the predimension $\delta(X)$ in the new structure $(M, L_1 \cup L_2)$ is a simple linear combination of d_1 and d_2, in fact uniquely determined by the free fusion principle.

6.12 Now we consider the new class of structures $\mathcal{H}_\delta$ consisting of all the $(M, L_1 \cup L_2)$ satisfying the *Hrushovski inequality*:

$$\delta(X) \geq 0 \text{ for any finite } X \subseteq M.$$

The next clever idea is to choose in the class $\mathcal{H}_\delta$ a structure that is algebraically closed in the class. A way of defining the notion of algebraic (*existential*) closedness in a class is well known in model theory. The prototypes are algebraically closed fields, differentially closed fields, algebraically closed difference fields considered above, and many others.

To define algebraically closed objects in $\mathcal{H}_\delta$, Hrushovski first introduces the notion of *strong embedding* $\mathbf{A} \leq_\delta \mathbf{B}$ in the class. This means that $\mathbf{A} \subseteq \mathbf{B}$ and for every finite $X \subseteq \mathbf{A}$,

$$\min\{d(Y) : X \subseteq Y, \text{ for finite } Y \subseteq A\} = \min\{d(Y) : X \subseteq Y, \text{ for finite } Y \subseteq B\},$$

that is, all dependencies between elements of $\mathbf{A}$ occurring in $\mathbf{B}$ can be detected already in $\mathbf{A}$.

A structure $\mathbf{M}^\sharp$ is said to be algebraically closed in $\mathcal{H}_\delta$ if any finite quantifier-free type over $\mathbf{M}^\sharp$ realized in a strong extension of $\mathbf{M}^\sharp$ is already realized in $\mathbf{M}^\sharp$.

Provided that $\mathcal{H}_\delta$ satisfies certain conditions, any two $\mathcal{H}_\delta$-algebraically closed structures are elementarily equivalent, and often their common theory is stable and even ω-stable. In the latter case, if $\mathbf{M}^\sharp$ is such an $\mathcal{H}_\delta$-algebraically closed structure, $\mathbf{M}^\sharp$ *becomes a homogeneous pregeometry* with the (combinatorial) dimension ∂ defined as follows:

$$\partial(X) = \min\{d(Y) : X \subseteq Y, \text{ for finite } Y \subseteq M\}.$$

6.13 Although at this step of Hrushovski's construction we have obtained a new homogeneous pregeometry, our aim is not yet achieved. The structure $\mathbf{M}^\sharp$ is not strongly minimal. Typically $\mathbf{M}^\sharp$ is *quasiminimal* in the following sense: the structure $\mathbf{M}^\sharp$ is uncountable but every definable subset $S \subseteq M$ is either countable or a complement of a countable one. So at the last stage of the construction one applies to $\mathbf{M}^\sharp$ a very delicate method called *collapse*: it chooses, following one of continuum many procedures μ inside $\mathbf{M}^\sharp$, a substructure $\mathbf{M}^\sharp_\mu$ with a smaller domain, which is strongly minimal. Remarkably, the pregeometry of $\mathbf{M}^\sharp_\mu$ agrees with the pregeometry of $\mathbf{M}^\sharp$, that is, the notion of dependence in the substructure is the same as in the ambient structure. In particular, the predimenision and notions of dimension in $\mathbf{M}^\sharp_\mu$ are defined exactly as in 6.12.

Thus we get a continuum many new strongly minimal nonlinear structures and pregeometries.

6.14 The discovery of the new strongly minimal structures in 1988 was an obvious challenge to the views and hopes expressed in 5.1 and 6.6. The success with the classification of Zariski geometries mitigated the disappointment, but nevertheless, the question whether the new structures are mathematical pathologies or a part of a bigger picture remained.

6.15 **Schanuel's conjecture.** A crucial breakthrough came with the following observation.

Let the original class $\mathcal{H}$ in 6.11 be the class of algebraically closed fields F of characteristic 0 and suppose we want to add a new function, called suggestively ex, to the field. We want the new function to be a homomorphism between the two group structures on F, that is,

$$\mathrm{ex}\,(x_1 + x_2) = \mathrm{ex}\,x_1 \cdot \mathrm{ex}\,x_2. \tag{9}$$

The free fusion principle uniquely determines then that the predimension δ for this class has to be

$$\delta(X) = \mathrm{trd}\,(X \cup \mathrm{ex}\,X) - \mathrm{ldim}\,X, \quad \text{for any finite } X \subseteq F,$$

where $\mathrm{ldim}\,X$ is the dimension of the **Q**-vector space generated by X. Now observe that the Hrushovski inequality of 6.12 is equivalent to

$$\mathrm{trd}\,(x_1, \ldots, x_n, \mathrm{ex}\,x_1, \ldots, \mathrm{ex}\,x_n) \geq n, \ \text{ for linearly independent} x_1, \ldots, x_n,$$

which is exactly the Schanuel conjecture for the exponentiation ex $=$ exp, F $=$ **C**, the central conjecture of transcendental number theory.

Variations of Schanuel's conjecture, e.g., for elliptic functions, are also well known and indeed can be written in the form of Hrushovski's inequality. It looks credible that the Hrushovski inequality properly applied is just the most general form of a Schanuel-type conjecture.

6.16 **Pseudoexponentiation.** In the particular case of the class $\mathcal{H}(\mathrm{ex})$ described above this author has carried out the steps 6.11 and 6.12 of Hrushovski's construction (with some modifications). The resulting class of structures called *algebraically closed fields with pseudoexponentiation,* ACFExp, has the following properties:

(i) ACFExp *is axiomatizable by an explicit list of (not first-order) formulas, stating* (*a*) *the validity of Schanuel's conjecture and*

(*b*) *that any system of n independent exponential-polynomial equations in n variables that does not directly contradict Schanuel's conjecture has a regular zero, but not more than countably many;*

(ii) ACFExp *is categorical in uncountable powers κ, that is, for every such κ there is a unique, up to isomorphism, algebraically closed field with pseudoexponentiation of cardinality κ;*

(iii) *An algebraically closed field with pseudoexponentiation carries a homogeneous pregeometry, in particular, any bijection between two bases of the pregeometry can be extended to an automorpism of the field with pseudoexponentiation.*

A consequence of the theorem is that Schanuel's conjecture is consistent with the field-theoretic algebra. The categoricity statement (ii) and homogeneity statement (iii) strengthen this further on: Not only is Schanuel's conjecture consistent, but along with other axioms, it also makes the algebra of the structure uniquely nice.

These simple arguments suggest the following.

6.17 **Conjecture.** *The unique algebraically closed field with pseudoexponentiation of power the continuum is isomorphic to* $(\mathbf{C}, +, \cdot, \exp)$, *the complex field with exponentiation.*

Clearly this conjecture implies Schanuel's conjecture. But there is also the part (b) in the axioms of ACFExp, which leads to the formulation of a new conjecture:

$(\mathbf{C}, +, \cdot, \exp)$ *is algebraically closed as a field with exponentiation.*

The precise meaning of the assumption (b) can be found in the original paper. We present here a theorem supporting the conjecture, that is the statement of the theorem is a formal corollary of the conjecture.

Theorem (W. Henson and L. Rubin, 1983) *Let $f(x)$ be a term in one variable in the language* $(+, \cdot, \exp)$ *and constant symbols for complex numbers. Assume that $f(x)$ is not of the form $e^{g(x)}$, where $g(x)$ is another such term. Then the equation $f(x) = 0$ has a solution in* $\mathbf{C}$.

6.18 **A test for Schanuel's conjecture.** The model-theoretic interpretation of Schanuel's conjecture has the advantage of the utmost generality. We can, for example, look for the simplest version of a Schanuel-like conjecture with the hope to test its validity. (Note that no natural version of a Schanuel-like conjecture has been proven so far.)

Apparently the easiest form of a Schanuel-like conjecture is for an analytic function $f(x)$ on $\mathbf{C}$ that satisfies no functional equation. In this case the Hrushovski inequality must have the form (8), Section 6.12. Does such a function exist? If yes, is the structure $(\mathbf{C}, +, \cdot, f)$ algebraically closed in the appropriate sense?

Both questions have positive answers. A. Wilkie has shown that an entire analytic function given as

$$f(x) = \sum_{n \geq 0} \frac{x^n}{a_n}$$

with a_n very rapidly increasing integers (e.g., $a_n = 2^{2^{n!}}$) satisfies the Hrushovski inequality. P. Koiran proved that the structure is algebraically closed.

7 Other Languages and Nonelementary Model Theory

The second-order languages such as L$_2$Reals proved unsuitable for a model-theoretic analysis, so various other, *more* tamer, extensions of first-order languages were considered. Among the most natural ones are the languages $L_{\lambda,\mu}$, for cardinal numbers λ and μ, which allow quantification over sequences of variables of length $< \mu$ and Boolean operations over sets of formulas of cardinality $< \lambda$.

These languages can be further enhanced by allowing, say, the quantifier $\mathcal{Q}$, which in expressions of the form $\mathcal{Q}x\, P(x)$ has the meaning "there exists at least $\aleph_1$-many x such that $P(x)$."

The main difficulty in studying these languages is the failure of any form of the compactness theorem.

Some progress in the study of these languages was achieved in the 1960s and 1970s, but further attempts, in particular in the spirit of classification theory of Sections 5 and 6, led to a complete rethinking of the approach to non-first-order model theory. Shelah introduced the new concept of abstract elementary classes, which is not based on any class of logic formulas.

7.1 **Definition.** Given cardinals λ and μ and an alphabet L, $L_{\lambda,\mu}(L)$ is the smallest collection of formulas that contain all atomic L-formulas in the variables v_α, $\alpha < \mu$, and closed under taking $\neg$, applying universal quantifiers to a string of variables $\forall v_{i_1} \cdots \forall v_{i_\alpha} \cdots P$, applying existential quantifiers to a string of variables $\exists v_{i_1} \cdots \exists v_{i_\alpha} \cdots P$, and applying disjunction $\bigvee_\alpha P_\alpha$ or conjunction $\bigwedge_\alpha P_\alpha$ to fewer than λ formulas.

The interpretation of $L_{\lambda,\mu}(L)$-formulas in L-structures is defined along the same lines as that for first-order formulas.

A formula of the language $L_{\infty,\mu}(L)$ is a formula of the language $L_{\lambda,\mu}(L)$, for some λ.

The language $L^{\mathcal{Q}}_{\infty,\mu}(L)$ is obtained by allowing the use, along with formulas of $L_{\infty,\mu}(L)$, of the quantifier $\mathcal{Q}$, with the interpretation explained above.

An example of the possible use of these languages is the axiomatization in 6.16. The axioms in (i)(a) require $L_{\omega_1,\omega}$, and in (b), $L^{\mathcal{Q}}_{\omega_1,\omega}$.

The following is one of the basic results about infinitary languages; compare with the Ehrenfeucht–Fraisse criterion.

7.2 **Theorem (C. Karp)** *Two L-structures* **A** *and* **B** *are* $L_{\infty,\omega}(L)$*-equivalent if and only if there is a back-and-forth system between* **A** *and* **B** (*definition 3.14*).

When $\mathbf{A}$ and $\mathbf{B}$ are countable we have a corollary that $\mathrm{L}_{\omega_1,\omega}(L)$-equivalence amounts to an isomorphism between the structures. A stronger result is the following categoricity result.

Theorem (D. Scott). *Given a countable L and a countable L-structure $\mathbf{A}$, there is an $\mathrm{L}_{\omega_1,\omega}(L)$-sentence $\Sigma(\mathbf{A})$ true in $\mathbf{A}$ and such that any countable model of $\Sigma(\mathbf{A})$ is isomorphic to $\mathbf{A}$.*

Note how this theorem emphasizes the special effect of categoricity in small cardinals, or cardinals small compared to the cardinality of the whole language, the set of all formulas. For a first-order language exactly the same statement holds when we replace "countable" by "finite." In fact, this very effect explains why the categoricity in uncountable cardinalities has given an impetus to the richest part of modern model theory, the first-order stability theory (Section 5).

7.3 **Löwenheim–Skolem theorems for $\mathrm{L}_{\lambda,\mu}$ and other languages.** The situation here is much more complex than for the first-order languages. The downward Löwenheim–Skolem holds but in a restricted form. Say, *for a countable L, an infinite L-structure $\mathbf{A}$, an infinite cardinal $\kappa \leq \operatorname{card} \mathbf{A}$, and each $\mathrm{L}_{\omega_1,\omega}$-sentence P that holds in $\mathbf{A}$ there is an L-substructure $\mathbf{B} \subset \mathbf{A}$ such that $\mathbf{B} \models P$.*

The proof uses the Skolem functions much in the same way as in the first-order case, see 3.4.

But the analogue of the upward theorem is not true. There are $\mathrm{L}_{\omega_1,\omega}$-sentences that have models but not higher than a certain cardinality. For example, in the language of arithmetic extended by a unary predicate N and a binary predicate ϵ we can state in the form of an $\mathrm{L}_{\omega_1,\omega}$-sentence Q that the predicate N defines the subset N of the model such that $(N, +, \cdot, 0, 1)$ is a standard arithmetic;

if $x \epsilon y$ holds then $x \in N$ and $y \notin N$; moreover,

$\forall y_1, y_2 \notin N\, (y_1 = y_2 \leftrightarrow \forall x \in N\, x \in y_1 \leftrightarrow x \in y_2)$.

Clearly this sentence has models at most of cardinality $2^{\aleph_0}$.

One can extend this method to obtain sentences with models of cardinalities bounded by $2^{2^{\aleph_0}}, 2^{2^{2^{\aleph_0}}}, \ldots$

For the general $\mathrm{L}_{\lambda,\mu}$-language the situation is even more complex.

7.4 **Categoricity for $\mathrm{L}_{\omega_1,\omega}$ in uncountable cardinals.** This problem was first attacked by J. Kiesler in the 1970s, in an attempt to extend the Morley theory to $\mathrm{L}_{\omega_1,\omega}$. Kiesler proved that the main results go through provided one can establish the fact that models of an $\mathrm{L}_{\omega_1,\omega}$-sentence categorical in an uncountable cardinal are homogeneous, which is of course the case for first-order languages. But shortly after Kiesler's work appeared, counterexamples to this assumption were found. More recently, examples of uncountably categorical $\mathrm{L}_{\omega_1,\omega}$-sentences with nonhomogeneous uncountable models were found in the context of mainstream mathematics.

7.5 **Example.** Consider the structure on the complex numbers

$$\mathbf{C}^e = (\mathbf{C}, +, p^{(3)}), \text{ where } p^{(3)}(x, y, z) \equiv e^x + e^y = e^z.$$

Notice that the subgroup $2\pi i\mathbf{Z}$ is definable in $\mathbf{C}^e$ as

$$\{v \in \mathbf{C} : \forall x, y, z \;\; e^x + e^y = e^z \leftrightarrow e^x + e^y = e^{z+v}\}.$$

Now, if we introduce a definable set $\mathbf{C}^* = \mathbf{C}/2\pi i\mathbf{Z}$ and a definable canonical homomorphism $\exp : \mathbf{C} \to \mathbf{C}^*$ we get an equivalent representation of the structure as a two-sorted structure $(\mathbf{C}, \; \mathbf{C}^* \cup \{0\})$ with the additive group structure $(\mathbf{C}, +)$ on the first sort, the field structure $(\mathbf{C}^* \cup \{0\}, \cdot, +)$ on the second sort, and exp mapping the first sort into the second sort. We can describe this structure by an $\mathrm{L}_{\omega_1,\omega}$-sentence Σ saying that:

- $(\mathbf{C}, +)$ is a divisible torsion-free group;
- $\mathbf{C}^* \cup \{0\}$ with respect to $+$ and $\cdot$ is an algebraically closed field of characteristic 0;
- the kernel of exp is an infinite cyclic group.

It takes a nontrivial algebra (theory of fields) in combination with model theory to prove that Σ has a unique, up to isomorphism, model in every uncountable cardinality. But any such model is not homogeneous.

7.6 **Abstract elementary classes.** Shelah, who has been in the forefront of studies in non-first-order model theory, was the first to realize that the syntactic specification of non-first-order languages has little relevance to model theory, and the more important are algebraic characteristics of classes of models, which eventually depend more on the meaning of specific axioms than the syntax of the language. This resulted in the following definition.

A class of L-structures K equipped with a notion of "strong submodel" $\preccurlyeq$ is said to be an *abstract elementary class* (AEC) if the class K and class of pairs satisfying the binary relation $\preccurlyeq$ are each closed under isomorphism and satisfy the following conditions:

(a) If $\mathbf{A} \preccurlyeq \mathbf{B}$ then $\mathbf{A} \subseteq \mathbf{B}$.
(b) $\preccurlyeq$ is a partial order on K.
(c) If $\{\mathbf{A}_i : i < \delta\}$ is a $\preccurlyeq$-increasing chain in K closed under limits, then:
  (i) $\mathbf{A}_\delta = \bigcup_{i<\delta} \mathbf{A}_i \in K$;
  (ii) for each $j < \delta$, $\mathbf{A}_j \preccurlyeq \mathbf{A}_\delta$;
  (iii) if each $\mathbf{A}_i \preccurlyeq \mathbf{B} \in K$ then $\mathbf{A}_\delta \preccurlyeq \mathbf{B}$.
(d) If $\mathbf{A}, \mathbf{B}, \mathbf{C} \in K$, $\mathbf{A} \preccurlyeq \mathbf{B}$ $\mathbf{B} \preccurlyeq \mathbf{C}$, and $\mathbf{A} \subseteq \mathbf{C}$ then $\mathbf{A} \preccurlyeq \mathbf{C}$.
(e) There is a (Löwenheim–Skolem) cardinal number $LS(K)$ such that if $\mathbf{A} \subseteq \mathbf{B} \in K$, there is an $\mathbf{A}' \in \mathbf{K}$ with $\mathbf{A} \subseteq \mathbf{A}' \preccurlyeq \mathbf{B}$ and $\operatorname{card} A' \leq \operatorname{card} A + LS(K)$.

7.7 **Examples.**

(a) Any first-order axiomatizable class of L-structures with respect to $\preccurlyeq$, the elementary embedding, is AEC.
(b) The class of models of the $\mathrm{L}_{\omega_1,\omega}$-sentence Σ in 7.5 with respect to the embedding $\subseteq$ is AEC.
(c) The class $\mathcal{H}_\delta$ emerging in Hrushovski's construction with respect to the strong embedding, see 6.12, is AEC.

(d) The class of fields with pseudoexponetiation is AEC with respect to the strong embedding corresponding to the Schanuel predimension $\delta(X) = \operatorname{trd} X - \operatorname{ldim} X$; see 6.16.

The theory of abstract elementary classes brought model theory closer to the tradition of abstract algebra but enriched with the vast technology of classification theory. The most powerful results of the theory are the following.

7.8 **Theorem (S. Shelah).** *There is a Hanf number μ (not computed but depending only on the Löwenheim–Skolem number $LS(K)$) such that if an AEC K has arbitrarily large models and satisfies the amalgamation property and the joint embedding property for its models, then provided that K is categorical in a successor cardinal larger than μ, it is categorical in all larger cardinals.*

In a more specific situation we have the following.

7.9 **Theorem (S. Shelah).** *Assume the mild set-theoretic assumptions $2^{\aleph_n} < 2^{\aleph_{n+1}}$ for all natural n. Let Σ be an $\mathrm{L}_{\omega_1,\omega}$-sentence that is categorical in $2^{\aleph_n}$ for every n. Then Σ has a unique model in every infinite cardinal.*

For further reading on the subject of infinitary languages and AEC see J. Baldwin's book [14].

Suggestions for Further Reading

I.–II. Introduction to Formal Languages. Truth and Deducibility

[1] Cori, René and Lascar, Daniel, *Mathematical logic*, A course with exercises. Part I, Propositional calculus, Boolean algebras, predicate calculus, Translated from the 1993 French original by Donald H. Pelletier, With a foreword to the original French edition by Jean-Louis Krivine and a foreword to the English edition by Wilfrid Hodges, Oxford University Press, Oxford, 2000, xx+338

[2] Cori, René and Lascar, Daniel, *Mathematical logic*, A course with exercises. Part II, Recursion theory, Gödel's theorems, set theory, model theory, Translated from the 1993 French original by Donald H. Pelletier, With a foreword to the original French edition by Jean-Louis Krivine and a foreword to the English edition by Wilfrid Hodges, Oxford University Press, Oxford, 2001, xx+331

III.–IV. The Continuum Problem and Forcing. The Continuum Problem and Construcitble set

[3] Kunen, Kenneth, *Set theory, Studies in Logic and the Foundations of Mathematics*, **102**, An introduction to independence proofs, Reprint of the 1980 original, North-Holland Publishing Co., Amsterdam, 1983, xvi+313

[4] Smullyan, Raymond M. and Fitting, Melvin, *Set theory and the continuum problem, Oxford Logic Guides*, **34**, Oxford Science Publications, The Clarendon Press Oxford University Press, New York, 1996, xiv+288

[5] Hrbacek, Karel and Jech, Thomas, *Introduction to set theory, Monographs and Textbooks in Pure and Applied Mathematics*, **220**, Third, Marcel Dekker Inc., New York, 1999, xii+291

[6] Jech, Thomas, *Set theory, Springer Monographs in Mathematics*, The third millennium edition, revised and expanded, Springer-Verlag, Berlin, 2003, xiv+769

V. Recursive Functions and Church's Thesis

[7] Boolos, George S. and Burgess, John P. and Jeffrey, Richard C., *Computability and logic*, **Fifth**, Cambridge University Press, Cambridge, 2007, xiv+350

[8] Rogers, Jr., Hartley, *Theory of recursive functions and effective computability*, **Second**, MIT Press, Cambridge, MA, 1987, xxii+482

VI. Diophantine Sets and Algorithmic Undecidability

[9] Matiyasevich, Yuri V., *Hilbert's tenth problem, Foundations of Computing Series*, Translated from the 1993 Russian original by the author, With a foreword by Martin Davis, MIT Press, Cambridge, MA, 1993, xxiv+264

[10] Li, Ming and Vitányi, Paul, *An introduction to Kolmogorov complexity and its applications, Graduate Texts in Computer Science*, **Second**, Springer-Verlag, New York, 1997, xx+637

VII. Gödel's Incompleteness Theorem

[11] Smullyan, Raymond M., *Gödel's incompleteness theorems, Oxford Logic Guides*, **19**, The Clarendon Press Oxford University Press, New York, 1992, xvi+139

VIII. Recursive Groups

[12] Higman, Graham and Scott, Elizabeth, *Existentially closed groups, London Mathematical Society Monographs. New Series*, **3**, Oxford Science Publications, The Clarendon Press Oxford University Press, New York, 1988, xiv+156

IX. Constructive Universe and Computation

[13] Kitaev, A. Yu. and Shen, A. H. and Vyalyi, M. N., *Classical and quantum computation, Graduate Studies in Mathematics*, **47**, Translated from the 1999 Russian original by Lester J. Senechal, American Mathematical Society, Providence, RI, 2002, xiv+257

X. Model Theory

[14] J.Baldwin, *Categoricity in Abstract Elementary Classes*, **245** Pages accepted by the American Mathematical Society, University Lecture Notes, accessible on http://www.math.uic.edu/~jbaldwin/pub/AEClec.pdf

[15] S.Buechler, *Essential stability theory, Perspectives in mathematical logic*, Berlin–London, Springer, 1996

[16] D.Marker, *Model theory: an introduction, Graduate texts in mathematics*, **217**, New York–London, Springer, 2002

[17] Shelah, S., *Classification theory and the number of nonisomorphic models, Studies in Logic and the Foundations of Mathematics*, **92**, **Second**, North-Holland Publishing Co., Amsterdam, 1990, xxxiv+705

Index